SPACE WEATHER OF THE HELIOSPHERE:
PROCESSES AND FORECASTS

IAU SYMPOSIUM 335

COVER ILLUSTRATION: TWO SEASONS IN THE CYCLE OF THE SUN

This is an enlarged part of the symposium poster, showing two seasons in the activity cycle of the Sun to represent the changes in space weather, against the backdrop of the Exeter cathedral.

The underlying inspiration for the poster logo comes from both the symposium topic, space weather, where the Sun is center-stage, and the conference location in Devon, UK. Since the rise of seaside resorts with the arrival of the railways in the 19th century, Devon's economy has been heavily reliant on tourism. The sunburst logo design is inspired by vintage British touristic posters, and is reminiscent of 19th century Art Deco sunburst. The two-season logo is reminiscent of a 1922 London Underground Museum Poster by British artist Albert E. Fruin.

The solar images in the poster are based on two events, respectively in period of high solar activity (2002 January 4 blasting Coronal Mass Ejection (CME), right forefront) and low solar activity (2010 July 11 solar eclipse, left background). The 'blasting CME' image is from the 'Best Of SOHO' image gallery courtesy of SOHO/EIT and SOHO/LASCO consortium (SOHO is a project of international cooperation between ESA and NASA). In this composite image, the solar disk from an EIT image in EUV light (304Å) was enlarged and superimposed on a LASCO C2 coronagraph image. The low-activity Sun is a composite of coronal images courtesy of NASA SDO/AIA and from the ESA 'Space in Images' gallery courtesy Institut d'Astrophysique de Paris (CNRS & UPMC), S. Koutchmy/J. Mouette. In this composite, the solar disk is made of SDO/AIA combined images at three wavelengths in the EUV (211, 193, 171Å) taken at about the same time as a reconstructed white-light image of the extended corona taken from the ground, in Atoll Hao, French Polynesia.

The landscape of the Exeter Cathedral in the poster is based on a photograph by Guy Edwards.

IAU SYMPOSIUM PROCEEDINGS SERIES

INTERNATIONAL ASTRONOMICAL UNION
UNION ASTRONOMIQUE INTERNATIONALE

International Astronomical Union

SPACE WEATHER OF THE HELIOSPHERE: PROCESSES AND FORECASTS

PROCEEDINGS OF THE 335th SYMPOSIUM OF THE INTERNATIONAL ASTRONOMICAL UNION HELD AT UNIVERSITY OF EXETER, UNITED KINGDOM JULY 17–21, 2017

Edited by

CLAIRE FOULLON
University of Exeter, UK

and

OLGA E. MALANDRAKI
IAASARS/National Observatory of Athens, Greece

CAMBRIDGE
UNIVERSITY PRESS

CAMBRIDGE UNIVERSITY PRESS
University Printing House, Cambridge CB2 8BS, United Kingdom
1 Liberty Plaza, Floor 20, New York, NY 10006, USA
10 Stamford Road, Oakleigh, Melbourne 3166, Australia

First published 2018

Printed in the UK by Bell & Bain, Glasgow, UK

Typeset in System LaTeX 2ε

A catalogue record for this book is available from the British Library Library of Congress Cataloguing in Publication data

This journal issue has been printed on FSC^{TM}-certified paper and cover board. FSC is an independent, non-governmental, not-for-profit organization established to promote the responsible management of the world's forests. Please see www.fsc.org for information.

ISBN 9781107192409 hardback
ISSN 1743-9213

Table of Contents

Part 7. Forecasting Models

Preface

The Sun is an active and variable star. Instabilities and non-stationary processes connected to the solar magnetic field and its evolutionary mechanisms modify its radiative and particle output on different time scales, from seconds to the evolutionary scale of the star. Solar activity affects interplanetary space and planetary environments. The physical state of space associated with short-term solar activity is called space weather, while long-term activity of our star determines space climate. The multi-disciplinary IAU Symposium 335 on 'Space Weather of the Heliosphere: Processes and Forecasts' was held at the University of Exeter, UK in July 17-21 2017. It linked various aspects of research in solar, heliospheric and planetary physics, emphasizing cross-disciplinary developments.

The ability to understand, monitor and forecast the space weather is of paramount importance to our high-technology society. Sudden changes in space weather, due to eruptive events on the Sun such as coronal mass ejections (CMEs) and solar flares, can impact the technology we rely on every day. In June 2015 the World Meteorological Organisation (WMO) endorsed space weather activities. In addition, in October 2015 the United States President's Office of Science and Technology Policy released the U.S.'s comprehensive Strategy and Action Plan for space weather, titled 'Space Weather: Understanding Potential Impacts and Building Resilience'. This release was a major step in the USA's determination to address civil societal issues related to all aspects of space weather. Also released that year was a lengthy international road map document by C. Schrijver et al. titled 'Understanding space weather to shield society: A global road map for 2015-2025 commissioned by COSPAR and the ILWS'. Thus, the Symposium subject was timely and strongly linked to our modern society's technological and economical developments. Moreover, the human exploration vision and ongoing efforts to send manned missions into space are putting increasing demands for space weather predictability beyond our Earth's environment. The effects of interplanetary space weather on planets or moons in our Solar System, as well as the space weather experienced by comets and asteroids, are enriching and stretching our understandings. Our scientific community has the responsibility to use its knowledge to support tackling societal problems; one of the roles that the community can play is to identify the research questions that have to be answered in order to improve forecasts.

The IAU Symposium 335 had received the support from 5 out of 9 Divisions across IAU, including the Coordinating Division E Sun and Heliosphere. It was one of the 9 approved by the IAU Executive Committee in May 2016. The symposium brought together scientific experts from various pertinent disciplines to the meeting from all over the world around the topic of Space Weather. We would like to emphasize the strongly international character of the symposium and that it was the first IAU symposium organised on this topic. We welcomed 185 participants from 30 different countries and 21 accompanying persons, exhibitors or public lecturer. Thanks to IAU and 14 cosponsors listed on page *xxi* of these Proceedings, we supported 47 scientists from around the world to attend and present their work. Particular noteworthy was the relatively high (36.8%) proportion of women attendees, one of the highest ever encountered at a major international meeting in our field.

The overall scientific program involved a total of 204 abstracts. Of those presented abstracts, 122 were posters and 82 were in the oral program. The oral program consisted of 26 invited talks by leaders in the field and 56 oral contributed presentations. 10 chairs led the sessions and provided lively question and answer sessions. The scientific program consisted of 8 sessions, scheduled over 4.5 days. Each session topic was explored by 2 to 4 invited talks, contributed talks and poster sessions. All oral presentations were plenary, except for the last theme where we had a common plenary and a split in 2 round tables. Thanks to the rich representation and optimal (scientific, regional, gender) diversity of the SOC, balance was respected in the selection of the 26 invited speakers and the general scientific program. The symposium therefore gave a balanced international overview of the general advances in space weather, focusing on the key topics of solar drivers and activity levels (session 1); Solar wind and heliosphere (2); Impact of solar wind, structures and radiation on and within terrestrial and planetary environments (3 and 4); Long-term trends and predictions for space weather (5); Challenges and strategy plans for Earth and the heliosphere (6); Forecasting models (7); Space weather monitoring, instrumentation, data and services (8). This volume contains a representative sample of 83 reviewed papers from all sessions whilst respecting the order of the scientific program.

Besides the main sessions, the oral program included a town hall session and round-table discussions, plenary summaries, and a few opening and closing talks. In the last few years, we have seen transformations of the UK political landscape, all in response to space weather being on the national risk register (e.g. the Met Office centre). The Symposium opened with a Welcome Introduction by M. Hapgood, chair of the Space Environment Impact Expert Group (SEIEG) that advises the Cabinet Office on Space Weather. Space weather is increasingly recognised as an international challenge faced by several communities. In Session 6, the United Nations Expert Group on Space Weather led a town hall session on 'Strategy for Developing an International Framework for Space Weather Services (2018-2030)'. On the last day, Session 8 included two parallel round tables RT1 'Data Handling and Assimilation' and RT2 'Relationships with the 'civil' society', with the findings summarized in the plenary by invited speakers. The ability to understand, monitor and forecast the space weather of the Earth and the heliosphere is of paramount importance for our high-technology dependent society and for the current rapid developments in our knowledge and exploration (robotic and human) of the Solar System. Space weather is not just important at Earth and this symposium brought a vast range of expertise together, discussing the activity and winds of other Suns and the effects on planets and other objects of the solar system such as Pluto.

To foster the excitement of younger colleagues in presenting their work, two competitions took place. A poster competition was organised with judges assigned among the senior participants; the poster winners were: in joint fourth, R. Chiba and V. Pant; in joint second, N. Hussain and M. Korsos; and in first place, S. Mahajan. An independent committee drawn from the SOC singled out K. Tlatova for her research, oral presentation and overall performance, and nominated her for a IAGA Young Scientist Award. Congratulations to all of them.

Special thanks are expressed to all SOC members for their help in organising the sessions. The editors take this opportunity to thank the following reviewers, for their valuable assistance in improving the papers: Z. Benkhaldoun, F. Berrilli, N.B. Crosby, S. Dalla, S. Dasso, B. Fleck, C. Foullon, I. Hannah, C. Jackman, D. Knipp, J. Luhmann, O.E. Malandraki, M. Owens, P. Reiff, K. Shiokawa, N. Srivastava, M. Thompson, I. Usoskin, E. Verwichte, J. Wang and D. Webb. The international services that address space weather disaster risks require a rapid dissemination of current research. May the IAU 335 Symposium and these companion Proceedings covering interdisciplinary topics and attracting a variety of interested parties become a crucial reference at this moment to help the international space weather community prosper.

Claire Foullon and Olga E. Malandraki,
Exeter, UK and Athens, Greece, January 15, 2018

THE ORGANIZING COMMITTEE

Scientific

C. Foullon (chair, UK)
O.E. Malandraki (co-chair, Greece)
Z. Benkhaldoun (Morocco)
F. Berrilli (Italy)
A. Bhardwaj (India)
A.S. Brun (France)
N.B. Crosby (Belgium)
S. Dasso (Argentina)
A. Donea (Australia)
H. Haubold (Austria)
H. Opgenoorth (Sweden)
P. Reiff (USA)
K. Shiokawa (Japan)
I. Usoskin (Finland)
J. Wang (China Nanjing)
D. Webb (USA)

Local

C. Foullon (chair, UoE)
M. Berger (co-chair, UoE)
D. Jackson (co-chair, MO)
M. Baldwin (UoE)
A. Mills (UoE)
D. Strange (NLO)

Acknowledgements

The symposium is sponsored and supported by the IAU Divisions E (Sun and Heliosphere, coordinating), B (Facilities, Technologies and Data Science), C (Education, Outreach and Heritage), D (High Energy Phenomena and Fundamental Physics) and G (Stars and Stellar Physics); other supporters include the Royal Astronomical Society (Geophysics: RAS G), UK Solar Physics community (UKSP), UK Sun-Planetary scientific community: Magnetosphere, Ionosphere and Solar-Terrestrial (MIST).

The Local Organizing Committee operated under the auspices of the
University of Exeter (UoE),
Meteorological Office (MO),
and
Norman Lockyer Observatory (NLO).

With thanks to the extended LOC team: E. Clarke, P. Hodges, M. Weber, A. Hillier, C. Fagilde, A. Finley, D. Griffin, G. Hawkes, R. Kieokaew, D. Miller, S. Strawbridge, D. Acreman, D. Barker, P. Whickam, C. Faunch, G. Poulton, C. Robson (UoE), S. Bingham (MO) and E. Verwichte (University of Warwick).

Funding by the
International Astronomical Union (IAU),
US Air Force Office of Scientific Research (AFOSR),
Royal Astronomical Society (RAS),

European Space Agency (ESA),
SCOSTEP activity 'Variability of the Sun and Its Terrestrial Impact' (VarSITI),
Science & Technology Facilities Council (STFC) Astronomy Division,
European Geophysical Union (EGU),
University of Exeter College of Engineering, Mathematics and Physical Sciences,
Committee on Space Research (COSPAR),
International Association of Geomagnetism and Aeronomy (IAGA),
Institute of Physics (IOP) Plasma Physics Group,
Journal Frontiers in Astronomy and Space Sciences,

and donation support from
Cambridge University Press (CUP),
Oxford University Press (OUP)
and
DK publisher,

are gratefully acknowledged.

Welcome Address by the Organizing Committee

Dear Colleague,

The University of Exeter is proud to host the IAU Symposium 335 on Space Weather of the Heliosphere: Processes and Forecasts. It is a pleasure to report that 193 participants from 31 different countries are expected to attend the event, sharing progress in all key topics of the scientific program.

Space weather is increasingly recognised as an international challenge faced by several communities. The ability to understand, monitor and forecast the space weather of the Earth and the heliosphere is of paramount importance for our high-technology society and for the current rapid developments in knowledge and exploration within our Solar System. This 5-day Symposium aims to further our understanding of space weather by linking various aspects of research in solar, heliospheric and planetary physics, and by putting particular emphasis on cross-disciplinary developments, merging different communities, learning from interplanetary comparisons and discussing some of the most current challenges at the international level via various forums, including invited and contributed talks and posters, but also a Town Hall and two Round Tables.

Exeter, UK, offers world-leading expertise in weather research and forecasting, and is rapidly expanding its efforts in space weather. Thus Exeter provides an ideal combination of communities to host the Symposium. What is most special about the history of space weather for our conference is that 'the first recorded impact of space weather on civilisation ' can be placed here, in the South West (when the former Great Western Railway (GWR) company was operating): on 18th of October 1841, the 10.5 PM train between Starcross and Exeter, had been delayed by 16 minutes, on the account of a very intense magnetic disturbance, at a time when telegraphs were means of communication to give clearance for railway switching (Cade 2013). Scientific excursions and local exhibitions are planned throughout the week, and will reveal all local treasures and facilities surrounding us.

The heart of the city offers a wealth of history, interesting architecture, cafés, pubs and restaurants for spending the evenings. Situated in the heart of Devon, Exeter is the perfect base to explore the South West of England, a scenery part of the world we hope our social activities will help you to discover.

As modern society becomes increasingly dependent on ground- and space-based technology, it also becomes increasingly vulnerable to the effects of space weather. Not only is it important for the public to better understand space weather but it is also an area of astronomy and space physics that inspires the public with beautiful phenomena such as auroras. We will promote our field of astronomy and space weather science, by giving you the opportunity to engage about space weather with our local young people (students), teachers and the general public in an active parallel education/public outreach program.

We thank you all for your dedication and your participation.
Best wishes on behalf of the Local and Scientific Organising Committees,
Dr Claire Foullon, Chair of IAUS335 LOC & SOC
Exeter, 9 July 2017

References

Cade, W. B., III (2013), The First Recorded Space Weather Impact?, Space Weather, 11, 489, doi:10.1002/swe.20091.

CONFERENCE PHOTOGRAPH

Participants

Franklin **Aldas Garces**, EPN, Ecuador franklinaldas@hotmail.com
Tanja **Amerstorfer**, AAS - Space Research Institute, Austria tanja.amerstorfer@oeaw.ac.at
Mahender **Aroori**, Osmania University, India mahenderastro@gmail.com
Ruth **Bamford**, STFC Rutherford Appleton Laboratory, UK ruth.bamford@stfc.ac.uk
Dipankar **Banerjee**, Indian Institute of Astrophysics, India dipu@iiap.res.in
Luke **Barnard**, University of Reading, UK l.a.barnard@reading.ac.uk
Ciaran **Beggan**, British Geological Survey, UK ciar@bgs.ac.uk
Zouhair **Benkhaldoun**, Oukaimeden Observatory- Cadi Ayyad University, Morroco zouhair@uca.ma
Mitchell **Berger**, University of Exeter, UK m.berger@exeter.ac.uk
Francesco **Berrilli**, University of Rome Tor Vergata, Italy francesco.berrilli@roma2.infn.it
Binod **Bhattarai**, St. Xavier's College, Nepal binod.bhattarai@hotmail.com
Prantika **Bhowmik**, IISER Kolkata, India prantika.tupur@gmail.com
Suzy **Bingham**, Met Office, UK suzy.bingham@metoffice.gov.uk
Mario **Bisi**, STFC Rutherford Appleton Laboratory, UK Mario.Bisi@stfc.ac.uk
Shaun **Bloomfield**, Northumbria University, UK shaun.bloomfield@northumbria.ac.uk
Gert **Botha**, Northumbria University, UK gert.botha@northumbria.ac.uk
Aziza **Bounhir**, UCA University, Morroco a.bounhir@uca.ma
Dave **Brain**, University of Colorado Boulder, USA david.brain@lasp.colorado.edu
Graziella **Branduardi-Raymont**, University College London, UK g.branduardi-raymont@ucl.ac.uk
Matthew **Browning**, University of Exeter, UK m.k.m.browning@exeter.ac.uk
Allan Sacha **Brun**, CEA-Saclay, France sacha.brun@cea.fr
Jack **Carlyle**, ESA/ESTEC, Netherlands jack.carlyle@esa.int
Sophie **Chabanski**, Royal Belgian Institute for Space Aeronomy, Belgium sophie.chabanski@aeronomie.be
Heon-Young **Chang**, Kyungpook National University, South Korea hyc68@naver.com
Nai-Hwa **Chen**, Korea Astronomy and Space Science Institute, South Korea nhchen@kasi.re.kr
Rimpei **Chiba**, Vienna Institute of Technology, Austria chiba.r.ab@m.titech.ac.jp
Arnab **Choudhuri**, Indian Institute of Science, Bangalore, India arnab@physics.iisc.ernet.in
Hebe **Cremades**, UTN-FRM and CONICET, Argentina hebe.cremades@frm.utn.edu.ar
Norma Bock **Crosby**, Belgian Institute for Space Aeronomy, Belgium Norma.Crosby@aeronomie.be
Xinghua **Dai**, National Astronomical Observatories- Chinese Academy of Sciences, China xhdai@nao.cas.cn
Alisson **Dal Lago**, INPE, Brazil alisson.dallago@inpe.br
Silvia **Dalla**, University of Central Lancashire, UK sdalla@uclan.ac.uk
M. Chantale **Damas**, CUNY/Queensborough Community College, USA mdamas@qcc.cuny.edu
Sanja **Danilovic**, Stockholm University, Sweden sdani@astro.su.se
Sergio **Dasso**, IAFE, Argentina sergio.dasso@gmail.com
Erwin **De Donder**, Royal Belgian Institute for Space Aeronomy, Belgium erwin.dedonder@aeronomie.be
Victor **De la Luz**, SCiESMEX UNAM, Mexico vdelaluz@igeofisica.unam.mx
Stefano **Della Torre**, INFN - Sez. Milano-Bicocca, Italy stefano.dellatorre@mib.infn.it
Mitsue **Den**, National Institute of Information & Comm. Technology, Koganei, Japan den@nict.go.jp
Mark **Dierckxsens**, BIRA-IASB, Belgium Mark.Dierckxsens@aeronomie.be
Melinda **Dsa**, Wigner Research Institute for Physics, Hungary dosa.melinda@wigner.mta.hu
Heather **Elliott**, Southwest Research Institute, USA helliott@swri.edu
Sean **Elvidge**, University of Birmingham, UK s.elvidge@bham.ac.uk
Robertus **Erdelyi**, University of Sheffield, UK robertus@sheffield.ac.uk
Paulo Roberto **Fagundes**, UNIVAP, Brazil fagundes@univap.br
Adam **Finley**, Physics and Astronomy University of Exeter, UK af472@exeter.ac.uk
Bernhard **Fleck**, ESA, USA bfleck@esa.nascom.nasa.gov
Andrzej **Fludra**, STFC, UK andrzej.fludra@stfc.ac.uk
Dominique **Fontaine**, CNRS - LPP, France dominique.fontaine@lpp.polytechnique.fr
Roberta **Forte**, Universit di Roma Tor Vergata, Italy roberta.forte@roma2.infn.it
Claire **Foullon**, University of Exeter, UK c.foullon@exeter.ac.uk
Merlin **Fox**, World Scientific Publishing, UK mfox@wspc.co.uk
Tim **Fuller-Rowell**, Univ. of Colorado, USA tim.fuller-rowell@noaa.gov
Stephen **Fuselier**, Southwest Research Institute, USA sfuselier@swri.edu
Tadhg **Garton**, School of Phyiscs, Trinity College Dublin, Dublin, Ireland, Republic of Ireland gartont@tcd.ie
Frederick **Gent**, Aalto University, Finland fred.gent.ncl@gmail.com
Sarah **Glauert**, British Antarctic Survey, UK sagl@bas.ac.uk
Nat **Gopalswamy**, NASA, USA nat.gopalswamy@nasa.gov
Romaric **Gravet**, LPC2E- University of Orleans- CNRS, France romaric.gravet@cnrs-orleans.fr
Daniel **Griffin**, University of Exeter, UK djg211@exeter.ac.uk
Pavel **Gritsyk**, Sternberg Astronomical Institute - MSU, Russia pgritsyk@gmail.com
Jordan **Guerra Aguilera**, Trinity College Dublin, Republic of Ireland jordan.a.guerra@tcd.ie
Norbert Gyula **Gyenge** , The University of Sheffield- SP2RC, UK n.g.gyenge@sheffield.ac.uk
Iain **Hannah**, University of Glasgow, UK iain.hannah@glasgow.ac.uk
Yongqiang **Hao**, Peking University, China yong.hao@gmail.com
Mike **Hapgood**, Science and Technology Facilities Council, UK mike.hapgood@stfc.ac.uk
Richard **Harrison**, RAL Space, UK richard.harrison@stfc.ac.uk
Gareth **Hawkes**, University of Exeter, UK gh378@exeter.ac.uk
Han **He**, National Astronomical Observatories- Chinese Academy of Sciences, China hehan@nao.cas.cn
Andrew **Hillier**, University of Exeter, UK a.s.hillier@exeter.ac.uk
Richard **Horne**, British Antarctic Survey, UK R.Horne@bas.ac.uk
Xin **Huang**, National Astronomical Observatories- Chinese Academy of Sciences, China xhuang@bao.ac.cn
Hugh **Hudson**, Space Sciences Lab, University of California, Berkeley CA, USA, UK hhudson@ssl.berkeley.edu
Ching Pui **Hung**, Institut de Physique du Globe de Paris / CEA-Saclay-AIM, France hung@ipgp.fr
Neal **Hurlburt**, LMATC, USA hurlburt@lmsal.com
Junga **Hwang**, Korea Astronomy and Space Science Institute, South Korea jahwang@kasi.re.kr
Nadia **Imtiaz**, TPD/PINSTECH, Pakistan nhussain@ualberta.ca
Karine **Issautier**, LESIA, France karine.issautier@obspm.fr
Caitriona **Jackman**, University of Southampton, UK c.jackman@soton.ac.uk
David **Jackson**, Met Office, UK david.jackson@metoffice.gov.uk
Daniel **Johnson**, Northumbria University, UK daniel5.johnson@northumbria.ac.uk

Navin Chandra **Joshi**, Kyung Hee University, South Korea njoshi98@gmail.com
Christina **Kay**, NASA GSFC, Republic of Ireland christina.d.kay@nasa.gov
Ralf **Keil**, ESA/ESOC, Germany ralf.keil@esa.int
Barry **Kellett**, STFC RAL Space, UK barry.kellett@stfc.ac.uk
Olga **Khabarova**, Institute of terrestrial magnetism (IZMIRAN)- Moscow, Russia olik3110@gmail.com
Rungployphan **Kieokaew**, University of Exeter, UK rk359@exeter.ac.uk
Jung Hee **Kim**, KASI, South Korea rskim@kasi.re.kr
Rok-Soon **Kim**, Kyungpook National University, South Korea hanulkkum@naver.com
Josuha **Knight**, Space Operations, RAF High Wycombe, UK Joshua.Knight138@mod.gov.uk
Delores **Knipp**, University of Colorado Boulder, USA delores.knipp@colorado.edu
Irina **Knyazeva**, Pulkovo Observatory, Russia iknyazeva@gmail.com
Petra **Kohutova**, University of Warwick, UK p.kohutova@warwick.ac.uk
Marianna **Korsos**, University of Sheffield, UK m.korsos@sheffield.ac.uk
Galina **Kotova**, Space Research Institute of RAS, Russia kotova@iki.rssi.ru
Garyfallia **Kromyda**, Aristotle University of Thessaloniki, Greece glkromyda@gmail.com
Maria **Kuznetsova**, NASA Community Coordinated Modeling Center, USA Maria.M.Kuznetsova@nasa.gov
Timo **Laitinen**, University of Central Lancashire, UK tlmlaitinen@uclan.ac.uk
Mai Mai **Lam**, University of Southhampton, UK mml@bas.ac.uk
KD **Leka**, NWRA, USA leka@nwra.com
Stefania **Lepidi**, Istituto Nazionale di Geofisica e Vulcanologia, Italy stefania.lepidi@ingv.it
Sanjay **Limaye**, University of Wisconsin, USA SanjayL@ssec.wisc.edu
Jon **Linker**, Predictive Science Inc., USA office@predsci.com
Mike **Lockwood**, Univrsity of Reading, UK m.lockwood@reading.ac.uk
Konstantina **Loumou**, University of Glasgow, UK k.loumou.1@research.gla.ac.uk
Janet **Luhmann**, SSL University of California- Berkeley, USA jgluhman@ssl.berkeley.edu
Sushant **Mahajan**, Georgia State University, USA mahajan@astro.gsu.edu
Olga **Malandraki**, National Observatory of Athens- IAASARS, Greece omaland@noa.gr
Ian **Mann**, University of Alberta, Canada imann@ualberta.ca
Petrus **Martens**, Georgia State University, USA martens@astro.gsu.edu
Valentin **Martinez Pillet**, National Solar Observatory, USA vmpillet@nso.edu
Helen **Mason**, University of Cambridge, UK hm11@damtp.cam.ac.uk
Aoife **McCloskey**, Trinity College Dublin- Ireland, Republic of Ireland mccloska@tcd.ie
Lee-Anne **McKinnell**, South African National Space Agency (SANSA), South Africa lmckinnell@sansa.org.za
Nigel **Meredith**, British Antarctic Survey, UK nmer@bas.ac.uk
Mauro **Messerotti**, INAF-Astronomical Observatory of Trieste, Italy messerotti@oats.inaf.it
Mark **Miesch**, HAO/NCAR, USA miesch@ucar.edu
Daniel **Miller**, University of Exeter , UK dm473@exeter.ac.uk
Diana **Morosan**, Trinity College Dublin, Republic of Ireland morosand@tcd.ie
Pradiphat **Muangha**, mahidol University, Thailand fhoone@hotmail.com
Karin **Muglach**, NASA GSFC, USA karin.muglach@nasa.gov
Milton **Munroe**, Retired Academic, High Wycombe, UK MMunroe300@aol.com
Sophie **Murray**, Trinity College Dublin, Republic of Ireland sophie.murray@tcd.ie
Kalevi **Mursula**, University of Oulu, Finland kalevi.mursula@oulu.fi
Irina **Myagkova**, Skobeltsyn Institute of Nuclear Physics - MSU, Russia irina@srd.sinp.msu.ru
Dibyendu **Nandi**, Center for Excellence in Space Sciences India- IISER Kolkata, India dnandi@iiserkol.ac.in
Gianluca **Napoletano**, University of L'Aquila, Italy gian.luke@hotmail.it
Ralph **Neuhaeuser**, AIU U Jena, Germany rne@astro.uni-jena.de
Ljubomir **Nikolic**, Natural Resources Canada, Canada ljubomir.nikolic@canada.ca
Naoto **Nishizuka**, NICT, Japan nishizuka.naoto@nict.go.jp
Nariaki **Nitta**, Lockheed Martin Solar and Astrophysics Laboratory, USA nitta@lmsal.com
Tomoya **Ogawa**, Kitasato University, Japan ogawa@kitasato-u.ac.jp
Terrance **Onsager**, NOAA Space Weather Prediction Center, USA terry.onsager@noaa.gov
Aleksandra **Osipova**, Central Astronomical Observatory - RAS, Pulkovo, Russia osisasha@gmail.com
Mathew **Owens**, University Of Reading, UK m.j.owens@reading.ac.uk
Vema Reddy **Panditi**, Indian Institute of Astrophysics, India vemareddy@iiap.res.in
Vaibhav **Pant**, Indian Institute of Astrophysics, India vaibhav@iiap.res.in
Jinhye **Park**, Kyung Hee University, South Korea jinhye@khu.ac.kr
Barbara **Perri**, CEA Saclay- AIM, France barbara.perri@cea.fr
Rui **Pinto**, IRAP - Institute of Research in Astrophysics and Planetology, France rui.pinto@irap.omp.eu
Dave **Pitchford**, SES, Germany dave.pitchford@ses.com
Arik **Posner**, NASA, USA arik.posner@nasa.gov
Gabby **Provan**, University of Leicester, UK gp31@le.ac.uk
Patricia **Reiff**, Rice University, USA reiff@rice.edu
Guenther **Reitz**, DLR (retired), Germany guenther.reitz@dlr.de
Davide **Rozza**, INFN and University of Milano-Bicocca- Italy, Italy davide.rozza@mib.infn.it
Chris **Russell**, UCLA, USA ctrussel@igpp.ucla.edu
Danislav **Sapundjiev**, Royal Meteorological Institute of Belgium, UK dasapund@meteo.be
Sabrina **Savage**, NASA Marshall Space Flght Ctr, USA sabrina.savage@nasa.gov
Camilla **Scolini**, KU Leuven, Belgium camilla.scolini@kuleuven.com
Michael **Sharpe**, Met Office, UK michael.sharpe@metoffice.gov.uk
Olga **Sheiner**, Radiophysical Research Institute, Russia oasheyner@gmail.com
Daikou **Shiota**, ISEE- Nagoya University, Japan shiota@isee.nagoya-u.ac.jp
Andrew **Sibley**, Met Office, UK andrew.sibley@metoffice.gov.uk
Hector **Socas-Navarro**, Instituto de Astrofisica de Canarias, Spain hsocas.iac@gmail.com
Nandita **Srivastava**, Udaipur Solar Observatory- PRL, India nandita@prl.res.in
Antoine **Strugarek**, CEA, France antoine.strugarek@cea.fr
Alexei **Struminsky**, Space Research Institute, Russia astrum@iki.rssi.ru
Bill **Swalwell**, UCLAN, UK bswalwell@uclan.ac.uk
Baolin **Tan**, National Astronomical Observatories of Chinese Academy of Sciences, China bltan@nao.cas.cn
Jianfei **Tang**, Xinjiang Astronomy Observatory, Chinese Academy of Sciences, China jftang@xao.ac.cn
Manuela **Temmer**, University of Graz, Austria manuela.temmer@uni-graz.at
Michael **Thompson**, National for Atmospheric Research, USA mjt@ucar.edu
Andrei **Tlatov**, Kislovodsk mountain astronomical station, Russia atlatov@gmail.com
Kseniya **Tlatova**, Kislovodsk mountain astronomical station, Pulkovo observatory, Russia k.tlatova@mail.ru
Yurdanur **Tulunay**, Middle East Technical University Ankara Turkey, Turkey ytulunay@ae.metu.edu.tr
S. Ersin **Tulunay**, ODTU-METU, Turkey ersintul@metu.edu.tr

Esa **Turunen**, University of Oulu, Finland esa.turunen@oulu.fi
Pauli **Visnen**, University of Oulu, Finland pauli.vaisanen@oulu.fi
Christine **Verbeke**, KU Leuven, Belgium christine.verbeke@kuleuven.be
Erwin **Verwichte**, University of Warwick, UK Erwin.Verwichte@warwick.ac.uk
Aline **Vidotto**, Trinity College Dublin, Republic of Ireland aline.vidotto@tcd.ie
Nicole **Vilmer**, LESIA-Paris Observatory, France nicole.vilmer@obspm.fr
Vincenzo **Vitale**, ASI & INFN, Italy vincenzo.vitale@roma2.infn.it
Mikhail **Vokhmyanin**, Saint Petersburg State University, Russia m.vokhmyanin@spbu.ru
Huaning **Wang**, Natioanl Astronomical Observatories - CAS, China hnwang@bao.ac.cn
David **Webb**, Boston College/ISR, USA david.webb@bc.edu
Maria **Weber**, University of Exeter, UK mweber@astro.ex.ac.uk
Marion **Weinzierl**, Durham University, UK marion.b.weinzierl@gmail.com
Elisabeth **Werner**, Uppsala University, Sweden elisabeth.werner.1453@student.uu.se
Matthew **West**, Royal Observatory of Belgium, UK matthew.west@observatoire.be
Yan **Yan**, National Astronomical Observatories of China, China yyan@bao.ac.cn
Yihua **Yan**, CAS Key Lab of Solar Activity- National Astronomical Observatories, China yyh@nao.cas.cn
Donghe **Zhang**, Peking University, China zhangdh@pku.edu.cn
Valentina **Zharkova**, Northumbria University, UK valentina.zharkova@northumbria.ac.uk

Part 1. Solar Drivers and Activity Levels

Space Weather of the Heliosphere: Processes and Forecasts
Proceedings IAU Symposium No. 335, 2017
C. Foulon & O.E. Malandraki, eds.

doi:10.1017/S1743921317007955

Statistical Analysis of Individual Solar Active Regions

Garyfallia Kromyda[1] and Loukas Vlahos[2]

[1]Department of Physics, Aristotle University of Thessaloniki GR-52124 Thessaloniki, Greece
email: glkromyda@gmail.com

[2]Department of Physics, Aristotle University of Thessaloniki GR-52124 Thessaloniki, Greece
email: vlahos@astro.auth.gr

Abstract. In the last decades, numerous observational and computational studies have shown that the global flare distribution is a power-law with a slope less than 2. In these studies, active regions are treated as statistically indistinguishable. To test this, we identify and separately analyze the flares produced by ten individual active regions (2006-2016). In five regions, we find a single power-law distribution, with a slope of $a < 2$. In the other five, we find a broken double power-law distribution, with slopes $a_1 < 2$ and $a_2 > 2$.

Keywords. Sun: corona, flares, X-rays, radiation mechanisms: thermal, methods: statistical, data analysis

1. Introduction

A possible coronal heating scenario was proposed by Parker (1988). Parker envisioned a corona full of magnetic flux tubes of all sizes, where small reconnection events are happening along each flux tube at a high occurrence rate. These small events release energy on the order of $10^{20} - 10^{24}$ erg and are called nano-flares. The energy released heats the corona, which radiates some of this energy away, predominantly in soft X-rays. The solar X-ray flare distribution follows a power-law form:

$$f(E) = f_o E^{-\alpha}, \quad \alpha > 0 \tag{1.1}$$

where $f(E)dE$ is the fraction of flares releasing an amount of energy between E and $E+dE$ per unit time. As shown in Hudson (1991) the total energy released per unit time by the ensemble of flares is:

$$E_{tot} = \int_{E_{min}}^{E_{max}} f(E)EdE = f_o \left[\frac{E^{2-\alpha}}{2-\alpha}\right]_{E_{min}}^{E_{max}}, \quad \alpha \neq 2 \tag{1.2}$$

- If $\alpha < 2$: the larger flares of the distribution are the source of coronal heating.
- If $\alpha > 2$: nanoflares are the source of energy for the corona.

Up to today, all observational studies have found distributions with slopes $1.7 < \alpha < 1.9$, contrary to what Parker's nano-flare heating theory would require.

The theory that is able to explain the power-law flare distribution is Self-Organized Criticality, pioneered by Bak *et al.* (1987, 1988), and adapted to solar flares by Lu & Hamilton (1991) and Lu *et al.* (1993) (see recent review by Aschwanden *et al.* 2016). An active region is treated as a system of energy storing nodes which are driven slowly (the energy injection rate is much slower than the energy release rate) and weakly (the amounts of energy injected into the system are much smaller than the amounts stored at any node) and exchange energy with each other. If the value at a node exceeds a critical

Active Region	Activity Period	# AR flares all flares	# X	# M	# C	# B	q-values
10848	14 - 29/1/2006	83/99	-	-	7	75	0.84
10930	3 - 18/12/2006	167/183	4	5	46	112	0.91
10960	31/5 - 13/6/2007	102/125	-	10	16	76	0.82
10963	7 - 18/7/2007	79/94	-	-	18	61	0.84
10978	3 - 19/12/2007	79/119	-	-	9	66	0.66
11029	24/10 - 1/11/2009	64/71	-	-	9	53	0.90
11158	11 - 21/2/2011	82/124	1	7	65	9	0.66
11271	13 - 30/8/2011	76/130	-	-	11	65	0.58
12192	14 - 31/10/2014	134/144	6	34	94	-	0.93
12529	4 - 21/4/2016	78/88	-	1	30	47	0.89

Figure 1. The selected active regions of this study

threshold, an instability able to also destabilize neighboring nodes sets in, and energy is released, known as an avalanche. The distribution of the magnitude of the avalanches for a system in the SOC state is a power-law. The avalanche time-series have a '1/f' power spectrum, with slopes between 1 and 3, indicative of long temporal correlations.

2. Data Collection

In this study we conduct a statistical analysis of the flares produced by 10 individual active regions, analyzing the flares of each active region separately. We search for active regions that produced more than 60 flares during their lifetime on the visible side of the Sun. To identify them we use the SolarSoft Latest Events Archive, where each flare record is assigned to an active region and is accompanied by a flare locator image. This search produced 10 active regions, shown in Figure 1. For each of these, we construct a flare-list, where each flare record has a start, stop and peak time, a position and a magnitude.

We start from the GOES soft X-ray time-series (long channel at $1 - 8\mathring{A}$ and short channel at $0.5 - 4\mathring{A}$), which show all flares irrespective of their active region origin. To construct time-series that show the flares of a single active region, we use the start, stop and peak times from the active region flarelist. We constructed an analysis algorithm in the IDL programming language that implements the following:

(a) Fills up data gaps in the observed time-series using interpolation of a linear model between the original flux value of the last point before the gap and the original flux value of first point after the gap. These gaps are not during flares. Their correction is necessary in order for the algorithm to correctly match times from the flare-lists to observations in the time-series.

(b) Corrects for duplicate flares (13 duplicate flares in total) by counting them only one time.

(c) If the start, stop and peak times from the flare-lists are plotted on the time-series, occasional discrepancies are revealed. The algorithm determines new start, stop and peak times for the flares, for the short and the long channel separately.

(d) Background subtraction for each flare (for the two channels separately). The algorithm:

- creates a background flux profile using the Wheatland (2010b) method.
- creates an alternative background flux profile by interpolating a linear fit between the flux at the start and the flux at the end of the flare.

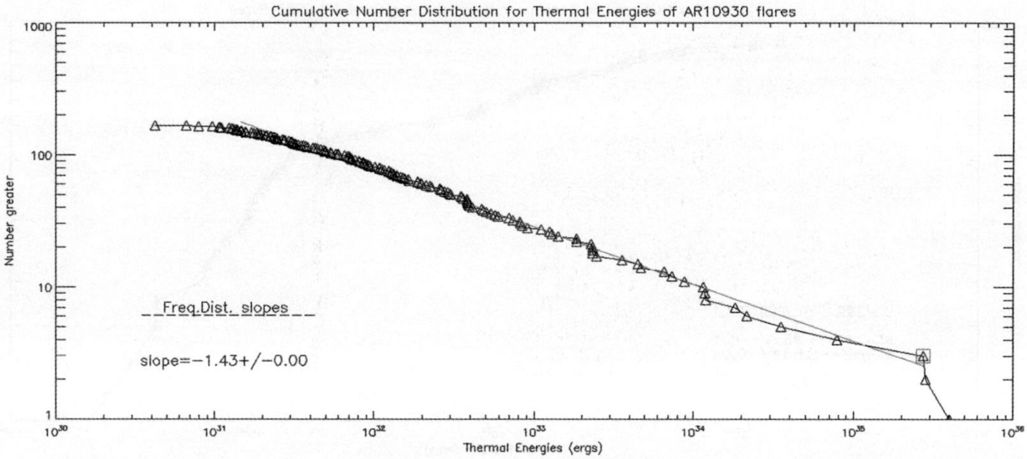

Figure 2. Cumulative Distribution of flare thermal energy events for active region 10930

- chooses the most appropriate background profile for each flare separately. This is the one that when subtracted, makes the peak flux of the flare profile greater. The two peak flux values do not differ by much, but using this criterion the most visually reasonable flare profile is chosen.

Using the two channels and the goes_chianti_tem SolarSoft routine, the temperature (T) and emission measure time-series (EM) corresponding to the flare timeseries are created. The thermal energy E acquired by plasma due to a flare can be found by assuming thermal equilibrium of a fully ionized hydrogen plasma, so that $E = 3kT\,\mathrm{EM}/n$. Using the calculated T and EM, and $n = 10^{10}\,cm^{-3}$, thermal energy time-series can be calculated.

3. Analysis and Results

Using the thermal energy time-series of each active region, and the thermal energy events corresponding to the region's flares, we construct cumulative distributions, as shown in Figure 2 and 3. For each energy value on the x-axis, the y-axis value corresponds to the number of flares with energy greater than the x-axis value.

For all 10 active regions (ARs), the powerlaw at the lowest energy limit, around 10^{31} erg, is less than 2, against what nano-flare heating requires. Furthermore, only half of the ARs show a single power-law. The other half show broken power-laws, with a first part having a slope less than 2, and a second part having a slope greater than 2. Only the single slope ARs can be considered true SOC systems. Their slope values are in agreement with the values predicted by Lu & Hamilton (1991). The double slope ARs cannot be considered true SOC systems. As a result we claim that the corona as a whole may be in a SOC state, but not all ARs can reach a SOC state during their flare-producting lifetime.

A possible explanation for broken power-laws is the existence of strong drivers, based on Georgoulis & Vlahos (1996) and Charbonneau (2001) simulations. According to Charbonneau (2001) strong drivers, where the energy injected at any time is comparable to the energy stored in system nodes, hinder the creation of large clusters. Thus scale-free, true SOC systems cannot exist, and mid-sized flares are favored, giving rise to broken power-laws. In the future, we plan to test the validity of the above claims, using Hinode vector magnetograms, since a magnetogram-driven simulation can clarify this point.

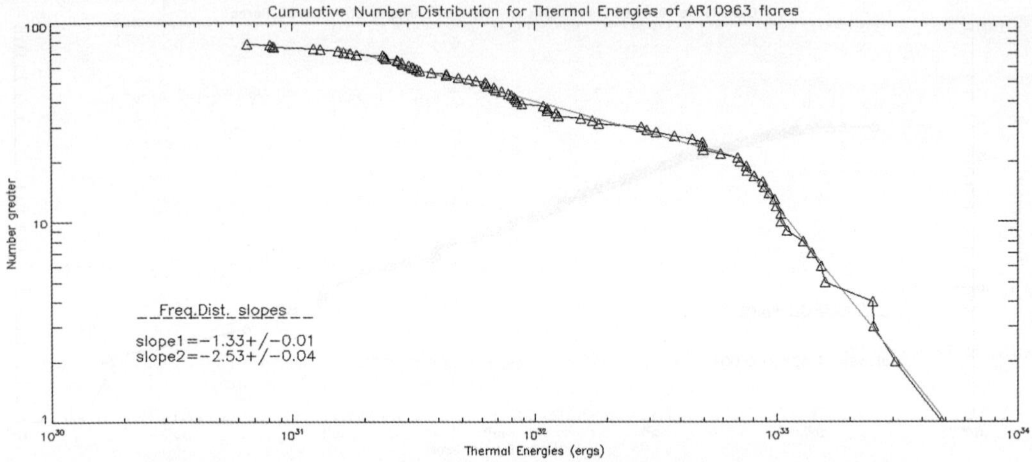

Figure 3. Cumulative Distribution of flare thermal energy events for active region 10963

No correlation is found in scatter plots of waiting time versus the energy of the previous or the following flare. Flares in individual ARs seem to be independent events in time. Such events have Poisson distributions, and their slopes in log-lin axis give us the average waiting time of the AR. This is inversely proportional to the average driving rate of the AR. We find one slow and one fast average waiting time, implying the existence of time-varying drivers in ARs. In general, the waiting time distributions are well-fitted by doublewise Poisson distributions.

The power spectrum of flare time-series of both channels sometimes show clear power-laws, other times show power-laws with fluctuations. The slopes average to 1.7. The auto-correlation functions (ACF) of the flare time-series show short correlation times. All these indicate that some type of self-organization exists in all ARs, but ARs are not typical SOC systems.

The power spectrum of background time-series of both channels (data points bewteen start and end of all flares in the original GOES timeseries are replaced with a linear fit between flux at start and flux at end of the flares) show clear powerlaws in all cases. The slopes average to 2.1. The ACFs of the background time-series show long correlation times. All these indicate that the background corona could be a SOC system in its own right, even though we cannot yet observationally identify the nano-flares it produces.

References

Aschwanden, M. *et al.* 2016, *Space Science Reviews*, 198, 47

Bak, P., Tang, C. & Wiesenfeld, K. 1987, *Physical Review Letters* vol. 59, p. 381-384

Bak, P., Tang, C. & Wiesenfeld, K. 1988, *Physical Review A*, vol. 38, p. 364-374

Charbonneau, P. 2001, *Solar Physics*, 203, 321-353

Freeland, S. L. & Handy, B. N. 1998, *Solar Physics*, 182, Issue 2, 497-500

Georgoulis, M. & Vlahos, L. 1996, *Astrophysical Journal*, 496, L135

Hudson, H. S. 1991, *Solar Physics*, 133, 357-369

Lu, E. T. & Hamilton, R. J. 1991, *ApJ*, 380, 89-92

Lu, E. T., Hamilton, R. J., McTiernan J. M. & Bromund, K. R. 1993, *ApJ*, 412, 841-852

Parker, E. N. 1988, *ApJ*, 330, 474-479

Vlahos, L., Georgoulis, M., Kluiving, R. & Pascos, P. 1995, *Astronomy and Astrophysics*, 299, 897

Wheatland, M. S. 2010, *ApJ*, 710, 1324-1334

Space Weather of the Heliosphere: Processes and Forecasts
Proceedings IAU Symposium No. 335, 2017
C. Foullon & O.E. Malandraki, eds.

doi:10.1017/S1743921317007670

Magnetic Activity Discrepancies of Solar-Type Stars Revealed by *Kepler* Light Curves

Han He[1,*], **Huaning Wang**[1], **Yan Yan**[1] **and Duo Yun**[1]

[1]CAS Key Laboratory of Solar Activity, National Astronomical Observatories, Chinese Academy of Sciences, Beijing, China
*email: hehan@nao.cas.cn

Abstract. Magnetic activity information is concealed in the shape of stellar light curves owing to the process of rotational modulation. We developed approaches to extract magnetic activity characteristics from stellar light curves, and applied the method to a solar-type star observed with *Kepler* space telescope and also to the Sun for comparison. The result reveals distinct magnetic activity discrepancies between the solar-type star and the Sun. (1) The light-curve periodicity of the solar-type star is generally stronger than that of the Sun. (2) For the solar-type star, when the range of light-curve fluctuation is larger, the periodicity is also higher; while for the Sun, only during the solar minima with minimal range of fluctuation, the light curves show some periodicity. We propose that on the solar-type star, it is the large-scale magnetic field that leads to the light curves with both high periodicity and large range of fluctuation.
Keywords. stars: activity, stars: magnetic field, stars: rotation, stars: solar-type, Sun: activity

1. Motivation

Magnetic activity properties of solar-type stars can be revealed by the shape of stellar light curves as demonstrated by the light curve of the Sun (Shapiro *et al.* 2016). The dark or bright magnetic features (such as spots and faculae) on the surface of a star cross the stellar disk along with the stellar rotation and cause regular or irregular fluctuation of light curves. This process is known as rotational modulation (Debosscher *et al.* 2011). By careful investigation on the light curves of the Sun, it can be found that the fluctuation range of the solar light curves is larger during solar maxima and smaller during solar minima, and such fluctuation range is in phase with the variation of the sunspot number. This property indicates that the rotational modulation of the Sun is dominated by sunspots. In addition, a wavelet or Fourier analysis of the solar light curves shows that the fluctuation of the light curves is more regular (periodic) during solar minima and irregular during solar maxima (Lanza *et al.* 2003). The irregular light curves during solar maxima can be understood by the rapid evolution of sunspots; while the possible magnetic features that cause the more regular light-curve modulation during solar minima need further investigation since sunspots are rarely seen during solar minima. He *et al.* (2015) compared the light curve of the Sun with the simultaneously observed photospheric magnetograms and found that it is the bright faculae associated with the enhanced network magnetic field that lead to the regular modulation. The spatial size of facula regions on the surface of the Sun is much larger than sunspots, and they are more stable and have a longer lifetime than sunspots. Thus the light curves modulated by the facula regions during solar minima is more regular and periodic than the light curves dominated by the sunspots during solar maxima.

From the analysis carried out on the light curves of the Sun, it can be seen that the shapes of the solar light curves show two kinds of cyclic variation: the cyclic variation of the range of fluctuation and the cyclic variation of the degree of periodicity. Both the two kinds of variations reflect the magnetic activity characteristics of the Sun. We want to know if these activity properties also exist on the solar-type stars other than the Sun. The *Kepler* space telescope (Borucki *et al.* 2010; Koch *et al.* 2010) observed a large number of solar-type stars in its field and provided continuous light curves of these solar-type stars for about four years. We developed a method to infer the general properties of the magnetic field of a star through the study of its light curve, and we applied our method to both a solar-type star observed with *Kepler* space telescope and the Sun for comparison. The result of our analysis reveals distinct magnetic activity discrepancies between the solar-type star and the Sun. In Section 2, we briefly explain the analytical method for the light curves. In Section 3, we present the result of our analysis.

2. Method

We aim to employ two mathematical measures to quantitatively evaluate the time variation of the fluctuation range and the degree of periodicity of a light curve. The original light-curve data are normalized by the median value to yield the relative flux expression. The definitions of the two measures are based on the relative flux data.

The fluctuation range of a light curve is evaluated by using the rms algorithm (e.g, García *et al.* 2010; Chaplin *et al.* 2011). After calculating the rms of the light curve using the relative flux data, we multiply the rms value by a factor of $2\sqrt{2}$ to reflect the intuitive range between crest and trough of the light curve. For example, given a sine curve $A\sin(\omega t)$, the rms is $\frac{A}{\sqrt{2}}$, while $2\sqrt{2}\cdot\text{rms} = 2\text{A}$, which is just the true range between crest and trough. Since the fluctuation of a real light curve is generally not uniform, we call this quantity the effective range (R_{eff}) of the light curve.

The degree of periodicity of a light curve is evaluated by using the autocorrelation algorithm (e.g., Chatfield 2003). Firstly, the autocorrelation function $\rho(h)$ is derived for the whole light curve with the time lag h from 0 to $N-1$, where N is the total number of the data points in the light curve. Then the average value of $|\rho(h)|$ for the first half of the autocorrelation function ($h \leqslant \frac{N}{2}$) is calculated. We call this quantity the autocorrelation index (i_{AC}) of the light curve. Larger i_{AC} value means higher (i.e., stronger) periodicity. We do not use the second half of the autocorrelation function to evaluate i_{AC} because if $h > \frac{N}{2}$, not all data points in the light curve are utilized to calculate $\rho(h)$, thus the second half of the autocorrelation function cannot reflect the global property of the light curve.

Here, we only give a brief description of the two measures. The detailed definitions and formulas can be found in He *et al.* (2015).

3. Result

We applied the two measures of light curve, R_{eff} and i_{AC}, to a solar-type star observed with *Kepler* (*Kepler* object ID: KIC 10864581). This star was selected from a superflare star catalog compiled by Shibayama *et al.* (2013) using the *Kepler* observations. The light-curve data of *Kepler* is arranged by quarters, in which quarters 2 to 16 (Q2–Q16) are full-length quarters (1 quarter $\sim$ 3 months). We evaluated the two measures for the light curves of the solar-type star quarter by quarter from Q2 to Q16, and obtained the time variation information of the two measures with the quarters. Before deriving the two measures, the noises and outliers in the original light curve data had been filtered

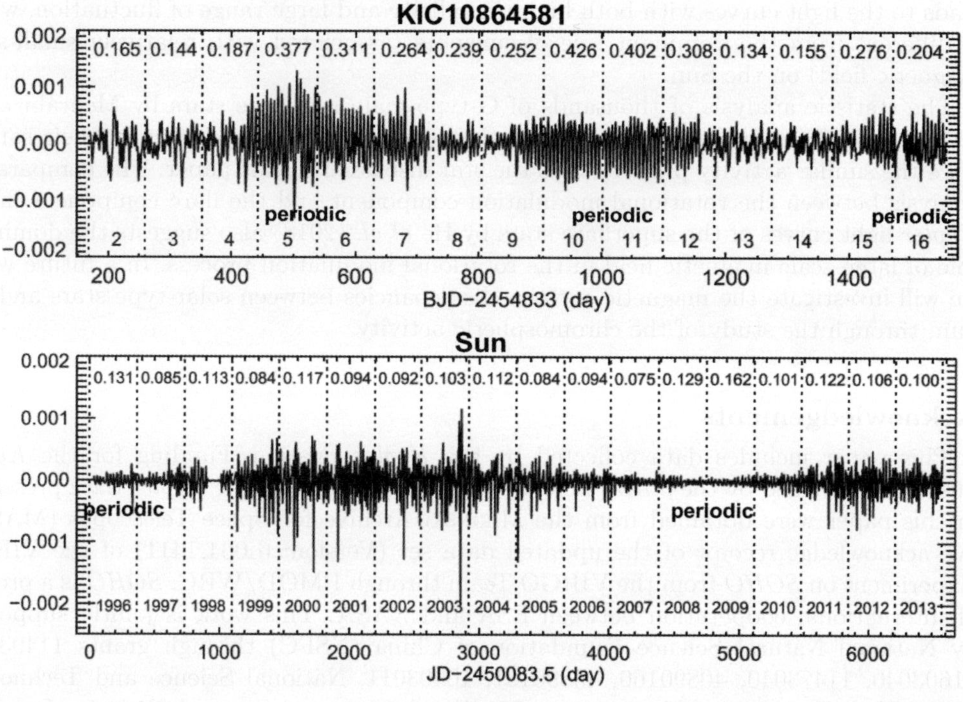

Figure 1. Top panel: Processed light curves (Q2–Q16) of the solar-type star (KIC 10864581) observed with *Kepler*. The quarter numbers are given below the light curves. The decimal numbers above the light curves are $i_{\rm AC}$ values of each quarter. Bottom panel: Processed light curves of the Sun (1996–2013) observed with *SOHO*. The year numbers are given below the light curves. The decimal numbers above the light curves are $i_{\rm AC}$ values of each year.

out. The processed light curves of the solar-type star (in relative flux expression) are shown in the top panel of Figure 1.

For comparison, we also applied the two measures to the light curves of the Sun. The solar light curves employed in this paper are from the total solar irradiance (TSI) data during 1996–2013, which were observed by the VIRGO instrument (Fröhlich *et al.* 1997) aboard the *SOHO* spacecraft (Domingo *et al.* 1995). The long-term trend in the TSI data was filtered out before calculating the two measures since it is irrelevant with the rotational modulation (Fröhlich & Lean 2004). The noises in the light curves were also filtered out. The two measures were evaluated year by year based on the processed light curves which are shown in the bottom panel of Figure 1.

Figure 1 demonstrates that the light curves of the solar-type star do have cyclic variation of fluctuation range and of periodicity, and the dynamic range (upper limit value of $R_{\rm eff}$) of the light curves has the same order of magnitude as the Sun (about 10^{-3}).

However, the result of the analyses also reveals distinct magnetic activity discrepancies between the solar-type star and the Sun. (1) The light-curve periodicity of the solar-type star is generally stronger than the one of the Sun, which is indicated by the relatively higher $i_{\rm AC}$ values of the solar-type star than the $i_{\rm AC}$ values of the Sun (see Figure 1). (2) For the solar-type star, when the range of light-curve fluctuation is larger, the periodicity is also higher; while for the Sun, only during the solar minima with minimal range of fluctuation, the light curves show some periodicity (see 'periodic' labels in Figure 1).

Based on the above result, we propose that on the solar-type star, it is the large-scale magnetic field (e.g., Vidotto *et al.* 2014) that dominates the rotational modulation and leads to the light curves with both high periodicity and large range of fluctuation, which is different from the dominant role of sunspots (associated with relatively small-scale magnetic field) on the Sun.

The statistic analysis of thousands of G-type main sequence stars by Mehrabi *et al.* (2017) based on the *Kepler* data shows that most of the G-type stars in their sample have the similar activity properties as the star discussed in this paper. The comparative analysis between the rotational modulation component and the flare component in the *Kepler* light curves of the superflare stars by He *et al.* (2018) also suggests the dominant role of large-scale magnetic field in the rotational modulation process. In a future work, we will investigate the magnetic activity discrepancies between solar-type stars and the Sun, through the study of the chromospheric activity.

Acknowledgements

This paper includes data collected by the *Kepler* mission. Funding for the *Kepler* mission is provided by the NASA Science Mission directorate. The *Kepler* data presented in this paper were obtained from the Mikulski Archive for Space Telescopes (MAST). We acknowledge receipt of the updated data set (Version: 6_004_1411) of the VIRGO Experiment on *SOHO* from the VIRGO Team through PMOD/WRC. *SOHO* is a project of international cooperation between ESA and NASA. This work is jointly supported by National Natural Science Foundation of China (NSFC) through grants 11403044, 11603040, 11473040, 40890160, 40890161, 10803011, National Science and Technology Basic Work Program of China (grant 2014FY120300), and Strategic Priority Research Program on Space Science, Chinese Academy of Sciences (grant XDA04060801).

References

Borucki, W. J., Koch, D., Basri, G., *et al.* 2010, *Science*, 327, 977
Chaplin, W. J., Bedding, T. R., Bonanno, A., *et al.* 2011, *ApJ* (Letters), 732, L5
Chatfield, C. 2003, *The analysis of time series: An introduction*, 6th Edition (Chapman & Hall)
Debosscher, J., Blomme, J., Aerts, C. & De Ridder, J. 2011, *A&A*, 529, A89
Domingo, V., Fleck, B. & Poland, A. I. 1995, *Solar Phys.*, 162, 1
Fröhlich, C. & Lean, J. 2004, *A&AR*, 12, 273
Fröhlich, C., Crommelynck, D. A., Wehrli, C., *et al.* 1997, *Solar Phys.*, 175, 267
García, R. A., Mathur, S., Salabert, D., *et al.* 2010, *Science*, 329, 1032
He, H., Wang, H. & Yun, D. 2015, *ApJS*, 221, 18
He, H., Wang, H., Zhang, M., *et al.* 2018, *ApJS*, in press, arXiv:1705.09028
Koch, D. G., Borucki, W. J., Basri, G., *et al.* 2010, *ApJ* (Letters), 713, L79
Lanza, A. F., Rodonò, M., Pagano, I., Barge, P. & Llebaria, A. 2003, *A&A*, 403, 1135
Mehrabi, A., He, H. & Khosroshahi, H. 2017, *ApJ*, 834, 207
Shapiro, A. I., Solanki, S. K., Krivova, N. A., Yeo, K. L. & Schmutz, W. K. 2016, *A&A*, 589, A46
Shibayama, T., Maehara, H., Notsu, S., *et al.* 2013, *ApJS*, 209, 5
Vidotto, A. A., Gregory, S. G., Jardine, M., *et al.* 2014, *MNRAS*, 441, 2361

Space Weather of the Heliosphere: Processes and Forecasts
Proceedings IAU Symposium No. 335, 2017
C. Foullon & O.E. Malandraki, eds.

doi:10.1017/S1743921317007992

Variation of Quiet Sun Radiation during Solar Cycles 23 and 24

Mahender Aroori, G. Yellaiah and K. Chenna Reddy

Department of Astronomy, Osmania University, Hyderabad, Telangana, India
email: mahenderastro@gmail.com

Abstract. Radio observations play a very important role in understanding the structure of the solar atmosphere. In this paper the quiet sun component of the solar radio emission has been investigated using data obtained from the Solar Indices Bulletin, National Geophysical Data Centre. By statistical method, the quiet sun component is estimated for 84 successive basic periods containing three solar rotations each using data obtained at different frequencies. From the quiet sun component we estimate the brightness temperature in each observing frequency.

Keywords. Sun: Radio radiation - Sun: Quiet sun radiation - Sun: Chromosphere, etc.

1. Introduction

Since the discovery of solar radio waves by Hey (1946), Southworth (1945) and Reber (1944), radio emission from the Sun can be broadly classified into two categories: disturbed emission and undisturbed emission. The undisturbed emission is further divided into two classes, the quiet sun radiation and the slowly varying component. The slowly varying component is due to localized active regions on the solar disk such as sunspots, the sunspot associated magnetic field, coronal mass ejections, solar flares, filaments and prominences. If the emission due to these active regions in the solar atmosphere is subtracted or removed, we get quiet sun radiation. Study of the quiet sun radiation enables us to know the size of the corona, electron density and electron temperature in different layers of the solar atmosphere. Measurement of the quiet sun emission provides accurate information about the source region of the radio emission. It also provides detailed information about the brightness temperature, magnetic fields and physical nature of the various sources inside the solar structure (Kundu 1965). The various frequency bands ranging from 410 MHz to 15400 MHz originate from different layers of the solar atmosphere starting from upper chromosphere to lower chromosphere.

2. Data

Sunspot number (SSN) and solar radio flux (F) at different frequencies during solar cycles 23 and 24 (1995-2013) have been collected from the Solar Indices Bulletin, National Geophysical Data Centre, Boulder published from the Department of Commerce, U.S. The radio flux is given in Solar Flux Unit (SFU), a measure of energy received per unit time per unit frequency interval. One SFU is equals to $10^{-22} W/m^2 Hz$.

3. Procedure and Analysis

The present paper provides the variation of the quiet sun component during solar cycles 23 and 24 (1995-2013). The period under investigation has been divided into 84 periods each of 81 days consisting of three successive Carrington rotations. The quiet sun

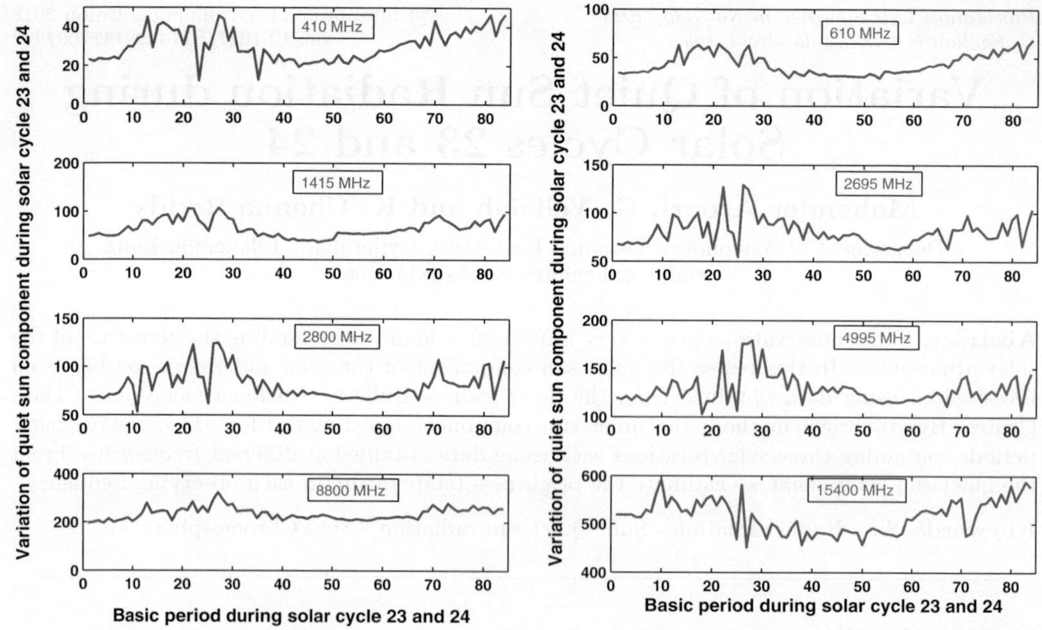

Figure 1. Variation of quiet sun radiation with basic periods at different frequencies(410, 610, 1415, 2695, 2800, 4995, 8800 and 15400 MHz) during solar cycle 23 and 24.

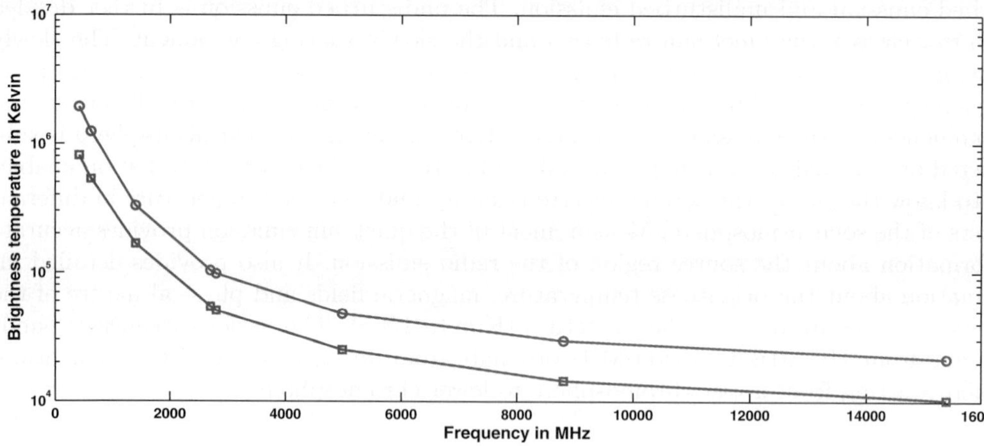

Figure 2. Variation of brightness temperature with frequency during solar cycle 23 (blue squares) and solar cycle 24 (red circles). To avoid data overlapping, cycle 24 data was shifted two times on y-axis.

component for each frequency under the present study is determined from the regression of scattered plot between radio flux and sunspot number. The generalised equation can be written as

$$F = A + B \times (SSN) \tag{3.1}$$

where A and B are arbitrary constants. The value of constant A contributes to the radio flux when the SSN is taken as zero; this simply gives the quiet sun component. The corresponding variations in brightness temperature has been determined by using the relation (Pawsey and Yabsley 1949)

$$T_b = \frac{Sc^2}{2k\nu^2\Omega}, \tag{3.2}$$

where k is the Boltzmann's constant (1.38×10^{-23} J / K), Ω is the solid angle subtended by the sun (where mean solid angle 6.8×10^{-5} Sr is taken from Pawsey and Yabsley 1949), ν is the frequency (MHz) and T_b is in degree Kelvin.

4. Results and Conclusion

Figure 1 shows the variation of quiet sun component at different frequencies averaged over 3 Carrington rotations, comprising 84 basic periods during the solar cycles 23 and 24 (1995-2013). The systematic fall and rise of quiet sun radiation connected with the 11-year solar cycle in Figure 1 shows some significant fluctuations in the course of passing from one basic period to another, similar to earlier studies (Zieba and Gula 1976). The quite sun component during solar minimum period is found to be stable at lower frequencies, consistent with the reports of Lokanadham & Subramanian (1986). Whereas at higher frequencies showing slight variation, while there is considerable variations during solar maximum period at all the frequencies. During the solar maximum period, the quiet sun component shows considerable differences at all the frequencies, and it remained more or less constant at lower frequencies with slight variations at higher frequencies during the solar minimum period. The number of disintegrating centres (break up of the solar activity centers) are so great that synchrotron radiation generated in their magnetic fields (during the maximum of the solar activity) could possibly have caused the differences. The decreasing density of electrons in the solar atmosphere causes a decline of thermal radiation especially on short waves (Kundu 1965). As outlined by Zieba & Gula (1976), the variation of the basic component with the phase of solar cycle depends on two basic processes namely variation of electron density in solar atmosphere and decay of vanishing magnetic fields. Mutual dependence of these two effects decides the actual shape of the quiet sun spectra. From Figure 2 the brightness temperature is decreasing with increasing frequency, this shows clearly the thermal nature of the quiet sun component at all frequencies.

5. Acknowledgement

We acknowledge the National Geophysical Data Centre (NGDC) Boulder, Department of Commerce, U.S for open source of data. And also we acknowledge the IAU for financial support and local hospitality. One of the authors, AM, is grateful to Honorary Chairman, SWAEROS for travel support.

References

Hey, 1946, *Nature*, 157, 47
Kundu, M. R., *Solar Radio Astronomy* 1965, *John Wiley & Sons publications*
Lokanadham, B. & Subramanian, P. K. 1986, *Astrophysics and Space Science*, 119, 27-32
Pawsey, J. L. & Yabsley, D. E., 1949, *Australian Journal of Scientific Research*, Vol. 2, p. 98
Reber, G. 1944, *Astrophys. J.*, 100, 279
Southworth, G. C. 1945, *J. Franklin Inst.*, 239, 285
Zieba, S. & Gula, R., 1976, *Acta Astronomica*, Vol. 26 No. 1

Space Weather of the Heliosphere: Processes and Forecasts
Proceedings IAU Symposium No. 335, 2017
C. Foullon & O.E. Malandraki, eds.

doi:10.1017/S1743921317007980

Temperature of Source Regions of ^{3}He-Rich Impulsive Solar Energetic Particle Events

N.-H. Chen[1], **R. Bučík**[2,3] **and R.-S. Kim**[1,4]

[1]Korea Astronomy and Space Science Institute, Daejeon, Republic of Korea
email: nhchen@kasi.re.kr

[2]Institut für Astrophysik, Georg-August-Universität Göttingen, Göttingen, Germany
[3] Max-Planck-Institut für Sonnensystemforschung, Göttingen, Germany
[4] University of Science and Technology, Daejeon, Republic of Korea

Abstract. Impulsive solar energetic particle (SEP) events originate from the energy dissipation process in small solar flares. Anomalous abundances in impulsive SEP events provide an evidence on unique, yet unclear, acceleration mechanism. The pattern of heavy-ion enhancements indicates that the temperature of the source plasma that is accelerated is low and not flare-like. We examine the solar source of the ^{3}He-rich SEP event of 2012 November 20 using Solar Dynamics Observatory (SDO)/ Atmospheric Imaging Assembly (AIA) images and investigate its thermal variation. The examined event is associated with recurrent coronal jets. The Differential Emission Measure (DEM) analysis is applied to study the temperature evolution/distribution of the source regions. Preliminary results show that the temperature of the associated solar source is ranged between 1.2-3.1 MK.

Keywords. Sun: UV radiation, particle emission, corona

---

1. Introduction

Impulsive solar energetic particles (SEP) events, discovered more than 40 years ago, are understood to come from the energy dissipation process in solar flares. One remarkable feature of impulsive SEP events is the enhanced abundance of ^{3}He isotopes and other heavier ions compared to corona. The in-situ observations of ion composition may provide important information on a possible nature of the sources as well as a clue on mechanism of particle acceleration on the sun.

It has been known from earilier observations that element abundance enhancement is an increasing function of mass-to-charge ratio (A/Q) of ions in impulsive SEP events (e.g., Reames *et al.* 2014; 2015a,b). An examination of theoretical A/Q profiles for various ion species at equilibrium temperature T indicates that any subtle variations in the temperature of source plasma could play a significant role in the properties of SEP events detected later. A narrow temperature range of 2-4 MK in the source region has been inferred to be responsible for the observed element abundances (Reames *et al.* 2014), though some studies (Mason *et al.* 2002, 2016) have proposed even lower temperatures ($<$ 1.5 MK). A direct measurement of charge states for low energy ions is consistent with the ambient plasma temperature (DiFabio *et al.* 2008). The higher charge states for high-energy ions are suggested to be due to charge stripping after acceleration. Thus the temperature of the source region in impulsive SEP events is uncertain as well as whether the temperature solely determines which element is enhanced during the acceleration.

With improved remote sensing observations (such as SDO/AIA) now, we are able to derive the detailed temperature distribution in active regions and thus to investigate the

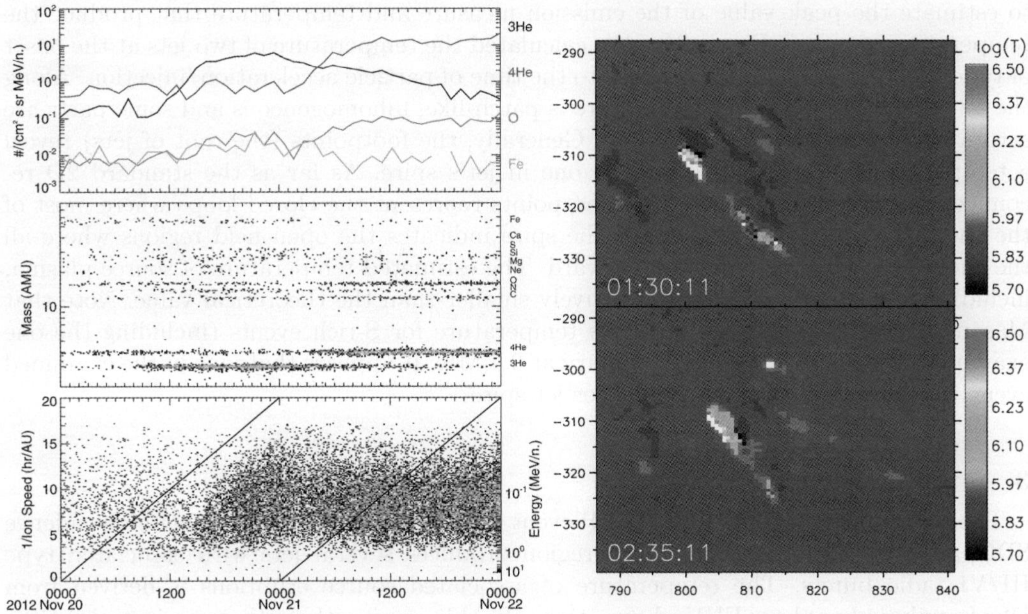

(a) (b)

Figure 1. (a) ^{3}He-rich SEP event. The panels from top to bottom are: 1 hr Intensity profile of ^{4}He, ^{3}He, O and Fe in energy range 0.23 - 0.32 MeV/nucleon; Mass spectrogram in energy range 0.4-10 MeV/nucleon and Inverted ion velocity-time spectrogram in mass range 10-70 amu. (b)Peak temperature map of the event source. The colorbar aside gives the magnitude of temperature. The selected subfigures are around the onset time of the accompanied type III burst.

conditions for the enhancement mechanism in ^{3}He-rich impulsive SEP events. In this work, we report for the first time the derived temperature distribution of the source region in one ^{3}He-rich impulsive SEP event. It is shown in section 2. The discussion and summary is given in section 3.

2. Observations and data analysis

The event under study was observed on 2012 November 20 (see Figure 1a). This prominent event started around 10:00 UT at 0.23-0.32 MeV/nucleon with a large ^{3}He enrichment and was intervened at the beginning of November 21 by a large-gradual SEP event with a normal coronal composition. Its source region has been identified as active region NOAA 11617 based on our identification method, including the connection analysis via the interplanetary and coronal field model and spatial-temporal coincidence on the EUV multi-wavelengths observation with the type III radio bursts (Chen *et al.* 2015). The associated solar eruption in the source region consists of two jets, accompanied by strong type III radio bursts with the onset at 01:29 and 02:32 UT, showing a deep drift below 0.1 MHz (see Figure 2 in Chen *et al.* 2015). Besides, the event is characterized by an enormously enhanced S/O ratio, ~5 times higher than typical ^{3}He-rich events S/O ratio (Mason *et al.* 2016).

To examine the temperature distribution at the source region, we utilized the Differential Emission Measure (DEM) analysis using six EUV channels (94, 131, 171, 93, 211, 335 Å) of SDO/AIA observations (Aschwanden *et al.* 2013). A Gaussian function is used to estimate the peak value of the emission measure and temperature that produce the temperature map (see Figure 1b). We calculated the temperature of two jets at the onset of type III bursts which may be close to the time of particle acceleration/injection. Along the jet, the distribution of temperature is patch-like, inhomogeneous and some parts are saturated (indicated by white color). Generally, the footpoints (the root of jets) reveal a temperature value higher than the one in jet's spire. As far as the standard 2D reconnection model is concerned, the footpoints represent the closed loops where most of the thermal energy is confined and the spire indicates the open field regions where all the energy would be transported outward. The estimated temperature of source plasma, including footpoint and spire, is relatively smaller than the theoretical value. Note that Mason *et al.* (2016) inferred that the temperature for S-rich events (including the one examined) is cooler with lower boundary at 0.4 MK. It appears that ions in the examined event were accelerated along the cooler jet spire.

3. Summary

The solar source of the ^{3}He-rich SEP event has been studied by analyzing high cadence SDO/AIA observations. The source region produces several jets with associated type III/VI radio bursts. The temperature of associated source eruptions is derived from DEM analysis based on EUV observations. In this events, the jet's spire is cooler than the footpoint regions. The estimated temperature of both jets, including footpoint and spire, ranges from 1.25 to 3.1 MK which is smaller than the standard range of 2-4 MK inferred from elemental abundance observations (Reames *et al.* 2014). In future works, we plan to characterize the temperature of the source regions in other (eg. ^{3}He-rich, Fe-rich) SEP events.

Acknowledgement

The work of R. Bučík is supported by the Deutsche Forschungsgemeinschaft under grant BU 3115/2-1.

References

Aschwanden, M. J., Boerner, P., Schrijver, C. J. & Malanushenko, A. 2013 *Solar Phys.*, 283 5A
Bučík, R., Innes, D. E., Mall, U., Korth, A., Mason, G. M. & Gómez-Herrero, R. 2014 *ApJ*, 786, 71B
Chen, N.-H., Ip, W.-H. & Innes, D. 2013 *ApJ*, 769 96C
Chen, N.-H, Bučík R., Innes, D. E. & Mason, G. M. 2015, *A&A*, 580A 16C
DiFabio, R., Guo, Z., Möbius, E., Klecker, B., Kucharek, H., Mason, G. M. & Popecki, M. 2008, *ApJ*, 687, 623
Mason, G. M., Nitta, N. V., Wiedenbeck, M. E. & Innes, D. E. 2016, *ApJ*, 823 138
Mason, G. M., Mazur, J. E. & Dwyer, J. R. 2002, *ApJ*, 565, L51
Reames, D. V., Cliver, E. W. & Kahler, S. W. 2014, *Solar Phys.*, 289, 3817-3841
Reames, D. V., Cliver, E. W. & Kahler, S. W. 2015a, *Solar Phys.*, 290, 1761-1774
Reames, D. V. 2015b, *Space Sci. Rev.* 194, 202

Space Weather of the Heliosphere: Processes and Forecasts
Proceedings IAU Symposium No. 335, 2017
C. Foullon & O.E. Malandraki, eds.

doi:10.1017/S174392131701136X

Spatial Inhomogeneity in Solar Faculae

A. Elek[1,3], N. Gyenge[1,2], M. B. Korsós[2,1] and R. Erdélyi[2,3],

[1]Debrecen Heliophysical Observatory (DHO), Konkoly Observatory,
Research Centre for Astronomy and Earth Sciences
Hungarian Academy of Sciences, Debrecen, P.O.Box 30, H-4010, Hungary
email: elek.anett@csfk.mta.hu

[2]Solar Physics and Space Plasmas Research Centre (SP^{2}RC),
University of Sheffield, Hicks Building, Hounsfield Road, Sheffield S3 7RH, England (UK)
[3]Dept. of Astronomy, Eötvös L. University, Pázmány P. sétány 1/A, Budapest, H1117,Hungary

Abstract. In this paper, we investigate the inhomogeneous spatial distribution of solar faculae. The focus is on the latitudinal and longitudinal distributions of these highly localised features covering ubiquitously the solar surface. The statistical analysis is based on white light observations of the Solar and Heliospheric Observatory (SOHO) and Solar Dynamics Observatory (SDO) between 1996 and 2014. We found that the fine structure of the latitudinal distribution of faculae displays a quasi-biennial oscillatory pattern. Furthermore, the longitudinal distribution of photospheric solar faculae does not show homogeneous behaviour either. In particular, the non-axisymmetric behaviour of these events show similar properties as that of the active longitude (AL) found in the distribution of sunspots. Our results, preliminary though, may provide a valuable observational constrain for developing the next-generation solar dynamo model.

1. Motivation

The connection of the small-scale magnetic field of solar photospheric faculae and the global magnetic field is studied, e.g., in the context of solar cycle dependence (Ermolli *et al.* 2014, Chatterjee *et al.* 2016) or the latitudinal distribution (Makarov *et al.* 1994). These studies highlight an important question, as it is not obvious, whether photospheric faculae should follow the global characteristics of active regions (ARs). ARs, as widely accepted, are the surface appearance of the solar global dynamo operating at the tachocline at the bottom of the convection zone. Global dynamo models work relatively well to describe the measured low-latitude global magnetic field. However, such global dynamos may struggle to adequately capture the physics of high-latitude magnetic phenomena, not to mention those that appear outside of ARs, or magnetic phenomena that may be generated by local dynamo, close to the solar surface. Since it is unclear whether the local dynamo itself indeed exists, it is even more unclear whether there is an interaction of the global and local dynamos. Therefore, studying the statistical properties of the highly localised photospheric faculae on spatial and temporal time scales well beyond their own extent and lifetime may provide further observational constrains to dynamo theorems. Valid and very interesting questions that should be answered are, such as e.g. (i) do faculae show indeed the butterfly distribution; (ii) how to account for the high-latitude distribution of faculae (Deng *et al.* 2016); (iii) is there a longitudinal pattern, supporting the mounting evidence of active longitudes (AL), of solar faculae; (iv) are all faculae the result of the global 11-year cycle dynamo, or, could there be a second dynamo operating (e.g. a local, surface one); (v) could actually the statistical analysis of spatio-temporal properties of faculae provide evidence towards the coupling of global and local dynamos; - just to raise a few questions that may justify why to pay so much attention to faculea.

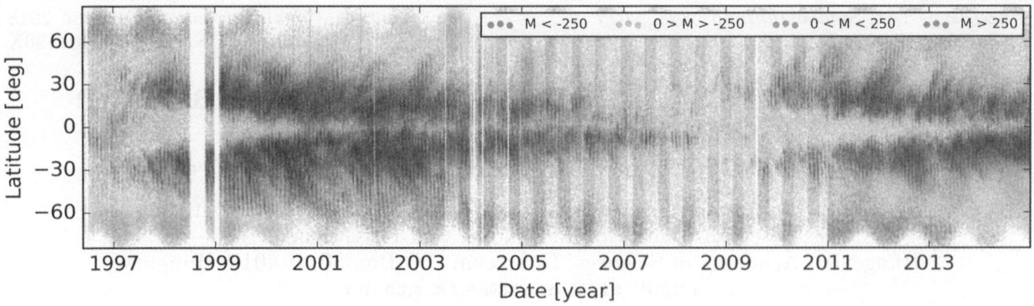

Figure 1. The latitudinal distribution of solar faculae based on SOHO and SDO observations. The dot colours refer to the associated magnitude of magnetic field of a given facula. Features, where $A < 2$, are omitted from the SDO data.

2. Solar Faculae Catalogue

We used the solar faculae database by the Debrecen Heliophysical Observatory (for details see Baranyi *et al.* 2016 and Győri *et al.* 2017). The wightlight observations are provided by the Solar and Heliospheric Observatory (SOHO) and Solar Dynamics Observatory (SDO) satellites. The catalogue contains information about the date of a facula event, area A in millions of solar hemisphere, position in Carrington coordinate system (latitude B and longitude L) and magnetic field strength (M). The database provides hourly cadence between 1996 and 2014. Our statistical sample includes all the observed solar faculae, except those data which failed one of the following criteria: (i) the longitudinal distance from the central meridian must be closer than 75 degrees and (ii) the absolute magnetic field strength must be less than 500 G. Both criteria are introduced due the errors in magnetograms. We distinguish four sub-samples based on the magnetic field strength (i) M: $M > 250$, (ii) $0 < M < 250$, (iii) $0 > M > -250$ and (iv) $M > -250$.

3. Latitudinal and Longitudinal Distribution of Solar Faculae

Figure 1 displays the latitude-time diagram of our statistical sample. The general shape of this diagram is the well-known butterfly-diagram. By following the global latitudinal migration towards the equator there seems to be a polar ward migration in narrow bands that repeats itself. Figure 1 shows that this migration is not always parallel, resulting in varying periodicity as function of latitude. The fine structure of the distribution reveals itself the negative and positive weak magnetic fields ($|B| < 250$) alternate.

The upper panel of Figure 2 shows the longitudinal distribution of solar faculae while the bottom panel plot shows the faculae longitudinal distance from the AL. The position of AL is defined in Gyenge *et al.* (2017). Essentially, the AL is the most enhanced longitudinal activity in a certain Carrington rotation, based on sunspot data, i.e the longitudinal distribution of sunspot groups is non-homogeneous. The parameter $\delta\phi$ represents the distance between the AL and the faculae in Carrington Phases. If $\delta\phi = 0$, the longitudinal position of the faculae is same as the longitudinal position of the AL, meanwhile $\delta\phi = 0.5$ corresponds to a 180 degrees shift. We summarised the faculae in each $\delta\phi = 0.05$ width bins. The obtained total counts N are converted to standard deviation units $(N - \mu(N))/\sigma(N)$, displayed by the vertical axis.

We conclude that he latitudinal migration shows the presence of the 11-year solar cycle. The fine structure also reveals a quasi-biennial oscillation - not highlighted in the displayed figures, however, a time series analysis reveals it. We determined this oscillation as follows: we cut the sunspot latitude-time diagram into 5-degree segments and analysed

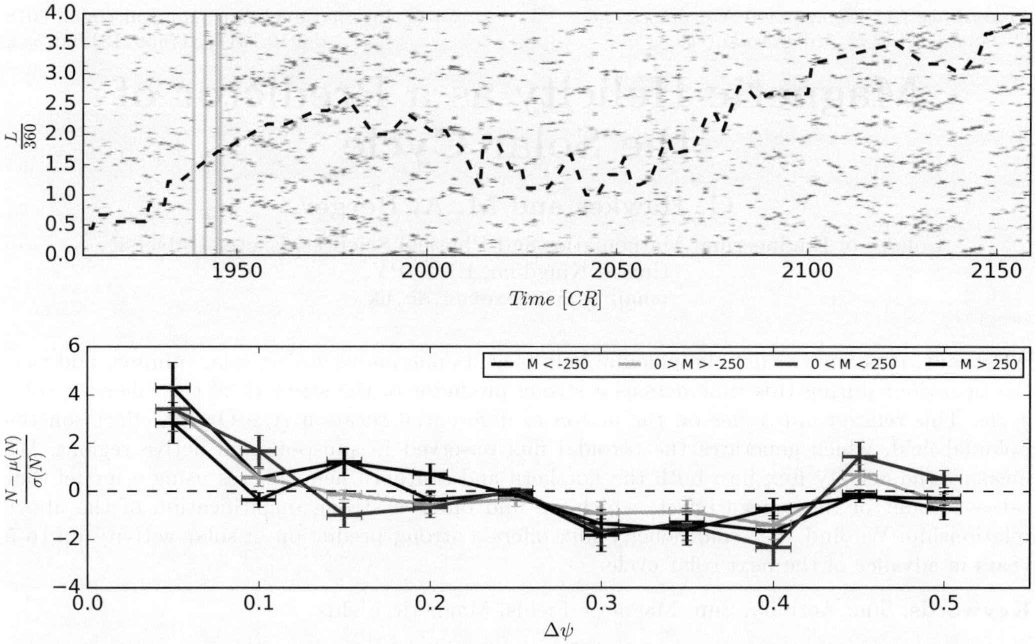

Figure 2. Upper: The longitudinal distribution of solar faculae. The dashed line indicates the migration of AL based on sunspot groups. Bottom: The distance of the AL migration and the solar faculae. The coloured lines represent the magnetic field strength. The error bars in x are due the differential rotation, while y are given by the square root of the certain bin.

the temporal variation histograms of the total magnetic field based on each segments. These histograms show periodic recurrent features with 1-2 years oscillation period. The periodicity may be dependent on latitude as the narrow bands are not parallel as Figure 1 suggests. This behaviour could mean that there is a second dynamo (or a string of dynamos) operating beside the one generating the global, 11-yr long solar cycle. The longitudinal distribution of solar faculae clearly shows that the vast majority of faculae occurred near AL. Faculae with strong magnetic field strength show similar behaviour as to that of the longitudinal distribution of sunspot groups. However, faculae with weak magnetic field strength also show AL-like features. We may therefore provide evidence for the manifestation of the non-axisymmetric solar dynamo at high latitudes, suggesting that the dipolar component of the global magnetic field plays a crucial role on the behaviour of AL.

Acknowledgements

RE is grateful to STFC, RE, NG and MBK acknowledge the Royal Society (UK) for the support received. AE is partially supported by Hungarian Solar Physics Foundation.

References

Baranyi, T., Győri, L. & Ludmány, A. 2016, *Sol. Phys.*, 291, 3081-3102
Chatterjee S., Banerjee D. & Ravindra B. 2016, *ApJ*, 827, 87
Deng, L. H., Li, B., Xiang, Y. Y. & Dun, G. T. 2016, *AJ*, 151, 2
Ermolli, I., Shibasaki, K., Tlatov, A. & van Driel-Gesztelyi, L. 2014, *Space Sci. Rev*, 186, 105-135
Gyenge, N., Singh, T., Kiss, T. S., Srivastava, A. K. & Erdélyi, R. 2017, *ApJ*, 838, 18
Győri, L., Ludmány, A. & Baranyi, T. 2017, *MNRAS*, 465 (2), 1259–1273
Hale, G. E. 1922, *MNRAS*, 82, 168
Makarov, V. I. & Makarova, V. V. 1994, *Sol. Phys.*, 163, 127, 267–289

Space Weather of the Heliosphere: Processes and Forecasts
Proceedings IAU Symposium No. 335, 2017
C. Foullon & O.E. Malandraki, eds.

doi:10.1017/S174392131700758X

Magnetic Helicity as a Predictor of the Solar Cycle

G. Hawkes and M. A. Berger

College of Engineering, Mathematics and Physical Sciences, Exeter University, United Kingdom, EX4 4PY,
email: gh378@exeter.ac.uk

Abstract. It is known that the poloidal field is at its maximum during solar minima, and that the behaviour during this time acts as a strong predictor of the strength of the following solar cycle. This relationship relies on the action of differential rotation (the Omega effect) on the poloidal field, which generates the toroidal flux observed in sunspots and active regions. We measure the helicity flux into both the northern and southern hemispheres using a model that takes account of the omega effect, which we find offers a strong quantification of the above relationship. We find that said helicity flux offers a strong prediction of solar activity up to 5 years in advance of the next solar cycle.

Keywords. Sun: Activity, Sun: Magnetic Fields, Magnetic Fields

1. Overview

Solar activity and its associated phenomena and drivers are known to have wide ranging effects on the heliosphere, and (for example) how cosmic rays pass through said regions, a process described in (for example) Ferreira & Potgieter (2004). There have been many attempts to make predictions of the solar activity cycle, which itself tends to be quantified by either sunspot/active region number or area. Prediction methodologies can be split into three subsets: extrapolation methods, precursor methods and model-based predictions, see Munoz-Jaramillo, Balmaceda & Deluca (2013). There have been many attempts to refine predictions of the solar cycle. One notable example is the work of Choudhuri, Chatterjee & Jiang (2007), which uses a mean field dynamo model. In this paper, they give a prediction of the strength of cycle 24.

Magnetic helicity has even recently been used as a proxy for solar eruptions, see Pariat *et al.* (2016), although this is admittedly restricted to singular events, rather than over the whole solar body. Considering only helicity flow across a boundary, Berger & Ruzmaikin (2000) gives the rate of change of helicity with respect to time as,

$$\frac{dH_{\mathcal{V}}}{dt} = 2\oint_{\mathcal{V}} (\mathbf{A}_P \cdot \mathbf{v}) B_n d^2x - 2\oint (\mathbf{A}_P \cdot \mathbf{B}) v_n d^2x, \tag{1.1}$$

where B_n is the component of the magnetic field normal to the sun's surface, $\mathbf{A}_P$ is the potential field and $\mathbf{v}$ is the plasma surface velocity. The expressions for B_r and $\mathbf{A}_P$ are expanded using a spherical harmonic model (see Berger & Ruzmaikin (2000)). The time variability of the model comes from the spherical harmonic co-efficients, which we use to calculate helicity flux from 1976 onwards.

Our estimations for helicity generation is based upon differential rotation alone (the Ω effect) - the zonal velocity of the solar body decreases as we move away from the equatorial slice towards the polar regions. The velocity field $\mathbf{v}$ is an analytical expression, which can be also be found in Berger & Ruzmaikin (2000). We therefore consider only the first integral of equation (1.1).

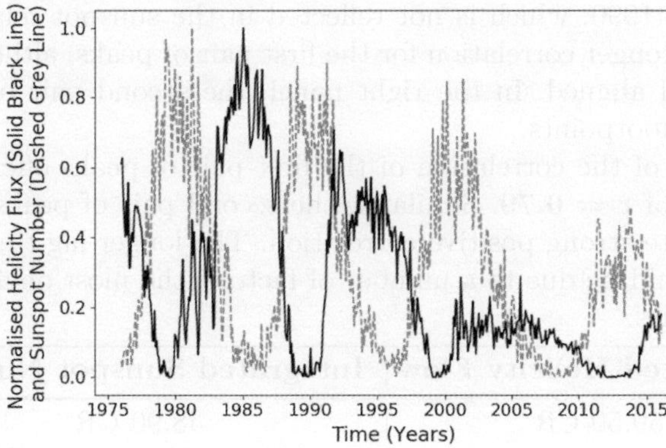

Figure 1. Normalised Helicity Flux (solid line) plotted against Normalised Sunspot Number (dashed line) for the period of 1976 to 2017

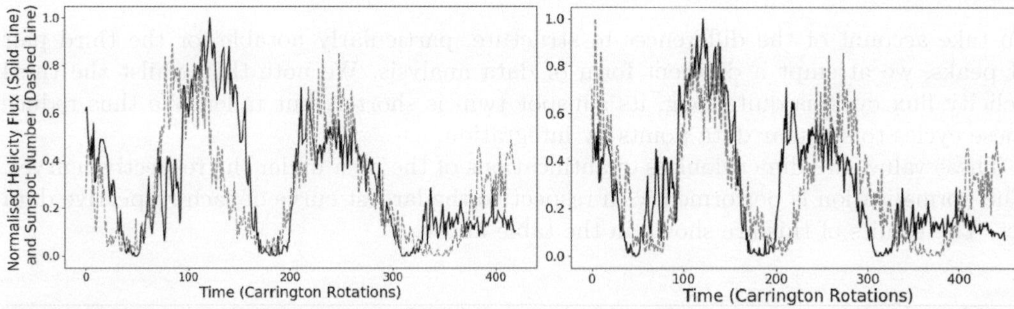

Figure 2. Normalised Helicity Flux (solid lines) plotted against Normalised Sunspot Number (dashed lines) with Helicity shifted forwards by 6.75 years (left) and 5.6 years (right)

Helicity flow for the solar body is modelled as shown in the aforementioned paper, with positive helicity flowing from the northern corona, through the northern and southern hemispheres (in that order) and out into the southern corona.

Sunspot number has frequently been used as a solar activity proxy, indicating regions of increased magnetic activity. Sunspots will often appear in pairs, of opposing polarity.

2. Data and Analysis

In Figure 1 we see Normalised helicity flux through the northern hemisphere plotted against monthly sunspot number over a period of $\sim 1976-2017$. Both data sets are averaged over a Carrington Rotation (27.28 days) - see Acknowledgements for data sources. During this period we observe three complete helicity cycles, and nearly four complete sunspot number cycles. Disregarding incomplete cycles, and the first sunspot cycle, we see clear similarities between each helicity cycle and the sunspot cycle that follows. We see the sunspot peaks decreasing approximately in line with the trend of the helicity flow peaks, with some phase shift (which is estimated below).

In Figure 2 we attempt to optimise the phase shift of the two underlying cyclic behaviours. On the left panel we have plotted we have an ~ 6.75 year phase shift on the left, whilst on the right we have an ~ 5.6 year phase shift. Both have their strong and weak points. For the former, we see a good correlation of foot points, and excellent correlation for the second pair of peaks. The first pair of peaks is less well aligned (in the region between the footpoints). This is could be due to the sudden drop observed in the helicity flux around the year 1990, which is not reflected in the sunspot relation. The 5.6 year phase shift gives a stronger correlation for the first pair of peaks, although the footpoints are no longer as well aligned. In the right panel, the second pair of peaks is less well aligned between the footpoints.

The maximisation of the correlation of the first pair of peaks offered a Pearson Correlation Co-efficient of $\mathbf{r} = \mathbf{0.79}$. Similarly, the second pair of peaks offered $\mathbf{r} = \mathbf{0.89}$, both of which indicate strong positive correlation. The longer lag time between the second pair of peaks could be due to a number of factors, the most obvious of which is the amplitude of the cycles.

Cycle	**Integrated Helicity Flow**	**Integrated Sunspot Number**	**Ratio**
1	59.50 CR	48.90 CR	74%
2	42.39 CR	46.79 CR	90.6%
3	17.5 CR	22.79 CR	76.8%

To take account of the differences in structure, particularly notable for the third pair of peaks, we attempt a different form of data analysis. We note that whilst the third helicity flux cycle is quite long, its sunspot twin is shorter, but taller. We thus reduce these cycles to singular data points by integration.

These values are dimensionless quantifications of the area under the respective curves. The normalisation is performed with respect to the largest curve of each respective data set. The results of this are shown in the table above.

3. Conclusions

We have provided strong indication of a relationship between the solar cycles of helicity flux and the corresponding sunspot number cycle. Our analysis indicates that there is a phase shift between the two mechanisms of approximately $5 - 7$ years. We aim to collect more data to further verify our hypothesis, and make direct comparisons with the predictive capability of the polar field alone (given the presence of B_n in our flux equation).

4. Acknowledgements

The Wilcox Solar Observatory data used in this study to calculate magnetic helicity flux was obtained via the web site http://wso.stanford.edu, courtesy of J.T. Hoeksema. Sunspot Number Source: WDC-SILSO, Royal Observatory of Belgium, Brussels. G Hawkes would also like to thank STFC for their funding under grant ST/N504063/1.

References

Choudhuri, A., Chatterjee, P. & Jiang, J. 2007, *Phys. Rev. Lett.*, 98, 13
Berger, M. A. & Field, G. B. 1984, *J. Fluid. Mech*, 147, 133
Berger, M. A. & Ruzmaikin, A 2000, *J. Geophys Research*, 105, 10481
Ferreira, S. E. S. & Potgieter, M. S. 2004, *ApJ*, 603, 744
Munoz-Jaramillo, A., Balmaceda, L. A. & Deluca, E. E. 2013, *Phys. Rev. Lett.*, 111(4)
Pariat, E., Leake, J. E., Valori, G., Linton, M. G., Zuccarello, F. P. & Dalmasse, K. 2016, *A & AR*, 631, 976

Space Weather of the Heliosphere: Processes and Forecasts
Proceedings IAU Symposium No. 335, 2017
C. Foullon & O.E. Malandraki, eds.

doi:10.1017/S1743921317010997

Simulating the Dynamics of Coronal Plasma Condensations

Petra Kohutova and Erwin Verwichte

Centre for Fusion, Space and Astrophysics, Department of Physics, University of Warwick, CV4 7AL, Coventry, UK
email: `p.kohutova@warwick.ac.uk`

Abstract. We present numerical MHD simulations of the dynamics of cool plasma condensations in a coronal loop. We address 2 mechanisms for how coronal rain leads to the excitation of coronal loop oscillations. We find that the combined effect of pressure gradients in the coronal loop plasma and magnetic tension force resulting from changes in magnetic field geometry explains observed sub-ballistic motion of coronal rain and longitudinal oscillations of the individual condensations. We also find that the condensations can excite sustained, small amplitude, vertically polarised transverse loop oscillations.

Keywords. Magnetohydrodynamics (MHD), Sun: corona, Sun: oscillations, Sun: magnetic fields

1. Introduction

Otherwise hot and diffuse solar corona contains numerous cool and dense plasma structures in the form of prominences and coronal rain. Coronal rain consisting of down-falling cool plasma condensations is a phenomenon occurring in footpoint-heated coronal loops as a result of thermal instability. Recent high resolution observations have shown that coronal rain is much more common than previously thought, suggesting its important role in the chromosphere-corona mass cycle. Due to its origin, coronal rain also provides us with physical insight into the atmospheric thermal cycle and into prominence formation and evolution.

2. The model and numerical setup

We use Lare2d (Arber *et al.* 2001), a 2.5D shock-capturing Lagrangian remap code that solves MHD equations on a staggered Cartesian grid. Thermal conduction and radiative loss terms are not included in the energy equation. The density variation between the loop and the background medium is given by the symmetric Epstein profile. We assume a realistic temperature profile representative of an atmosphere consisting of a cool chromosphere, transition region layer and hot corona. The vertical density profile for the non-isothermal stratified atmosphere is determined by numerically solving for a hydrostatic pressure balance. For analysis of the condensation dynamics a long rectangular domain is used, corresponding to a straightened coronal loop (Figure 1). This configuration uses uniform vertical magnetic field and reduced gravity to account for the semicircular model of the loop. A condensation, represented by a Gaussian density enhancement with chromospheric temperature is introduced near the loop apex and let to evolve. For analysis of the excitation of vertical loop oscillations a square domain is used using a current free magnetic field configuration to model a coronal arcade (Figure 1). A cool and dense extended condensation region is located at the loop apex.

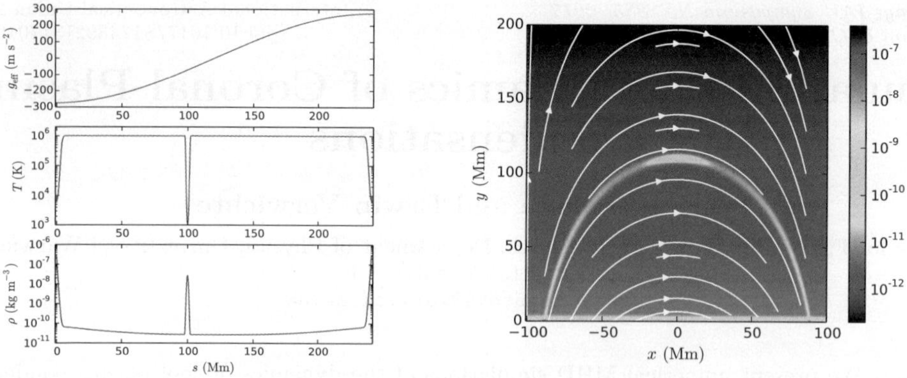

Figure 1. Left: Initial configuration of the straightened loop. Plots show the variation of effective gravity, temperature and density along the centre of the domain. Right: Initial density configuration of the coronal loop with cool condensation region. The white lines show the B-field direction.

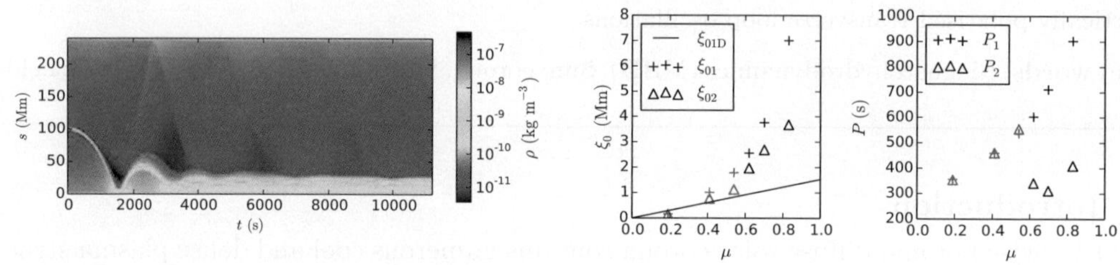

Figure 2. Left: Time distance plot of the density evolution along the centre of the loop. Right: Amplitudes and periods of the oscillations for different condensation region masses. Solid line shows amplitudes predicted by 1D model by Verwichte *et al.* (2017).

3. Results

3.1. *Condensation dynamics*

The downward motion of the condensation becomes increasingly sub-ballistic for low condensation region densities and large magnetic field values. If the density or magnetic field cross the threshold value, the condensation does not follow simple free fall motion but is instead found to oscillate, until it eventually settles in an equilibrium position (Figure 2). Analysis of the evolution of the individual forces acting on the condensation suggests that the oscillatory motion is caused by the combination of two forces: gas pressure force resulting from the compression of the underlying loop plasma confined below the blob and magnetic tension force resulting from bending of the magnetic field lines. The problem can therefore be approximated as an MHD piston aligned with the magnetic field. We consider the dynamics of the plasma blob as a perturbation of the equilibrium where pressure forces above and below the blob match the gravity on the blob. For small-amplitude perturbation a dispersion relation is derived that relates the field-aligned oscillation period of the rain blob with the blob-plasma density ratio, equilibrium position and density scale-height (Kohutova & Verwichte 2017a). Although in high resolution solar observations the majority of coronal rain condensations are seen to fall directly towards the solar surface, individual blobs are sometimes observed to longitudinally oscillate up and down before falling (Kohutova & Verwichte 2016, Verwichte *et al.* 2017). The mechanism investigated above is therefore likely responsible for this type of observed motion.

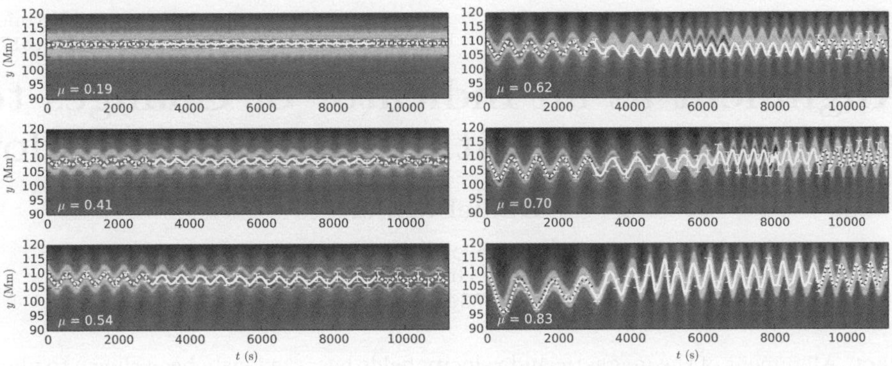

Figure 3. Time-distance plots of the loop apex position for different condensation region masses. Oscillation period decreases after condensations fall towards the surface.

3.2. *Vertical oscillations*

The presence of the dense condensation region displaces the axis of the loop downwards, which triggers vertical loop oscillations with the oscillation periods and amplitudes increasing with the overall condensation mass (Figure 2). The amplitudes predicted by 1D mechanical model by Verwichte *et al.* (2017) agree with simulation amplitudes for lowest masses of the condensation region, while at higher masses the two diverge and the mechanical model underestimates the amplitudes. This is as expected given that the assumptions made by the mechanical model (point-line condensation, constant loop length) only agree with the simulated scenario for low condensation region masses. As the condensations fall towards the solar surface under the influence of gravity, the distribution of the mass along the loop changes. This change in the loop density profile leads to a change in the period of the fundamental kink mode of the loop (Figure 3). The ratio of the initial to final oscillation period also increases with increasing condensation mass. The simulated period ratio can be compared with values deduced from 1D model for fundamental kink mode frequency of a long thin loop with longitudinally varying density:

$$\frac{\mathrm{d}^2\xi}{\mathrm{d}s^2} + \frac{\omega^2}{C_\mathrm{k}^2}\xi = 0$$

where ξ is the loop displacement and $C_\mathrm{k}(s)$ reflects the corresponding density profile, reaching good agreement for high condensation region masses (Kohutova & Verwichte 2017b). The first observational evidence of excitation of vertical coronal loop oscillations by coronal rain was shown in Verwichte & Kohutova (2017). The observations presented therein also showed a change of the oscillation period caused by the drainage of the coronal loop mass, in agreement with simulation results presented in this work.

References

Arber, T. D., Longbottom, A. W., Gerrard, C. L. & Milne, A. M. 2001, *J Comput Phys.*, 171, 151

Kohutova, P. & Verwichte, E. 2016, *ApJ*, 827, 39

Kohutova, P. & Verwichte, E. 2017a, *A&A*, 602, A23

Kohutova, P. & Verwichte, E. 2017b, *A&A*, submitted

Verwichte, E., Antolin, P., Rowlands, G., Kohutova, P. & Neukirch, T. 2017, *A&A*, 598, A57

Verwichte, E. & Kohutova, P. 2017, *A&A*, 601, L2

Space Weather of the Heliosphere: Processes and Forecasts
Proceedings IAU Symposium No. 335, 2017
C. Foullon & O.E. Malandraki, eds.

doi:10.1017/S1743921317008663

Alignment as an Indicator of Changes to Modal Structure within the Roberts Flow

Daniel Miller

University of Exeter,
Exeter, Great Britain
email: dm473@exeter.ac.uk

Abstract. Alignment of the magnetic and velocity fields has previously been shown to play a role within nonlinear dynamo theory (Cameron and Galloway 2006), MHD turbulence (Matthaeus *et al.* 1980) and mean field theory (Yokoi 2013). What has not been previously examined is whether it is beneficial to examine alignment within kinematic dynamo theory. I show how measurements of alignment within kinematic dynamo theory for the Roberts flow can indicate a change in the structure of the magnetic field.

Keywords. MHD, Kinematic Dynamo Theory

1. Introduction

The evolution of an incompressible, electrically conducting, fluid and the magnetic field that it couples with is governed by the equation of motion for the fluid (1.1) and the induction equation (1.2). At the start of the dynamo the magnetic field is weak. The Lorentz force term ($(\mathbf{B}\cdot\nabla)\mathbf{B}$) within equation (1.1) can thus be neglected and the fluid evolves independently from the magnetic field. The initial growth of a seed magnetic field can therefore be examined by solving (1.2) using a prescribed flow $\mathbf{u}$. As the induction equation is linear solutions will be of the form $\mathbf{B}(x,y,z)\exp(\sigma t)$. Amplification of the seed magnetic field therefore requires $\Re\{\sigma\} > 0$.

One example of a flow that produces dynamo action and has been studied extensively is the Roberts flow (Roberts 1972) which is shown in equation (1.3). The flow amplifies the magnetic field by exponentially stretching along the null lines of the flow. The flow results in a magnetic field whose structure depends upon the wavenumber in the z direction, k_z, and it is for this reason that I use it to examine alignment.

$$\frac{\partial \mathbf{u}}{\partial t} = -\nabla P + (\mathbf{B}\cdot\nabla)\mathbf{B} - (\mathbf{u}\cdot\nabla)\mathbf{u} + \nu\nabla^2\mathbf{u} + \mathbf{F} \tag{1.1}$$

$$\frac{\partial \mathbf{B}}{\partial t} = (\mathbf{B}\cdot\nabla)\mathbf{u} - (\mathbf{u}\cdot\nabla)\mathbf{B} + \eta\nabla^2\mathbf{B} \tag{1.2}$$

$$\mathbf{u} = \sqrt{\frac{3}{2}}\left[\cos(y), \sin(x), \sin(y) + \cos(x)\right] \tag{1.3}$$

The quantity that I use to measure alignment is shown within equation (1.4) and is the volume average of $|\cos(\theta)|$ where θ is the angle between $\mathbf{u}$ and $\mathbf{B}$ at each point. $\mathbf{AH}_c$ therefore measures alignment over the entire domain.

$$\mathbf{AH}_c = \frac{k_z}{8\pi^3}\int_0^{\frac{2\pi}{k_z}}\int_{-\pi}^{\pi}\int_{-\pi}^{\pi}\frac{|\mathbf{u}\cdot\mathbf{B}|}{|\mathbf{u}||\mathbf{B}|}dV \tag{1.4}$$

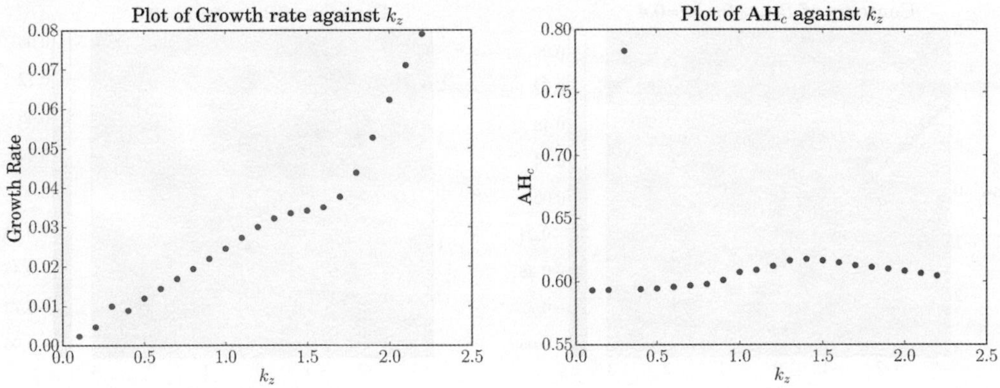

(a) (b)

Figure 1. $\mathbf{AH}_c$ highlights the difference between $k_z = 0.3$ and the other k_z which is not obvious from the plot of the growth rate

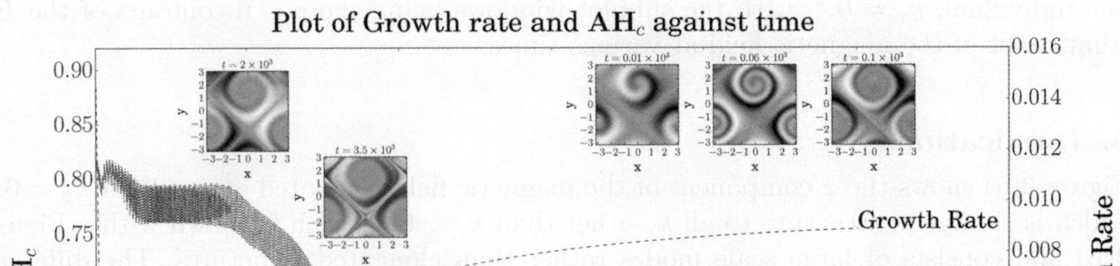

Figure 2. $k_z = 0.4$ run showing how $\mathbf{AH}_c$ changes in response to a change in the structure of **B**. Light indicates strong positive field dark is strongly negative. Position indicates time where the plots are taken except those in the top right which are at times $t = 10, 60, 100$ respectively.

2. Overview

Due to the z independence of $\mathbf{u}$ and the linear nature of (1.2) it can be shown that the magnetic field may be rewritten in the form shown within equation (2.1). The wavenumber in the z direction, k_z is therefore a free parameter.

$$\mathbf{B}(x, y, z, t) = 2\Re\{\tilde{\mathbf{B}}(x, y) \exp(ik_z z + \sigma t)\} \tag{2.1}$$

I solve (1.2) within a doubly periodic geometry for prescribed flow (1.3) and $\eta = 1/1000$. I use a spectral code which makes use of an Adams-Bashforth: Adams-Moulton predictor-corrector scheme for timestepping. I begin at $k_z = 0.1$ with **B** as a random seed field and then increase k_z in steps of 0.1. Each subsequent k_z is then initialised using the rescaled magnetic field of the previous k_z for computational efficiency. Figure 1(a) shows a plot of the growth rate vs k_z and is in good agreement with published results (Roberts 1972). Figure 1(b) shows a plot of the time asymptotic $\mathbf{AH}_c$ against k_z. Figure 2 shows

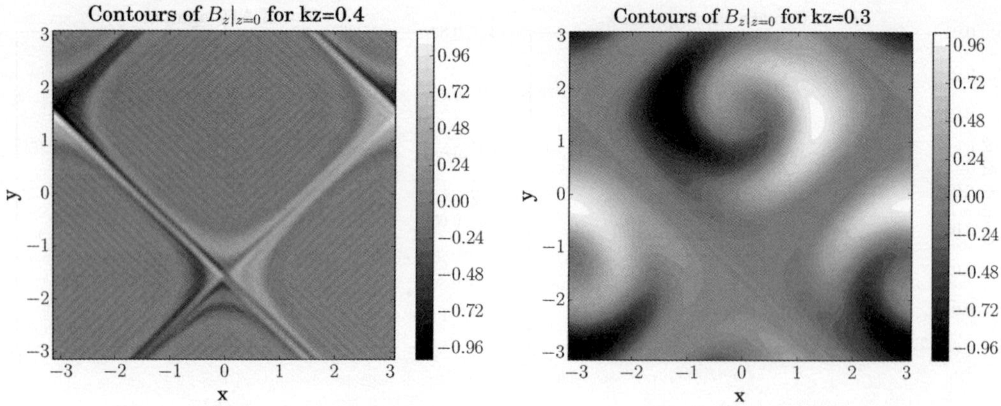

(a) $z = 0$ contour for $k_z = 0.4$ (b) $z = 0$ contour for $k_z = 0.3$

Figure 3. Figure shows that the structure of B_z for $k_z = 0.3$ is different to that of all other k_z

one individual, $k_z = 0.4$, with the subplot windows being the $z = 0$ contours of the B_z component of the magnetic field at various times.

3. Implications

Figure 3(a) shows the z component of the magnetic field evaluated at $z = 0$ for $k_z = 0.4$ which is similar in structure to all k_z other than $k_z = 0.3$ which is shown within Figure 3(b) and consists of large scale modes rather than elongated structures. The different modal structure that $k_z = 0.3$ has compared to all other k_z examined is clearly seen by the large change in value of $\mathbf{AH}_c$ shown within Figure 1(b) whereas examination of the growth rate alone would not perhaps highlight the special nature of the magnetic field structure of $k_z = 0.3$. This suggests that the calculation of $\mathbf{AH}_c$ may be a useful method to identify differing magnetic field structures as k_z is varied.

In Figure 2 we see that as time evolves the magnetic field leaves the center region of the flow and becomes exponentially stretched along the null lines of the flow forming into a seperatrix like structure. Correspondingly as the magnetic field structure transitions from a large scale modal structure to one that is more seperatrix like the value of $\mathbf{AH}_c$ drops to a value in line with that obtained for other seperatrix structures. Time series of $\mathbf{AH}_c$ thus enable us to see that the magnetic field is restructuring itself without needing to examine plots of the magnetic field directly.

In Summary I have shown via numerical simulations that the measure of alignment $\mathbf{AH}_c$ between the flow and magnetic field within kinematic dynamo simulations can be a useful method for highlighting differences in magnetic field structure. This suggests that $\mathbf{AH}_c$ may be a useful diagnostic tool within kinematic dynamo theory.

I would like to thank EPSRC for a PhD studentship and the University of Exeter. I would also like to thank my supervisor Dr Joanne Mason for many useful discussions.

References

Cameron, R. & Galloway, D. 2006, *Mon. Not. R. Astron. Soc.*, 365, 735-746

Matthaeus, W. H., Montgomery, D. C., Wan, M & Servidio, S. 2012, *Journal of Turbulence*, 13, 37

Roberts, G. O. 1972, *Philosophical Transactions Of The Royal Society A*, 271, 1216

Yokoi, N. 2013, *Geophys. Astrophys. Fluid Dyn*, 107,114-184

Space Weather of the Heliosphere: Processes and Forecasts
Proceedings IAU Symposium No. 335, 2017
C. Foullon & O.E. Malandraki, eds.

doi:10.1017/S1743921317008821

Hard X-Ray Emission of Solar Flares Measured by Lomonosov Space Mission

Irina N. Myagkova[1], **Andrey V. Bogomolov**[1], **Vitaly V. Bogomolov**[1,2], **Sergey I. Svertilov**[1,2], **Sergey Yu. Bobrovnikov**[1], **Ivan V. Yashin**[1], **Mikhail I. Panasyuk**[1,2], **Vladimir V. Kalegaev**[1], **Vasily L. Petrov**[1] **and Svetlana S. Shishkova**[2]

[1]Skobeltsyn Institute of Nuclear Physics of Lomonosov Moscow State University, 1(2), Leninskie gory, Moscow, 119991, Moscow, Russia, email: irina@srd.sinp.msu.ru

[2]Physical Department of Lomonosov Moscow State University, 1(2), Leninskie gory,119991, Moscow, Russia

Abstract. Solar hard X-ray and gamma-ray emission was measured by BDRG instrument, the part of set of instruments operated on board the Russian satellite Lomonosov from April 2016 until now (solar-synchronous orbit with altitude 490 km, inclination of 97.6 degrees). Lomonosov measurements (11 flares with the X-ray energy more than 10 keV, and more than half of them have class in soft X-rays less than C2) were compared to the data obtained by RHESSI and Fermi space observatories as well as the Nobeyama Radioheliograph operating at the same time. The quasi-periodicity with different periods were found in some of them.

Keywords. Sun: Flares, Sun: X-rays, gamma rays, Instrumentation: detectors

1. Introduction

Measurements of neutral emission of solar flares (X- and gamma-rays produced at the Sun), their time history and spectral shape, and comparing them with microwave emission of solar flare can provide us with the most direct important information about particle injection and acceleration processes in solar flares. Besides that, in the context of space weather, the data about the hardness of solar hard X-ray (HXR) and gamma-ray emission are important for estimating the possible damage that may be caused by a given flare to technical systems.

It is well known that the thermal flare emission of hot plasma produces soft X-rays (SXR). The most common classification of solar flares is based upon the energy of the thermal plasma defining its class in SXR according to observations on board GOES satellites (Garcia, 1994). The non-thermal emission of solar flares hard X-ray and gamma-emission is the result of charged particle interaction with the solar atmosphere. It is the superposition of electron bremsstrahlung continuum and gamma-ray line emission (see, e.g., Fletcher *et al.* 2011 and references therein).

2. Instrumentation

The Block of Detector of the Roentgen and Gamma emission (BDRG) instrument includes three identical detector units mounted on the spacecraft payload platform in such a way that their axes are shifted at 90 degrees from each other. Each detector has cosine angular dependence of sensitive area. The central axis, relatively to which the detector axes are inclined, is directed toward to the local Zenith. Thus, the total field of view of all three detectors is about 2 sr. Each BDRG detector unit is similar to the

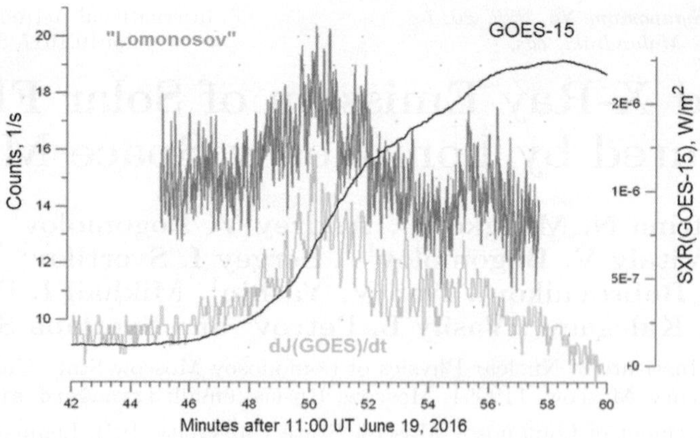

Figure 1. The time profiles of HXR emission measured during solar flare near noon UT June, 19, 2016 (C1.7) on board Lomonosov

Detector of the Roentgen and Gamma Emission (DRGE) detectors used for solar flare measurements on board Vernov satellite (Myagkova *et al.* 2016). It consists of optically coupled thin (0.3 cm) NaI(Tl) and considerably thick (1.7 cm) CsI(Tl) crystals. The diameter of both scintillators is 13 cm; NaI(Tl) is placed on top of the CsI(Tl) crystal, and both crystals are viewed by the same Photomultiplier tube Hammamatsu R877. Working ranges are 0.01-0.65 MeV for NaI(Tl) and 0.05-3 MeV for CsI(Tl). In this way, the NaI(Tl) works as the main detector for hard X-ray timing while the CsI(Tl) is used as an active shield against background gammas, and it can also detect gammas with energies up to a few MeV.

Information from BDRG contains a number of data frames formed continuously and a set of different data frames formed by trigger algorithms with characteristic times of 10 ms (fast), 1 s (slow) and 20 s (super-slow). Continuous data contain 3 main types of frames: monitoring (count rate in 8 energy channels for NaI(Tl) and CsI(Tl) from each BDRG detector, every 0.1 s), spectrum (724 channel spectra for NaI(Tl) and CsI(Tl) separately), and event mode (primary values of two scintillation components for a fixed number of events combined with time data). Trigger data contain the same set of different data frames, similar to the regular frames, with better time resolution (up to 0.001s).

3. Experimental data

The duty cycle for the detection of solar flares on board Lomonosov satellite was about 35-40 % as a result of its orbit parameters, so many flares were missed. However, 11 flares with HXR-emission were detected by Lomonosov during June-August 2016.

In Figure 1, HXR-emission produced in the June 19, 2016 solar flare of C1.7 GOES class (start at 11:44 UT, maximum at 11:58 UT), observed by Lomonosov is shown as an example. For comparison, the black line shows the change in SXR (GOES-15), and the gray line is the SXR flux derivative. It is clearly seen that the time profile of HXR and the derivative of SXR are strongly similar, in accordance with the Neupert effect (Neupert, 1968). Data on the solar HXR emission observed during the weak flares of C-class in soft X-rays can help solve the problem of so-called “cold” flares (e.g., Fleishman *et al.* 2011).

Lomonosov is a polar low-altitude satellite, thus during about a third of observational time the satellite crosses the Earth's radiation belts (ERB). Moreover, even in the rather suitable regions the observed increase of count rates may be caused by both solar flare radiation and electron precipitations. A possible way to specify the source of the

Table 1. The catalog of solar flares observed by Lomonosov satellite in June - August, 2016.

No	Day/month of 2006	Time, start-max-end SXR, hh:mm:ss (GOES)	Class SXR	Time, start-end HXR hh:mm (Lomonosov)	QPP	Other data	AR
1.	19/06	11:04-11:58-12:08	C1.7	11:44:20-11:57:40	+	GOES	12558
2.	09/07	09:01-09:05-09:05	B9.8	13:01:15-13:04:00	-	GOES	12564
3.	10/07	00:53-00:59-01:03	C8.6	00:56:20-00:59:40	+	F	12564
4.	18/07	08:09-08:23-08:32	C4.4	08:11:00-08:16:10	+	F	12567
5.	20/07	23:02-23:09-23:55	C5.0	23:04:45-23:05:30	-	F, N	12567
6.	21/07	00:42-00:46-00:50	M1.2	19:00:53-19:01:45	+	F, N	12567
7.	07/08	05:28-05:35-05:40	C1.3	05:33:55-05:34:10	+	F	12571
8.	07/08	10:18-10:24-10:29	C1.6	10:22:15-10:24:50	-	F	-
9.	11/08	08:34-08:38-08:42	B5.6	08:36:20-08:37:30	-	R	12574
10.	14/08	19:29-19:36-19:39	C1.1	19:34:00-19:35:15	-	R	12578
11.	15/08	00:17-00:23-00:29	C1.1	00:19:20-00:22:50	+	R	12578

detected signal is to compare time profiles observed on Lomonosov with the results of other experiments. We used the following data which is publicly available.

Characteristics of the flares detected on Lomonosov as well as SXR ones (GOES) are presented in Table 1: the day and the month, moments of the start, the maximum and the end of the SXR measured by GOES-15 satellite, flare class in SXR, the beginning and the end of the HXR emission with the energy more than 10 keV measured on board Lomonosov satellite, QPP - quasi-periodic pulsations (+ if they were detected), and the list of experiments in which, in addition to Lomonosov, the flare emission was detected: F - Fermi GBM ftp://legacy.gsfc.nasa.gov/glast/data/gbm/triggers/2016; N - Nobeyama http://solar.nro.nao.ac.jp/norh; R - RHESSI http://hesperia.gsfc.nasa.gov/rhessi3.

4. Summary

1. The time resolution of 0.1 seconds allows us to investigate in detail the fine temporal structure of HXR weak solar flares.

2. More than half of the HXR flares observed by Lomonosov had the SXR class lower than C2. It suggests that in these events solar electrons are accelerated without heating.

3. Experiment on board Lomonosov demonstrated that solar hard X-ray emission can be recorded during flares of the B-class in SXR.

This study was supported by the Russian Science Foundation, grant no.16-17-00098

References

Fleishman, G. D., Kontar, E. P., Nita, G. M. & Gary, D. E. 2011, *Astrophys. J. Lett.*, 731, L19

Fletcher, L., Dennis, B. R., Hudson, H. S., Krucker, S., Phillips, K., Veronig, A., Battaglia, M.,Bone, L., Caspi, A., Chen, Q., Gallagher, P., Grigis, P. T., Ji, H., Liu, W., Milligan, R. O. & Temmer, M. 2011, *Space Sci. Rev.*, 159, 19

Garcia, H. A. 1994, *Solar Phys.*, 154, 275

Myagkova, I. N., Bogomolov, A. V., Kashapova, L. K., Bogomolov, V. V., Svertilov, S. I., Panasyuk, M. I., Kuznetsova, E. A. & Rozhkov, G. V. 2016, *Solar Phys.*, 291, 3439

Neupert, W. M. 1968, *Astrophys. J. Lett.*, 153, L59

Space Weather of the Heliosphere: Processes and Forecasts
Proceedings IAU Symposium No. 335, 2017
C. Foullon & O.E. Malandraki, eds.

doi:10.1017/S1743921317008924

On the Dynamics of the Largest Active Region of the Solar Cycle 24

Ranadeep Sarkar[1], Nandita Srivastava[1] and Sajal Kumar Dhara[2]

[1]Udaipur Solar Observatory, Physical Research Laboratory,
Badi Road, Udaipur 313001, India
email: ranadeep@prl.res.in
email: nandita@prl.res.in

[2]Istituto Ricerche Solari Locarno, Universita della Svizzera Italiana,
CH-6605 Locarno Monti, Switzerland
email: sajal@irsol.ch

Abstract. We have studied the dynamics of the solar active region (AR) NOAA 12192 using full-disc continuum images and the vector magnetograms observed by the Helioseismic and Magnetic Imager (HMI) onboard Solar Dynamics Observatory (SDO). AR 12192 is the largest region of the solar cycle 24. It underwent a noticeable growth and produced 6 X-class, 22 M-class and 53 C-class flares during its disc passage. But the most peculiar fact of this AR is that it was associated with only one CME in spite of producing several X-class flares. In this work, we present the area evolution of this giant sunspot group during the first three rotations when it appeared as AR 12172, AR 12192 and AR 12209, respectively. We have also attempted to make a comparative study of the flare-related photospheric magnetic field and Lorentz force changes for both the eruptive and non-eruptive flares produced by AR 12192.

Keywords. Photosphere, sunspots, flares, coronal mass ejections (CMEs).

1. Introduction

During the maximum phase of solar cycle 24, the largest solar active region of the cycle appeared on the solar disc as NOAA 12192 in October 2014. It completed its passage across the visible solar disc from October 17 to 30 and became the largest sunspot group in 24 years since NOAA 6368 in November 1990 (Thalmann *et al.* 2015). As the solar cycle 24 was a comparatively weak cycle (Jiang *et al.* 2015), the huge size of AR 12192 was quite unexpected.

Interestingly, the sunspot group AR 12192 was the return of AR 12172 from the previous rotation. It returned back to the next two rotations as NOAA 12209 and NOAA 12237, respectively, as shown in Figure 1. In the previous cycle also, it was non-eruptive in nature and produced 12 C-class and 1 M-class flares during its disc passage from September 21 to October 3. After disappearing behind the limb on October 3, it was seen in GONG farside images (Figure 2). From October 4 to October 16 it crossed the far side of the Sun and evolved to grow as a giant sunspot group and on October 17, it appeared as AR 12192 on the eastern limb of visible solar disc. However, it started showing pre-signatures of its appearance from October 14 when it was behind the eastern limb. Hinode flare catalog reveals two M-class flares on October 14 and 3 C-class and 1 M-class flare on October 16 at the location S14 E88. These strong flares behind the eastern limb and the appearance of a big spot in GONG far side image (Figure 2) on October 14, were the distinct signatures of an active region group approaching over the eastern limb.

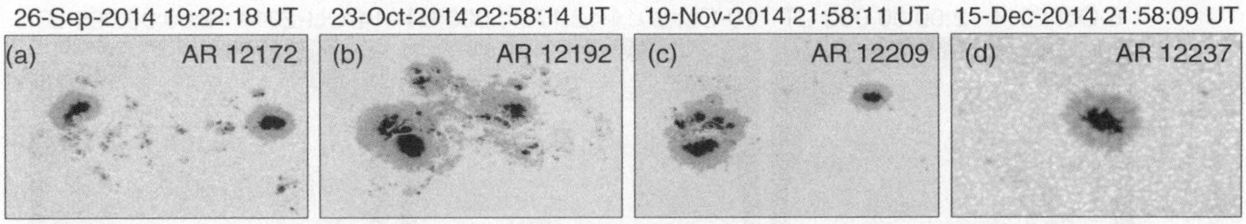

Figure 1. Images show the SDO continuum intensity map of AR 12192 (b) in its previous rotation (a), and subsequent rotations (c) and (d), respectively. The heliographic coordinates of the ARs illustrated in the above panels are S11W05, S12W08, S13W06 and S14E15 respectively from left to right.

From October 17 to 30, AR 12192 crossed the visible solar disc and produced 6 X-class flares, 22 M-class flares, and 53 C-class flares. On October 24, when it was close to the central meridian it produced one X3.1 class flare which set a record in flare energy for an event without a CME (RHESSI science nugget no.239, http://sprg.ssl.berkeley.edu/~tohban/wiki/index.php/A_Record-Setting_CMEless_Flare). At that time, the location of the AR was S16W21, which was an ideal location to produce an Earthward-directed halo CME event. However, no major CME was observed. Only one small CME was launched during an M4.0 class flare on October 24, which was the only CME produced by this active region during the disc passage.

On October 31, AR 12192 disappeared behind the western limb and returned back in the next rotation on November 13 as NOAA 12209 with a reduced size. From November 13 to November 26 NOAA 12209 was visible on the solar disc and produced 16 C class flares. After that, it went behind the visible solar disc and further reduced to a very small size and again became visible on the solar disc for the last time in the next rotation from December 10 to 22 as NOAA 12237.

In this work, we have studied the evolution of AR NOAA 12192 and its relation with the flare and CME productivity.

2. Data analysis

The evolution of AR 12192 during its disc passage was well captured by the observations from the Helioseismic and Magnetic Imager (HMI) (Schou *et al.* 2012) and the Atmospheric Imaging Assembly (AIA) (Lemen *et al.* 2012) onboard Solar Dynamics Observatory (SDO) (Pesnell *et al.* 2012). To study the umbral and penumbral area variation for all the three rotations of AR 12192 we have used the full-disc continuum images observed in the Fe I line at 6173 Å with a spatial scale of $0.5''$ per pixel and temporal scale of 12 minutes. To study the photospheric magnetic field evolution of AR 12192, we have used the 12 minutes cadence HMI vector magnetogram series from the version of Space weather HMI Active Region Patches (SHARP; Turmon *et al.* 2010). For identifying the umbral-penumbral and penumbral-quiet Sun boundaries we have calculated the cumulative histogram (Pettauer & Brandt 1997, Mathew *et al.* 2007) of the intensity of each pixel's brightness within the cutout which encloses the whole active region as well as the immediately surrounding quiet Sun region. Before calculating the cumulative histogram, we have normalized all the pixels of the cutout region by the median of brightness value of a $10{\times}10$ pixel^2 quiet Sun region surrounding the active region. The cumulative histograms were computed for all the days during the disc passage of AR 12192 from 2014 October 21 to 2014 October 26 when its longitude was within $\pm75°$ from the central meridian. We have found the normalized intensity threshold for umbral-penumbral boundary is 0.53 and that of penumbral-quiet Sun boundary is 0.90. These

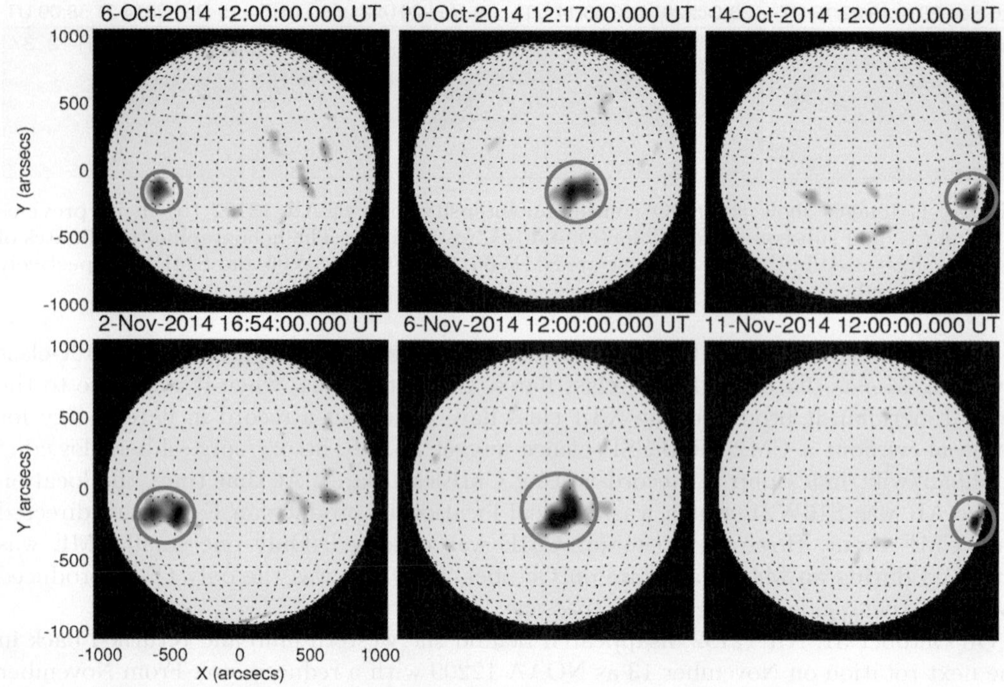

Figure 2. The upper panels show the appearance of AR 12192 (indicated by the red circles) in GONG far-side images in between the first and second rotation. The bottom panels show the same in between the second and third rotation.

values were then used to calculate the umbral and penumbral area evolution during the first three rotations of the AR. As the area during the 4th rotation was too small it was not included in the analysis. To calculate the true area of the sunspot groups, we have used the algorithm introduced by Cakmak (2014). The variation in area of AR 12192 for the first three rotations is illustrated in Figure 3.

3. Summary and Discussions

AR 12192 underwent a gradual growth starting from the previous disc passage during which the umbral and penumbral area increased from ~40 to ~85 MSH (Millionths of Solar Hemisphere) and ~300 to ~500 MSH respectively. The further growth of the AR occurred on the backside of the visible solar disc. Finally, when it appeared as AR 12192 in the second rotation, both the penumbral and umbral area became almost 6 times larger than the previous rotation. Also the major dynamics of this AR occurred during this rotation when the umbral area grew from ~500 to ~650 MSH and the penumbral area increased from ~3200 to ~4000 MSH. This noticeable growth in both the umbral and penumbral area reflected in the magnetic energy content of AR 12192, as it produced six high energetic X-class flares within an interval of one to two days during its disc passage. It implies that the huge size of AR 12192 may be the reason of its sufficient energy storage to trigger the recurrent high energetic flares. In the third rotation, the active region area decreased by almost ten times of that from the previous rotation. Notably, in all the three rotations the penumbral to umbral area ratio remained almost constant (approximately 5.5), implying that both the umbra and penumbra evolved almost in a similar fashion during the whole evolution period.

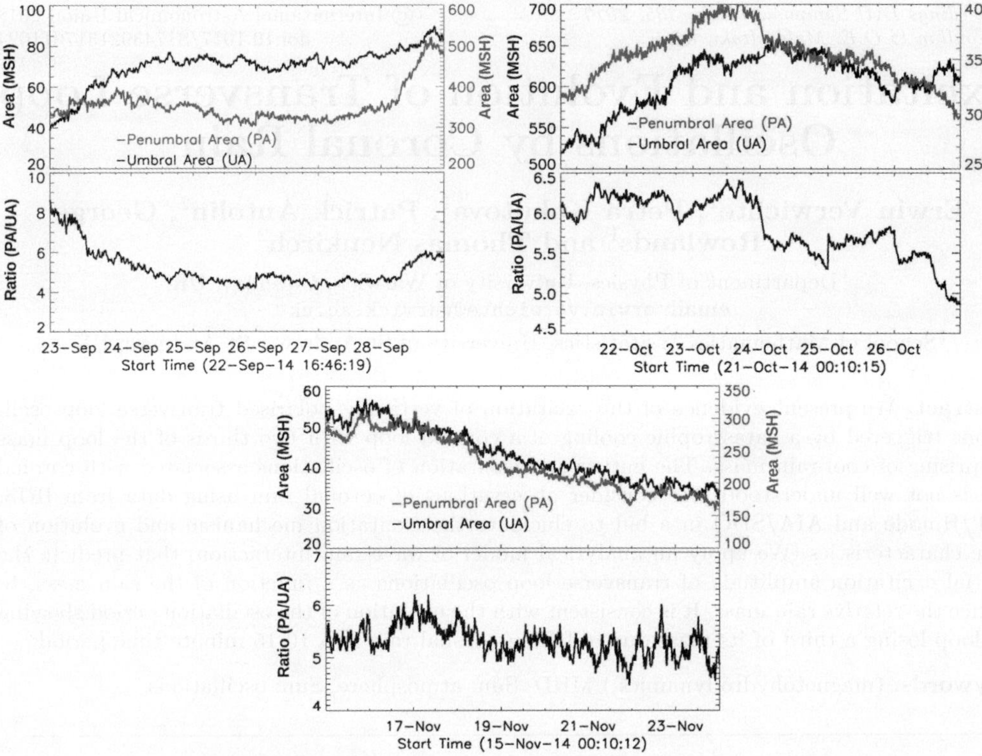

Figure 3. Area variation of the AR for the first three rotations

From the analysis of HMI vector magnetograms, we have studied the four non-eruptive major flares which occurred near the core region of AR 12192 and one eruptive flare that occurred away from the core region. We found abrupt and permanent changes in the horizontal magnetic field near the magnetic neutral line along with significant changes in the radial component of Lorentz force. Importantly, the above changes were significantly less in case of the four confined flares than the eruptive one. The distinct changes in the magnetic properties for both the eruptive and non-eruptive flares reveal that the large flares leave significant magnetic imprints on the solar photosphere which may carry the information of the nature of eruptions. A detail comparative study of the magnetic characteristics associated with both the eruptive and non-eruptive flares produced by AR 12192 is presented in Sarkar & Srivastava (2018).

References

Cakmak, H. 2014, *Experimental Astronomy*, 37, 539

Jiang, J., Cameron, R. H. & Schussler, M. 2015, *ApJL*, 808, L28

Lemen, J. R., Title, A. M., Akin, D. J., *et al.* 2012, *Sol. Phys.*, 275, 17

Mathew, S. K., Martínez Pillet, V., Solanki, S. K. & Krivova, N. A. 2007, *Astronomy & Astrophysics*, 465, 291

Pesnell, W. D., Thompson, B. J. & Chamberlin, P. C. 2012, *Sol. Phys.*, 275, 3

Pettauer, T. & Brandt, P. N. 1997, *Sol. Phys.*, 175, 197

Sarkar, R. & Srivastava, N. 2018, *Sol. Phys.*, 293, 16

Schou, J., Scherrer, P. H., Bush, R. I., *et al.* 2012, *Sol. Phys.*, 275, 229

Thalmann, J. K., Su, Y., Temmer, M. & Veronig, a. M. 2015, *ApJL*, 801, L23

Turmon, M., Jones, H. P., Malanushenko, O. V. & Pap, J. M. 2010, *Sol. Phys.*, 262, 277

Space Weather of the Heliosphere: Processes and Forecasts
Proceedings IAU Symposium No. 335, 2017
C. Foullon & O.E. Malandraki, eds.

doi:10.1017/S1743921317011024

Excitation and Evolution of Transverse Loop Oscillations by Coronal Rain

Erwin Verwichte[1], Petra Kohutova[1], Patrick Antolin[2], George Rowlands[1] and Thomas Neukirch[2]

[1]Department of Physics, University of Warwick, Coventry, UK
email: erwin.verwichte@warwick.ac.uk

[2]School of Mathematics & Statistics, University of St Andrews, St Andrews, UK

Abstract. We present evidence of the excitation of vertically polarised transverse loop oscillations triggered by a catastrophic cooling of a coronal loop with two thirds of the loop mass comprising of cool rain mass. The nature and excitation of oscillations associated with coronal rain is not well understood. We consider observations of coronal rain using data from IRIS, SOT/Hinode and AIA/SDO in a bid to elucidate the excitation mechanism and evolution of wave characteristics. We apply an analytical model of wave-rain interaction, that predicts the inertial excitation amplitude of transverse loop oscillations as a function of the rain mass, to deduce the relative rain mass. It is consistent with the evolution of the oscillation period showing the loop losing a third of its mass due to falling coronal rain in a 10-15 minute time period.

Keywords. (magnetohydrodynamics:) MHD, Sun: atmosphere, Sun: oscillations

1. Introduction

Coronal rain are dense cool plasma blobs that form near the top of coronal loops and fall towards the solar surface guided by the magnetic field (e.g. Schrijver 2001, Antolin *et al.* 2010). They are caused by catastrophic cooling of coronal plasma in response to localised heating increasing the amount of radiating in a loop (e.g. Müller *et al.* 2005). Coronal rain frequently exhibits transverse oscillatory motions (Antolin & Verwichte 2011, Kohutova & Verwichte 2016). These oscillations typically have periods in the order of a few to tens of minutes with projected displacement amplitudes of hundreds of to a thousand km. The physical mechanism for excitation of such oscillation in quiescent coronal loops is unclear. Verwichte *et al.* (2017) demonstrated that the presence of coronal rain through its concentrated mass may excite transverse loop oscillations and thus provide an additional seismological tool to determine the fraction of the rain mass relative to the total loop mass. This mechanism could explain the excitation of some rain oscillations such as the short-period oscillations measured by Kohutova & Verwichte 2016).

The approximate expressions found by Verwichte *et al.* (2017) are based on an analytical model where rain blobs are point masses, which displace the loop by its gravity. For loop length, L, inclination angle of the loop plane with respect to the photospheric normal, θ and fraction of coronal rain mass relative to the total loop mass, m/M, the expected displacement amplitudes of the excited oscillation, for fundamental horizontally and vertically polarised modes, are given by

$$\xi_{0,\wedge} = (9.6 \pm 0.7)\,10^{-3}\sqrt{\frac{2\theta}{\pi}\frac{m}{M}}\,L\ ,\ \ \xi_{0,\perp} = (5.0 \pm 0.6)\,10^{-3}\sqrt{1-\frac{2\theta}{\pi}\frac{m}{M}}\,L\ . \qquad (1.1)$$

These expressions are in agreement with full MHD simulations in the limit of small rain masses but underestimate the amplitude for larger masses (Kohutova & Verwichte 2017a,

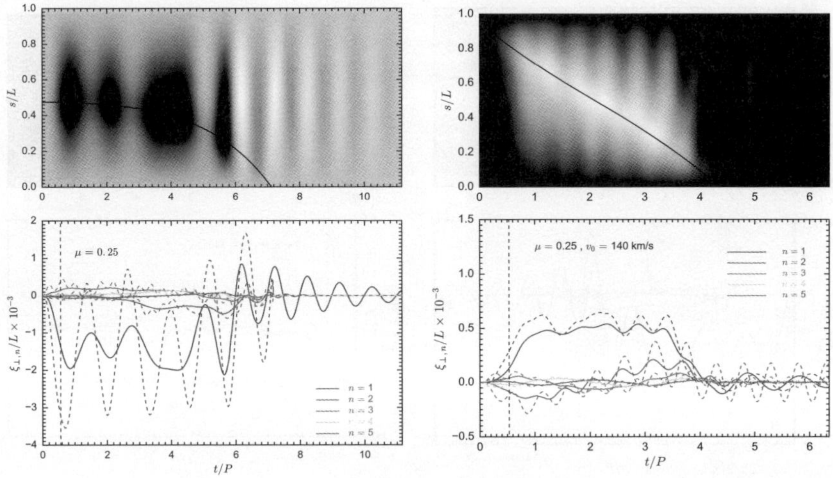

Figure 1. Left: Excitation of vertical oscillations due to a finite mass rain blob ($m/M = 0.2$) released near the loop top. The rain mass is exponentially ramped up over a typical time of 100s. Top: Displacement along the loop as a function of time with blob path superimposed. Bottom: Amplitudes of the five first wave mode harmonics as a function of time. Dashed lines are undamped modes (from Verwichte *et al.* 2017). Right: Same but for a rain blob launched into the loop from a foot point at a speed of 140 km/s. This case highlights the effect of wave excitation via the centrifugal force.

2017b). Excitation of vertical oscillations may also be achieved through the centrifugal force from fast-moving rain, with speeds in excess of $\sqrt{g_\odot R} \sim 100$ km/s, displacing the loop outwards (see Figure 1).

2. Observation

We present observations by IRIS in Si IV and Hinode/SOT in Ca II of a vertically polarised transverse oscillation in a coronal loop of length 80 Mm with coronal rain on the 27th August 2014. From 08:42 UT cold plasma condensations are forming at the loop top that quickly fill up to 60% of the loop length. At the same time a fundamental vertically polarised transverse oscillation becomes visible as periodic up and down motions of the condensations (see Figure 2). A tried and tested semi-automated method is used to determine the loop top position as a function of time at two time intervals, and we fit an undamped sinusoidal signal with a background trend (Verwichte *et al.* 2004, Verwichte *et al.* 2009). Critically, for the two time intervals we find different displacement amplitudes and periods, i.e. $\xi_{0,1} = 0.22{\pm}0.07$ Mm, $P_1 = 2.6{\pm}0.1$ min, and $\xi_{0,2} = 0.15{\pm}0.07$ Mm, $P_2 = 2.1{\pm}0.1$ min. From inserting these values into Eq. (1.1) the rain mass ratios at the two time intervals are deduced to be (Verwichte & Kohutova 2017): $(m/M)_1 = 0.58^{+0.4}_{-0.3}$, $(m/M)_2 = 0.38^{+0.4}_{-0.2}$. 58% of the loop mass is rain at the start of the event when condensations are seen forming along 60% fraction of the loop length. This result is similar to what had been found to explain the short-period horizontally polarised transverse oscillations by Kohutova & Verwichte (2016) and consistent with the finding of Antolin *et al.* (2015) that coronal rain may make up most of the loops mass. Towards the end of the visible oscillation, the mass fraction has dropped down to 40%. The evolution of period independently holds information on the rain mass. Assuming the hot loop mass does not change, we deduce from the period ratio that $(m/M)_2 = 1 - (P_1/P_2)^2[1 - (m/M)_1] = 0.34^{+0.7}_{-0.2}$, which independently confirms the earlier found value.

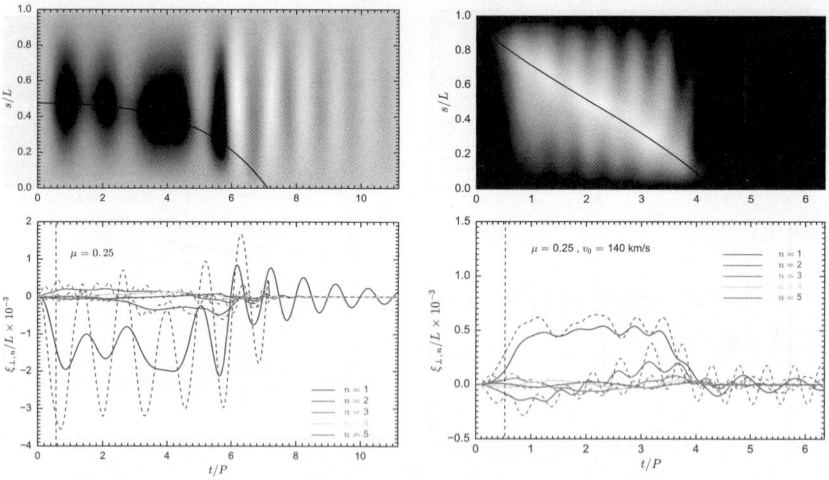

Figure 2. Top left: Partial FOV of the IRIS Si IV observations at three times. Bottom left: Time-distance cuts perpendicular (top) and parallel (bottom) to the loop axis in the two IRIS data sets. Right top: Loop position obtained from IRIS as a function of time. Right middle: Loop displacement as a function of time obtained by subtracting the linear profile from the loop position time series. The red and green curves are fitted sinusoidal curves with the corresponding parameters listed. Right bottom: Position of individual rain blobs as distance relative to the approximate loop top as a function of time (from Verwichte & Kohutova 2017).

Eq. (1.1) assumes an instantaneous appearance of the rain blob. In reality rain blobs form on a radiative cooling time-scale $\tau_{\rm rad} = (k_{\rm B}/(\gamma-1)\,\chi)n_e^{-1}T^{1-\alpha}$, where we have assumed a piece-wise linear radiation profile as a function of temperature. For a typical coronal loop with temperature $T = 10^6$ K and mass density $\rho = 10^{-11}$ kg/m^{-3}, we obtain $\tau_{\rm rad} = 5$ mins. We have used SDO/AIA images and the DEM inversion method by Hannah & Kontar (2012) to estimate the temperature and density as a function of time during the condensation phase of the thermal limit cycle. We obtain $\tau_{\rm rad}$=500 s, i.e. the same order of magnitude but larger. We may model this gradual appearance by ramping up the rain mass on the same time-scale (see Figure 1). However, this does not take into account pressure effects at low masses. Full MHD simulations with self-consistent thermal instability are required to answer this question more definitively.

References

Antolin, P., Shibata, K. & Vissers, G. 2010, *ApJ*, 716, 154
Antolin, P. & Verwichte, E. 2011, *ApJ*, 736, 121
Antolin, P., Vissers, G., Pereira, T. M. D., Rouppe van der Voort, L. & Scullion, E. 2015, *ApJ*, 806, 81
Hannah, I. G. & Kontar, E. P. 2012, *A&A*, 539, A146
Kohutova, P. & Verwichte, E. 2016, *ApJ*, 827, 39
Kohutova, P. & Verwichte, E. 2017a, *A&A*, 602, A23
Kohutova, P. & Verwichte, E. 2017b, *A&A*, 606, A120
Müller, D. A. N., De Groof, A., Hansteen, V. H. & Peter, H. 2005, *A&A*, 436, 1067
Schrijver, C. J. 2001, *Sol. Phys.*, 198, 325
Verwichte, E., Nakariakov V. M., Ofman, L. & Deluca, E. E. 2004, *Sol. Phys.*, 223, 77
Verwichte, E., Aschwanden, M. J., Van Doorsselaere, T., Foullon, C. & Nakariakov, V. M. 2009, *ApJ*, 698, 397
Verwichte, E., Antolin, P., Rowlands, G., Kohutova, P. & Neukirch, T. 2017, *A&A*, 598, A57
Verwichte, E. & Kohutova, P., E. 2017, *A&A*, 601, L2

The Northcott Theatre auditorium, main venue for the oral program (speaker: H. Cremades).

Part 2. Solar Wind and Heliosphere

Space Weather of the Heliosphere: Processes and Forecasts
Proceedings IAU Symposium No. 335, 2017
C. Foullon & O.E. Malandraki, eds.

doi:10.1017/S1743921317008067

Gamma-Ray Solar Flares and In Situ Particle Acceleration

Alexei Struminsky
Space Research Institute,
Profsoyuznaya st., 84/32, 117997, Moscow, Russian Federation
email: astrum@iki.rssi.ru

Abstract. At present two concurrent paradigms of solar energetic particle (SEP) origin exist: acceleration directly in the flare site or by the shock wave of coronal mass ejection (CME). Active discussions on a relative role of flares and coronal mass ejections for SEP acceleration and propagation are continuous until now. In my opinion only future observations of solar high energy γ–emission with better spectral, spatial and temporal resolution may clarify this issue. In my report I discuss possible signatures of the flare and shock acceleration processes. What is a picture provided by the current instruments? What can we expect to observe with a perfect instrument in high energy gamma rays in one or another case on a time scale of impulsive and long decay flare phases?

Keywords. Sun: activity, Sun: flares, Sun: x-rays and gamma-rays, Sun: particle emission

1. Introduction

A task to measure γ–radiation was set up just from the beginning of space era. A paper of Chupp (1964), possibly, was the first, where estimates for intensities of solar γ–rays and neutrons had been presented. Later Edward Chupp (1927-2017) became a principal investigator for the γ-ray spectrometer experiments on the OSO-7 and the Solar Maximum Mission satellites, which made pioneering observations of solar flare γ–rays and neutrons. Reading his reviews (Chupp, 1964; Chupp, 1971; Chupp, 1984; Chupp, 1996; Chupp & Ryan, 2009) one can follow the history of solar γ–ray observations from first proposals to the Fermi Gamma Ray Observatory (GRO).

Milestones of the solar γ–ray observations attributed to famous solar events and particular missions are: August 1972, OSO -7 - solar nuclear γ–lines (Chupp *et al.* 1973); June 3, 1982, SMM - pion decay solar γ–emission (Forrest *et al.* 1986), two phases of proton acceleration (Ramaty *et al.* 1987), first γ–flare statistics (Cliver *et al.* 1989); June 1991, Compton GRO (Kanbach *et al.* 1993, Rank *et al.* 2001, Murphy *et al.* 1997), GAMMA-1 (Akimov *et al.* 1996) – long duration solar γ–ray flares, GRANAT – coronal γ–ray source (Ramaty *et al.* 1997); July 23, 2002, RHESSI - precise γ–line spectroscopy (Lin *et al.* 2003) and γ–ray images (Hurford *et al.* 2003); Fermi GRO – images of pion decay emission(Ackermann *et al.* 2014), M class γ–flares (Ackermann *et al.* 2012), confirmation of long duration γ–flares (Ajello *et al.* 2014). The lastest reviews on the subject are Lin (2011), Vilmer *et al.* (2011) and Raymond *et al.* (2012).

High energy particles accelerated in flares interact in the solar atmosphere producing bremsstrahlung (electrons), γ–line ($>$ 30 MeV protons) and pion decay ($>$ 144 MeV protons) emission. Their fluxes should be rather high and they need to interact in a thick target in order to generate gamma emission observable at 1 AU. There are two possible sites of the thick target in the Sun, the photosphere (large density) and the corona (low density and, possibly, large time of trapping). A nice scheme of the flare gamma-ray diagnostics is presented in Chupp (1983). Studying spectral and temporal characteristics of this emission it is possible to separate cases of several acceleration episodes or trapping, to evaluate their characteristic time and to estimate a total energy of interacting particles. Spatial–temporal observations of γ–emission may allow to determine a position and size of the source, separate different sources, and to track the evolution of sources in time. In an ideal case the γ–ray observations may provide an injection function of solar protons into the interplanetary space.

What should the injection function of solar protons (> 30 MeV for γ–lines and > 144 MeV for pion production) look like based on their measurements near the Earth? Solar proton events (Kurt *et al.* 2004, Belov *et al.* 2005) at the Earth well correlate with M–X class solar flares with coordinates E10–W120 (the nominal Earth connecting field line is at W60). Generally a release time of solar protons we may not estimate better than ~ 10 min keeping in mind all possible uncertainties of a detector's background and a length of magnetic field line. To reproduce temporal profiles of proton intensity we need to consider several injection episodes with total duration from ~ 10 min up to several hours (Struminsky, 2003, 2013a). Prolonged injection of solar protons into the interplanetary space within ~ 60 degree is necessary, which may be provided by flare acceleration and coronal propagation or by CME–driven shock wave acceleration. Arguments between flare and shock acceleration protagonists are continuing, one of the last examples is a polemic between Grechnev *et al.* (2015), who are in favor of flare acceleration, and Cliver (2016), who is a strong supporter of shock acceleration. Recent reviews on the topic are Aschwanden (2012), Reames (2013), Desai & Giacalone (2016), Bazilevskaya (2017), Klein & Dalla (2017).

Below I discuss possible signatures of the flare and shock acceleration processes. What we might expect to observe by a perfect instrument in high energy γ–rays in one or another case on a time scale of minutes (impulsive phase) and hours (long decay phase)?

2. Spectral, temporal and spatial characteristics of solar γ–flares

Let us first consider statistical results of solar X–ray (SXR) and γ-ray observations in proton events. Solar proton events (Kurt *et al.* 2004, Belov *et al.* 2005) at the Earth well correlate with M-X class solar flares. Since plasma–emitting SXR is heated by electrons accelerated in solar flares (Neupert, 1968) it is reasonable to suggest that solar proton events should correlate well with hard X-ray (HXR) events. If electrons and ions are accelerated nearly simultaneously (Forrest & Chupp 1983, Ackermann *et al.* 2012) then fluences of HXR and γ–rays should correlate. However from the statistical analysis of the 2.223 MeV line fluence and the fluence > 50 keV it follows that such a correlation exists only when a threshold in the production of ions is reached (Shih *et al.* 2009). A similar relationship had been discussed at the time of SMM by Cliver *et al.* (1994). This ultimately suggests that while the acceleration of protons above 30 MeV is closely related to the acceleration of relativistic electrons, the acceleration of subrelativistic electrons is only proportional to the acceleration of relativistic electrons and ions when a given threshold of high energy particles is reached Vilmer *et al.* (2011). Possibly, there are two acceleration processes, one producing proportional quantities of relativistic electrons and ions, the other one producing mostly subrelativistic electrons (Frost & Dennis 1971).

The SMM data suggest that progressive hardening (soft-hard-harder, SHH, the Kiplinger effect) of HXR spectrum is a diagnostic of high-energy electron and proton acceleration (Kiplinger, 1995). This is not a manifestation of the big flare syndrome (Kahler, 1982). Grayson *et al.* (2009) checked the Kiplinger effect using RHESSI and demonstrated a statistically significant dependence of SHH and SEP observations. This a link that is unexplained in the standard scenario of SEP acceleration at the CME shock front and encourages further investigation of the mechanisms which could be responsible (Grayson *et al.* 2009). A critical view on the Kiplinger effect is presented by Kahler (2012).

A hardening of γ–spectrum is visible in individual events and it should be the characteristic of solar proton events. Ramaty *et al.* (1987) presented evidence of two stages of pion production on June 3, 1982. In the first stage pions were produced by particles with relatively soft spectrum and later in the second stage by particles with much harder spectrum. Kiener *et al.* (2006) analyzing INTEGRAL data on gamma-emission of the 2003 October 28 flare distinguished three phases: A – only continuum emission, B – nuclear gamma-lines, C – decay. According to Coronas–F observations (Kuznetsov *et al.* 2011) the pion production occurred in the B and C phases. Pion decay emission was observed by Coronas–F also in the flares of November 4, 2003 and January 20, 2005 (Kurt *et al.* 2010). If electrons and protons are accelerated by the same processes, then a temporal structure of γ–emission on time scale of several seconds should exist within each phase corresponding to a fine spatial structure of coronal loops (Zimovets *et al.* 2013). A zero time for the solar flare on December 6, 2006, which had showed clearly preflare, impulsive and decay phases, was defined by Struminsky & Zimovets (2010) as the beginning of 15.4 GHz microwave emission. Struminsky (2013b) deduced a zero time for the pion flares superimposing their temperature time profiles with that of the event on December 6, 2006. It appeared that a visible pion production in all events started 4-5 minutes after the zero time. These 4-5 minutes is a characteristic time for acceleration of relativistic protons (spectrum hardening). The acceleration mechanism responsible for this spectra hardening is still unknown.

Gamma line and pion decay emission are observed on a time scale of several hours during some solar flares. High energy solar protons, which produced this emission, might be accelerated during the impulsive phase and subsequently trapped, or they might be accelerated continuously during entire flare. Mandzhavidze & Ramaty (1992) showed that a model in which the particles were accelerated during the impulsive phase and subsequently trapped in coronal loops could explain the emission detected at the late phase of the flare on June 1, 1991. According to Kocharov *et al.* (1994) a continuous and simultaneous acceleration of protons and relativistic electrons at the gradual phase of the flare on June 15, 1991 gives a natural explanation of the data. Akimov *et al.* (1996) provided additional evidence that the gamma-ray and other emissions observed well after the impulsive phase of the flare on June 15, 1991 was initiated by the prolonged non–stationary particle acceleration rather than prolonged trapping. The observations of burst-on-tail (BOT) of the flare on October 23, 2003 presented by Zimovets & Struminsky (2012) showed distinctly that episodes of electron acceleration may occur well after CME lift–off. It is unknown whether protons have been accelerated during the BOT.

The brightest and longest solar high energy γ–flare to date was detected by the Fermi LAT on March 7, 2012. Ajello *et al.* (2014) showed that the fluxes and spectra of the high-energy γ–rays evolve differently during the impulsive phase and the sustained emission. Also they noticed correlations and some differences between the fluxes and spectral indexes of the protons required for the production of high-energy γ–rays and SEP protons seen at 1 AU. From these results Ajello *et al.* (2014) suggested that the high-energy γ–rays are most likely produced by protons (rather than electrons) accelerated in the corona (rather than in the associated CME shock) continuously during the entire flare.

Spatial observations of solar γ–ray are controversial. Comparisons of imaged and spatially integrated fluences (events on October 28, 29 and November 2, 2003) show that in all cases most, if not all, of the emission was confined to compact sources with size scales of tens of arcseconds or smaller that are located within the flare active region (Hurford *et al.* 2006). Thus, the ions producing γ–rays appear to be accelerated by the flare process and not by a widespread shock driven by a fast coronal mass ejection. The 28 October event yielded the first such image to show double–footpoint γ-ray line sources. These footpoint sources straddled the flaring loop arcade but were displaced from the corresponding 0.2–0.3 MeV electron-bremsstrahlung emission footpoints. As with the previously studied event on July 23, 2002 (Hurford *et al.* 2003), this implies spatial differences in acceleration and/or propagation between the flare-accelerated ions and electrons. An erupting flux rope can act as a trigger of energy release in two–ribbon flares. Its successive interaction with different loops of a parent active region can lead to apparent motion of HXR sources and to a series of HXR pulsations (Kuznetsov *et al.* 2016). Similar motion and pulsations can be observed for γ–ray sources in a case of flare acceleration.

Ackermann *et al.* (2014) and Ajello *et al.* (2014) showed that during most of the long-duration emission the high energy γ–rays appear to come from the same active regions responsible for the flare emission. Observations of γ-emission during flares behind the limb (occulted) show that a γ-source can be larger than an active region. Cliver *et al.* (1993) sugested that protons produced the γ–emission on the visible disk during the flare behind the limb on September 29, 1989 were accelerated by CME–driven shock wave. Ramaty *et al.* (1997) provided evidence for γ–ray production in the corona during the giant flare behind the east limb on June 1, 1991. Ackermann *et al.* (2017) presented Fermi GRO/LAT spatially resolved observations of four flares behind the limb. They believed that the HXR emission was due to electron bremsstrahlung from a coronal thin-target loop-top, but the $>$ 100 MeV γ–rays to a pion production in the photoshere by the CME–accelerated protons precipitating downward.

3. Discussion and conclusions

Nuclear line and pion decay γ-rays are emitted on a time scale from several minutes up to several hours during some solar flares but their statistics are very poor. Accelerated protons ($>$ 30 MeV, $>$ 144 MeV) interact in the photosphere and the corona, and might be be injected into the interplanetary space, but spatial properties of the injection function is unclear from γ-ray observations. Consequently we still do not know what is accelerating mechanism of solar cosmic rays, what is a source of interplanetary (propagating) solar protons. In my opinion only future observations of solar high energy γ–emission with better spectral, spatial and temporal resolution will answer these questions. We need special telescopes for continuous observations of solar hard X-ray and γ–emissions with better sensitivity, energetic, temporal and spatial resolution aboard spacecraft in deep space, at the Lagrangian points, or in eccentric orbits with high apogees.

In a case of protons accelerated in flares we expect to observe γ–emission (lines, pion decay) multiple photospheric footpoint sources (target density $\sim 10^{17}$cm^3) and weak coronal sources (target density $\sim 10^9 - 10^{12}$ cm^3) with spatial scale up to active region size (L $\sim$ 10 degree=0.17R), its correlation with HXR and microwave emission, and fine temporal structure ($\sim$ 1 min, $\sim$ 10 sec). In a case of shok–accelerated protons that would be decreasing γ–emission from unstable photospheric footpoints (target density $\sim 10^{17}$cm^3), weak moving and extending coronal source (target density $< 10^9$ cm^3) $VT = 2500kms^{-1} \times 600s = L \sim 2.1R = 120°$), without fine temporal structure and not correlating with HXR and microwave emissions.

Lin *et al.* (2010) presented a concept of the Solar Eruptive Event (SEE) 2020 mission, which was unfortunately not supported. Comparing the spectral, spatial, and temporal behavior of ions and electrons requires X-ray and γ–ray imaging spectroscopy from < 10 keV to > 15MeV. For future progress we need high–resolution γ–ray imaging with a factor of > 10 increase in sensitivity and < 10 arcsec resolution compared to RHESSIs 35 arcsec resolution. A step in this direction is a development of the Gamma Ray Imager/Polarimeter for Solar flares (GRIPS). This instrument incorporates key technological improvements over the current state of the art at HXR/γ–ray energies. Dunkan *et al.* (2016) described GRIPS's first Antarctic long-duration fight in January 2016, and presented preliminary calibration and scientific results.

Last results in the high energy γ rays (> 100 MeV) were obtained by the Fermi GRO/LAT, but its sensitivity and spatial resolution are not enough to distinguish between interacting flare–accelerated ions that are localized to the flare loops and those ions accelerated by the CME shock that precipitate to form a larger diffuse source. Chupp *et al.* (1998) and Chupp *et al.* (2003) proposed two techniques which may locate the emission regions of high-energy neutral emissions to arcminute or better spatial accuracy and also determine their energy spectra. I do not know any development in this direction from that time.

Solar Probe Plus and Solar Orbiter may give a key for the problen of solar cosmic ray origin. Both spacecrafts are designed for measurements close to the Sun, where interplanetary scattering and transport effects are significantly reduced, allowing to discriminate between different acceleration sites and mechanisms and to isolate the contributions of numerous physical processes (Desai & Giacalone 2016).

The author thanks the Russian Foundation for Basic Research (grant 16-02-00328) for providing a travel support to attend the Space Weather IAU Symposium No. 335.

References

Ackermann, M., Ajello, M., Allafort, A., *et al.* 2012, *ApJ*, 745, 144

Ackermann, M., Ajello, M., Albert, A., *et al.* 2014, *ApJ*, 787,15

Ackermann, M., Allafort, A., Baldini, L., *et al.* 2017, *ApJ*, 835, 219

Ajello, M., Albert, A., Allafort, A., *et al.* 2014, *ApJ*, 789,20

Akimov, V., Ambroz, P., Belov, A., *et al.* 1996, *Solar Phys.*, 166, 107

Aschwanden, M. 2012, *Space Sci. Revs*, 171, 3

Bazilevskaya, G. 2017, *JPhCS*, 798, 012034

Belov, A., Garcia, H., Kurt, V., Mavromichalaki, H. & Gerontidou, M. 2005, *Solar Phys.*, 229, 135

Chupp, E. 1964, in: W. M. Hess (eds), *The Physics of Solar Flares, Proceedings of the AAS-NASA Symposium held 28-30 October, 1963 at the Goddard Space Flight Center, Greenbelt, MD.* (Washington, DC: National Aeronautics and Space Administration, Science and Technical Information Division), p. 445

Chupp, E. 1971, *Space Sci. Revs*, 12, 486

Chupp, E., Forrest, D., Higbie, P., Suri, A., Tsai, C. & Dunphy, P. 1973, *Nature*, 241, 333

Chupp, E. 1983, *Solar Phys.*, 86, 383

Chupp, E. 1984, *ARAA*, 22, 359

Chupp, E. 1996, *AIP-CP*, 374, 3

Chupp, E., Aprile, E., Bertsch, D., Dunphy, P. & Ramaty, R. 1998, *Ad. Sp. Res.* 21, 333

Chupp, E. L., Trottet, G., Dunphy, P. P. & Rieger, E. 2003, in:T. Kajita, Y. Asaoka, A. Kawachi, Y. Matsubara and M. Sasaki (eds), *Proc.the 28th ICRC. July 31-August 7, 2003.* Trukuba, Japan. Under the auspices of IUPAP. p. 3171

Chupp, E. & Ryan, J. 2009, *Res. A&A*, 9, 11

Cliver, E., Forrest, D., Cane, H., Reames, D., McGuire, R., von Rosenvinge, T., Kane, S. & MacDowall, R. 1989, *ApJ*, 343, 953

Cliver, E., Kahler, S. & Vestrand, W. 1993, in D.A. Leahy, R. B. Hicks, and D. Venkatesan (eds),

23rd International Cosmic Ray Conferenceheld 19-30 July, 1993 at University of Calgary, Alberta, Canada. Singapore: World Scientific, 3, 91
Cliver, E., Crosby, N. & Dennis, B. 1994, *ApJ*, 426, 767
Cliver, E. 2016, *ApJ* 832, 128
Desai, M. & Giacalone, J. 2016, *Living Rev. Solar Phys.* 13, 3
Duncan, N., Saint-Hilaire, P., Shih, A., *et al.* 2016, *Proceedings of the SPIE* 9905, 99052Q
Forrest, D. & Chupp, E. 1983, *Nature* 305, 291
Forrest, D., Vestrand, W., Chupp, E., Rieger, E. & Cooper, J. 1986, *Ad. Sp. Res.*, 6, 115
Frost, K. & Dennis, B. 1971, *ApJ*, 165, 655
Grayson, J., Krucker, S. & Lin, R. 2009, *ApJ*, 707, 1588
Grechnev, V., Kiselev, V., Meshalkina, N. & Chertok, I. 2015, *Solar Phys.*, 290, 2827
Hurford, G., Schwartz, R., Krucker, S., Lin, R., Smith, D. & Vilmer, N. 2003, *ApJ* (Letters) 595, L77
Hurford, G., Krucker, S., Lin, R., Schwartz, R., Share, G. & Smith, D. 2006, *ApJ* (Letters) 644, L93
Kahler, S. 1982, *JGR*, 87, 3439
Kahler, S. 2012, *ApJ* 747, 66
Kanbach, G., Bertsch, D., Fichtel, C., Hartman, R., Hunter, S., Kniffen, D., Kwok, P., Lin, Y., Mattox, J. & Mayer-Hasselwander, H. 1993, *A&AS*, 97, 349
Kiplinger, A. 1995, *ApJ*, 453, 973
Kiener, J., Gros, M., Tatischeff, V. & Weidenspointner, G. 2006, *A&A*, 445, 725
Klein, K.–L. & Dalla, S. 2017, *Space Sci. Revs* on line first, 27 July
Kocharov, G., Chuikin, E., Kovaltsov, G., Usoskin, I. & Kocharov, L. 1994, *AIP-CP*, 294, 45
Kurt V., Belov A., Mavromichalaki H. & Gerontidou, M. 2004, *Annales Geophysicae* 22, 2255
Kurt, V., Yushkov, B., Kudela, K. & Galkin, V. 2010, *Cosmic Res.*, 48, 70
Kuznetsov, S., Zimovets, I., Morgachev, A. & Struminsky, A. *Solar Phys.*, 291, 3385
Kuznetsov, S., Kurt, V., Yushkov, B., Kudela, K. & Galkin, V. 2011, *Solar Phys.*, 268, 175
Lin, R., Krucker, S., Hurford, G., *et al.* 2003, *ApJ* (Letters) 595, L69
Lin, R., Caspi, A., Krucker, S., *et al.* 2010, *astro-ph.SR*, arXiv:1311.5243v1
Lin, R. 2011, *Space Sci. Revs*, 159, 421
Mandzhavidze, N. & Ramaty, R. 1992, *ApJ* (Letters) 396, L111
Murphy, R., Share, G., Grove, J., Johnson, W., Kinzer, R., Kurfess, J.,Strickman, M. & Jung, G. V. 1997, *ApJ*, 490, 883
Neupert, W. 1968, *ApJ* (Letters) 153, L59
Raymond, J., Krucker, S., Lin, R. & Petrosian, V. 2012, *Space Sci. Revs*, 173, 197
Ramaty R., Mandzhavidze N., Barat C. & Trottet G. 1997, *ApJ*, 479, 458
Ramaty, R., Murphy, R. & Dermer, D. 1987, *ApJ* (Letters) 316, L41
Rank, G., Ryan, J., Debrunner, H., McConnell, M. & Schonfelder, V. 2001, *A&A*, 378, 1046
Reames, D. 2013, *Space Sci. Revs*, 175, 53
Shih, A., Lin, R. & Smith, D. 2009, *ApJ* (Letters) 698, L152
Struminsky, A. 2003, *Astron. Rep.* 47, 916
Struminsky, A. & Zimovets, I. *Astron. Lett.* 36, 430
Struminsky, A. 2013a, *JPhCS* 409, 012148
Struminsky, A. 2013b, *Geomagn. & Aeron.* 409, 012148
Vilmer, N., MacKinnon A. & Hurford, G. 2011, *Space Sci. Revs*, 159, 167
Zimovets, I. & Struminsky, A. 2012, *Solar Phys.* 281, 749
Zimovets, I., Kuznetsov, S. & Struminsky, A. 2013, *Astron. Lett.*, 39, 267

Space Weather of the Heliosphere: Processes and Forecasts
Proceedings IAU Symposium No. 335, 2017
C. Foullon & O.E. Malandraki, eds.

doi:10.1017/S1743921317009681

The Relationship between Long-Duration Gamma-Ray Flares and Solar Cosmic Rays

Hugh S. Hudson[1,2]

[1]Space Sciences Laboratory, University of California, Berkeley CA USA 94720
email: hhudson@ssl.berkeley.edu

[2]SUPA School of Physics and Astronomy, University of Glasgow, UK

Abstract. A characteristic pattern of solar hard X-ray emission, first identified in SOL1969-03-30 by Frost & Dennis (1971), turns out to have a close association with the prolonged high-energy gamma-ray emission originally observed by Forrest *et al.* (1985). This identification has become clear via the observations of long-duration γ-ray flares by the *Fermi*/LAT experiment, for example in the event SOL2014-09-01. The distinctive features of these events include flat hard X-ray spectra extending well above 100 keV, a characteristic pattern of time development, low-frequency gyrosynchrotron peaks, CME association, and gamma-rays identifiable with pion decay originating in GeV ions. The identification of these events with otherwise known solar structures nevertheless remains elusive, in spite of the wealth of EUV imagery available from SDO/AIA. The quandary is that these events have a clear association with SEPs in the high corona, and yet the gamma-ray production implicates the photosphere itself, despite the strong mirror force that should focus the particles *away* from the Sun We discuss the morphology of these phenomena and propose a solution to this problem.

Keywords. Sun: flares, Sun: X-rays, gamma rays, Sun: coronal mass ejections (CMEs), Sun: particle emission

1. Introduction

Flares and CMEs involve substantial particle acceleration, and we can see the consequences of this in several forms. Radio-astronomical techniques allow us to remote-sense some populations of energetic electrons in the corona, but such information at high energies often involves the transport of particles to regions of high density where emission can form. For electrons, the basic diagnostic radiation is bremsstrahlung, while for accelerated ions it is a variety of nuclear processes, such as the inelastic scattering of high-energy primary particles on ambient nuclei.

The processes involved with particle acceleration in the corona remain ill-understood, mainly because of the weak diagnostic power of the direct emissions (e.g., Krucker *et al.* 2010) and the difficulty of interpretation of the radio signatures (e.g. Kundu 1965). Both electrons and ions contribute to solar particle events detectable in the heliosphere, but the identification of these emissions with specific structures in the corona, as identified for example by soft X-ray or EUV imaging, remains largely unclear. In the heliosphere, for example, we have a clear association of particle acceleration with interplanetary shock waves, but we have little information about how these shocks form, and how they attain their energy.

Recently these problems have become especially prominent because of the detection of many long-duration events by the *Fermi*/LAT experiment, at energies above 20 MeV. Share *et al.* (2017) coined the term "sustained gamma-ray events" (SGREs) to imply a source population possibly distinct from that of the normal flare γ rays. Their detected

Table 1. Properties of coronal high-energy radiations.

Hard X-rays	**High-energy γ-rays**
SOL1969-03-30[1]	SOL1982-06-03[2]
Coronal origin (by occultation)[3]	Coronal origin (by inference)[4]
Hard spectrum, $J \sim (h\nu)^{-2}$	Pion decay signature
Duration tens of minutes	Duration up to many hours
Association with type II/IV burst[5]	Association with type II/IV burst
Drifting cm-wave source[6]	Neutrons
Association with SEPs	Association with SEPs
Large angular scale[7]	Unknown structure

Notes: [1]Frost & Dennis 1971. [2]Forrest *et al.* 1985. [3]Hudson 1978. [4]Chupp *et al.* 1985. [5]Smerd 1970. [6]Énomé & Tanaka 1971. [7]Krucker & Lin 2008.

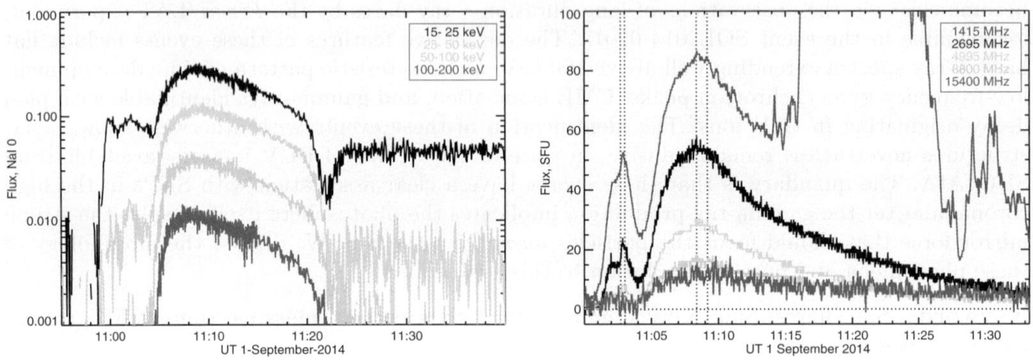

Figure 1. SOL2014-09-01, observed in hard X-rays by *Fermi*/GBM (log scale, left) and by the San Vito radio observatory (linear scale, right). The light curves match extremely well, except at longer radio wavelengths at which a distinct spectral component appears.

photon spectra can extend above 1 GeV and display the characteristics of pion decay, again implying GeV particle energies. Table 1 summarizes the properties of such events, based on the "discovery" events in hard X-rays (Frost & Dennis, 1971) and γ-rays (Forrest *et al.* 1985). In principle any sustained γ-ray emission could come either from continued particle acceleration or from the gradual interaction of trapped particles. Mandzhavidze & Ramaty (1992) favored the latter, as further discussed here, and noted a requirement for the mirror ratio to exceed 10 and the ambient density to lie below 5×10^{11} cm^{-3}, "quite reasonable" conditions. Klein *et al.* (2017) also discuss the mirror force in this context, though Plotnikov *et al.* (2017) do not.

Recently a "Rosetta Stone" event appeared (Pesce-Rollins *et al.* 2015; Ackermann *et al.* 2017) , in which both a Frost-Dennis hard X-ray source and sustained high-energy γ-ray emission occurred together, almost coincident in time but with the γ rays remaining detectable for almost two hours. This event, SOL2014-09-01, as shown in Figure 1, matches the HXR/radio properties in Table 1 quite well. As inferred previously for Frost-Dennis events, the extreme flatness of the HXR spectrum and its association with a matching non-thermal microwave spectrum, with low turnover frequency, strongly implicates relativistic electrons in the corona. In this event *Fermi*/LAT also detected high-energy γ-radiation.

2. The mirror force and particle numbers

The association of Frost-Dennis/SGRE events with SEPs naturally suggests the idea that the SEPs themselves return to the Sun and interact to produce the HXR and γ-ray emission (Pesce-Rollins *et al.* 2015; De Nolfo *et al.* 2016, Share *et al.* 2017, Ackermann *et al.* 2017). This interpretation, however, immediately seems counterintuitive. The conventional explanation of SEPs describes them as the result of diffusive shock acceleration in a bow wave driven by the CME. Such a shock would form in the middle corona, where a relatively weak magnetic field might favor shock formation. This very weakness of the field implies a strong mirror force that would tend to "focus" the particles away from the Sun, especially on open field structures. The probability for an accelerated particle to fall in the loss cone defined by collisions at photospheric height would be given by a probability $P = 1 - \sqrt{1 - B/B_0}$, where B is the ambient field intensity and B_0 the intensity at the precipitation footpoints. Note that a similar mirror force would apply to a diffusive SEP propagation, but that in any case we really do not have any quantitative knowledge of the actual geometry of the coronal magnetic field. For the SolarSoft PFSS field extrapolation (Schrijver & DeRosa 2003) at the time of SOL2014-09-01, we find that $\bar{P} = \Delta\Omega/4/\pi \approx 0.003$ for closed field, based on the mean field strengths at source surface at 2.5 $R_\odot$ and the photosphere. As an independent crude estimate, a dipole field originating 0.1 $R_\odot$ below the photosphere would have a mirror ratio $P = 0.0008$.

A more basic question about the association between SEPs and SGREs comes from the total numbers of particles needed. Mewaldt *et al.* (2005) suggest particle fluences of order 0.1-10 protons $(\mathrm{cm}^2\ \mathrm{sr\ MeV})^{-1}$ at 500 MeV, for a set of five well-observed events. This estimate includes rough approximations for multiple counting and for longitude dependence. Converted to energy fluxes, assuming a 10% propagation cone, these fluences imply total particle energies of $2 \times 10^{24-26}$ erg above 500 MeV. Ackermann *et al.* (2017) estimate at least $2.5 \times 10^{24} - 10^{27}$ erg for the energy required to sustain the occulted SGRE events they study. So, even not considering the mirror force, there is a substantial number problem: more particles appear to be required to produce the γ-rays than even the large energy fluxes in the interplanetary SEPs.

3. A "Lasso" scenario

Based on the above, the suggestion that SEP particles can somehow produce the SGRE emissions seems highly unlikely. I propose a "Lasso" scenario to enable this relationship, in which distended but closed magnetic fields would form a noose that could capture some of the SEPs, not allowing their escape into the heliosphere. In this picture the particle accelerator deposits particles on both open and closed field structures. The closed-field domain might have a large spatial scale, perhaps extending to several $R_\odot$ in height. This then bodily retracts, as a part of the recovery of the corona to the CME eruption, and thereby transporting the non-thermal particles by advection into denser regions. In this process one would also expect the betatron process and first-order Fermi acceleration to increase the particle energies and effectively to multiply their number relative to the SEPs themselves (the Compton-Getting effect). This idea in a sense modernizes the original suggestion by Elliot (1969) for particle storage in the low-beta corona. The Lasso idea just imagines that closed field can ensnare some of the Elliot particles and bring them down towards the Sun, where increased density and/or actual precipitation can produce the nuclear interactions needed. Post-CME inward flows on large spatial scales, as observable by LASCO, are not unusual (Sheeley *et al.* 2004, 2014), as required by the heliospheric regulation of open solar flux. Indeed the readily-available LASCO movies show hints of an inflow at the N flank of the CME in the SOL2014-09-01 event, in a manner consistent with the Lasso picture.

Do we actually require the lasso action (retraction of large-scale magnetic flux bundles) to find sufficient energetic particles? A static field volume at a sufficiently low density might in principle contain 500 MeV protons long enough to generate pions and γ-ray continuum *in situ* via trapping in turbulence. The relativistic Bethe-Bloch cross-section for collisional losses is about 3×10^{-23} cm^2 (Olive *et al.* 2014), which for density 10^8 cm^{-3} translates into a time scale of five hours; for these parameters the Alfvén speed is about 3000 km/s. Thus a "coronal thick target" scenario could provide an alternative to the Lasso picture suggested here.

4. Conclusions

The events described here (the Frost-Dennis and SGRE phenomena) appear to form a distinctive pattern, and one that probably happens fairly commonly during active periods. The feebleness of the diagnostic radiations in hard X-rays and γ-rays makes their detection difficult, but their energetic significance is profound. The Lasso picture itself, if analyzed theoretically, would definitely need to take into account the feedback between particle and plasma dynamics, since the energy content of the particles can be large. This means that a test-particle description of the collapsing trap (e.g., Somov & Kosugi 1997; Eradat Oskoui & Neukirch, 2014) may not capture its physics self-consistently.

The essential novelty of the Lasso picture is twofold: first, substantial particle acceleration must take place on closed field lines, implying that the CME shock (if that is the accelerator) must penetrate such regions; and second, that the post-event dynamics of flux bundles containing high-energy particles can facilitate their interaction with the lower atmosphere. The magnetic retraction would help out with the number problem via the additional acceleration terms. In such a case the centroid position of the γ-ray source might lie well above the limb, given turbulent trapping; if loss-cone dynamics were important, the sources would be in the lower solar atmosphere. If the CME-driven shock accelerates the particles, it must have developed in the deep atmosphere or else have flanks extending into closed-field regions not participating in the eruption. In neither case would we expect substantial γ-ray source motion. If SEPs could return to the Sun from well-connected open fields, somehow overcoming the mirror force (e.g. Jin *et al.* 2017; Plotnikov *et al.* 2017), this might not be the case.

Acknowledgements

I am grateful for extensive discussion of these and related ideas in the ISSI team "Energetic Ions: The Elusive Component of Solar Flares," chaired by Alec MacKinnon and with the participation of Federico Benvenuto, Christina Cohen, Nicole Duncan, Luis Fernandez Menchero, Säm Krucker, Alec MacKinnon, Melissa Pesce-Rollins, Albert Shih, Paulo Simões, Rami Vainio, Nicole Vilmer, and Pietro Zucca. Georgia De Nolfo and Jim Ryan helped by reading the first version, and I also benefited from discussions with Frederic Effenberger and Ludwig Klein.

References

Ackermann, M., Allafort, A., Baldini, L., *et al.* 2017, *The Astrophysical Journal*, 835, 219

Chupp, E. L., Forrest, D. J., Vestrand, W. T., Dubrunner, H., Flueckiger, E. O., Cooper, J. F., Kanbach, G., Reppin, C. & Share, G. H. 1985, *International Cosmic Ray Conference*, 4 de Nolfo, G. A., Boezio, M., Bruno, A., Christian, E. R., Martucci, M., Mergè, M., Mocchiutti, E., Munini, R., Ricci, M., Ryan, J. M., Share, G. H. & Stochaj, S. 2016, *AGU Fall Meeting Abstracts*

Elliot, H. 1969, in *Solar Flares and Space Research*, ed. C. de Jager & Z. Svestka, 356

Enomé, S. & Tanaka, H. 1971, in *IAU Symposium*, Vol. 43, Solar Magnetic Fields, ed. R. Howard, 413–416

Eradat Oskoui, S. & Neukirch, T. 2014, *Astron. Astrophys.*, 567, A131

Forrest, D. J., Vestrand, W. T., Chupp, E. L., Rieger, E., Cooper, J. F. & Share, G. H. 1985, *International Cosmic Ray Conference*, 4

Frost, K. J. & Dennis, B. R. 1971, *Astrophys. J.*, 165, 655

Hudson, H. S. 1978, *Astrophys. J.*, 224, 235

Jin, M., Petrosian, V., Liu, W. & Omodei, N. 2017, in *AAS/Solar Physics Division Meeting*, Vol. 48, AAS/Solar Physics Division Meeting, 3.03

Klein, K.-L., Tziotziou, K., Zucca, P., Valtonen, E., Vilmer, N., Malandraki, O. E., Hamadache, C., Heber, B. & Klener, J. 2017, *Springer Astrophysics and Space Science Library*

Krucker, S., Hudson, H. S., Glesener, L., White, S. M., Masuda, S., Wuelser, J.-P. & Lin, R. P. 2010, *Astrophys. J.*, 714, 1108–1119

Krucker, S. & Lin, R. P. 2008, *Astrophys. J.*, 673, 1181

Kundu, M. R. 1965, *Solar radio astronomy*, New York: Interscience Publication

Mandzhavidze, N. & Ramaty, R. 1992, *Astrophys. J. Lett.*, 396, L111–L114

Mewaldt, R. A., Cohen, C. M. S., Mason, G. M., Haggerty, D. K., Looper, M. D., Vourlidas, A., Desai, M. I., Giacalone, J., Labrador, A. W., Leske, R. A. & Mazur, J. E. 2005, in ESA Special Publication, Vol. 592, Solar Wind 11/SOHO 16, Connecting Sun and Heliosphere, ed. B. Fleck, T. H. Zurbuchen & H. Lacoste, 67

Olive, K. A., *et al.* 2014, *Chinese Physics C*, 38, 090001

Pesce-Rollins, M., Omodei, N., Petrosian, V., Liu, W., Rubio da Costa, F., Allafort, A. & Chen, Q. 2015a, *Astrophys. J. Lett.*, 805, L15

Pesce-Rollins, M., Omodei, N., Petrosian, V., Liu, W., Rubio da Costa, F., Allafort, A. & Fermi-LAT Collaboration. 2015b, in *International Cosmic Ray Conference*, Vol. 34, 34th International Cosmic Ray Conference (ICRC2015), 128

Plotnikov, I., Rouillard, A. P. & Share, G. H. 2017, *Astron. Astrophys.*, 608, A43

Schrijver, C. J. and De Rosa, M. L. 2003, *Sol. Phys.*, 212, 165–200

Share, G. H., Murphy, R. J., Tolbert, A. K., Dennis, B. R., White, S. M., Schwartz, R. A. & Tylka, A. J. 2017, arXiv:1711.01511

Sheeley, Jr., N. R. & Wang, Y.-M. 2014, *Astrophys. J.*, 797, 10

Sheeley, Jr., N. R., Warren, H. P. & Wang, Y.-M. 2004, *Astrophys. J.*, 616, 1224

Smerd, S. F. 1970, *Proceedings of the Astronomical Society of Australia*, 1, 305

Somov, B. V. & Kosugi, T. 1997, *Astrophys. J.*, 485, 859

Space Weather of the Heliosphere: Processes and Forecasts
Proceedings IAU Symposium No. 335, 2017
C. Foullon & O.E. Malandraki, eds.

doi:10.1017/S1743921318000443

Acceleration of Solar Wind Particles Passing through the 3D Heliospheric Current Sheet

Valentina V. Zharkova

Department of Mathematics, Physics and Electrical Engineering, University of Northumbria, Newcastle upon Tyne, NE1 8ST, UK
email: valentina.zharkova@northumbria.ac.uk

Abstract. Additional acceleration of protons and electrons passing through a 3D reconnecting current sheet (RCS) of the solar corona and heliosphere is investigated with PIC approach. The simulation confirms spatial separation of electrons and protons and generation of a polarisation electric field induced by separated particles. In the heliospheric current sheet with a weak magnetic field there are two populations of particles: transit and bounced. The transit particles (both protons and electrons) are accelerated to high energies while the bounced electrons fail to reach the midplane. Instead they form an electron cloud of horseshoe or medallion types at some distance D from its midplane, which is larger for bigger guiding field magnitudes. These energetic electrons and protons appearing near the HCS boundaries can be a great danger for the satellites crossing the sector boundaries.

Keywords. Sun: flares, plasmas, acceleration of particles, magnetic field

1. Introduction

The interplanetary magnetic field (IMF) in the heliosphere is divided into two sectors of the open magnetic field lines with the opposite magnetic field polarities separated by the equatorial plane providing a magnetic field reversal known as the sector boundary (SB). The latter is considered to be the heliospheric current sheet (HCS), which governs the magnetic field reversals from and toward the Sun along Parker's spiral (see http://wso.stanford.edu/gifs/HCS.html).

Simulations of particle trajectories were carried out by using a test particle approach in a 3D magnetic configuration with the guiding field emulating a simple case of magnetic reconnection for a constant reconnection electric field (Zharkova and Gordovskyy 2004) or for those enhanced near the magnetic X-nullpoint due to anomalous resistivity (Wood and Neukirch 2005). Zharkova and Gordovskyy (2004) and Zharkova and Agapitov (2009) showed that the trajectories of particles with the opposite charges (electrons or protons) can be either fully symmetric or strongly asymmetric towards the midplane of the RCS depending on the ratio between the magnetic field components.

As a result, some fraction of the released magnetic energy is transformed into kinetic energy of accelerated particles. The energy spectra of these particles also depend on a magnetic field topology, electric field strength and the magnitude of a transverse magnetic field. Accelerated particles gain energies up to 100 keV for the electrons and up to a few tens MeV for the protons (Zharkova and Gordovskyy 2005; Wood and Neukirch 2005; Zharkova and Agapitov 2009).

Recent study with particle-in-cell (PIC) of particle acceleration in the 3D heliospheric current sheet with low magnitudes of magnetic field relevant to the interplanetary magnetic field at the Earth orbit (Zharkova and Khabarova 2012) allowed to provide the plausable interpretation of the energies and pitch angle distributions of electrons and protons after their passage through the HCS.

2. Description of the model

A PIC approach was used for the simulation of particle acceleration in a 3D reconnecting current sheet with simple magnetic field topology and ambient particle densities relevant to the HCS.

The simulation domain is a small part of the RCS (see model in Siversky and Zharkova (2009) where periodic boundary conditions are used to contain the full trajectories of accelerated particles. The main component B_z depends on x as follows: $B_z(x) = -B_{z0}\tanh\left(\frac{x}{L_x}\right)$. Similar to Zharkova and Gordovskyy (2004), the B_x component is assumed to be constant inside the simulation domain, i.e. $B_x = -B_{x0}$. The guiding (out-of-plane) magnetic field B_y is maximal in the midplane and vanishes outside the RCS: $B_y(x) = B_{y0}\text{sech}\left(\frac{x}{L_x}\right)$. Particles are considered accelerated by the reconnection electric field E and to generate their own electric and magnetic fields as described by Siversky and Zharkova (2009).

Simulations are carried out for the following current sheet parameters: the main component of the magnetic field $B_{z0} = 10^{-3}$ T, the current sheet half-thickness $L_x = 1$ m, the drifted electric field $E_{y0} = 250$ V/m and the guiding, B_{y0}, and transverse, B_{x0}, components of the magnetic field are selected to range from $(0.1 - 10) \times 10^{-4}$ T to cover acceleration at the various parts of the RCS.

3. Results

The PIC simulations have shown that the induced magnetic field $\tilde{\mathbf{B}}$ is much smaller than the background $\mathbf{B_z}$. On the other hand, the electric field $\tilde{\mathbf{E}}_\mathbf{x}$, called a polarization field, induced by the separation of electrons from protons is much larger than the reconnection (drifted) field E_y as pointed out by Zharkova and Agapitov (2009) and Siversky and Zharkova (2009). The polarization electric field $\tilde{E}_x$, which is perpendicular to the current sheet, leads to a local non-neutrality of the plasma, which becomes stronger if B_{x0} decreases, or B_{y0} increases.

Electrons entering from the $x < 0$ semi-space reveal the dynamics similar to those of the 'transit' protons, e.g. drifting towards the midplane, becoming accelerated and ejected to the $x > 0$ semi-space. However, the polarization field $\tilde{E}_x(x)$, which extends beyond the current sheet and has a component parallel to the magnetic field B_z decelerates the ejected electrons and returns them back to the point where they entered the acceleration region (see Figure 1a). While transit electrons are accelerated at the midplane, ejected from it into the interplanetary space without return because of a weak magnetic field.

The observations of solar wind energetic electrons with energies above 2 keV often reveal that the locations of their motion reversals do not always coincide with the locations (and times) of the magnetic field reversals in the HCS, or sector boundary crossings (SBC) (Kahler and Lin 1994, 1995). According to Crooker *et al.* (2004a,b), the mismatch between the SBC allegedly indicated by a change of electrons pitch angles occuring in one location (see Figure 1b) and the change of a sign and magnitude of the IMF occurring in another location is caused by entangling of the large-scale magnetic field.

However, these discrepancies in spatial observations of the electrons in a vicinity of the HCS can be explained in much more simple and logical manner by rather dynamic

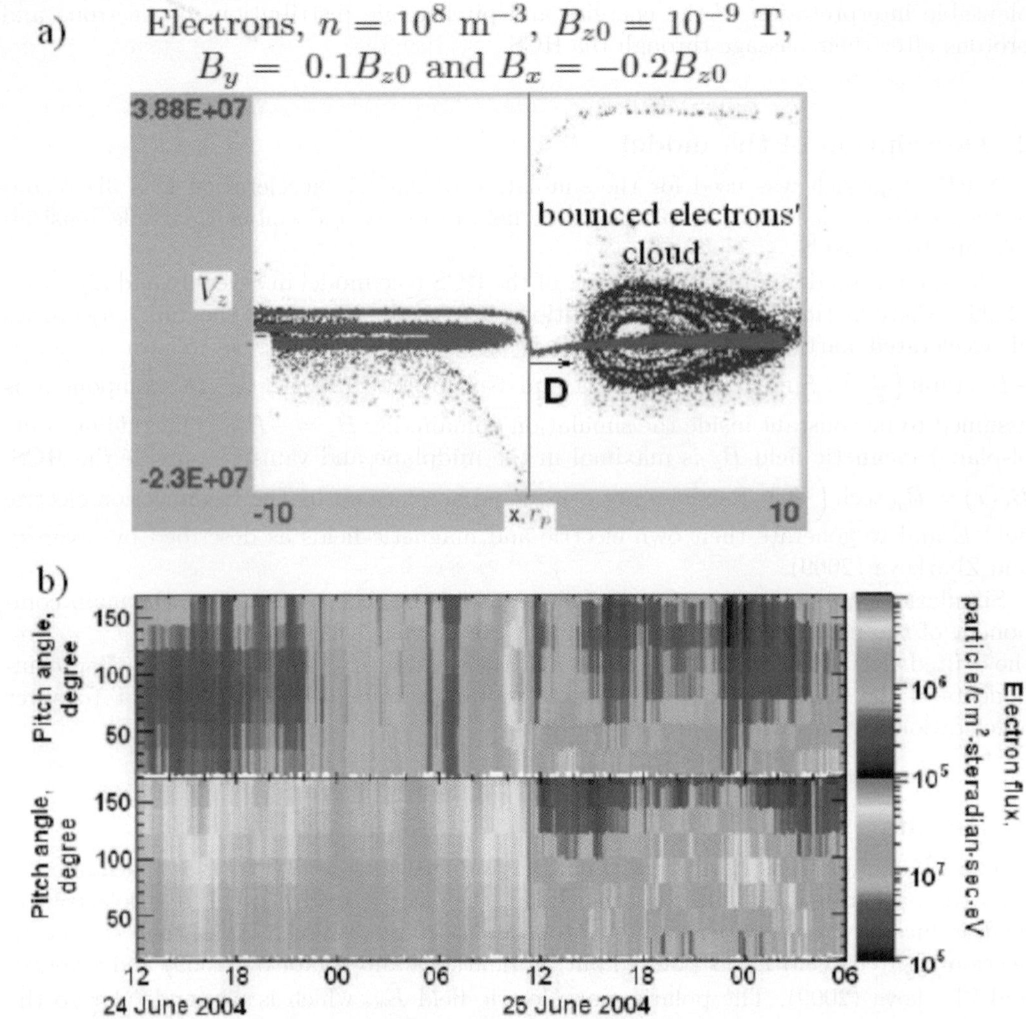

Figure 1. Comparison of a) accelerated electron (black dots) and proton (blue dots) distributions simulated for strong guiding field with PIC with respect to the current sheet midplane (X=0) and b) observed energy and pitch angle distributions of electrons passed through the HCS. Note that simulated distributions of the bounced electrons (for X> 0) do not reach the midplane but turned back at some distance D from it. This causes the pitch angle distributions shown in b) with the variations of a pitch angle from 0 to 180° similar to the simulated variation in a).

physical conditions in the HCS caused by the ongoing magnetic reconnection and correspondence of SBC to the HCS midplane. In this case electrons become accelerated as shown in Figure 1a to form the transit electrons leaving the current sheet with energies of a few hundreds of eV as measured, and the bounced electrons with energies of a few tens of eVs, which are turning by 180 degrees from their initial motion (towards the SB, or midplane of the HCS) from the HCS at some distance D from its midplane (SB for HCS).

Remarkably, the population of bounced electrons forms either the horseshoe or medallion shapes for weaker IMF magnitudes of 10^{-9} nT (see Figure 1a). This resembles the pitch angle distributions often observed in the interplanetary space (Figure 1b). The distance D, at which the electrons change their pitch angles is larger for a bigger guiding field magnitude that fits rather well the observations (see Table 1 in Zharkova and Khabarova 2012).

The analysis of the measurements near the SBC from the Interball-1 spacecraft and Wind SWE data (1995–2005) revealed rather peculiar profiles of the solar wind velocities V across the HCS (see Figure 5 in Zharkova and Khabarova 2012). The unexplained decrease of the solar wind velocity observed 1-3 days before the SBC and a sharp increase of this velocity 1-3 days after the SBC is defined by the polarisation electric field. Only in about 50% of cases the ion velocity dostributions can be assigned to a presence of the co-rotating interaction region (CIR) approaching the HCS, observed though after the SBC. Hence, the other 50% cases are to be associated with the particles accelerated during their sector boundary crossing. The observed asymmetric velocity profiles of the solar wind ions during their crossing of the SB are likely to be caused by the effect of an asymmetric polarisation electric field induced by the separation of accelerated electrons and protons with respect to the current sheet midplane, or SB (Zharkova and Khabarova 2012).

4. Conclusions

The PIC simulation has shown that accelerated particles produce a strong polarisation field, $\tilde{E}_x$, caused by the separation of accelerated particles across the current sheet. This process results in formation of peculiar velocity profiles of ions following the polarisation electric field and of electron clouds in the form of a 'horse shoe' or 'medallion' well before the electrons can reach the current sheets. The clouds disappear only when protons gain sufficient energy to leave the model RCS while dragging electrons with them.

References

Crooker, N. U., Forsyth, R., Rees, A., Gosling, J. T. & Kahler, S. W. 2004a, *Journal of Geophysical Research (Space Physics)*, 109, A06110

Crooker, N. U., Huang, C.-L., Lamassa, S. M., *et al.* 2004b, *Journal of Geophysical Research (Space Physics)*, 109, A03107

Gosling, J. T., Eriksson, S. & Schwenn, R. 2006a, *Journal of Geophysical Research (Space Physics)*, 111, A10102

Gosling, J. T., McComas, D. J., Skoug, R. M. & Smith, C. W. 2006b, *Geophysics Research Letters*, 33, L17102

Kahler, S. & Lin, R. P. 1994, *Geophysics Research Letters*, 21, 1575

Kahler, S. W. & Lin, R. P. 1995, *Solar Physics*, 161, 183

Lin, R. P., Ko, Y. K., Sui, L., *et al.* 2005, *Astrophysical Journal*, 622, 1251

Priest, E. & Forbes, T. 2000, *Magnetic Reconnection* (Magnetic Reconnection, by Eric Priest and Terry Forbes, pp. 612. ISBN 0521481791. Cambridge, UK: Cambridge University Press, June 2000.)

Siversky, T. V. & Zharkova, V. V. 2009, *Journal of Plasma Physics*, 75, 619

Verboncoeur, J. P. Langdon, A. B. & Gladd, N. T. 1995, *Comp. Phys. Comm.*, 87, 199

Wood, P. & Neukirch, T. 2005, *Solar Physics*, 226, 73

Zharkova, V. V. & Gordovskyy, M. 2004, *Astrophysical Journal*, 604, 884

Zharkova, V. V. & Gordovskyy, M. 2005, *Monthly Notices of the Royal Astronomical Society* **356**, 1107–1116.

Zharkova, V. V. & Agapitov, O. V. 2009, *Journal of Plasma Physics*, 75, 159

Zharkova, V. V. & Khabarova, O. V. 2012, *Astrophysical Journal*, Vol. 752, 35

Zharkova, V. V. & Khabarova, O. V. 2015, *Annales Geophys.*, 33, 457

Space Weather of the Heliosphere: Processes and Forecasts
Proceedings IAU Symposium No. 335, 2017
C. Foullon & O.E. Malandraki, eds.

doi:10.1017/S1743921317010882

Pursuing Forecasts of the Behavior and Arrival of Coronal Mass Ejections through Modeling and Observations

Hebe Cremades[1,2]

[1]Universidad Tecnológica Nacional– Facultad Regional Mendoza, CEDS
[2]Consejo Nacional de Investigaciones Científicas y Técnicas (CONICET)
Rodríguez 243, 5500, Mendoza, Argentina
email: hebe.cremades@frm.utn.edu.ar

Abstract. Sophisticated instrumentation dedicated to studying and monitoring our Sun's activity has proliferated in the past few decades, together with the increasing demand of specialized space weather forecasts that address the needs of commercial and government systems. As a result, theoretical and empirical models and techniques of increasing complexity have been developed, aimed at forecasting the occurrence of solar disturbances, their evolution, and time of arrival to Earth. Here we will review groundbreaking and recent methods to predict the propagation and evolution of coronal mass ejections and their driven shocks. The methods rely on a wealth of data sets provided by ground- and space-based observatories, involving remote-sensing observations of the corona and the heliosphere, as well as detections of radio waves.

Keywords. Sun: coronal mass ejections (CMEs), solar-terrestrial relations, interplanetary medium

1. Introduction

The first hints of Sun-Earth connection were revealed in the mid-eighteenth century, as Olof Hiorter and Anders Celsius noticed perturbations in a geomagnetic needle in connection with aurorae. A century later, these magnetic disturbances were found to have the same frequency and phase as sunspots (Sabine 1852). Although several groundbreaking discoveries related with solar physics and solar-terrestrial relations followed in the years to come (*e.g.* Carrington 1859, Hale 1908, Fairfield & Cahill 1966), it was not until the beginning of the space age when coronal mass ejections (CMEs) were discovered. Their existence had been proposed to explain geomagnetic disturbances (see summary in Burlaga *et al.* 1991), so that CMEs were soon associated with magnetic interplanetary structures originating at the Sun (*e.g.* Schwenn 1983, Sheeley *et al.* 1985).

Ground- and space-based observatories dedicated to Heliophysics proliferated in the following years, enabling in-depth studies of the Sun's interior and atmosphere, as well as of its impact in the interplanetary medium, Earth, and other planets. Several aspects of CMEs came to light, regarding *e.g.* their source regions, kinematics, and morphology. In spite of significant progress on these matters, to date it has not been possible to foresee when and where on the Sun the next CME might take place. This limitation implies that forecasting must be based *sine qua non* with the occurrence of a CME, *i.e.* only once a CME has erupted, its arrival at Earth and degree of influence may be forecasted. The identification of Earth-directed CMEs has proven to be non-trivial, since it is mainly based on observations from white-light coronagraphs. Because of the Thomson scattering effect (*e.g.* Vourlidas & Howard 2006), Earth-directed CMEs may be at times undetected by coronagraphs located in the Sun-Earth line. This problem was addressed by Cremades *et al.* (2015a), who investigated the production of CMEs from an active region crossing central meridian, and found twice as many CMEs traveling towards Earth when using coronagraphs located $\sim 90°$ away from the Sun-Earth line, implying a significant number of missed alarms. In addition, a subset of events dubbed "stealth CMEs" (Robbrecht *et al.* 2009) may not leave perceptible imprints on low coronal images, complicating even more the identification of potentially geoeffective CMEs.

Once a CME is detected, a good assessment of the three-dimensional (3D) trajectory and size serves to determine *if* it will arrive at Earth, while knowledge on its propagation profile is crucial to ascertain *when* it will arrive. Furthermore, information on *how* the associated magnetic fields are configured is key to determine whether the Earth's magnetosphere is to be disrupted and to what extent. In the next sections, I will briefly discuss tools and models that have been proposed to address these aspects.

2. Determining propagation direction and size

Soon after the discovery of CMEs, there was some speculation on whether CMEs could be best described as planar loops (Trottet & MacQueen 1980) or 3D structures (*e.g.* Howard *et al.* 1982, Crifo *et al.* 1983), with the latter being quickly accepted. Likewise, it did not take long to recognize that CMEs and their trajectories appear projected in the plane of the sky, thus hindering proper assessment of their 3D characteristics. This is particularly critical for events that propagate close to the Sun-Earth line, which are certainly the most threatening for the space weather at Earth. To overcome the problem for these events, particularly halo CMEs, cone models based on coronagraph observations were proposed before (*e.g.* Howard *et al.* 1982) and during the SOHO era beginning in the mid-1990s (SOHO: Solar and Heliospheric Observatory; Domingo *et al.* 1995). Some of these include Zhao *et al.* (2002), Michałek *et al.* (2003), Xie *et al.* (2004), Cremades & Bothmer (2005), Zhao (2008), and Na *et al.* (2013). The outcome of these models in terms of propagation direction and size is particularly useful to feed models of propagation and arrival time of CMEs and shocks (see Section 4).

Other ways of finding out the 3D propagation direction based on single-viewpoint observations rely on characteristics of the low-coronal environment at the time of eruption. This is the case of the methods proposed by Cremades *et al.* (2006) and Gopalswamy *et al.* (2009), which consider coronal holes as playing a key role in the deflection of CMEs from their source regions. Kay *et al.* (2013) developed a more sophisticated data-driven model of the coronal background that also takes into account CME properties to predict deflection for the first solar radii of propagation.

The launch of the STEREO mission in 2006 (Solar-Terrestrial Relations Observatory; Kaiser *et al.* 2008) enabled new ways of finding these CME attributes while minimizing uncertainties. The most widespread methods are either based on the tie-pointing/triangulation reconstruction (*e.g.* Temmer *et al.* 2009, Mierla *et al.* 2009, Srivastava *et al.* 2009, Liu *et al.* 2009, Liewer *et al.* 2011), polarization ratio (*e.g.* Moran *et al.* 2010, deKoning & Pizzo 2010) or forward modeling (*e.g.* Thernisien *et al.* 2009, Wood *et al.* 2010). The first method requires at least two viewpoints, and that the same parcel of coronal material can be discerned in stereoscopic images, which is not always straightforward. With the polarization ratio technique, however, observations from one viewpoint are enough, although multiple views are helpful to constrain the problem. Forward modeling approaches fit an *ad hoc* 3D density distribution to achieve visual agreement with data. Best results are obtained when applied to observations from three different viewpoints, while solutions are strongly undetermined for single vantage point observations. Unfortunately, these tools are often unappropriately used, because their limitations are overlooked and an untrained eye may find various erroneous solutions that fit the same CME. For a comparison of these and other reconstruction techniques, see Mierla *et al.* (2009, 2010).

The interpretation of white-light images for size assessment must be taken with care, given that the outermost rim of a CME may represent the shock (Vourlidas *et al.* 2013), especially for the case of CMEs seen as halos. It should be noted that events that succeed to be seen as halo CMEs in past and present coronagraphs are indeed "special" (Lara *et al.* 2006), faster and wider than average (*e.g.* Shen *et al.* 2013). In fact, full limb halo CMEs –halo CMEs completely surrounding a coronagraph's occulter but at the same time originating close to the solar limb– may be geoeffective (Gopalswamy *et al.* 2007, Cid *et al.* 2012), while a fairly bright and wide CME seen at the limb may not show up as a halo when changing vantage point by $\sim 90°$ (Cremades *et al.* 2015a).

3. Determining flux rope orientation

A period of negative Bz component of the interplanetary magnetic field at 1 AU is well known for disrupting the Earth's magnetic field. This can be provided by parts of the helical magnetic flux rope(s) embedded within interplanetary CMEs (ICMEs) or by the interplanetary magnetic field compressed ahead of the ICME. While the latter is more difficult to predict, requiring complex high-resolution dynamic 3D models of the ambient solar wind, potential geoeffectiveness of a white-light CME may be partially ascertained based on the knowledge of the configuration of its magnetic flux rope (see Vourlidas *et al.* 2013 for a discussion on the fraction of CMEs embedding flux ropes).

Bothmer & Schwenn (1998) recognized different magnetic field configuration patterns in Helios 1 and 2 in-situ data sets, in agreement with the helical magnetic flux rope picture. The magnetic configuration of filaments could be related to that of the so-called magnetic clouds detected in-situ, having four possible types, resulting from two possible directions of the field axis and two possible values of magnetic helicity or handedness. Although for the examined time interval they found that the axes of the flux ropes were generally close to the ecliptic plane, later studies identified other orientations (*e.g.* Mulligan *et al.* 2000, Huttunen *et al.* 2005, Lepping *et al.* 2006). The apparent relationship between the magnetic configuration of interplanetary flux ropes and that of their solar sources (*e.g.* Yurchyshyn 2008, Marubashi *et al.* 2015) has motivated a number of studies toward magnetic field prediction at 1 AU (Savani *et al.* 2015, 2017, Kay *et al.* 2017).

As stated above, CME deflection from their solar sources and rotation of their main axis are common effects in the first stages of eruption. While various approaches have been proposed to tackle the issue of deflection (see Section 2), only few studies address rotation (*e.g.* Vourlidas *et al.* 2011, Kay & Opher 2015, Kay *et al.* 2017). Instead of deducing CME rotation with respect to their source regions, at times it can be more convenient to determine flux rope orientation from white-light observations of CMEs, given that at heights of a few solar radii they have experienced most of these effects to eventually behave self-similarly. However, this is not an easy task, since orientations of the main axis of reconstructed CMEs have large uncertainties (Thernisien *et al.* 2009), which in turn may have large impact on the predictions of in-situ conditions through simulations of ICMEs (*e.g.* Savani *et al.* 2017, Kay *et al.* 2017). Stronger constraints on latitude, longitude, and tilt of the main axis of white-light CMEs can be achieved with a proper combination of vantage points and CME propagation direction. As shown by Cremades and Bothmer (2004), the projected morphology exhibited by CMEs depends on their propagation direction and orientation with respect to the vantage point (Figure 1a). If all remote-observing spacecraft as well as the CME trajectory approximately lie in the same plane, *i.e.* the ecliptic, to find the orientation of the CME's main axis will involve

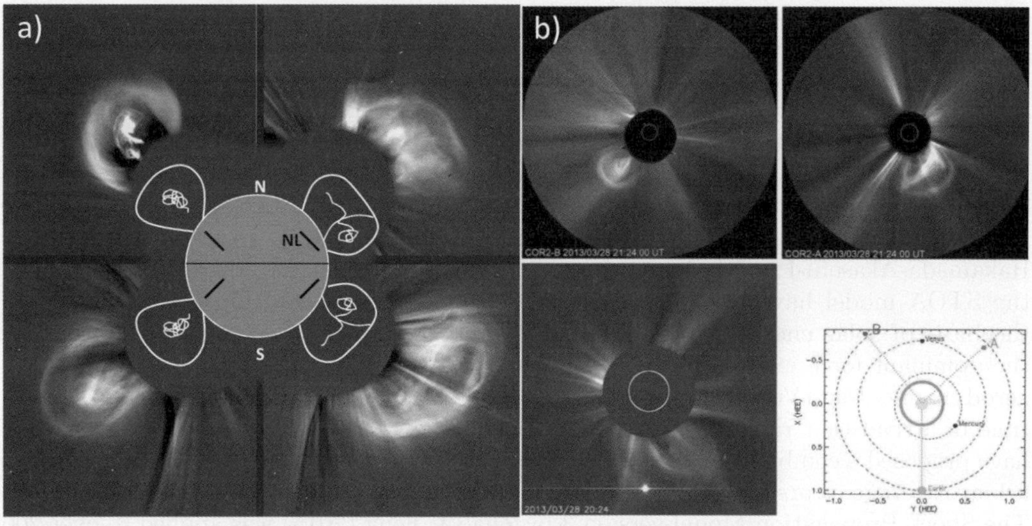

Figure 1. a) The 3D configuration scheme and four sample CMEs in agreement with it. The main symmetry axes of the CMEs to the East are oriented along the line of sight (axial view), while those to the West are aligned with the plane of the sky (lateral view). The black lines on the solar disk represent the neutral lines of the related source regions. Adapted from Cremades and Bothmer (2004). b) Axial (top-left) and lateral (top-right) views of the same CME, made possible from observations in quadrature respectively from STEREO B and A, and because of a nearly polar propagation and favorable orientation of the CME's main axis with respect to the vantage points. The bottom-left image is an intermediate view provided by SOHO. Bottom-right: location of the STEREO spacecraft relative to the Sun and Earth, with the CME traveling away from the ecliptic represented by the bold circle surrounding the Sun. From Cabello *et al.* (2016).

inherent uncertainties. Unfortunately, this is the case for most Earth-directed CMEs. To help reduce these uncertainties, a configuration in quadrature between vantage points and CME propagation direction is required. This problematic has been explained by Cabello *et al.* (2016), who presented the first simultaneous observation of both flux rope views for the same CME, which was possible for an event propagating nearly perpendicular to the ecliptic plane (Figure 1b). The upcoming Solar Orbiter mission will enable analogous observations for Earth-directed CMEs through observations from above the solar poles.

4. Determining time of arrival

Most of the forecasting models naturally focus on the forecasting of arrival times of CMEs and/or their associated shocks, provided that a CME with a propagation direction component toward Earth has been ejected. Inspired by the classification proposed by Zhao & Dryer (2014), forecasting models are here sorted into three classes: (i) empirical, data assimilative models, (ii) physics-based models, and (iii) magnetohydrodynamic (MHD) models. Empirical, data assimilative models rely on typically simple analytic expressions fed with parameters obtained from solar, coronal, or interplanetary detections. Examples of these are those by Gopalswamy *et al.* (2001, 2005), Wang *et al.* (2002), Srivastava & Venkatakrishnan (2004), Xie *et al.* (2006), and Michalek *et al.* (2008). Other approaches rely on the concept of radial and expansion speed of CMEs (Schwenn *et al.* 2005, Mäkelä *et al.* 2016). Prediction techniques may for instance also profit from interplanetary scintillation information (*e.g.* Jackson *et al.* 1998, Manoharan 2006), from heliospheric imaging (*e.g.* Sheeley *et al.* 2008, Rouillard *et al.* 2008, Davis *et al.* 2011, Colaninno *et al.* 2013, Möstl *et al.* 2014, Rollett *et al.* 2016), or from tomographic considerations (*e.g.* Jackson *et al.* 2011). Low-frequency type II radio emissions can be useful to forecast CME shock arrival times, assuming that the emission takes place at the Earth-directed shock sector and that at those distances (farther than 30 solar radii) the shock propagates with nearly constant speed. Only one third of L1 shocks are associated to low-frequency type II events, however shock arrival times based on these emissions have errors of less than ± 6 h for 85 % of the events (Cremades *et al.* 2015b).

Among the physics-based models the Shock Time of Arrival (STOA; Dryer & Smart 1984), the Interplanetary Shock Propagation Model (Smith & Dryer 1990), and the Hakamada-Akasofu-Fry (HAF) version 2 (Fry *et al.* 2003) stand out. Improvements to the STOA model have been proposed by Qin *et al.* (2009) and Liu & Qin (2012) on the basis of solar energetic particles and X-rays. The drag-based model assumes that the dominant force exerted on ICMEs is the magnetohydrodynamical equivalent of the aerodynamic drag (Vršnak *et al.* 2007, 2013). The aforementioned models have errors in time of arrival that range from ~ 6 to 12 h. Shi *et al.* (2015) and Hess & Zhang (2015) have proposed a combination of stereoscopic measurements with the drag-based model, to achieve average errors of 8.6 h for 16 events, and under 3.5 h for seven events respectively. The Shock Propagation Model version 3 by Zhao & Feng (2015) was applied to over 200 Earth-directed events to yield an average absolute error of ~ 9 h.

Several of the most renowned MHD models are routinely used for space weather prediction purposes. Enlil may be constrained by the Wang-Sheeley-Arge (WSA) method and/or a cone model (Odstrcil *et al.* 2004, 2005, Xie *et al.* 2013). CORHEL uses a coronal model (Magnetohydrodynamics outside A Sphere–MAS or WSA) combined with a heliospheric model (MAS or Enlil) (Linker *et al.* 2009). The Alfven-Wave driven solar wind Model (AWSoM), part of the Space Weather Modeling Framework, is based on the BATS-R-US code (Tóth *et al.* 2005, 2012, Van der Holst *et al.* 2014) and can be used in combination with the Eruptive Event Generator – Gibson and Low (Jin *et al.* 2017). The HAF version 3 combined with the 3D MHD model by Wu *et al.* (2007, 2011), and the Solar-InterPlanetary Conservation Element/Solution Element MHD model by Feng *et al.* (2007, 2010) are other prediction models of widespread use.

The inherent uncertainties of input parameters, the limited knowledge on how kinematics and morphology behave for different events, and the difficulties in determining the real solar wind conditions in 3D space and time are the main factors that affect the performance of existing models (Zhao & Dryer 2014). This includes interactions between transients and with other structures (*e.g.* Dasso *et al.* 2009, Liu *et al.* 2012, Lugaz *et al.* 2013, Temmer *et al.* 2014). According to Burlaga (2002), two out of three ICMEs are complex events, which implies that realistic simulations require a high level of complexity.

5. Final remarks

In the past three decades, diversity and progress of Heliophysics-dedicated space missions and ground-based observatories have enabled enormous advancement and increasing complexity of techniques and models aimed at forecasting CME behavior and arrival. However, the various techniques and models have different caveats, pros and cons, *e.g.* with respect to simplicity, running time, lead time, data assimilation, and number of events on which they have been validated. At the present time, the maximum benefit can only be achieved by recognizing which methods or models perform best under different circumstances, not only to avoid false and missed alarms, but also to obtain good proxies of arrival time. Furthermore, coronagraphs and EUV imagers, particularly those offset from the Sun-Earth line, are compelling to identify Earth-directed events and forecast their time of arrival.

Acknowledgements

HC is member of the Carrera del Investigador Científico (CONICET). She thanks the invitation and financial aid provided by the SOC and LOC of this IAU Symposium. She appreciates support from Argentine grants PICT 2012-973 (ANPCyT), PIP 2012-01-403 (CONICET) and UTI4035TC (UTN). The author is grateful to Fernando López for help with this manuscript. This proceeding is dedicated to the memory of Rainer Schwenn.

References

Bothmer, V. & Schwenn, R. 1998, *Annales Geophysicae*, 16, 1
Burlaga, L. F. E. 1991, *Physics and Chemistry in Space*, 21, 1
Burlaga, L. F., Plunkett, S. P. & St. Cyr, O. C. 2002, *J. Geophys. Res.*, 107, 1266
Cabello, I., Cremades, H., Balmaceda, L. & Dohmen, I. 2016, *Solar Phys.*, 291, 1799
Carrington, R. C. 1859, *Mon. Not. R. Astron. Soc.*, 20, 13
Cid, C., Cremades, H., Aran, A., Mandrini, C., *et al.* 2012, *J. Geophys. Res.*, 117, A11102
Colaninno, R. C., Vourlidas, A. & Wu, C. C. 2013, *J. Geophys. Res.*, 118, 6866
Cremades, H. & Bothmer, V. 2004, *A&A*, 422, 307
Cremades, H. & Bothmer, V. 2005, *IAU Symp. No 226, Dere, K., Wang, J. & Yan, Y. (eds.)*
Cremades, H., Bothmer, V. & Tripathi, D. 2006, *Adv. Space Res.*, 38, 461
Cremades, H., Mandrini, C. H., Schmieder, B. & Crescitelli, A. M. 2015a, *Solar Phys.*, 290, 1671
Cremades, H., Iglesias, F. A., St. Cyr, O. C., Xie, H., *et al.* 2015b, *Solar Phys.*, 290, 2455
Crifo, F., Picat, J. P. & Cailloux, M. 1983, *Solar Phys.*, 83, 143
Dasso, S., Mandrini, C., Schmieder, B., Cremades, H., *et al.* 2009, *J. Geophys. Res.*, 114, A02109
Davis, C. J., de Koning, C. A., Davies, J. A., Biesecker, D., *et al.* 2011, *Space Weather*, 9, S01005
Dryer, M. & Smart, D. F. 1984, *Adv. Space Res.*, 4, 291
de Koning, C. A. & Pizzo, V. J. 2010, *AGU Fall Meeting Abstracts*
Domingo, V., Fleck, B. & Poland, A. I. 1995, *Solar Phys.*, 162, 1
Fairfield, D. H. & Cahill, Jr., L. J. 1966, *J. Geophys. Res.*, 71, 155
Feng, X., Zhou, Y. & Wu, S. T. 2007, *ApJ*, 655, 1110
Feng, X., Yang, L., Xiang, C., Wu, S. T., *et al.* 2010, *ApJ*, 723, 300
Fry, C. D., Dryer, M., Smith, Z., Sun, W., *et al.* 2003, *J. Geophys. Res.*, 108, 1070
Gopalswamy, N., Lara, A., Yashiro, S., Kaiser, M. L., *et al.* 2001, *J. Geophys. Res.*, 106, 29207
Gopalswamy, N., Lara, A., Manoharan, P. K. & Howard, R. A. 2005, *Adv. Space Res.*, 36, 2289
Gopalswamy, N., Yashiro, S. & Akiyama, S. 2007, *J. Geophys. Res.*, 112, A06112
Gopalswamy, N., Mäkelä, P., Xie, H., Akiyama, S., *et al.* 2009, *J. Geophys. Res.*, 114, A00A22
Hale, G. E. 1908, *ApJ*, 28, 315
Hess, P. & Zhang, J. 2015, *ApJ*, 812, 144
Howard, R. A., Michels, D. J., Sheeley, N. R. & Koomen, M. J. 1982, *ApJ* (L), 263, L101
Huttunen, K. E. J., Schwenn, R., Bothmer, V. & Koskinen, H. 2005, *Ann. Geophys.*, 23, 625
Jackson, B. V., Hick, P. L., Kojima, M. & Yokobe, A. 1998, *J. Geophys. Res.*, 103, 12049
Jackson, B. V., Hamilton, M. S., Hick, P. P., *et al.* 2011, *J. Atmos. Sol-Terr. Phys.*, 73, 1317
Jin, M., Manchester, W. B., van der Holst, B., Sokolov, I., *et al.* 2017, *ApJ*, 834, 173
Kaiser, M. L., Kucera, T. A., Davila, J. M., St. Cyr, O. C., *et al.* 2008, *Space Sci. Rev.*, 136, 5
Kay, C. & Opher, M. 2015, *ApJ* (L), 811, L36
Kay, C., Opher, M. & Evans, R. M. 2013, *ApJ*, 775, 5
Kay, C., Gopalswamy, N., Reinard, A. & Opher, M. 2017, *ApJ*, 835, 117
Lara, A., Gopalswamy, N., Xie, H., Mendoza-Torres, E., *et al.* 2006, *J. Geophys. Res.*, 111, 6107
Lepping, R. P., Berdichevsky, D. B., Wu, C.-C., Szabo, A., *et al.* 2006, *Ann. Geophys.*, 24, 215
Liewer, P. C., Hall, J. R., Howard, R. A., *et al.* 2011, *J. Atmos. Sol-Terr. Phys.*, 73, 1173
Linker, J. A., Riley, P., Mikic, Z., Lionello, R., *et al.* 2009, *AGU Fall Meeting Abstracts*
Liu, H.-L. & Qin, G. 2012, *J. Geophys. Res.*, 117, A04108
Liu, Y., Luhmann, J. G., Lin, R. P., Bale, S. D., *et al.* 2009, *ApJ* (L), 698, L51
Liu, Y. D., Luhmann, J. G., Möstl, C., Martinez-Oliveros, J. C., *et al.* 2012, *ApJ* (L), 746, L15

Lugaz, N., Farrugia, C. J., Manchester, IV, W. B. & Schwadron, N. 2013, *ApJ*, 778, 20
Mäkelä, P., Gopalswamy, N. & Yashiro, S. 2016, *Space Weather*, 14, 368
Manoharan, P. K. 2006, *Solar Phys.*, 235, 345
Marubashi, K., Akiyama, S., Yashiro, S., Gopalswamy, N., *et al.* 2015, *Solar Phys.*, 290, 1371
Michałek, G., Gopalswamy, N. & Yashiro, S. 2003, *ApJ*, 584, 472
Michałek, G., Gopalswamy, N. & Yashiro, S. 2008, *Solar Phys.*, 248, 113
Mierla, M., Inhester, B., Marqué, C., Rodriguez, L., *et al.* 2009, *Solar Phys.*, 259, 123
Mierla, M., Inhester, B., Antunes, A., Boursier, Y., *et al.* 2010, *Ann. Geophys.*, 28, 203
Moran, T. G., Davila, J. M. & Thompson, W. T. 2010, *ApJ*, 712, 453
Möstl, C., Amla, K., Hall, J. R., Liewer, P. C., *et al.* 2014, *ApJ*, 787, 119
Mulligan, T., Russell, C. T. & Luhmann, J. G. 2000, *Adv. Space Res.*, 26, 801
Na, H., Moon, Y.-J., Jang, S., Lee, K.-S., *et al.* 2013, *Solar Phys.*, 288, 313
Odstrcil, D., Riley, P. & Zhao, X. P. 2004, *J. Geophys. Res.*, 109, A02116
Odstrcil, D., Pizzo, V. J. & Arge, C. N. 2005, *J. Geophys. Res.*, 110, A02106
Qin, G., Zhang, M. & Rassoul, H. K. 2009, *J. Geophys. Res.*, 114, A09104
Robbrecht, E., Patsourakos, S. & Vourlidas, A. 2009, *ApJ*, 701, 283
Rollett, T., Möstl, C., Isavnin, A., Davies, J. A., *et al.* 2016, *ApJ*, 824, 131
Rouillard, A. P., Davies, J. A., Forsyth, R. J., Rees, A., *et al.* 2008, *Geophys. Res.* (L), 35, L10110
Sabine, E. 1852, *Phil. Trans. R. Soc. Lond.*, 142, 103
Savani, N. P., Vourlidas, A., Szabo, A., Mays, M. L., *et al.* 2015, *Space Weather*, 13, 374
Savani, N. P., Vourlidas, A., Richardson, I. G., Szabo, A., *et al.* 2017, *Space Weather*, 15, 441
Schwenn, R. 1983, *Space Sci. Rev.*, 34, 85
Schwenn, R., dal Lago, A., Huttunen, E. & Gonzalez, W. D. 2005, *Ann. Geophys.*, 23, 1033
Shen, C., Wang, Y., Pan, Z., Zhang, M., *et al.* 2013, *J. Geophys. Res.*, 118, 6858
Sheeley, Jr., N., Howard, R. A., Michels, D., Koomen, M., *et al.* 1985, *J. Geophys. Res.*, 90, 163
Sheeley, Jr., N. R., Herbst, A. D., Palatchi, C. A., Wang, Y.-M., *et al.* 2008, *ApJ*, 675, 853
Shi, T., Wang, Y., Wan, L., Cheng, X., *et al.* 2015, *ApJ*, 806, 271
Smith, Z. & Dryer, M. 1990, *Solar Phys.*, 129, 387
Srivastava, N. & Venkatakrishnan, P. 2004, *J. Geophys. Res.*, 109, A10103
Srivastava, N., Inhester, B., Mierla, M. & Podlipnik, B. 2009, *Solar Phys.*, 259, 213
Temmer, M., Preiss, S. & Veronig, A. M. 2009, *Solar Phys.*, 256, 183
Temmer, M., Veronig, A. M., Peinhart, V. & Vršnak, B. 2014, *ApJ*, 785, 85
Thernisien, A., Vourlidas, A. & Howard, R. A. 2009, *Solar Phys.*, 256, 111
Tóth, G., Sokolov, I. V., Gombosi, T., Chesney, D., *et al.* 2005, *J. Geophys. Res.*, 110, A12226
Tóth, G., van der Holst, B., Sokolov, I., De Zeeuw, D., *et al.* 2012, *J. Comput. Phys.*, 231, 870
Trottet, G. & MacQueen, R. M. 1980, *Solar Phys.*, 68, 177
van der Holst, B., Sokolov, I. V., Meng, X., Jin, M., *et al.* 2014, *ApJ*, 782, 81
Vourlidas, A. & Howard, R. A. 2006, *ApJ*, 642, 1216
Vourlidas, A., Colaninno, R., Nieves-Chinchilla, T. & Stenborg, G. 2011, *ApJ* (L), 733, L23
Vourlidas, A., Lynch, B. J., Howard, R. A. & Li, Y. 2013, *Solar Phys.*, 284, 179
Vršnak, B. & Žic, T. 2007, *A&A*, 472, 937
Vršnak, B., Žic, T., Vrbanec, D., Temmer, M., *et al.* 2013, *Solar Phys.*, 285, 295
Wang, Y. M., Ye, P. Z., Wang, S., Zhou, G. P., *et al.* 2002, *J. Geophys. Res.*, 107, 1340
Wood, B. E., Howard, R. A. & Socker, D. G. 2010, *ApJ*, 715, 1524
Wu, C.-C., Fry, C. D., Wu, S. T., Dryer, M., *et al.* 2007, *J. Geophys. Res.*, 112, A09104
Wu, C.-C., Dryer, M., Wu, S. T., Wood, B. E., *et al.* 2011, *J. Geophys. Res.*, 116, A12103
Xie, H., Ofman, L. & Lawrence, G. 2004, *J. Geophys. Res.*, 109, 3109
Xie, H., Gopalswamy, N., Ofman, L., St. Cyr, O. C., *et al.* 2006, *Space Weather*, 4, 10002
Xie, H., St. Cyr, O. C., Gopalswamy, N., Odstrcil, D., *et al.* 2013, *J. Geophys. Res.*, 118, 4711
Yurchyshyn, V. 2008, *ApJ* (L), 675, L49
Zhao, X. P. 2008, *J. Geophys. Res.*, 113, 2101
Zhao, X. P., Plunkett, S. P. & Liu, W. 2002, *J. Geophys. Res.*, 107, 1223
Zhao, X. & Dryer, M. 2014, *Space Weather*, 12, 448
Zhao, X. H. & Feng, X. S. 2015, *ApJ*, 809, 44

Space Weather of the Heliosphere: Processes and Forecasts
Proceedings IAU Symposium No. 335, 2017
C. Foullon & O.E. Malandraki, eds.

doi:10.1017/S1743921317010857

Geomagnetic Consequences of Interacting CMEs of June 13-14, 2012

Nandita Srivastava[1], Zavkiddin Mirtoshev[2] and Wageesh Mishra[3]

[1]Udaipur Solar Observatory, Physical Research Laboratory,
Badi Road, Udaipur 313001, India
email: nandita@prl.res.in

[2]Department of Physics, Samarkand State University,
15 University Boulevard, Samarkand 140104, Uzbekistan
email: zmirtoshev@gmail.com

[3]CAS Key Laboratory of Geospace Environment, Department of Geophysics and Planetary Sciences, University of Science and Technology of China, Hefei 230026, China
email: wageesh@ustc.edu.cn

Abstract. We have studied the consequences of interacting coronal mass ejections (CMEs) of June 13-14, 2012 which were directed towards Earth and caused a moderate geomagnetic storm with Dst index ~ -86 nT. We analysed the in-situ observations of the solar wind plasma and magnetic field parameters obtained from the OMNI database for these CMEs. The in-situ observations show that the interacting CMEs arrive at Earth with the strongest ($\sim$ 150 nT) Sudden Storm Commencement (SSC) of the solar cycle 24. We compared these interacting CMEs to a similar interaction event which occurred during November 9-10, 2012. This occurred in the same phase of the solar cycle 24 but resulted in an intense geomagnetic storm (Dst ~ -108 nT), as reported by Mishra *et al.* (2015). Our analysis shows that in the June event, the interaction led to a merged structure at 1 AU while in the case of November 2012 event, the interacted CMEs arrived as two distinct structures at 1 AU. The geomagnetic signatures of the two cases reveal that both resulted in a single step geomagnetic storm.

Keywords. Coronal mass ejections, solar wind plasma, magnetic field, geomagnetic storm.

1. Introduction

Coronal mass ejections (CMEs) are large expulsions of matter and magnetic field from the Sun, moving with speeds ranging from few hundreds to few thousand kilometers per second. Their travel time from the Sun to the Earth ranges between 1 to 4 days. The number of CMEs launched from the Sun is about 5 per day around solar maximum (St. Cyr *et al.* 2000; Yashiro *et al.* 2004; Webb and Howard 2012). Sometimes, when CMEs occur in quick succession, particularly during solar maximum, they can interact or merge with each other if they propagate along the same direction through interplanetary space. If the successive CMEs are directed Earthward, their interaction can enhance the geoeffectiveness due to the extended period and enhanced strength of southward magnetic field causing intense geomagnetic storms (Wang *et al.* 2003; Farrugia and Berdichevsky, 2004; Farrugia *et al.* 2006).

In this paper, we present the geomagnetic consequences of interacting CMEs of June 13-14, 2012. This interaction led to the strongest SSC of solar cycle 24 and to a moderate intensity geomagnetic storm. In an earlier study, Mishra *et al.* (2015) reported a similar interaction event that occurred on November 9-10, 2012 and led to an intense geomagnetic storm. We present a comparison of the geomagnetic signatures of the two cases.

2. Interacting CME Events

June 13-14, 2012 CMEs: Two CMEs originating from the same active region NOAA 11504 were launched on June 13 and 14, 2012 separated by several hours. These CMEs has average speeds of 630 km s^{-1} and 990 km s^{-1} respectively and were observed to interact at a distance of 100 $R_\odot$ from the Sun. The interaction resulted in a geomagnetic storm of moderate intensity (Dst ~ -86 nT). The interaction phase and the interplanetary signatures have been investigated in detail and presented in an earlier paper (Srivastava *et al.* 2017, in press). From the examination of in-situ properties of these interacting CMEs, we found that it led to the Strongest Storm Commencement or SSC ($\sim$ 150 nT) of the present solar cycle. The SSC was unique because of its long duration of approximately 20 hours.

November 9-10, 2012 CMEs: A similar interaction event was observed in the same phase of solar cycle 24 when two CMEs with similar speeds 620 km s^{-1} and 910 km s^{-1} were launched a few hours apart on November 9 and 10, 2012 from NOAA AR 11608. These CMEs propagated in the same direction and were observed to interact at a distance of 35 $R_\odot$ from the Sun. The detailed study of this interaction event by Mishra *et al.* (2015) showed that it led to a strong geomagnetic storm with (Dst ~ -108 nT).

In the following section, we compare the in-situ and geomagnetic parameters of the interacting CMEs of June 13-14, 2012 with those of November 9-10, 2012.

3. In-situ and Geomagnetic Signatures

We have analysed in-situ observations of the solar wind plasma and interplanetary magnetic field recorded during the passage of the interacting CMEs of June 13-14, 2012 as observed by the ACE and Wind spacecraft located at the L1 point. These have been plotted in Figure 1. Here S1 and S2 mark the arrival of the interplanetary shocks associated with the preceding CME of June 13 (CME1) and the following CME of June 14 (CME2), respectively. Figure 1(a), (b), (c) and (g) show the variation in density, velocity, ram pressure and horizontal component of the Earth's magnetic field near the equator. The sudden rise in density and velocity associated with arrival of shock, S2, driven by CME2 gave rise to a strong ram pressure which contributed significantly to the rise of Sym-H index. The Sym-H index increased from around 40 nT to 150 nT within less than half an hour. The details of this strongest observed SSC in the solar cycle 24 is presented in Srivastava *et al.* (2017). Figure 1(g) measures the energy content of the symmetric ring current flowing in the magnetic equatorial plane of the Earth and its decreasing value shows the main phase of the geomagnetic storm. The intensity of the geomagnetic storm reached a peak value of ~ -86 nT and its main phase lasted for $\sim$ 16 hrs.

The AL index in Figure 1 (h) measures the westward (midnight sector) electrojet current and captures the auroral substorm process. We note that the fluctuation in southward component (Bz) of interplanetary magnetic field (IMF) in Figure 1(d) and in the y-component of electric field (IEFy) shown in Figure 1(e) in the shock-sheath region of CME1 have not intensified the AL index significantly. Despite the large value of southward IMF Bz and positive IEFy during the arrival of shock S2, their short duration resulted in a low substorm activity. The Bz and positive IEFy are well correlated with the AL index during the passage of CME2. Figure 1(f) shows the variation in the Polar Cap (PC) index or the variation in the ionospheric electric field over the polar region. We note that the region of CME1 and shock driven by CME2 could not strongly intensify the PC index. However, the smooth variation in Bz and IEFy in the region of CME2 is reasonably well correlated with the PC index and have intensified its value. Thus, we find

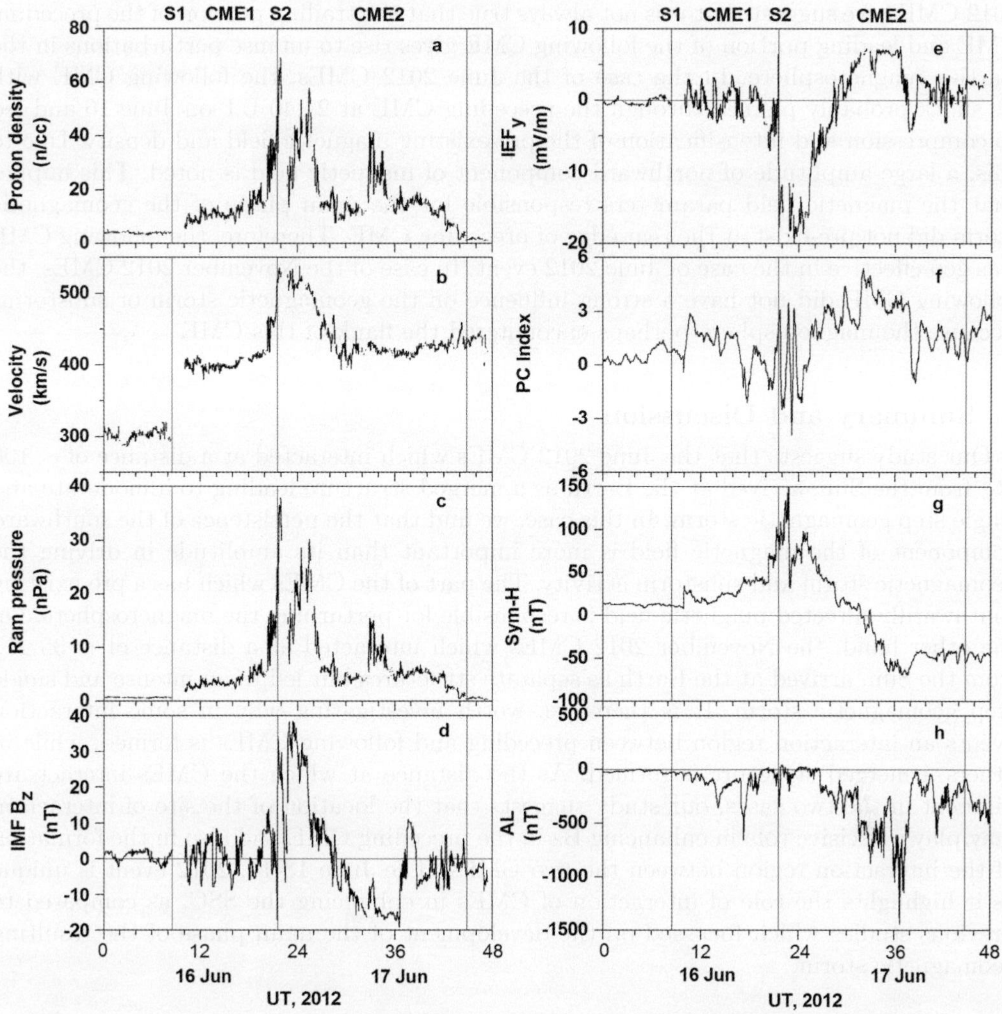

Figure 1. Variation of the in-situ solar wind and geomagnetic field parameters associated with the passage of interacting CMEs of June 13-14, 2012, obtained during June 16-18, 2012. Left panels show, (from top to bottom): Proton density (n/cc), Velocity (km s^{-1}), Ram pressure (nPa) and Bz component (nT). Right panels show, (from top to bottom), the 'y' component of the interplanetary electric field IEFy (mV/m), polar cap (PC) index, Sym-H (nT) and AL index (nT).

that the preceding CME did not strongly influence the triggering of substorms nor in the perturbation of the ionospheric electric field over the polar region. Figure (1) also shows that when the terrestrial magnetosphere encounters the passage of the trailing edge of the following CME, i.e. CME2 (excluding its rearmost part), intensification of the AL and PC indices took place.

In our earlier study of interacting CMEs of 2012 November 9-10 (Mishra *et al.* 2015), we found that the trailing edge of CME1 (November 9) and the interaction region identified between the preceding and following (November 10) CMEs played a direct role in the storm-time AL intensification. In this case, the peak amplitude of the PC index was nearly the same in the shock-sheath region of the preceding CME as in the interaction region. Therefore, comparing the interaction events of June 2012 CMEs and November

2012 CMEs, we suggest that it is not always true that the trailing portion of the preceding CME and leading portion of the following CME gives rise to intense perturbations in the Earth's magnetosphere. In the case of the June 2012 CMEs, the following CME with its shock probably pierced through the preceding CME at 21:40 UT on June 16 and led to compression and intensification of the pre-existing magnetic field and density. Due to this, a large amplitude of northward component of magnetic field is noted. This implies that the magnetic field parameters responsible for the main phase of the geomagnetic storm did not pre-exist at the rear edge of preceding CME. Therefore, the following CME was geo-effective in the case of June 2012 event. In case of the November 2012 CMEs, the following CME did not have a strong influence on the geomagnetic storm or substorms because the magnetosphere perhaps encountered the flank of this CME.

4. Summary and Discussion

Our study suggests that the June 2012 CMEs which interacted at a distance of ~ 100 $R_\odot$ from the Sun, arrived at the Earth as a merged structure leading to a moderate and single step geomagnetic storm. In this case, we find that the persistence of the southward component of the magnetic field is more important than its amplitude in driving the geomagnetic storm and substorm activity. The part of the CMEs which has a pre-existing southwardly directed magnetic field is responsible for perturbing the magnetosphere. On the other hand, the November 2012 CMEs which interacted at a distance of ~ 35 $R_\odot$ from the Sun, arrived at the Earth as separate structures but led to an intense and single step geomagnetic storm. It is therefore, worth investigating why in some interaction events an interaction region between preceding and following CMEs is formed, while in others a merged structure is formed. As the distance at which the CMEs interact are different in the two cases, our study suggests that the location of the site of interaction may play a decisive role in enhancing Bz in the preceding CME and also in the formation of the interaction region between the two CMEs. The June 13-14, 2012 event is unique as it highlights the role of interaction of CMEs in enhancing the SSC, as compared to previous studies which focussed on the development of the main phase of the resulting geomagnetic storm.

5. Acknowledgements

We acknowledge the use of in-situ data of the solar wind from ACE and WIND spacecraft from NASA (`http:cdaweb.gsfc.nasa.gov`). Work by Z. Mirtoshev was carried out at the Udaipur Solar Observatory, PRL, Udaipur during January-July 2017, under the NAM S&T Centre Research Training Fellowship for Developing Country Scientists (RTF-DCS) by the Department of Science and Technology, India.

References

Farrugia, C. J. & Berdichevsky, D. 2004, *Ann. Geophys.*, 22, 3679-3698
Farrugia, C. J., Jordanova, V. K., Thomsen, M. F., *et al.* 2006, *J. Geophys. Res.*, 111, 11104
Mishra, W., Srivastava, N. & Chakrabarty, D. 2015, *Solar Phys.*, 290, 527
Srivastava, N., Mishra, W. & Chakrabarty, D. 2017, *Solar Phys.*, in press
St. Cyr, O. C., Plunkett, S. P., Michels, D. J., *et al.* 2000, *J. Geophys. Res.*, 105, 18169
Wang, Y. M., Ye, P. Z. & Wang, S. 2003, *J. Geophys. Res. Space Physics*, 108, 1370
Webb, D. F. & Howard, T. A. 2012, *Living Rev. Solar Phys.*, 9, 3
Yashiro, S. N., Gopalswamy, G., Michalek, O. C., *et al.* 2004, *J. Geophys. Res. Space Physics*, 109, A07105

Space Weather of the Heliosphere: Processes and Forecasts
Proceedings IAU Symposium No. 335, 2017
C. Foullon & O.E. Malandraki, eds.

doi:10.1017/S1743921318000066

Effects of ICMEs on High Energetic Particles as Observed by the Global Muon Detector Network (GMDN)

A. Dal Lago[1], C. R. Braga[1], R. R. S. de Mendonca[1], M. Rockenbach[1], E. Echer[1], N. J. Schuch[2], K. Munakata[3], C. Kato[3], T. Kuwabara[4], M. Kozai[5], H. K. Al Jassar[6], M. M. Sharma[6], M. Tokumaru[7], M. Duldig[8], J. Humble[8], P. Evenson[9] and I. Sabbah[10]

[1]National Institute for Space Research (INPE), Avenida dos Astronautas-12227-010, São José dos Campos-SP, Brazil
email: alisson.dallago@inpe.br

[2]Southern Regional Space Research Center - CRS/INPE, Santa Maria, Brazil

[3]Shinshu University, Matsumoto, Japan

[4]Chiba University, Chiba City, Chiba, Japan

[5]Japan Aerospace Exploration Agency - JAXA, Sagamihara, Kanagawa, Japan

[6]Kuwait University, Kuwait City, Kuwait

[7]Institute for Space-Earth Environmental Research, Nagoya University, Japan

[8]University of Tasmania, Hobart, Tasmania, Australia

[9]Bartol Research Institute, University of Delaware, Newark, USA

[10]Department of Natural Sciences, College of Health Sciences, Public Authority for Applied Education and Training, Kuwait City, Kuwait

Abstract. The Global Muon Detector Network (GMDN) is composed by four ground cosmic ray detectors distributed around the Earth: Nagoya (Japan), Hobart (Australia), Sao Martinho da Serra (Brazil) and Kuwait city (Kuwait). The network has operated since March 2006. It has been upgraded a few times, increasing its detection area. Each detector is sensitive to muons produced by the interactions of ~50 GeV Galactic Cosmic Rays (GCR) with the Earth's atmosphere. At these energies, GCR are known to be affected by interplanetary disturbances in the vicinity of the earth. Of special interest are the interplanetary counterparts of coronal mass ejections (ICMEs) and their driven shocks because they are known to be the main origins of geomagnetic storms. It has been observed that these ICMEs produce changes in the cosmic ray gradient, which can be measured by GMDN observations. In terms of applications for space weather, some attempts have been made to use GMDN for forecasting ICME arrival at the earth with lead times of the order of few hours. Scientific space weather studies benefit the most from the GMDN network. As an example, studies have been able to determine ICME orientation at the earth using cosmic ray gradient. Such determinations are of crucial importance for southward interplanetary magnetic field estimates, as well as ICME rotation.

Keywords. interplanetary medium, cosmic rays, coronal mass ejections (CMEs)

1. Introduction

Ground-based detectors are optimal for detecting secondary particles resulting from the interaction of Galactic Cosmic Rays (GCR) with energies from tens to several hundreds of GeV with the Earth's atmosphere (Duldig 2000; McDonald 2000). At this energy interval, GCR are modulated by the solar activity and corresponding interplanetary magnetic field

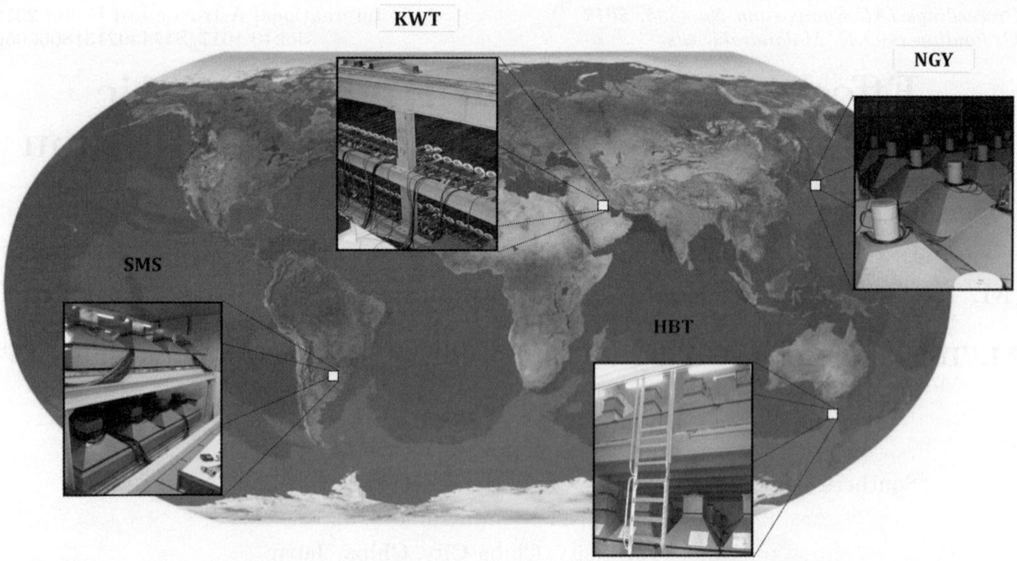

Figure 1. Current configuration of the Global Muon Detector Network (GMDN). From right to left, the detectors are: Nagoya, Japan (NGY), Hobart, Australia (HBT), Kuwait City, Kuwait (KWT) and Sao Martinho da Serra, Brazil (SMS).

(IMF) perturbations (Kuwabara *et al.* 2009). Among the main disturbances that drive geomagnetic activity at the Earth are the interplanetary counterparts of coronal mass ejections (ICMEs) (Gosling *et al.* 1991). They are the traveling disturbances originated at solar mass ejections, also referred to as coronal mass ejections (CMEs), as they are widely known because they are observed in the solar corona using coronagraphs (Schwenn 1996; Schwenn *et al.* 2005). ICMEs are the main drivers of geomagnetic storms for two main reasons, the first being the fact that their flux rope structures carry out-of-the ecliptic IMF for substantial amounts of time, sometimes up to ten hours or more. The second reason is the fact that they often drive interplanetary shocks, whose sheath compressed magnetic fields were found to be the most frequent sources of intense and very intense geomagnetic storms (Gonzalez *et al.* 2011; Szajko *et al.* 2013). In this paper, we shall present a brief description of some of the most important modulations produced by ICMEs on ground cosmic rays observed by the Global Muon Detector Network.

2. The Global Muon Detector Network (GMDN)

Since 2006, four muon detectors are in operation around the earth: Nagoya (Japan), Hobart (Australia), Sao Martinho da Serra (Brazil) and Kuwait city (Kuwait). Their locations were chosen to be as homogeneously distributed as possible. Since 1992, however, the Nagoya and Hobart detectors were already in operation as a two-hemisphere network. In 2001, a 4 m^2 prototype was installed in Sao Martinho da Serra in order to improve the coverage over the Atlantic. It was later upgraded to a 28m^2 detector. In 2006, a detector was installed in Kuwait city, after which all 4 instruments together were named the "Global Muon Detector Network (GMDN)". Since then, three detectors have had their detection area increased: Kuwait was upgraded to 25m^2, Hobart was upgraded to 16m^2 and Sao Martinho da Serra was upgraded to 32m^2. Nagoya detection area is 36m^2. Figure 1 shows the current distribution of GMDN detectors around the Earth.

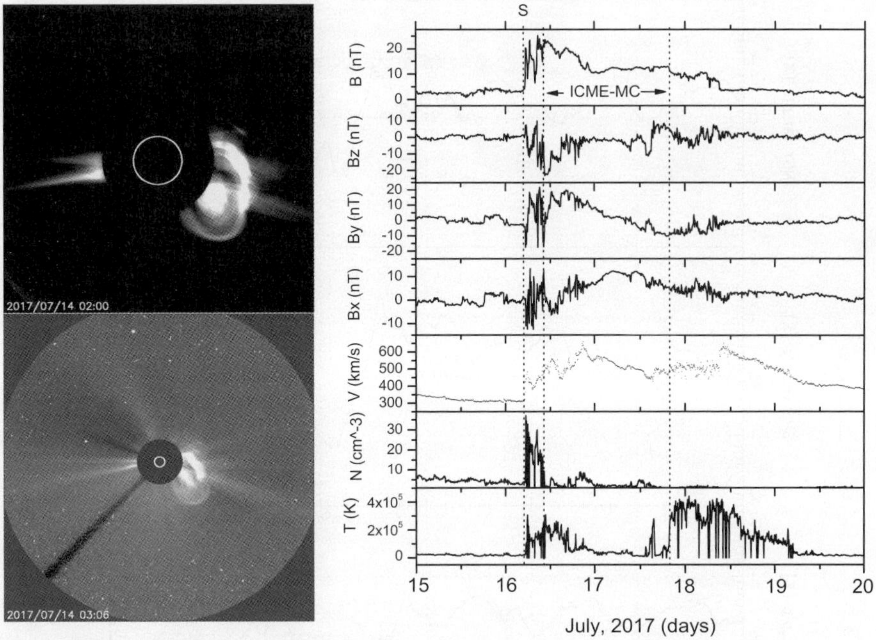

Figure 2. (left) Coronal mass ejection occurred on 14 July 2017, observed by the LASCO instrument aboard the SOHO space observatory. (right) From top to bottom, magnetic field intensity, its z, y and x components, solar wind speed, number density and temperature observed by ACE on the period from 15-19 July 2017. An interplanetary shock is observed in the first dotted line (from left to right), marked as "S". The ICME-magnetic cloud is located inside the second and third vertical lines. A magnetic field rotation can be identified in the "y" component of the magnetic field (By). One can see the abrupt changes in all parameters across the shock. A sheath compressed magnetic field is clearly seen between the shock and the ejecta (ICME). The ICME has a magnetic field which is few times stronger than the normal solar wind value. The only data available from ACE was level 1, which we used because our purpose was only to characterize the macroscopic features of the event. No quantitative calculations were made, which would require level 2 data.

3. Interplanetary coronal mass ejections

Coronal mass ejections (CMEs) are one of the most important solar energetic transients that drive Space Weather (Schwenn *et al.* 2005). They consist of large expulsions of solar material observed by coronagraphs. Figure 2 (left) shows a CME observed on 14 July 2017 by the Large Angle and Spectroscopic Coronagraph (LASCO), which is one instrument aboard the Solar and Heliospheric Observatory (SOHO). When pointed along the Sun-Earth line, they appear as "halos" in Earth-orbiting or L1 coronagraphs (Howard *et al.* 1982). Such CMEs can either be traveling away or towards the Earth. In the latter case, they are very likely to produce an interplanetary disturbance that has the potential to impact our planet. Such disturbances are the interplanetary counterparts of the CMEs, the so-called ICMEs. Some of them are ejecta-like structures, possessing a organized magnetic structure, called a magnetic clouds (MC) (Burlaga *et al.* 1981). Because of their high speeds, they often drive shocks in front of them. A sheath compressed magnetic field is produced in between the shock and the ICME ejecta. Figure 2 (right) shows an example of an ICME-MC observed on 15-19 July 2017 by the Advanced Composition Explorer (ACE). It is presented from top to bottom the magnetic field intensity, its z, y and x components, solar wind speed, number density and temperature.

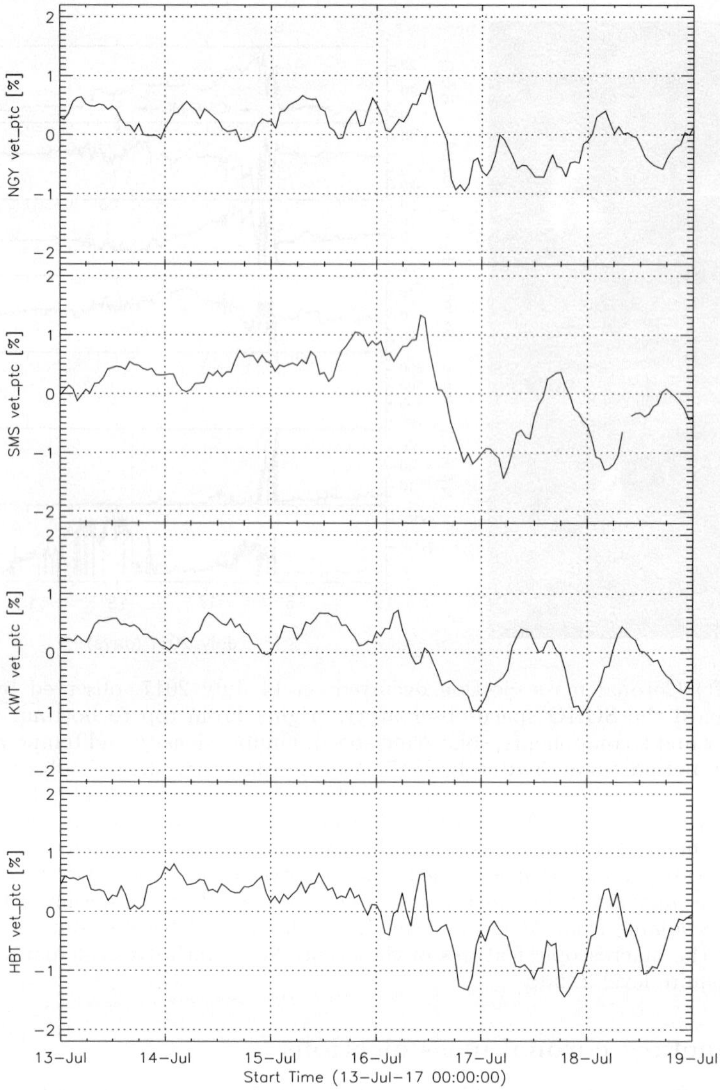

Figure 3. Pressure and temperature corrected vertical channels of all GMDN detectors, NGY (Japan), SMS (Brazil), KWT (Kuwait) and HBT (Hobart) during the passage of the ICME on 16 July 2017, related to the CME observed at the Sun on 14 July 2017. Clear Forbush decreases are observed on 16 July in all detectors.

4. Modulation

Before comparing ground cosmic ray intensity modulation due to extraterrestrial phenomena, it is necessary to take into account atmospheric effects, such as pressure and temperature. They produce a seasonal variation on the cosmic ray intensity observed by muon detectors. We recommend a careful reading of De Mendonca *et al.* (2016) for details on corrections of GMDN observations for atmospheric effects. Perhaps the most well known modulation that ICMEs impose in ground cosmic rays is the Forbush decrease (Forbush 1937; Simpson 1954). It consists of a global decrease of cosmic ray intensity observed during the passage of the Earth through interplanetary disturbances. Figure 3 shows the Forbush decrease observed by all four GMDN detectors, NGY (Japan),

SMS (Brazil), KWT (Kuwait) and HBT (Hobart) during the passage of the ICME on 16 July 2017, related to the CME observed at the Sun on 14 July 2017. Another type of modulation imposed by approaching ICMEs at Earth is the loss cone effect, which produces a precursory anisotropy. This effect occurs before the arrival of the ICME at the Earth, thus attracting attention to possible Space Weather applications towards forecasting arrival of interplanetary disturbances. Very good descriptions of this anisotropic modulation can be found in Nagashima *et al.* (1992), Munakata *et al.* (2000), Kuwabara *et al.* (2006), Fushishita *et al.* (2010), Rockenbach *et al.* (2011) and Braga *et al.* (2011). Munakata *et al.* (2000) reported many such loss cone precursors using observations from two muon detectors (Nagoya and Hobart). His overall conclusion was that such precursors could be detected up to 8 hours before the arrival of the ICME at Earth. Later work from Rockenbach *et al.* (2011) found similar results. As already mentioned in Figure 2 description, ICMEs often drive interplanetary shocks. As these shocks propagate, they sweep out galactic cosmic rays forming a depleted region behind them. Kozai *et al.* (2016) studied the cosmic ray density and gradient derived by GMDN observations during Forbush decreases of a number of CME-ICME shock events, originating in several positions at the solar disk. These authors found a distinct modulation of galactic cosmic rays, one related to the shock-sheath region and another due to the ICME flux rope.

5. Orientation of ICME flux ropes

It has been shown to be possible to determine the flux rope orientation of ICMEs using cosmic ray directional anisotropies observed by ground-based muons detectors. Remarkable results were reported by Kuwabara *et al.* (2009), where excellent agreement of ICME inclination was found when comparing a cosmic ray derivation and a more widely used Minimum Variance Analysis method, which is based on the Interplanetary Magnetic Field.

6. Summary

In this paper we presented a brief description of the Global Muon Detector Network (GMDN) and of some of the most important modulations produced by interplanetary counterparts of coronal mass ejections (ICMEs) on ground cosmic rays observed by the GMDN. In particular, we described the Forbush decrease occurred in 11-16 July 2017 observed by the detectors of the GMDN. Following on, we make reference to many previous studies related to the loss cone effect produced by ICMEs approaching the Earth. This particular effect has potential for Space Weather applications. We also pointed out some important results related to the estimate of the orientation of ICME-magnetic clouds using cosmic ray observations from GMDN.

7. Acknowledgements

A.D.L., E.E. and R.R.S.M thank CNPq for grants 304209/2014-7, 302583/2015-7 and 152050/2016-7. C.R.B. thanks FAPESP 2014/24711-6. We thank the SOHO/LASCO and ACE teams for the data used in this work. SOHO is a joint collaboration between ESA and NASA.

References

Braga *et al.* 2011, *32nd International Cosmic Ray Conference*, 10, 286, ISBN: 978-1-63439-138-2

Burlaga, L. F., E. Sittler, F. Mariani & R. Schwenn 1981, *J. Geophys. Res.*, 86, 6673
De Mendonca *et al.* 2016, *ApJ*, 830, 88
Duldig, M. 2000, *Space Science Review*, 93, 207
Forbush, S. E. 1937, *Physical Review*, 51, 12, 1108
Fushishita, A., *et al.* 2010, *ApJ*, 715, 2, 12391247
Gosling, J. T., McComas, D. J., Phillips, J. L. & Bame, S. J. 1991, *J. Geophys. Res.*, 96, A5, 7831
Gonzalez, W. D., Echer, E., Tsurutani, B. T., Clua de Gonzalez, A. L. & Dal Lago, A. 2011, *SSR*, 158, 69
Howard, R. A., *et al.* 1982, *ApJ*, 263, 2L, 101
Kozai, M., Munakata, K., Kato, C., Kuwabara, T., Rockenbach, M., Lago, A. Dal, Schuch, N. J., Braga, C. R., Mendonça, R. R. S., Jassar, H. K. Al, Sharma, M. M., Duldig, M. L., Humble, J. E., Evenson, P., Sabbah, I. & Tokumaru, M. 2016, *ApJ*, 825, 100
Kuwabara, T., *et al.* 2006, *Space Weather*, 4, 8, S08001
Kuwabara, T., *et al.* 2009, *JGR*, 114, A05109
McDonald, F. B. 2000, *Space Science Reviews*, 93, 263
Munakata, K., *et al.* 2000, *JGR*, 27457
Nagashima, K., *et al.* 1992, *Planetary and Space Science*, 40, 8, 1109
Rockenbach, M., *et al.* 2011, *GRL*, 38, L16108
Schwenn, R. 1996, *Astrophys. Space Sci.*, 243, 1, 187
Schwenn, R., Dal Lago, A., Huttunen, E. & Gonzalez, W. D . 2005, *Ann. Geophys.*, 23, AG/2004180, 1033
Simpson, J. A. 1954, *Physical Review*, 94, 2, 426
Szajko, N. S., Cristiani, G., Mandrini, C. H. & Dal Lago, A. 2013, *ASR*, 51, 10, 1842

Space Weather of the Heliosphere: Processes and Forecasts
Proceedings IAU Symposium No. 335, 2017
C. Foullon & O.E. Malandraki, eds.

doi:10.1017/S1743921318000285

Re-Acceleration of Energetic Particles in Large-Scale Heliospheric Magnetic Cavities

Olga V. Khabarova[1], **Olga E. Malandraki**[2], **Gary P. Zank**[3,4], **Gang Li**[3,4], **Jakobus A. le Roux**[3,4] **and Gary M. Webb**[3]

[1]Pushkov Institute of Terrestrial Magnetism, Ionosphere and Radio Wave Propagation of RAS (IZMIRAN), Moscow, Russia, email: olik3110@gmail.com
[2]IAASARS, National Observatory of Athens, Penteli, Greece, email: omaland@astro.noa.gr
[3]Center for Space Plasma and Aeronomic Research, University of Alabama in Huntsville, AL, USA
[4]Department of Space Science, University of Alabama in Huntsville, AL, USA

Abstract. Case studies show that some energetic particle flux enhancements up to MeV/nuc. observed at 1 AU cannot be treated as a consequence of particle acceleration at shocks or during flares. Atypical energetic particle events (AEPEs) are often detected during crossings of magnetic cavities formed by strong current sheets of various origins in the solar wind. Such cavities confine small-scale magnetic islands (SMIs) produced by magnetic reconnection. SMIs, in turn, trap and re-accelerate energetic particles according to predictions based on the theory of Zank *et al.* describing stochastic particle energization in the supersonic solar wind via numerous dynamically interacting SMIs. AEPEs possess energies that overlap SEP events and can be an important component in understanding space weather.

Keywords. Solar wind, particle acceleration, magnetic islands, current sheets

1. Introduction

Energetic particle flux enhancements (EPFEs) observed at 1 AU in the keV-MeV energy range are predominantly treated as a phenomenon that has a distant origin. Some EPFEs are classical solar energetic particle (SEP) events, produced by particle acceleration near the Sun, which result from either solar flares or diffusive particle acceleration at shocks associated with interplanetary coronal mass ejections (ICMEs). The same diffusive shock acceleration (DSA) mechanism is usually considered to be responsible for 1 AU EPFEs observed in quiet (non-flare) times and associated with long-lived corotating interaction regions (CIRs) or randomly occurring stream interaction regions (SIRs). In this case, interplanetary shocks are formed at CIR/SIR edges further from the Earth's orbit, with energetic particles reaching the Earth from 2–3 AU as they stream towards the Sun. In both cases, the source of accelerated particles is located far from the point of observation, and so it has long been thought that the processes in the solar wind that could accelerate particles locally to substantial energies do not exist. However, an increasing number of EPFEs detected at 1 AU cannot be explained by the dominant paradigm of nonlocal particle acceleration, suggesting instead that local particle acceleration or modulation of EPFE time-intensity profiles by local structures should be considered. The problem that observers typically face, for the interpretation of such events, is the observation of a time shift in variations of EPFE time-intensity profiles that corresponds to the propagation of the thermal solar wind, and not energetic particles from a distant source (Khabarova *et al.* 2015, 2016; Khabarova & Zank, 2017; Khabarova *et al.* 2017). These energetic particle flux enhancements, called "atypical energetic particle events"

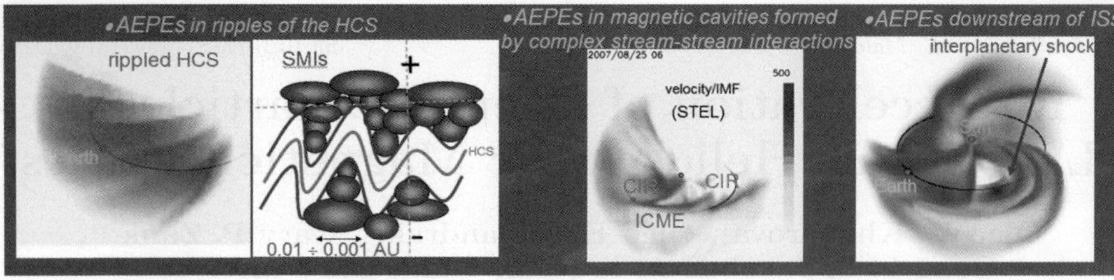

Figure 1. The three most common situations for the formation of magnetic cavities bounded by strong current sheets in the solar wind. The cavities are filled with small-scale magnetic islands (SMIs) created by magnetic reconnection. SMIs confine SMIs and trap and re-accelerate energetic particles up to several MeV, depending on the initial energy of seed particles. Left: magnetic cavities formed in ripples of the HCS. Middle: magnetic cavities produced by the interaction of CIRs and an ICME. Right: magnetic cavities formed by the interaction of an interplanetary shock with the HCS.

(AEPEs), are associated with additional mechanisms of particle acceleration that occur locally. In this paper we present examples of AEPEs that result from local particle acceleration in the solar wind, with up to MeV energies. Consequently, AEPEs should be regarded as another, if underestimated, hazardous factor in terms of the space radiation environment for space weather.

2. Atypical energetic particle events observed at 1 AU

Formation of magnetic cavities in the solar wind Recent studies of particle acceleration mechanisms in the heliosphere reveal the importance of stream-stream interactions and heliospheric current sheet (HCS) – stream interactions that often occur in the solar wind. These interactions can produce huge magnetic cavities that are bounded by strong current sheets. The cavities are usually filled with small-scale magnetic islands produced by magnetic reconnection (Wei *et al.* 2000; Khabarova *et al.* 2015, 2016). The entire system confines small scale magnetic islands and traps and re-accelerates energetic particles via a stochastic mechanism proposed by Zank *et al.* (2014, 2015) and le Roux *et al.* (2015, 2016). The mechanism of repeated particle interactions with dynamically interacting, reconnecting, merging, and contracting magnetic islands or flux ropes has been confirmed by numerous in situ observations (Khabarova *et al.* 2015, 2016, 2017). Stochastic particle acceleration by small-scale magnetic islands can yield significant particle energization, particularly when the seed particle population is initially energized by another mechanism such as, for example, DSA. As a result, crossings of magnetic cavity regions are often associated with unusual variations in the energetic particle flux up to several MeV/nuc. near the Earth's orbit.

Figure 1 illustrates some typical situations that are favourable for local particle acceleration and re-acceleration during the propagation of the solar wind from the corona to the Earth's orbit. Velocity and IMF plots reconstructed from interplanetary scintillation data (Jackson *et al.* 2013) show that the HCS often possesses a rippled form (left panel of Figure 1), and in situ observations confirm the occurrence of small-scale magnetic islands (SMIs) that fill the HCS ripples. Observations show that the heliospheric plasma sheet (HPS) region and its vicinity contain not only the embedded HCS, but also numerous small-scale structures, namely, SMIs and secondary current sheets produced presumably by magnetic reconnection. The rippled form of the HCS effectively confines SMIs and

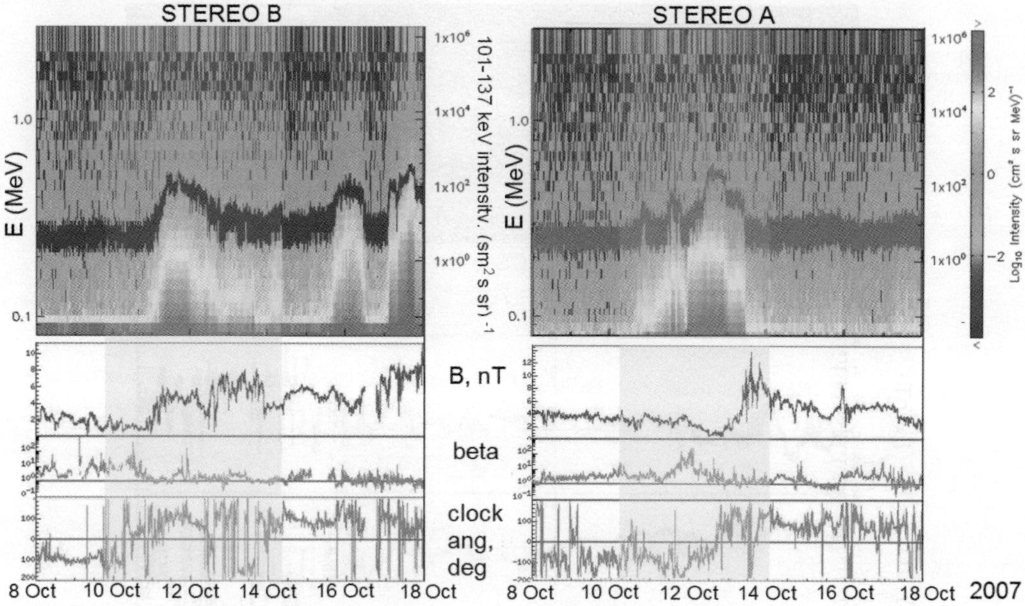

Figure 2. Crossing of the rippled HPS and associated AEPEs observed by STEREO B (left) and STEREO A (right). The separation angle is 36°. From top to bottom: energy flux of ions (omni-directional flux spectrogram) and variations in the 101–137 keV energy channel of the anti-Sun telescope; the IMF strength, the plasma beta and the IMF clock angle. The latter varies quickly back and forth many times as the spacecraft crosses numerous SMIs within the HCS ripples, and its more-or-less regular change across zero angle suggests an IMF sector crossing. The yellow stripes identify a region filled with SMIs within and around the HPS. HPS crossings are characterized by a very large plasma beta.

allows for additional particle acceleration. Magnetic cavities can also form through CIR-ICME interactions or by CIR-HCS interactions. The middle panel of Figure 1 shows the case of a CIR-ICME-CIR interaction that lead to the formation of a large magnetic cavity (Khabarova *et al.* 2016). Another example that creates magnetic cavities is interplanetary shocks (ISs). The right panel of Figure 1 illustrates an IS interacting with the HCS, which is transparent to the shocks, but is deformed. It has been shown that IS-HCS interaction intensifies magnetic reconnection and SMI production (Khabarova *et al.* 2016). The turbulent sheath region downstream of ISs may also be filled with SMIs (Zank *et al.* 2015).

AEPEs in ripples of the HCS Even non-disturbed current sheets in the solar wind represent multi-layer structures that have a tendency to reconnect and form magnetic islands (Wei *et al.* 2000; Lazarian et al, 2012; Malova *et al.* 2017). This characteristic makes it difficult to find signatures of simple Petschek-like magnetic reconnection near strong current sheets, including the HCS. However, the occurrence of reconnecting current sheets can be revealed by plasma and IMF signatures as well as particle behaviour, all of which are typical for multi-X-point magnetic reconnection (Zharkova & Khabarova, 2015; Khabarova *et al.* 2015, 2016).

Figure 2 shows STEREO multi-spacecraft observations of AEPEs during the crossing of a highly disturbed, rippled HPS filled with SMIs during October 2007. The HPS shape was slightly different when observed by STEREO Behind than at the STEREO Ahead spacecraft, which were separated by 35°. The upper panels of Figure 2 show omni-directional flux and energetic ion flux of 103–137 keV from the anti-Sun STEREO

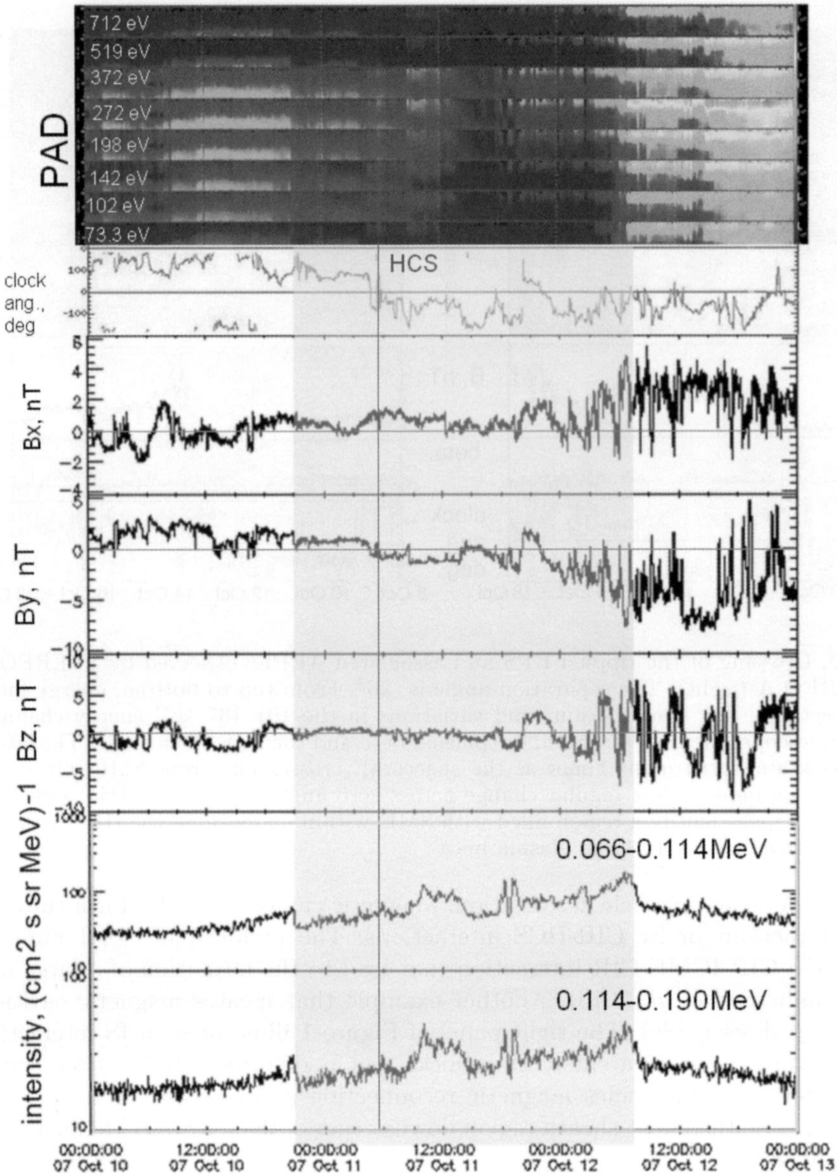

Figure 3. The heliospheric plasma sheet crossed by ACE. The yellow strip indicates both the HPS location and the occurrence of an energetic ion flux enhancement. A typical disorganized PAD (pitch-angle distribution) histogram of suprathermal electrons is seen during the HPS crossing. This can be explained by the rippled profile of the HCS/HPS and the occurrence of small- scale magnetic islands produced by magnetic reconnection in the region (see Figure 1). A corresponding energetic particle flux enhancement is shown in the bottom two panels. The vertical line shows the position of the HCS within the HPS.

telescopes. The time-intensity ion flux profiles are shifted with respect to each other, which corresponds to the peculiarities of the HPS rippled profile observed by the two spacecraft. The energetic ion flux demonstrates a clear anisotropy during the October 2007 AEPE event, i.e. time-intensity profiles were different for different viewing directions. Besides the main maximum of energetic ion flux indicated in Figure 2, additional

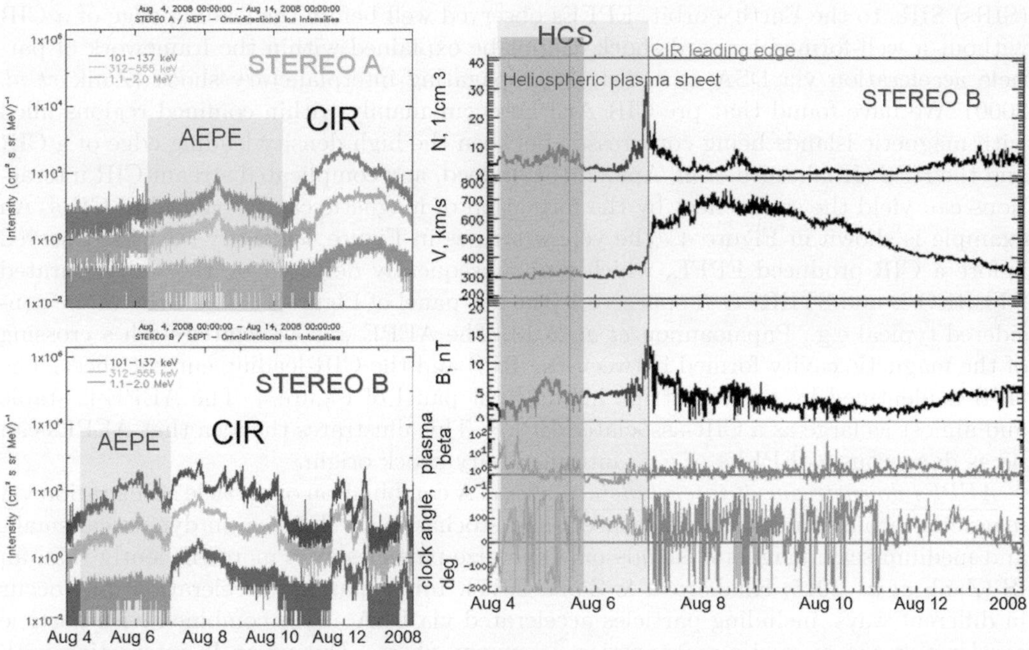

Figure 4. STEREO observations of a typical CIR (see Papaioannou *et al.* 2014), preceded by an AEPE (left panel). The separation angle between STEREO B and STEREO A was 65.5°. The AEPE is associated with a magnetic cavity formed by the heliospheric plasma sheet (HPS) (the green vertical stripe) and the leading CIR current sheet (right panel).

energetic ion flux enhancements related to pre-CIR deformation of the HCS and particle acceleration in the region filled with SMIs were detected by STEREO B after the main crossing of the HCS/HPS. However, the enhancements were not so pronounced in the STEREO A observations.

Details of the HCS crossing and corresponding AEPE observed by STEREO A in October 2007 were discussed in Zharkova & Khabarova (2015) in terms of confirmation of ongoing magnetic reconnection at the HCS embedded into the HPS. It was noted that the region filled with accelerated particles was much wider than expected from PIC modelling, which might be a signature of additional acceleration by dynamical magnetic islands surrounding the structure. Magnetic islands can be identified by the rotation of the IMF during the period of interest (not shown).

This particular crossing of the HCS was detected by the L1 spacecraft (WIND and ACE) on October 11, 2007 with a corresponding time shift (i.e., see Leif Svalgaard's list of sector boundary crossings: http://www.leif.org/research/sblist.txt). It is often thought that the behaviour of suprathermal electrons identifies the HCS as a boundary at which suprathermal elections change direction abruptly. However, there are some crossings that do not follow the dominant paradigm that suggests all energetic electrons are produced in the solar corona. Observations show that local processes can affect PADs dramatically (see the yellow strip in Figure 3). The occurrence of corresponding energetic flux enhancements associated with HPS crossings confirms the idea that SMIs bounded by strong current sheets can accelerate particles to at least 0.5 MeV.

AEPEs in magnetic cavities formed by a complex stream-stream interaction AEPEs are often observed before the arrival of regular CIRs or irregular stream interaction regions

(SIRs) SIRs to the Earth's orbit. EPFEs observed well before the leading edge of a CIR without a well-formed forward shock cannot be explained within the framework of particle acceleration via DSA at a distant propagating interplanetary shock (Zank *et al.* 2000). We have found that pre-CIR AEPEs occur mainly within confined regions filled with magnetic islands being compressed between the high-density leading edge of a CIR and the HPS (Khabarova *et al.* ApJ, 2016). Indeed, any complicated stream-CIR interactions can yield the same effect by the formation of magnetic cavities ahead of CIRs. An example is shown in Figure 4. The yellow stripes in Figure 4 identify AEPEs observed before a CIR-produced EPFE, which was subsequently detected by the well-separated STEREO B and STEREO A spacecraft (the left panel of Figure 4). The CIR can be considered typical e.g., Papaioannou *et al.* 2014. The AEPE was associated with a crossing of the magnetic cavity formed between the HPS and the CIR-leading current sheet. The HPS is identified by a green stripe in the right panel of Figure 4. The AEPE is stable and almost as large as a CIR-associated EPFE. This illustrates the idea that AEPEs can be as dangerous as EPFEs of e.g., interplanetary shock origin.

AEPEs downstream of interplanetary shocks A combination of particle acceleration via magnetic reconnection at current sheets and stochastic acceleration in dynamical small- and medium-scale magnetic islands may energize particles even more efficiently than an IS (Zank *et al.* 2015; Khabarova & Zank, 2017). Initial particle acceleration may occur in different ways, including particles accelerated via DSA at ISs combined with particle acceleration via magnetic reconnection at current sheets. During an IS interaction with the HCS (i) the IS pushes the HCS, so it is typically observed both before and after the passage of the IS, and (ii) it intensifies magnetic reconnection in the HPS. As a result, AEPEs are sometimes observed far beyond the turbulent wake of an IS, simultaneously with the HPS crossing (Khabarova *et al.* 2016). We have found observationally that particle energization at ISs ahead of an ICME can be much weaker than that of combined particle acceleration near reconnecting current sheets at the edges of a fragmented magnetic cloud within the ICME (Khabarova & Zank, 2017).

3. Summary and conclusions

This study emphasizes the need for a comprehensive systematic investigation of AEPEs produced by local particle acceleration in the solar wind. Observations show that AEPEs are not rare since magnetic cavities filled with dynamical magnetic islands are almost inevitably formed by the HCS on one side and current sheets at the edge of an ICME or CIR on the other. Energetic particles can be efficiently accelerated up to energies of several MeV within magnetic cavities that are filled with magnetic islands. AEPEs may themselves be as intense as ordinary EPFEs produced by DSA or flares. Therefore, AEPEs represent a serious and currently poorly recognized danger for spacecraft and astronauts.

The analysis of in situ multi-spacecraft data often shows that streams propagate and interact with current sheets in a very complicated fashion. This interaction frequently leads to the formation of magnetic cavities. In the presence of complicated interacting large-scale dynamical structures, relating data from distant spacecraft may offer little insight into understanding the large-scale topology of the region in which particle acceleration may occur, because even multi-point measurements cannot reconstruct the current form of a magnetic cavity, nor its origin and prior evolution. One approach that is useful utilizes interplanetary scintillation tomographic data to reconstruct the solar wind speed, density and interplanetary magnetic field, thereby providing some understanding of the 3-D structure of stream interactions that lead eventually to AEPEs.

Acknowledgements

STELab data used in Figure 1 are obtained from the official website http://smei.ucsd.edu/new_smei/data&images/data&images.html (thanks to B. Jackson and M. Bisi). The Solar Electron and Proton Telescope STEREO data were obtained from the http://www2.physik.uni-kiel.de/STEREO/index.php (University of Kiel, Germany). Other STEREO data are from the STEREO science center: http://stereossc.nascom.nasa.gov/data.shtml. This work is partly supported by the International Space Science Institute (ISSI) in the framework of International Team 405 entitled "Current Sheets, Turbulence, Structures and Particle Acceleration in the Heliosphere." GPZ acknowledges NSF, DOE number PHY – 1707247. OVK's work is partially supported by RFBR grants 16-02-00479, 17-02-00300 and 17-02-01328.

References

Jackson, B. V., Clover, J. M., Hick, P. P., *et al.* 2013, *SoPh*, 285, 151

Khabarova, O. V., Zank, G. P., Malandraki, O. E., *et al.* 2017, *Sun and Geosphere*, 12, 23

Khabarova, O. & Zank, G. P. 2017, *ApJ*, 843, 4, https://doi.org/10.3847/1538-4357/aa7686

Khabarova, O., Zank, G. P., Li, G., *et al.* 2015, *ApJ*, 808, 181

Lazarian, A., Vlahos, L., Kowal, G., *et al.* 2012, *SSRv*, 173, 557

le Roux, J. A., Zank, G. P., Webb, G. M. & Khabarova, O. 2015, *ApJ*, 801, 112

le Roux, J. A., Zank, G. P., Webb, G. M. & Khabarova, O. 2016, *ApJ*, 827, 47

Malova, H. V., Popov, V. Yu., Grigorenko, E. E., *et al.* 2017, *ApJ*, 834, 34

Papaioannou, A., Malandraki, O. E., Dresing, N., *et al.* 2014, *Astron. Astrophys.*, 569, A96, doi: 10.1051/0004-6361/201323336

Wei, F., Hu, Q., Schwen, R. & Feng, X. 2000, *Sci. China Ser. A-Math.*, 43, 629. https://doi.org/10.1007/BF02908775

Zank, G. P., Hunana, P., Mostafavi, P., *et al.* 2015, *ApJ*, 814, 137

Zank, G. P., le Roux, J. A., Webb, G. M., Dosch, A. & Khabarova, O. 2014, *ApJ*, 797, 28

Zank, G., Rice, W. K. M. & Wu, C. 2000, *J. Geophys. Res.*, 105, 25079

Zharkova, V. V. & Khabarova, O. V. 2015, *AnGeo*, 33, 457

Space Weather of the Heliosphere: Processes and Forecasts
Proceedings IAU Symposium No. 335, 2017
C. Foullon & O.E. Malandraki, eds.

doi:10.1017/S1743921317010122

Structure of the Power Spectral Density of Galactic Cosmic Ray Variation during 1953-2016

Pauli Väisänen, Ilya Usoskin and Kalevi Mursula
Space Climate Research Unit, University of Oulu, Finland
email: pauli.vaisanen@oulu.fi

Abstract. Fluxes of galactic cosmic rays (GCR) observed at 1 AU are modulated inside the heliosphere at different time scales. Here we study the properties of the power spectral density (PSD) of galactic cosmic ray variability using hourly data from 31 neutron monitors (NM) from 1953 to 2016. We pay particular attention to the reliability of the used datasets and methods. We present the overall PSD and discuss different parts of the spectrum and the related periodicities. We find significant spectral peaks at the periods of 11 years, 1.75 years, 155 days, 27 days and 24 hours and the harmonics of the latter two peaks. We calculate a power law slope of -1.79 ± 0.13 for the period range between 50 and 130 hours and a slope of -1.34 ± 0.17 for the period range between 40 days and 3.4 years (1 000 – 30 000 h).

Keywords. Space Climate, Cosmic rays, Heliospheric modulation

1. Introduction

The flux of galactic cosmic rays in the heliosphere is modulated by solar activity and, especially, by the changes in the heliospheric magnetic field (HMF). Irregularities in the HMF scatter cosmic ray particles, which is seen as a reduced intensity of GCR observed at Earth. These changes are monitored by the NMs which measure the secondary nucleons generated by the interaction of GCR with the Earth's atmosphere.

A common way to study the temporal variability and possible periodicities of a measured quantity is to examine its power spectral density (PSD), which shows the amount of power distributed at different frequencies. In addition to spectral peaks generated by systematic periodicities, the PSD typically exhibits a continuous power law behaviour which can be associated with turbulence, self-organized criticality and/or stochastic processes.

In this study we compute the PSDs of NM data using multiple NM stations. These PSDs are then analysed for different periodicities and power law behaviour.

2. Data, methods and parameters

We use NM data from 31 globally distributed stations listed in Figure 1 spanning the years 1953-2016, which correspond to roughly six solar cycles. Principles of the NM as a cosmic ray detector can be found elsewhere (e.g., Moraal *et al.* 2000; Simpson 2000; Usoskin *et al.* 2005). The full list of NMs and their operation periods is available at http://cidas.isee.nagoya-u.ac.jp/WDCCR/datacoverage_num.html.

Data were mostly obtained through the IZMIRAN data site at http://cr0.izmiran.ru/common/links.htm, which contains full-length datasets of NM data for most stations, containing also data from closed stations. Outliers, discontinuities and GLE events were removed using a median filter and visually examining the data. The median NM

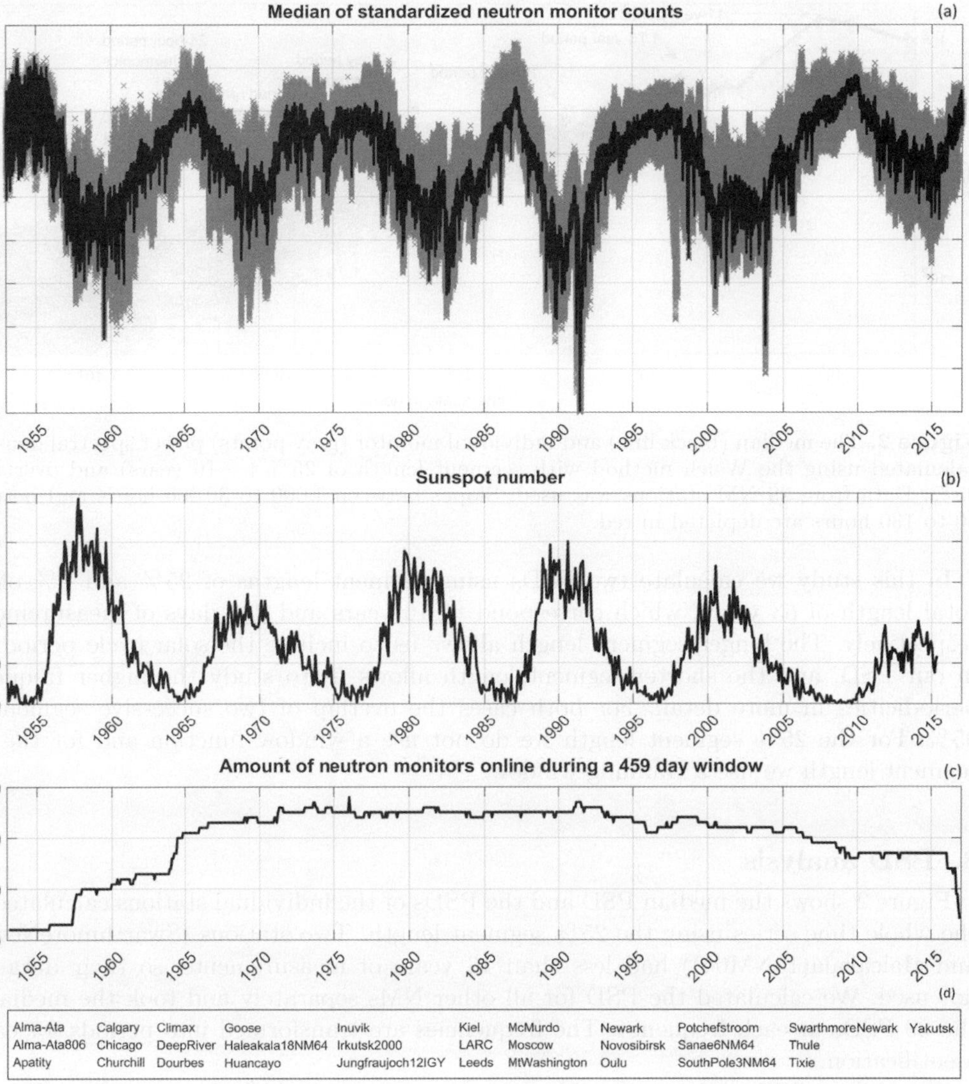

Figure 1. [a]: Median value (black line) of standardized count rates of 31 neutron monitors listed in panel [d]. [b]: Monthly sunspot number. [c]: Number of operating monitors during a 459-day (=17 solar rotations) time window.

count calculated from the standardized counts of all 31 NMs and the number of stations operating at each time are shown in Figure 1.

All the used datasets are linearly interpolated, detrended and standardized before calculating the PSD. For the PSD calculation we used the FFT with Welch method of averaging periodograms. This is a good PSD estimation method for data that includes data gaps (Munteanu *et al.* 2016). FFT transform of the signal is also computationally very fast. In the Welch method, the time series is first split into multiple, partly overlapping segments, and FFT and PSD are calculated for each segment. Then the segment PSDs are finally averaged to obtain a PSD with reduced noise. However, this process loses information at the low frequencies of the signal, depending on the used segment length.

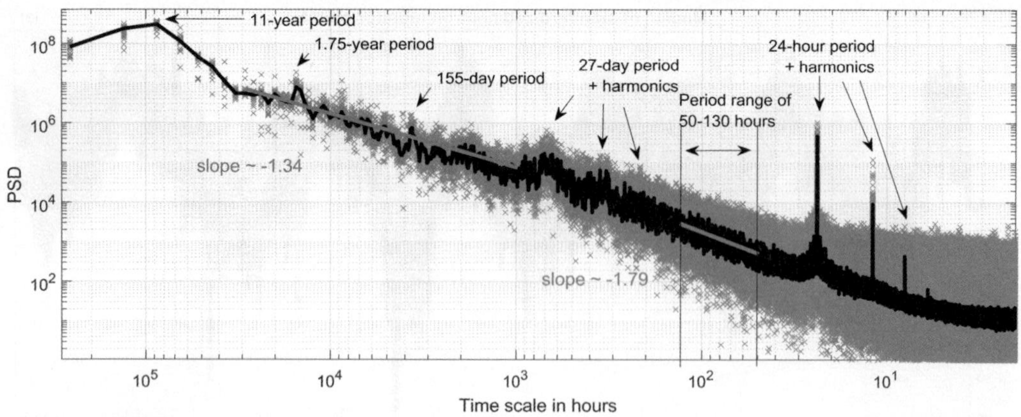

Figure 2. The median (black line) and individual monitor (gray points) power spectral densities calculated using the Welch method with segment length of 25 % ($\sim$ 16 years) and overlap of 95 %. Data from 29 NM stations was used. Slopes between 1 000 to 30 000 hours and between 50 to 130 hours are depicted in red.

In this study we calculate two PSDs using segment lengths of 25 % and 1 % of the total length of 63 years, which correspond to 16 years and 230 days of measurements, respectively. The longer segment length allows us to include the solar cycle periodicity in our PSD, and the shorter segment length allows us to study the higher frequency periodicities in more detail. For both cases the overlap of two successive segments is 95 %. For the 25 % segment length we do not use a window function and for the 1 % segment length we use a Hanning window.

3. PSD analysis

Figure 2 shows the median PSD and the PSDs of the individual stations calculated for the whole time series using the 25 % segment length. Two stations (SwarthmoreNewark and Haleakala(18NM64)) had less than 16 years of measurements, so their data was not used. We calculated the PSD for all other NMs separately and took the median of the 29 PSDs at each frequency. The frequencies are transformed into periods for easier identification.

The highest PSD peak is observed at the 11-year mark, which corresponds to the solar cycle. The solar cycle is the most prominent driver of change in the space conditions (space climate), and many other solar periodicities are caused or influenced by it.

A midterm quasiperiodicity is observed at around 1.75 years, identified earlier in some other studies on cosmic ray data (Valdés-Galicia *et al.* 1996), as well as in solar wind speed and geomagnetic activity (Mursula *et al.* 2003).

The Rieger periodicity, which has been suggested to be connected to the Rossby waves in the tachocline (Zaqarashvili *et al.* 2010) is observed at around 155 days. A weak second harmonic at 77 days is also visible.

Another solar activity related peak is at the 27-day period of solar rotation (see, e.g., Gil and Mursula, 2017, and references therein). Differential rotation and the evolution of active regions and coronal holes on the solar surface cause the 27-day peak to be wide. The 13.5-day second harmonic of the solar rotation can also be clearly observed, and, less clearly, also the 9-day third harmonic. The diurnal variation, which is stronger than the solar rotation period, is seen at the 24-hour mark. The 2nd and 3rd harmonics at 12 and 8 hours are also clearly noticeable.

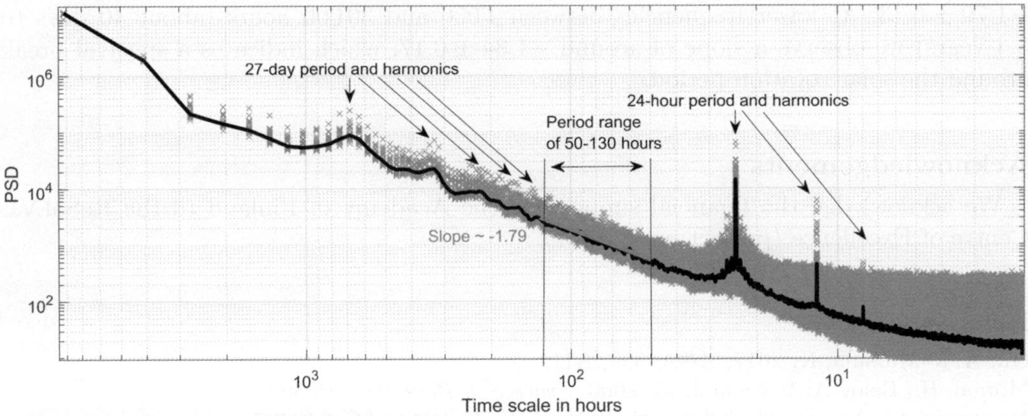

Figure 3. The median (black line) and the individual (gray points) power spectral densities calculated using the Welch method with the segment length of 1 % ($\sim$ 230 days), overlap of 95 % and a Hanning window using all 31 NM stations. Slope of -1.79 (between 50 to 130 hours) is depicted in red.

All the identified main peaks are significant at the confidence level better than 95%, determined by comparing the lower 95 % confidence bound of the peak PSD value to the upper confidence bound value of points before and after the peak.

At time scales less than 10 hours we can observe the PSD flatten to white noise.

Figure 2 also outlines the time scales between 50 and 130 hours which exhibits a range of periods least affected by power peaks and a clear power-law behaviour. This part of the spectrum is between the harmonics of the 27-day peak and the diurnal peak. The power of this period was earlier studied, e.g., by Kudela *et al.* (1991). We calculate the PSD slope for each monitor and obtain a mean slope of -1.79 and the standard deviation of ± 0.13 from the individual slopes. At the low frequency part of the spectrum between 1000 and 30000 hours we calculate a slope of -1.34 with a standard deviation of ± 0.17, which indicates a spectral break around the solar rotation period. This lower frequency part of the spectrum is, however, affected by spectral peaks.

Figure 3 gives a closer look at the high-frequency part of the power spectrum. Using a segment of 1 % of the total time, we obtain an smoother spectrum than in Figure 2. All the 31 NMs are now used. The spectrum is quite smooth below the 200-hour mark, but small peaks at the fourth and fifth harmonics of solar rotation period at 164 and 132 hours are the reason why 130 hours is used here as the lower cut-off frequency for this part. At the higher frequency end of this part we observe an increase of the spectrum towards the diurnal peak, which sets the selected 50 hours as the higher frequency cut-off for this part. For this PSD between 50 and 130 hours we calculate a mean slope of -1.79 ± 014.

4. Conclusions

We have calculated the power spectral densities of neutron monitor counts from 31 stations that span the years 1953-2016. Two power spectral densities were calculated, one covering all periods up to 16 years, the other for high frequencies above one year.

The strongest peaks in the power spectral densities are the 11-year, 1.75-year, 155-day, 27-day and 24-hour peaks. The second and third harmonics of 27-day and the 24-hour periodicities are also clearly present in the spectrum. The most clear power-law is at frequencies corresponding to time scales between 50 and 130 hours, where the slope was

-1.79 ± 0.13. At lower frequencies between 1 000 and 30 000 hours (about 40 days to 3.4 years) we observe a slope of around -1.34 ± 0.17, which indicates a spectral break around the solar rotation period.

Acknowledgements

We acknowledge the financial support by the Academy of Finland to the ReSoLVE Centre of Excellence (project no. 272157).

References

Gil, A. & Mursula, K. 2017, *A&A*, 599, A112

Moraal, H., Belov A. & Clem J. M. 2000, *Space Sci. Revs*, 93, 285-303

Munteanu, C., Negrea, C., Echim, M. & Mursula, K. 2016, *AG*, 34, 437

Mursula, K., Zieger, B. & Vilppola, J. H. 2003, *Solar Phys.*, 7, 123

Kudela, K., Ananth, A. G. & Venkatesan, D. 1991, *J. Geophys. Res.: Space Physics*, 96, 2156

Simpson, J. A., 2000, *Space Sci. Revs*, 93, 11-32

Usoskin, I. G., Alanko-Huotari, K., Kovaltsov, G. A. & Mursula, K., 2005 *J. Geophys. Res.*, 110, A12108

Valdés-Galicia, J. F., Pérez-Enríquez, R. & Otaola, J. A. 1996, *Solar Phys.*, 167, 409

Zaqarashvili, T. V., Carbonell, M., Oliver, R. & Ballester J. L. 2010, *ApJ*, 709, 749

Space Weather of the Heliosphere: Processes and Forecasts
Proceedings IAU Symposium No. 335, 2017
C. Foullon & O.E. Malandraki, eds.

doi:10.1017/S1743921317007268

Testing Models of the Fast Solar Wind using Spectroscopic and In Situ Observations

Andrzej Fludra[1] **and Enrico Landi**[2]

[1]RAL Space, STFC Rutherford Appleton Laboratory, Didcot, UK
email: andrzej.fludra@stfc.ac.uk

[2]University of Michigan, Ann Arbor, USA
email: elandi@umich.edu

Abstract. We present a new technique to study joint observations of EUV spectral line intensities and in situ charge states of the fast solar wind. We solve the time-dependent equation for ionization and recombination for a chosen element and calculate the charge state evolution along the open magnetic fields for elements such as C, O, Ne, Mg, Si and Fe. Comparing predicted spectral lines intensities above the limb and in situ charge states to observations from SOHO/SUMER and Ulysses/SWICS, we test how well the modelled thermodynamic parameters of the solar wind reproduce observations. We outline the application of this method to Solar Orbiter data.

Keywords. Sun: solar wind, Sun: UV radiation, methods: numerical

1. Introduction

Understanding the sources and acceleration mechanisms of the solar wind remains one of the major unsolved problems in solar physics. Further progress can be achieved by making joint observations with high-resolution EUV spectrometers and in situ heavy ion sensors that provide measurements of several ionization stages of heavy ions observed near the solar surface and in the heliosphere. The combination of these two different observables provides means of testing solar wind model predictions. In this paper we present a method of combining spectroscopic and in situ data sets, summarize results obtained from the application of this method to SOHO/SUMER and Ulysses/SWICS data and three solar wind models, and outline its application to future observations from the Solar Orbiter SPICE and SWA/HIS instruments.

2. Method

The method, developed by Landi *et al.* (2012a), consists of 5 steps:

Step 1: Obtain joint observations of high resolution EUV spectral line intensities measured in the low corona and in situ solar wind charge state distribution of the same solar wind plasma parcels.

Step 2: Establish the magnetic connectivity between the solar surface and the location of the in situ sensor. The magnetic field lines are tracked from the spacecraft to the Suns surface using 3D MHD global models (e.g. AWSoM, van der Holst *et al.* 2014) to determine the exact location of the source region of the wind plasma measured in-situ.

Step 3: Select the solar wind models to be tested and extract from them the solar wind parameters along the open magnetic field line: electron temperature, electron density, and the solar wind velocity.

Step 4: Solve ionization balance equations and calculate evolution of charge states. We use the Michigan Ionization Code (MIC - Landi *et al.* 2012b):

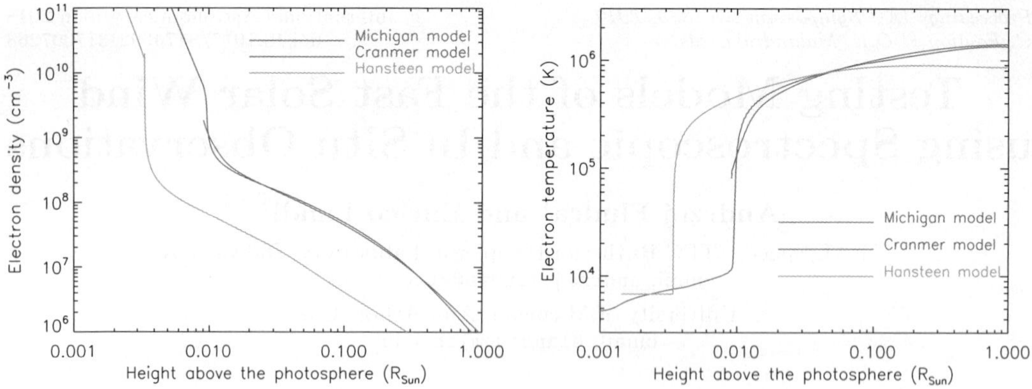

Figure 1. Electron density (left) and temperature (right) from the three models.

$$\frac{\partial y_m}{\partial t} = n_e(y_{m-1}C_{m-1}(T_e) + y_{m+1}R_{m+1}(T_e) - y_m(C_m(T_e) + R_m(T_e))) \tag{2.1}$$

where y_i is the fractional abundance of ionization stage i, and $C_i(T)$ and $R_i(T)$ are ionization and recombination rates from stage i, taken from CHIANTI.

Step 5: Calculate the local spectral line intensities using non-equilibrium MIC ion fractions and electron density and temperatures from the solar wind models; then integrate intensities along the line of sight. Results are compared with observations.

This procedure can also be used to build empirical models of the solar wind. In this case, all 5 steps of the procedure are repeated, except that in step 3, rather than using a theoretical model, we adopt user-selected wind parameters as a function of position along the magnetic field line. The comparison with observations provides information on what needs to be changed in the solar wind parameters: these are modified, and fed again into the procedure at step 4 until satisfactory agreement with observations is obtained.

3. Application to SOHO and Ulysses data

Landi *et al.* (2014) applied this procedure to compare predictions from three models of the fast solar wind to high resolution spectra recorded by SOHO/SUMER at the north polar limb and Ulysses/SWICS fast wind charge states. They considered the following models: (a) Hansteen and Leer (1995) - 1D, single temperature, ad-hoc heating function; (b) Cranmer *et al.* 2007 (coronal hole) - 1D, single temperature, wave-based plasma heating and wind acceleration; and (c) the AWSoM model (van der Holst *et al.* 2014) - 3D, 2-temperature, Alfven-wave plasma heating and wind acceleration. Predicted electron temperature and density are shown in Figure 1.

Results of steps 3, 4 and 5 in Figure 2 show that none of the models can fully reproduce observations. The Cranmer and Michigan models provide the best fit, with similar levels of disagreement. Possible sources of discrepancies include: (a) solar wind source region located in the corona rather than in the chromosphere, (b) presence of photoionization, (c) presence of supra-thermal electrons, (d) geometrical effects in the line-of-sight integration.

4. Application to Solar Orbiter

An unprecedented opportunity to extend this study to new data will become available after the launch of the Solar Orbiter mission (Müller *et al.* 2013). The spacecraft will travel close to the Sun (0.28 R_s at perihelion), climb out of the ecliptic up to 32 degrees, and carry out remote observations of the Sun and in situ measurements of fields and

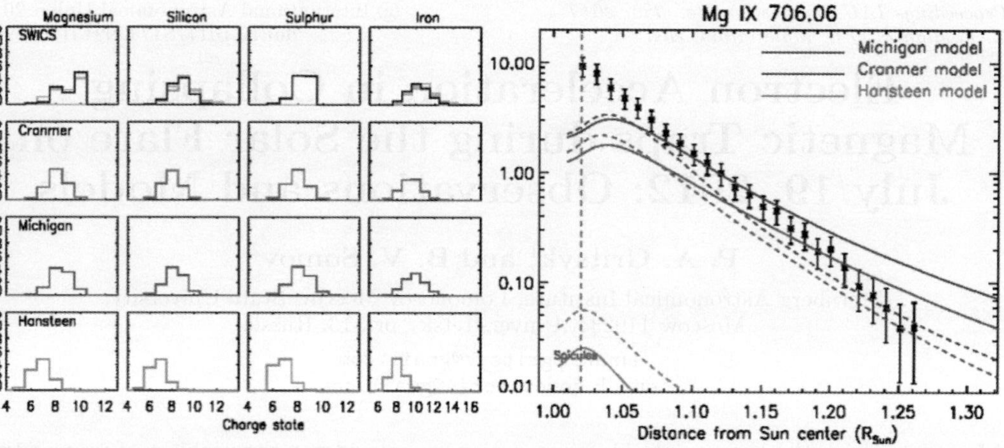

Figure 2. (a) Left: measured (SWICS, upper panel) and predicted charge state distributions. (b) Right: Full lines are intensities (phot cm^{-2} s^{-1} arcsec^{-2}) predicted using non-equilibrium ion fractions from MIC. Dashed lines: intensities predicted assuming ionization equilibrium. Error bars are observations from SUMER.

particles. The two instruments important for the extension of this study are the SPICE spectrometer (Fludra *et al.* 2013) and SWA/HIS ion sensor.

SPICE will measure 2D high resolution spectra and spectral images in the 70.4 - 79.0 nm and 97.3 - 104.9 nm ranges, which include emission lines from ions of H, C, O, N, Ne, S, Mg, Si and Fe, formed at temperatures from 10,000 K to 10 MK. SPICE will obtain the first ever out-of-ecliptic spectral observations of polar regions.

The SWA Heavy Ion Sensor (HIS) will measure the properties of ions heavier than He in the solar wind and interplanetary CMEs. The HIS instrument is a top-hat mass spectrometer capable of measuring energy, energy per charge, time-of-flight, azimuth angle, and elevation angle. The ions measured by HIS are He^{1-2+}, C^{2-6+}, N^{2-7+}, O^{2-8+}, Ne^{6-9+}, Mg^{6-12+}, Si^{6-12+}, S^{6-14+}, Fe^{3-24+}.

Several elements and ions will be observed jointly by both instruments. Therefore, when SPICE and SWA/HIS are used together, SPICE intensities will test the solar wind model predictions in the transition region and low corona, while HIS will test predictions of charge states in situ as described in Section 2 and 3. In addition, SPICE and HIS can measure the low-FIP/high-FIP abundance ratios (e.g., Si/C, Fe/O, S/N, Mg/Ne). Comparing SPICE abundance ratio maps with in situ SWA/HIS abundances may help confirm the magnetic connectivity of the solar wind observed at the location of the spacecraft to the wind sources on the Sun.

References

Cranmer, S. R., van Ballegooijen, A. A. & Edgar, R. J. 2007, *ApJS*, 171, 520
Fludra, A., *et al.* 2013, Proceedings of the SPIE Optical Engineering conference: Solar Physics and Space Weather Instrumentation V, Vol. 8862, 88620F
Hansteen, V. H. & Leer, E. 1995, *JGR*, 100, 21577
Landi, E., *et al.* 2012a, *ApJ*, 750, 159
Landi, E., *et al.* 2012b, *ApJ*, 744, 100
Landi, E., *et al.* 2014, *ApJ*, 790, 111
Müller, D., *et al.* 2013, *Sol. Phys.*, 285, 25
van der Holst, B., Sokolov, I. V., Meng, X., *et al.* 2014, *ApJ*, 782, 81

Space Weather of the Heliosphere: Processes and Forecasts
Proceedings IAU Symposium No. 335, 2017
C. Foullon & O.E. Malandraki, eds.

doi:10.1017/S1743921317008912

Electron Acceleration in Collapsing Magnetic Traps during the Solar Flare on July 19, 2012: Observations and Models

P. A. Gritsyk[1] and B. V. Somov[2]

Sternberg Astronomical Institute, Lomonosov Moscow State University,
Moscow 119234, Universitetsky pr., 13, Russia
[1]email: pgritsyk@gmail.com
[2]email: somov.boris@gmail.com

Abstract. Using the appropriate kinetic equation, we considered the problem of propagation of accelerated electrons into the solar corona and chromosphere. Its analytical solution was used for modelling the M7.7 class limb flare occurred on July 19, 2012. Coronal above-the-loop-top hard X-Ray source was interpreted in the thin-target approximation, the foot-point source - in the thick-target approximation with account of the reverse-current electric field. For the foot-point source we found a good accordance with the RHESSI observations. For the coronal source we also got very accurate estimate of the power-law spectral index, but significant differences between the modelled and observed hard X-ray intensities were noticed. The last discrepancy was solved by adding the coronal magnetic trap model to the thin target model. The former one implies that the trap collapses in two dimensions, locks and accelerates particles inside itself. In our report, we confirm an existence and high efficiency of the electron acceleration in collapsing magnetic traps during solar flares. Our new results represent (e.g. for RHESSI observations) the theoretical prediction of the double step particle acceleration in solar flares, when the first step is the acceleration in reconnection area and the second one – the acceleration in coronal trap.

Keywords. acceleration of particles, Sun: corona, flares, magnetic fields, X-rays

1. Introduction

The acceleration of charged particles to high energies remains a topical problem of modern astrophysics as applied to phenomena different in their scales and nature, for example, solar flares. The latter are of particular interest for two reasons. The first one is the studying and forecasting magnetic storms and other manifestations of space weather, which are directly related with accelerated particles. The second one is the studying the physical particle acceleration mechanisms, because they can be investigated most comprehensively with high spatial, temporal, and spectral resolutions. At present, the achieved accuracy of electromagnetic radiation detectors onboard spacecraft is so high that the observational data allow us to reconstruct a complete picture of particle acceleration and propagation in the solar atmosphere and, thus, to check whether the existing model assumptions are correct. The overall picture and scenario of a solar flare may be considered without exaggeration to have been well studied (Somov 2012; Somov 2013; Krucker *et al.* 2008) – magnetic reconnection plays a decisive role in this manifestation of solar activity.

Let's consider the particles acceleration process during solar flares (Somov & Kosugi 1997). The electrons and ions are pre-accelerated by the electric field in the reconnecting current layer (RCL in Figure 1). After this first step in the acceleration process, they end up in the coronal magnetic trap (trap in Figure 1), whose length and cross-sectional size (thickness) decrease rapidly. In such a collapsing trap the captured particles are

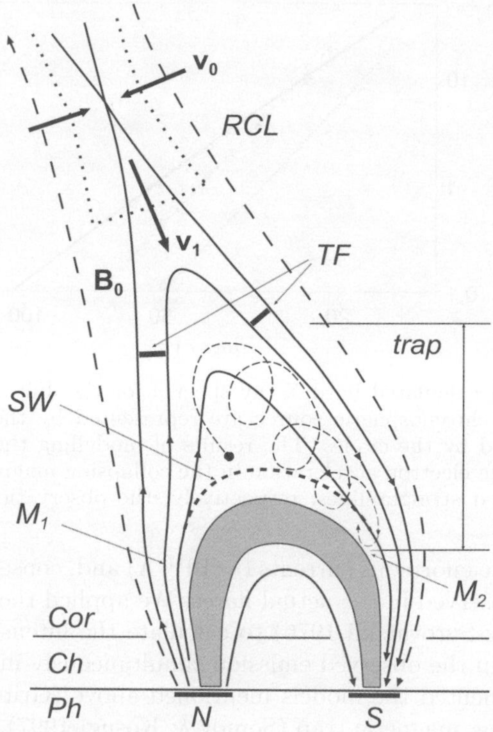

Figure 1. The coronal magnetic trap (trap) between the turbulent front (TF) and the magnetic mirrors (M1, M2). The dashed spiral indicates the trajectory of a trapped electron. SW is the shock wave separating the hot and cold plasmas. RCL is the high-temperature reconnecting current layer, the source of primarily accelerated electrons.

reflected from the shock wave or from the magnetic mirrors M1 and M2 in Figure 1. Inside the collapsing trap the charged particles acquire an additional acceleration through the first-order Fermi mechanism and betatron heating. Such a picture was called double step acceleration in Somov & Kosugi 1997 and has not yet been confirmed by convincing observations of flares, remaining predominantly a theoretical prediction. In this short report we present the new observations, which possible confirm the double step acceleration model.

2. Multi-wavelength observations of the flare and models

The July 19, 2012 flare was observed at the solar limb by the RHESSI, GOES, and SDO spacecraft (Liu 2013) with a high accuracy, making it suitable for investigating the magnitude and efficiency of the possible additional electron acceleration in the corona. The data on the spectrum, locations, and spatial scales of the coronal and chromospheric hard X-ray sources are described in detail in Krucker & Battaglia 2014 and Krucker *et al.* 2015. We calculated the energy flux density of the electrons accelerated in the reconnecting current layer (to fit the intensity of the hard X-ray spectrum observed in the chromosphere footpoint) or, more precisely, the fast electrons escaping from the current layer through the turbulent front (TF) in schematic Figure 1 within the framework of a self-consistent thick-target model with a reverse current (Gritsyk & Somov 2014). The need to consider the reverse current, which compensates for the electric current carried by a bundle of fast moving electrons, is obvious. If it did not exist, bundles of accelerated

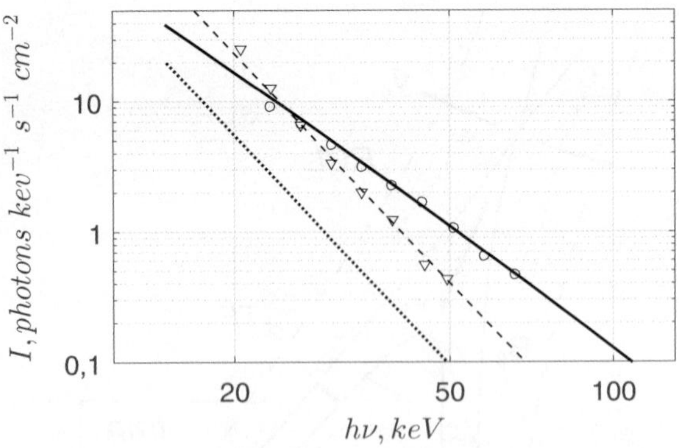

Figure 2. Observed and calculated hard X-ray spectra for the July 19, 2012 solar flare. The results of modelling the chromospheric source are represented by the straight solid line; the observations are indicated by the circles. The results of modelling the coronal source without and with allowance for the electron acceleration in the collapsing magnetic trap are represented by the dotted and dashed straight lines, respectively; the observations are indicated by the triangles.

electrons would generate enormous currents ($\sim 10^{17}$ A) and, consequently, huge magnetic fields, which are not observed in the actual flares. We applied the formalism of the thin-target model (Somov & Syrovatskii 1976) to calculate the intensity of the coronal hard X-ray source. To explain the observed emission simultaneously in the chromosphere and the corona, we supplemented the models mentioned above (Gritsyk & Somov 2017) by the model of a collapsing magnetic trap (Somov & Kosugi 1997).

The solid line in Figure 2 represents the hard X-ray spectrum for the chromospheric source of the July 19, 2012 flare that was calculated in the thick target model with a reverse current. It is clearly seen that the calculated X-ray spectrum in the chromosphere closely coincides with the observed one in both intensity and slope. The spectrum of the coronal source calculated in the thin-target approximation is also presented in Figure 2. Let's discuss the important conclusion of this report: it is fundamentally impossible to model the observed spectra of the coronal and chromospheric hard X-ray sources in terms of the classical flare model. Indeed, the difference of slope indexes of these spectra $\varphi_{Cor} - \varphi_{Ch} = 2$ always in the classical model, while according to the observations of the July 19, 2012 flare, $\varphi_{Cor} - \varphi_{Ch} \approx 1.6$. In the model with a reverse current the observed ratio of the slopes is naturally obtained for the observed flare parameters. In other words, the thick-target approximation with a reverse current not only accurately describes the X-ray spectrum of the chromospheric source, but also allows the slope of the spectrum in the corona to be properly estimated. The intensity of the coronal source (the dotted line in Figure 2), which is lower in the model than the observed one by a factor of ≈ 4.5, constitutes an exception. In our opinion, such a marked difference between the calculated and observed intensities, while the spectral slopes closely coincide, is a weighty argument for the idea of the existence of particle acceleration and its high efficiency in the collapsing magnetic trap formed by convergent magnetic field lines in the upper part of the flare loop (Figure 1). Such an observational picture was predicted theoretically by Bogachev & Somov (2007), but in the absence of accurate space experiments it has had no convincing confirmation until now.

3. Conclusions

Today, the study of solar flares in the context of space weather is of great practical importance. The key point in this issue is the process of primary energy release and related with it acceleration of charged particles to high energies. For studying the role of different acceleration mechanisms, we considered a model for the limb flare of July 19, 2012, for which highly accurate satellite observations (primarily from the RHESSI spacecraft) are available. This flare was chosen for our modelling due to the presence of a bright coronal hard X-ray source, which was observed synchronously with the sources located in the chromosphere near the flare loop footpoints. The flare model is based on the analytical methods proposed in earlier papers (Syrovatskii & Shmeleva 1972; Somov & Kosugi 1997; Somov & Bogachev 2003; Gritsyk & Somov 2014). The classical flare model based on the thick-target approximation was shown to be inapplicable for the description of solar flares with a coronal X-ray source, because it incorrectly predicts the slope of its spectrum.

Our attention was focused on investigating the coronal hard X-ray source that, in our opinion, is attributable to the existence of collapsing magnetic traps produced by magnetic reconnection. Convincing arguments for the existence of coronal traps and their decisive role in the particle acceleration in solar flares constitute the main result of this report. This result was obtained from general assumptions about the pattern of particle acceleration during solar flares, with the main one being the assumption about the existence of double step electron acceleration (Somov & Kosugi 1997). Owing to the existing and prospective space experiments, one might expect new confirmations of our conclusions.

The authors thank the referee for helpful comments. We also thank the Symposium Organizing Committee, International Astronomic Union and cosponsors for support and help. The work was also supported by Russian Foundation for Basic Research (16-02-00585A).

References

Bogachev, S. A. & Somov, B. V. 2007, *Astron. Lett.*, 33, 54

Gritsyk, P. A. & Somov, B. V. 2014, *Astron. Lett.*, 40, 499

Gritsyk, P. A. & Somov, B. V. 2017, *Astron. Lett.*, 43, 614

Krucker, S. & Battaglia, M. 2014, *ApJ*, 780, 107

Krucker, S., Battaglia, M., Cargill, P. J., Fletcher, L., Hudson, H. S., MacKinnon, A. L., Masuda, S., Sui, L., *et al.* 2008, *A & AR*, 16, 155

Krucker, S., Saint-Hilaire, P., Hudson, H. S., Haberreiter, M., Martinez-Oliveros, J. C., Fivian, M. D., Hurford, G., Kleint, L., *et al.* 2015, *ApJ*, 802, 19

Liu, R. 2013, *MNRAS*, 434, 1309

Somov, B. V. 2012, *Plasma Astrophysics. Part I: Fundamentals and Practice* (Springer Science, New York), p. 498

Somov, B. V. 2013, *Plasma Astrophysics, Part II: Reconnection and Flares* (Springer Science, New York), p. 504

Somov, B. V. & Bogachev, S. A. 2003, *Astron. Lett.*, 29, 621

Somov, B. V. & Kosugi, T. 1997, *ApJ*, 485, 859

Somov, B. V. & Syrovatskii, S. I. 1976, *Sov. Phys. Usp.*, 19, 813

Syrovatskii, S. I. & Shmeleva, O. P. 1972, *Soviet Astron.*, 16, 273

Space Weather of the Heliosphere: Processes and Forecasts
Proceedings IAU Symposium No. 335, 2017
C. Foullon & O.E. Malandraki, eds.

doi:10.1017/S174392131800056X

A Behavioural Model of the Solar Magnetic Cycle

Milton Munroe
Retired Academic, High Wycombe, UK
email: MMunroe300@aol.com

1. Introduction

All recent models of solar magnetic cycle behaviour assume that the Ω-effect stretches an existing poloidal magnetic field into a toroidal field using differential rotation (Featherstone and Miesch 2015). The α-effect recycles the toroidal field back to a poloidal field by convection and rotation and this is repeated throughout the cycle. Computer simulations based on that conceptual model still leave many questions unanswered. It has not resolved where the solar dynamo is located, what it is or what causes the differential rotation which it takes for granted. Does this paradigm need changing? The conceptual model presented here examines the sun in horizontal sections, analyses its internal structure, presents new characterizations for the solar wind and structures found and shows how their interaction creates rotation, differential rotation, the solar dynamo and the magnetic cycle.

2. Model Inputs and Interpretations

2.1. *Ions Heavier Than H^+ Travel Faster*

Researchers report that heavier positive solar wind ions (He^{+2}, O^{+6}, etc.) travel faster than the bulk plasma and Pierrard (2012) has derived a kinetic model for their velocity distribution. The data from Pierrard can be approximated by the expression

$$u^2 = U^2 - \frac{A - GMm}{2\pi m r} \tag{1}$$

where A = force of expansion, u = velocity of ion, m = mass of ion, r = distance from the sun, U = final velocity, G = Gravitational Constant, M = mass of the sun. Estimated values: U $\approx 5\text{x}10^5$ m/s, the Heliopause is $\approx 2.8\text{x}10^9$ m from the sun, A $\approx 1.2\text{x}10^{-5}$ Nm^2, GMm $\approx 2.21\text{x}10^{-7}$ Nm^2, A $\approx$ 55 x GMm. The solar wind would therefore easily expand into space against gravity.

2.2. *Electric Current in the Solar Wind*

Since heavier positive ions travel faster than the bulk plasma, the solar wind is an electric current with positive ions as the charge carriers. The sun is therefore a current source and no "return path" is required or necessary for this ion current to flow. Current flow is ideally radial from all points on the solar surface. The electric current density $\boldsymbol{J}_r$ at distance $\boldsymbol{r}$ from the Sun is

$$J_r = \sum_{i=1}^{n} \rho_i \delta v_i \text{ (summed across all heavy ion species; >50 ion species)} \tag{2}$$

where ρ_i is the volume charge density of heavy ion species "i" (<10% by volume cumulatively) and δv_i is the velocity difference between the heavy ion species "i" and the bulk plasma.

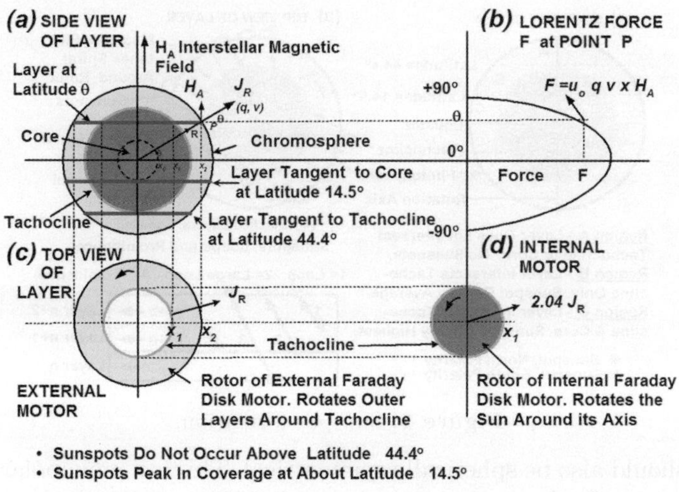

Figure 1. Typical Layer with 2 FD motors.

2.3. *The Solar Wind Rotates the Sun as Stacked Layers of Faraday Disk Motors*

Busby and Mason (2004) report that contours of constant angular velocity lie roughly on lines of constant latitude throughout the convection zone. The sun is sliced into horizontal layers, since all points on each layer rotate at the same speed throughout the zone. Ions which become the solar wind are assumed to maintain the electric current from the core to the surface. Photons are estimated to take about 175000 years to cross the 300000 km wide convection zone. Helium and Hydrogen ions may also take a comparable time. Electrons at this velocity create a current density of about 8.5x10^5 Amps/m^2 in 20 gauge copper wire. Since the sun is in the interstellar magnetic field, $\boldsymbol{H}_A$ (N-pole upwards), the ion current interacts with $\boldsymbol{H}_A$ and Lorentz forces rotate the layer anti-clockwise. Analysis shows that layers which intersect the tachocline contain 2 concentric Faraday Disk (FD) electric motors with 360° wrap-around brushes and the ions exit the solar surface to become the solar wind. Figure 1 shows a layer in the northern hemisphere which intersects the tachocline and identifies the 2 concentric FD motors embedded in that layer. These motors can be "run" to determine system performance.

2.4. *Rotation and Differential Rotation*

The Internal FD motor rotates the entire sun about the vertical axis and has as its rotor the radiative zone. The maximum torque (at equator) per unit area of solar surface is

$$T_{Max} = \pi\mu_0(H_D + H_A)R^3 J_r \text{ n.m/m}^2 \tag{3}$$

$\boldsymbol{H}_A$ is the ambient magnetic field, $\boldsymbol{H}_D$ is the dipole magnetic field created by the External FD motor and $\boldsymbol{J}_r$ is the current density of the solar wind at the solar surface. The External FD motor rotates the outer layers differentially around the tachocline. Its rotor extends from the tachocline to the solar surface. The latitude-dependent torque on an External FD motor is

$$T_\theta = 0.51\pi\mu_0 H_A R^3 J_r \cos^2\theta \text{ n.m/m}^2 \tag{4}$$

2.5. *The Solar Magnetic Field*

The apparent solar magnetic field has three components: (1) The ambient (interstellar) magnetic field $\boldsymbol{H}_A$, (2) A monopole component $\boldsymbol{H}_M$. Being an electric current, the solar wind carries its own magnetic field. Since the current is spherically symmetrical, its

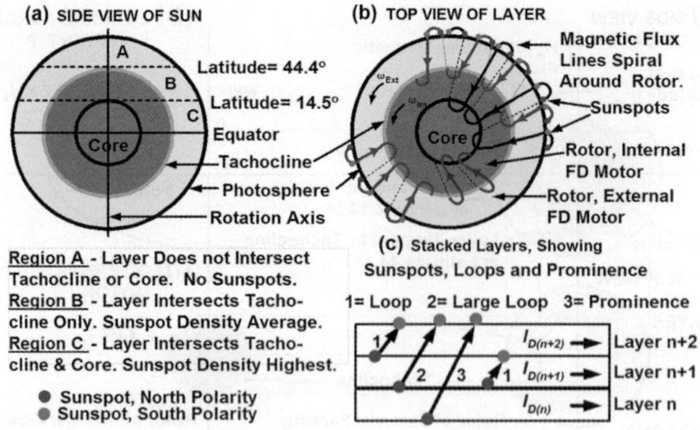

Figure 2. Sunspot formation.

magnetic field should also be spherically symmetrical. This is a monopolar magnetic field. The Ulysses spacecraft observed this (Schatten, 1999) and Dirac magnetic monopoles were observed in the laboratory (Ray et al, 2014). This shows that the monopolar magnetic field exists. This model, supported by the Ulysses data, shows that the sun is a special case and a term must be added to Maxwell's fourth equation to reflect that $\nabla.B \neq 0$. (3) A dipole component $\boldsymbol{H}_D$ created by rotation and induction and in accordance with Lenz's law opposes $\boldsymbol{H}_A$.

The resultant magnetic field has 4 sectors of alternate magnetic polarity separated by current sheets. It is formed in the vertical plane, but centrifugal forces and gravity twist it into a near-horizontal plane. This distortion persists throughout as folds in the heliospheric current sheet.

3. The System

3.1. *The Solar Magnetic Cycle, the Solar Dynamo and Sunspot Formation*

The Internal FD motor has as its magnetic field the dipole field of the External FD motor. Feedback produces oscillations in the speeds of the FD motors. Induced and conduction currents in the rotors also follow the harmonic oscillations. The operating magnetic fields of the FD motors form the solar magnetic cycle. The induced and conduction currents in the system across all the FD motors form the "Solar Dynamo". Figure 2 shows that sunspots are formed where the magnetic flux lines on the rotor of the External FD motor break the surface. Magnetic flux lines are spirals since the medium itself is moving.

3.2. *The Fast and Slow Solar Winds and the Solar Corona*

Solar wind ions exiting at the poles travel along open monopole field lines as the Fast Solar Wind. Ions exiting at lower latitudes become trapped in the dipole to form the corona and accumulate there until their gas pressure exceeds the magnetic pressure of the dipole, when they escape to form the Slow Solar Wind. The loss of kinetic energy heats up the corona to about $2\text{x}10^6$ K. Majahan *et al.* (2002) have also proposed this. The estimated heat capacity of the solar wind is $2\text{x}10^4$ to 10^5 J/kg.K. Trapping of solar wind ions by the dipole explains the shape and temperature of the corona and the greater heavy ion content of the Slow Solar Wind.

4. Discussion and Conclusions

4.1. *Implications of the Model*

If there were no heavy ions in the solar wind, the sun would have no magnetic field and no corona. Stars of this type would group by gravity (Globular Clusters). If there were no ambient magnetic field, the sun would not rotate and would be an ideal magnetic monopole. If the ambient magnetic field were larger then the magnetic cycle would be longer and deeper, enabling ions trapped in the corona to reach higher gas pressures. When the dipole field is decreasing, the gas pressure in the corona can exceed the magnetic pressure of the dipole, leading to a coronal event (CME or even Supernova). An FD motor would only reach equilibrium speed when $\boldsymbol{H}_D \approx \boldsymbol{H}_A$. If the sun were at the centre of the Galaxy where billions of stars contribute to $\boldsymbol{H}_A$, the sun would accelerate for billions of years without being able to reach equilibrium speed, since it cannot create a sufficiently large magnetic field to balance $\boldsymbol{H}_A$. Centrifugal forces would progressively flatten the convective zone followed by the radiative zone until the inner core is exposed. Immense magnetic forces would anchor the flattened tachocline and corona to the core and nuclear radiation and products would exit directly into space at the exposed poles (Quasars, Pulsars, Black Holes). Astronomers are searching for planets like ours in the Goldilocks zone of other stars. They should also consider whether the star is soon to become a Supernova.

4.2. *Conclusions*

This conceptual model of the solar magnetic cycle provides a sound and useful basis for further investigations. Starting from only first principles and with published data on the internal structure of the sun and composition of the solar wind, it predicts many aspects of behaviour that are characteristic of the sun and solar wind. The model shows: (1) heavier positive ions in the solar wind travel faster because of their greater mass; (2) the solar wind is an electric current of positive ions, with implications for Earth and space; (3) the solar magnetic field is an active field and not derived from a fossil field, with implications for the α- and Ω-effects; (4) the solar magnetic field has a monopole component, with implications for Maxwell's fourth equation.

Specifically, the model provides reasoned explanations for: (1) the rotation, differential rotation and cyclic magnetic activity of the sun and solar-type stars; (2) the Hale Magnetic Cycle and the solar dynamo; (3) the nature of sunspots. This model also accounts for (1) the empirically determined Gnevyshev – Ohl rule and (2) why the surface of the sun rotates slower than the interior and that a special case of this effect accounts for the anti-solar type rotation found in some stars (where the poles rotate faster than the equator; Featherstone & Miesch 2015).

Positive electricity in the solar wind is another form of electricity. Therefore investigating its properties would (1) improve the understanding of solar-terrestrial relations, (2) enable better protection from the potential negative effects on humans and conventional electric and electronic systems, and (3) potentially enable harnessing the solar wind as an energy source.

References

Busby, P. & Mason, J. 2004, *A&G*, 45, 47

Featherstone, N. A. & Miesch, M. S. 2015, *ApJ*, 804, 67

Majahan, S. M., Miklaszewski, R., Nikol'skaya, K. I. & Shatashvili, N. L. 2002, *Adv. Space Res*, 30, 571

Pierrard, V. 2012, in M. Lazar (ed), *Exploring the Solar Wind (In Tech)*, Ch.10

Ray, M. W., Ruokokoski, E., Kandel, S., Möttönen, M. & Hall, D. S. 2014, *Nature*, 505, 657

Schatten, K. H. 1999, in J. L. Burch, R. L. Carovillano, S. K. Antiochos (eds), *Sun-Earth Plasma Connections, Geophys Monogr Ser*, 109, 129

Space Weather of the Heliosphere: Processes and Forecasts
Proceedings IAU Symposium No. 335, 2017
C. Foullon & O.E. Malandraki, eds.

doi:10.1017/S1743921317010134

The Aging Solar Wind: a Break in Wind Evolution at Older Ages?

Dúalta Ó Fionnagáin and Aline A. Vidotto
School of Physics, Trinity College Dublin,
College Green, Dublin 2, Ireland
email: ofionnad@tcd.ie

Abstract. The current solar wind is well studied from remote observations and in situ measurements. However, we have very little information of the solar wind as it has evolved. We investigate the evolution of the solar wind by modeling the winds of solar analogues. By using X-ray temperatures as proxies for wind temperatures, we find that a break in behaviour occurs. At 2 Gyr there is a sharp decline in coronal temperatures, which results in a steep decay in mass loss rates for older stars. As the wind is responsible for stellar spin down, through angular momentum loss due to magnetised winds, our results suggest a decline in angular momentum loss for older stars. This agrees with recent observations which find anomalously high rotation rates in older stars. We also find that this evolution in the wind has adverse effects on the Earth's magnetosphere, with an Earth aged 100 Myr having a magnetosphere 3 Earth radii in size.

Keywords. Late-type stars; stars: winds, outflows; Sun: solar wind

1. Introduction

Winds are known to carry away mass and angular momentum from stars, causing them to spin down. These magnetised winds from solar analogues are assumed to be analogous to the aging solar wind. As these stars age, various properties evolve over time, such as rotation and magnetic activity. Magnetic stars tend to have more active winds, and as such, spin down faster as they lose angular momentum. van Saders *et al.* (2016) recently showed that a sample of more evolved stars, observed by Kepler, display unusually fast rotation rates. These do not agree with previously defined age-rotation relations (Skumanich 1972), and it is suggested that there is a weakening in magnetic braking of stars after rossby number ≈ 2.

The 'Sun in time' program considered the evolution of solar-type stars, examining properties such as X-ray flux (Güdel 2007), high energy flux (Ribas *et al.* 2005), magnetic activity (Guinan & Engle 2009). This provided inspiration to study stellar winds from solar analogues, and investigate how they evolve to what we see today from our Sun. Our sample is based off this study and shown in Table 1. We use a 1D stellar wind model (Parker 1958) to simulate the winds of solar analogues. We take X-ray derived coronal temperatures (Johnstone & Güdel 2015) and scale these down to find wind temperatures; an important input parameter for our model. The results presented in these proceedings are discussed in depth in Ó Fionnagáin & Vidotto (2018).

2. Break in wind behaviour

Recent works have suggested that there is a break in the rotation-activity behaviour of solar analogue stars at older ages. This includes a break in coronal temperature with

Table 1. Sample of stars used in the present study: This sample is largely similar to that used in The Sun In Time sample Güdel (2007). Values are mostly taken from Güdel (2007); Vidotto *et al.* (2014). The X-ray luminosity of the sun here is considered to be between activity maximum and minimum.

Star	M ($M_\odot$)	R ($R_\odot$)	P_{rot} (d)	Age (Gyr)	$log[L_X]$ (erg/s)	d (pc)
EK Dra	1.04	0.97	2.77	0.12	29.93	34.5
HN Peg	1.10	1.04	4.55	0.26	29.00	17.95
χ^1 Ori	1.03	1.05	4.83	0.5	28.99	186.0
π^1 UMa	1.00	1.00	5	0.5	28.97	14.36
BE Cet	1.09	1.00	12.4	0.6	29.13	20.9
κ^1 Cet	1.03	0.95	9.3	0.65	28.79	9.14
β Com	1.10	1.10	12.4	1.6	28.21	9.13
15 Sge	1.01	1.10	13.5	1.9	28.06	17.69
18 Sco	0.98	1.02	22.7	3.0	26.8	13.9
Sun	1.00	1.00	27.2	4.6	$\approx$27	1 AU
α Cen A	1.10	1.22	30	5.5	27.12	1.34
16 Cyg A	1.00	1.16	35	7.0	26.89	21.1

rotation for older stars, which we use to define our wind temperatures (Ó Fionnagáin & Vidotto 2018). There is also evidence for a break in magnetic energy in solar type stars after 2-3 Gyr, which can be seen in Vidotto *et al.* (2014) (Figure 2 within). Additionally, van Saders *et al.* (2016) found a break in angular momentum loss for stars older than $\sim$ 4 Gyr, Booth *et al.* (2017) found similar behaviour after $\sim$ 2 Gyr, with a steeper drop in X-ray luminosity from these stars. From our 1D hydrodynamical simulations of these solar type stars, we find that a steep decrease in mass loss rate has occurred for the Sun as it aged. Figure 1 shows that the mass loss rate of aging solar type stars breaks around 2 Gyr, leading to a steep decrease of $\dot{M} \propto t^{-3.89}$, compared to the relatively flat $\dot{M} \propto t^{-0.74}$ relation before 2 Gyr.

From our models we can derive the values of wind velocity, density and temperature at the vicinity of an Earth-like orbit. This allows us to calculate the ram pressure impinging on the Earth as it has aged. By equating the ram pressure and the magnetospheric pressure from Earth (Chapman & Ferraro 1930), we can calculate the magnetospheric standoff distance. We find that a young Earth orbiting a young Sun would have a significantly compressed magnetosphere, at an age of 100 Myr displaying a 3 R_E magnetosphere size.

3. Implications of an aging solar wind

The break in wind-age relation that we found has many consequences for the aging solar wind, affecting mass loss rates which indirectly affects angular momentum, planetary environments, Earth's magnetosphere and atmospheric loss:

• Older stars with lower-than-expected base wind temperatures will experience lower mass-loss rates from their winds (Figure 1). Since the wind is responsible for stellar spin down, through the loss of angular momentum in these magnetised winds, it is expected that this will also result in older stars exhibiting much faster rotation than expected from standard age-rotation relationships. This agrees well with asteroseismic observations presented by van Saders *et al.* (2016).

• Although the mass loss rates presented here result in a rapidly increase in mass loss rate in the past, it is not sufficient to account for the discrepancy in solar evolution

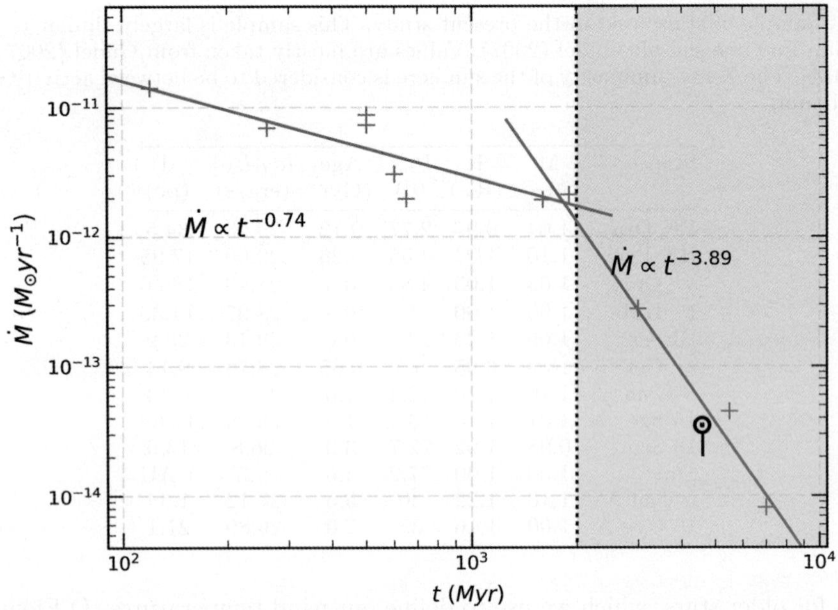

Figure 1. Calculated mass loss rates (*red crosses*) using our 1D wind model shown in this plot describe how the mass loss rate of the Sun would change as it ages. Included in our calculation is our estimation of the current solar mass loss rate ($\odot$) and the range of values calculated directly from observations (*black line*) (Wang 1998). Note the clear break into a rapidly declining mass loss rate regime (*grey shaded region*) at 2 Gyr.

models and evidence for liquid water on Earth in the past, also known as the faint young Sun problem (Feulner 2012). This is discussed further in Ó Fionnagáin & Vidotto (2018) in which we find a total mass loss since 100 Myr of $0.8\%M_{\odot}$.

• As these winds permeate and compose the interplanetary medium, they directly effect exoplanets. Our evolving winds from solar analogues can not only provide information on the environment surrounding exoplanets, but also on the environment around an evolving young Earth. We find that an Earth of 100 Myr possesses a magnetosphere of 3 Earth radii in size.

• We find that this decreased magnetosphere would cause the polar opening angle in the magnetosphere to increase in size dramatically. With closed field regions beginning at 55-60° latitude. This could possibly contribute to a significant boost in atmospheric loss.

Acknowledgements

DÓF acknowledges funding from a Trinity College Postgraduate Award through the School of Physics.

References

Booth, R. S., Poppenhaeger, K., Watson, C. A., Silva Aguirre, V. & Wolk, S. J., 2017, *Mon. Not. Royal Soc.*, 471, 1012

Chapman, S. & Ferraro, V. C. A., 1930, *Nature*, 126, 129

Feulner, G., 2012, *Reviews of Geophysics*, 50, RG2006

Güdel, M., 2007,*Living Reviews in Solar Physics*, 4, 3

Guinan, E. F. & Engle, S. G., 2009, in Mamajek E. E., Soderblom D. R., Wyse R. F. G., eds, IAU Symposium Vol. 258, The Ages of Stars. pp 395–408 (arXiv:0903.4148), doi:10.1017/S1743921309032050

Johnstone, C. P. & Güdel, M., 2015, *A&A*, 578, A129

Ó Fionnagáin, D. & Vidotto, S. S., 2018, *Mon. Not. Royal Soc.*, 476, 2465

Parker, E. N., 1958, *Astrophysical Journal*, 128, 664

Ribas, I., Guinan, E. F., Güdel, M. & Audard, M., 2005, *Astrophysical Journal*, 622, 680

Skumanich, A., 1972, *Astrophysical Journal*, 171, 565

Vidotto, A. A., *et al.* 2014, *Mon. Not. Royal Soc.*, 441, 2361

Wang, Y.-M., 1998, in Donahue R. A., Bookbinder J. A., eds, Astronomical Society of the Pacific Conference Series Vol. 154, Cool Stars, Stellar Systems, and the Sun. p. 131

van Saders, J. L., Ceillier, T., Metcalfe, T. S., Silva Aguirre, V., Pinsonneault, M. H., García, R. A., Mathur, S. & Davies, G. R., 2016, *Nature*, 529, 181

Part 3. Impact of Solar wind, Structures and Radiation on Magnetospheres

Space Weather of the Heliosphere: Processes and Forecasts
Proceedings IAU Symposium No. 335, 2017
C. Foullon & O.E. Malandraki, eds.

doi:10.1017/S1743921317009863

Comparison and Time Evolution of the Geomagnetic Cutoff at the ISS Position: Internal vs External Earth's Magnetic Field Models

Matteo J. Boschini[1,2], Stefano Della Torre[1], Massimo Gervasi[1,3], Davide Grandi[1,3], Giuseppe La Vacca[1,3], Simonetta Pensotti[1,3], Pier Giorgio Rancoita[2], Davide Rozza[1,3] and Mauro Tacconi[1,3]

[1]INFN Sezione di Milano Bicocca, I-20126 Milano, Italy
email: davide.rozza@mib.infn.it
[2]Cineca, Segrate, Italy
[3]Università degli Studi di Milano Bicocca, I-20126 Milano, Italy

Abstract. Our back-tracing code (GeoMagSphere) reconstructs the cosmic ray trajectories inside the Earth's magnetosphere. GeoMagSphere gets the incoming directions of particles entering the magnetopause and disentangles primary from secondary particles (produced in atmosphere) or even particles trapped inside the Earth's magnetic field. The separation of these particle families allows us to evaluate the geomagnetic rigidity cutoff. The model can be used considering the internal symmetric (IGRF-12) magnetic field only, or adding the asymmetric external one (Tsyganenko models: T89, T96 or TS05). A quantitative comparison among these models is presented for quiet (solar pressure $P_{dyn} < 4$ nPa) and disturbed ($P_{dyn} > 4$ nPa) periods of solar activity, as well as during solar events like flares, CMEs. In this analysis we focused our attention on magnetic field data in magnetosphere, from Cluster, and simulated cosmic rays for a generic detector on the ISS as for example AMS-02. We found that high solar activity periods, like a large fraction of the period covering years 2011-2015, are better described using IGRF+TS05 model. Results, i.e. the average vertical rigidity cutoff at the ISS orbit, are shown in geographic maps of $2° \times 2°$ cells.

Keywords. Cosmic rays, Earth's magnetic fields, Sun activity

1. Introduction

Cosmic rays (CRs) arriving at Earth interact with the geomagnetic field. At the International Space Station (ISS) orbit, depending on the geographic position, a combination of primary, secondary and trapped particles is present; a quick description regarding these families can be found in e.g. Bobik *et al.* (2006). Primary particles are CRs arriving from the space outside the magnetosphere (the region affected by the Earth's magnetic field). Secondary particles, observed at ISS altitude ($\sim$ 400 km above the surface), are produced mainly by primary CRs interactions in the atmosphere. A fraction of secondaries can be trapped in it increasing the population of the radiation belts. Trajectories of charged particles inside the magnetosphere depend on particle velocities and magnetic field strength. For a specific geographic region and incoming direction it is possible to detect particles, depending on their rigidities, coming from outside the magnetosphere (primaries), from the atmosphere (secondaries) or from the radiation belts (trapped). At each rigidity it is possible to determine the origin of these particles. The transition between secondary+trapped particles and primary ones is not sharp and it is called

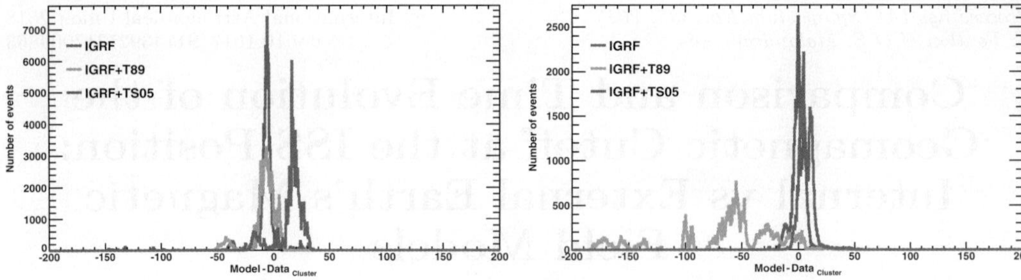

Figure 1. Amount of data as a function of the difference, regarding the component B_x of the magnetic field, between the model predictions (IGRF alone in red, IGRF+T89 in green and IGRF+TS05 in blue) and the Cluster satellite measurements during 2 - 7 November 2011 (left) and 5 - 9 March 2012 (right).

penumbra (see e.g. Cooke *et al.* 1991, Bobik *et al.* 2006). The rigidity value of the highest detected secondary particle is called *upper* (R_U) rigidity cutoff, above this value, only primary CR can reach the detector position. The rigidity value of the lowest primary particle is called *lower* (R_L) rigidity cutoff and, below it, only secondary (or trapped) particles can be detected, see e.g. Cooke *et al.* (1991). A method to establish the particle origin is to perform the so-called *back-tracing*, i.e. the reconstruction of particle trajectory back in time from the detector position up to the magnetopause (if particle is primary), or eventually the Earth's surface (in case of secondary particles).

2. GeoMagSphere model

The GeoMagSphere model (see e.g., GMS-website: http://www.geomagsphere.org/, Bobik *et al.* 2005, Bobik *et al.* 2006) performs the trajectory reconstruction inside the magnetosphere. Input parameters of the numerical code are: the particle mass and charge, its geographic position, time and arrival direction. Both particle charge and time are reversed for the back-tracing, in this way, as matter of fact, the propagation equation remains unchanged. GeoMagSphere adopts *IGRF* as internal magnetic field model (Finlay *et al.* 2010) and add an external magnetic field chosen among: *T89* (Tsyganenko 1989 and Peredo *et al.* 1993), *T96* (Tsyganenko 1995, 1996) or *TS05* (Tsyganenko and Sitnov 2005). The external border of the magnetosphere (the magnetopause) can be selected between the two models proposed by Sibeck *et al.* (1991) and Shue *et al.* (1997). One of the effects of adding the external field is to obtain an asymmetric magnetosphere as required by the interaction with the solar wind. A comparison between experimental magnetic field data, recorded by the Cluster satellite (Escoubet *et al.* 2001), and several models, with and without external field, is reported in Figure 1. The two periods selected are 2 - 7 November 2011 (left), considered quiet days without significant solar events, and 5 - 9 March 2012 (right) when large solar events (flare and CME) occurred. On the x-axis is reported the difference, regarding the component B_x of the magnetic field, between the model predictions and the experimental data. On the y-axis is reported the amount of data corresponding to that difference. The data analysed have a time bin of $\sim$ 4 s corresponding to the Cluster data acquisition. In both cases magnetic field strength data are better described by IGRF+TS05 fields, in particular during March 2012, as expected since TS05 was specifically developed to reproduce the magnetosphere during magnetic storms. It is also shown that the differences between the IGRF+T89 and IGRF+TS05 models are mostly negligible during quiet periods (see also Grandi *et al.* 2015).

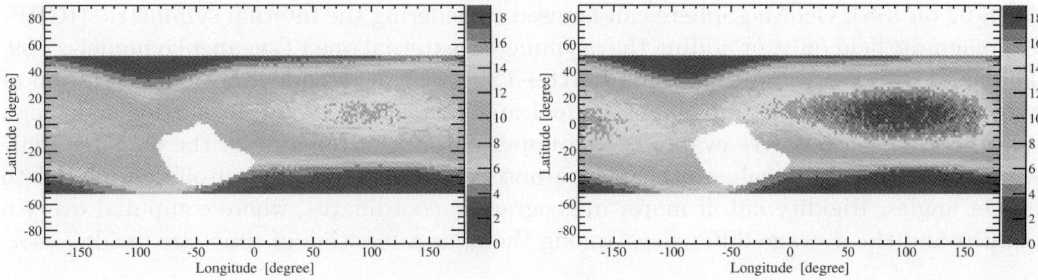

Figure 2. Average upper cutoff computed for the first six months of 2012 selecting particles arriving within 5° (left) and 20° (right) from the local zenith. The color scale represents the cutoff rigidity in GV. The white region in both maps is the South Atlantic Anomaly region excluded in this analysis.

3. Rigidity cutoff results

We then used GeoMagSphere, considering both IGRF and external fields, to evaluate the rigidity cutoff at the ISS orbit. Using TS05 as external field, we calculated R_{cut} for a given arrival direction (e.g., inside 5° from the local zenith), geographic position (e.g., at the ISS orbit) and time. We performed the back-tracing using the real time and position of the space station, considering simulated particles inside the acceptance cone of AMS-02 (Aguilar *et al.* 2015) (excluding the South Atlantic Anomaly region). We restricted our analysis to particles arriving inside a certain acceptance cone from the zenith at each given orbital position. We divided the geographic cutoff maps in $2° \times 2°$ cells and for each cell, the maximum rigidity for secondary (or trapped) particles identifies the upper cutoff (R_U). We analysed the cutoff widening the angle from the local zenith from 5° to 20°. In Figure 2 we reported the cutoff map regarding the first six months of 2012 selecting particles arriving within 5° (left) and 20° (right) from the local zenith. The white region inside the maps is the South Atlantic Anomaly as defined by the AMS-02 collaboration and, for this reason, excluded by the analysis. The magnetic latitudinal gradient is clearly shown. Figure 2 reports, as expected, that increasing the opening angle, more particles (e.g. secondaries with higher energy) can be observed in the detector, consequently the upper rigidity cutoff increase.

A time evolution of this cutoff was also calculated with a resolution of 6 months from July 2011 to June 2015. For each cell we evaluated the relative differences in R_U obtained in several time periods. The average differences, as a function of time, are inside $\sim 4\%$.

A deeper study was done on a specific day (i.e. March 7th, 2012) where a large solar event occurred (see e.g., Della Torre 2016 and Di Fino *et al.* 2014). For this day we calculated the vertical cutoff using both IGRF+T89 and IGRF+TS05 models. The rigidity cutoff difference obtained by the two models was computed as a function of the geomagnetic latitude (λ_{mag}). For almost 90% of the observation time the difference between the two models is within 5%, while in the remaining 10% of the time, when the detector is close to the polar regions, this difference increases up to 70%.

4. Conclusions

We developed a code (GeoMagSphere) to reconstruct charged particle trajectories inside the geomagnetic field. GeoMagSphere is able to separate primary, secondary and trapped particles using a back-tracing technique. In this way, we are able to evaluate the rigidity cutoff. This procedure is mandatory in order to get the effective primary cosmic ray flux for every detector in space (we applied it to the magnetic spectrometer

AMS-02 on ISS). GeoMagSphere can be used considering the internal symmetric (IGRF-12) magnetic field only, or adding the asymmetric external one (Tsyganenko models: T89, T96 or TS05). We found that using IGRF+TS05 we can provide a better description of experimental data, both for magnetic field and cosmic ray fluxes, in particular in high solar activity periods. We evaluated the upper cutoff as a function of the opening angle with respect to the local zenith and we observed higher rigidity cutoffs correlated to higher angles. Rigidity cutoff maps, in geographic coordinates, where computed every 6 months and the average differences among them, as a function of time, are inside $\sim 4\%$.

Acknowledgements

This work is supported by ASI (Agenzia Spaziale Italiana) under contract ASI-INFN I/002/13/0 and ESA (European Space Agency) contract 4000116146/16/NL/HK.

References

Aguilar, M., *et al.* (AMS Collaboration). 2015, *Phys. Rev. Lett.*, 114, 171,103

Bobik, P., Boschini, M., Grandi, D., *et al.* 2005, in *Washington DC American Geophysical Union Geophysical Monograph Series*, Vol. 155, The Inner Magnetosphere: Physics and Modeling, ed. T. I. Pulkkinen, N. A. Tsyganenko & R. H. W. Friedel, 301

Bobik, P., Boella, G., Boschini, M. J., *et al.* 2006, *Journal of Geophysical Research: Space Physics*, 111(A5)

Cooke, D. J., Humble, J. E., Shea, M. A., *et al.* 1991, *Il Nuovo Cimento C*, 14(3), 213–234

Della Torre, S. (AMS-02 Collaboration) 2016, in *25th European Cosmic Ray Symposium (ECRS 2016)* Turin, Italy - eConf C16-09-04.3

Di Fino, L., Zaconte, V., Stangalini, M., Sparvoli, R., Picozza, P., *et al.* 2014, *J. Space Weather Space Clim.*, 4, A19

Escoubet, C. P., Fehringer, M. & Goldstein, M. 2001, *Ann. Geophys.*, 19, 1197–1200

Finlay, C. C., Maus, S., Beggan, C. D., *et al.* 2010, *Geophysical Journal International*, 183(3), 1216

Grandi, D., Bertucci, B., Boschini, M., *et al.* 2015, In Proceedings of the 34th *International Cosmic Ray Conference (ICRC2015)* The Hague, The Netherlands)

Peredo, M., Stern, D. P. & Tsyganenko, N. A. 1993, *Journal of Geophysics Research*, 98, 15,343

Sibeck, D. G., Lopez, R. E. & Roelof, E. C. 1991, *Journal of Geophysics Research*, 96, 5489

Shue, J.-H., Chao, J. K., Fu, H. C., *et al.* 1997, *Journal of Geophysical Research*, 102, 9497

Störmer, C. 1956, *The polar aurora* Oxford University Press

Tsyganenko, N. A. 1989, *Planetary Space Science*, 37, 5–20

Tsyganenko, N. A. 1995, *Journal of Geophysics Research*, 100, 5599–5612

Tsyganenko, N. A. 1996, In E. J. Rolfe & B. Kaldeich, eds., *International Conference on Substorms*, vol. 389 of ESA Special Publication, 181

Tsyganenko, N. A. & Sitnov, M. I. 2005, *Journal of Geophysical Research (Space Physics)*, 110, A03208

Space Weather of the Heliosphere: Processes and Forecasts
Proceedings IAU Symposium No. 335, 2017
C. Foullon & O.E. Malandraki, eds.

doi:10.1017/S1743921317011620

How does the Sun Influence the Magnetospheres of Jupiter and Saturn?

Caitríona M. Jackman[1] and Christopher S. Arridge[2]

[1]Department of Physics and Astronomy, University of Southampton, Southampton, SO17 1BJ, UK
email: `c.jackman@soton.ac.uk`

[2]Department of Physics and Astronomy, Lancaster University, Lancaster, LA1 4YB, UK
email: `c.arridge@lancaster.ac.uk`

Abstract. Spacecraft have visited Jupiter and Saturn at all phases of the solar cycle and thus we have a wealth of data with which to explore both upstream parameters and magnetospheric response. In this paper we review upstream parameters including interplanetary magnetic field strength and direction, solar wind dynamic pressure, plasma beta and Mach number. We consider the impact of changing solar wind on dayside coupling via reconnection. We also comment on how solar UV flux variability over a solar cycle influences the plasma and neutral tori in the inner magnetospheres of Jupiter and Saturn, and thus estimate the solar cycle effects on internally driven magnetospheric dynamics. Finally we place our results in the context of the now complete set of data from the Cassini mission at Saturn and the current data streaming in from Juno at Jupiter, outlining future avenues for research.

Keywords. Jupiter, Saturn, solar wind, interplanetary medium, UV radiation, planets and satellites: general, molecular processes

---

The magnetospheres of Jupiter and Saturn are fascinating environments. Their dynamics are often said to be dominated by internal factors associated with strong planetary magnetic fields, rapid planetary rotation as an energy source, and internal plasma loading from moons and rings (e.g. Vasyliunas 1983; Hill 2017, and references therein). However, there is also evidence that Jupiter and Saturn respond to forcing from their external environment, the solar wind. Thus, characterising the upstream environment is critical to understand this internal vs. external balance. Both planets have been visited by multiple spacecraft. Saturn has hosted the orbiting Cassini spacecraft since 2004, completely changing our understanding of this beautiful planet, and Jupiter has recently seen the arrival of the Juno spacecraft. Figure 1 shows the variation of sunspot number with time from 1970 to 2017, with the times of spacecraft flybys and orbital tours. It illustrates that there has been spacecraft coverage at both Jupiter and Saturn at every phase of the solar cycle, and we have significant heritage datasets from which to draw a statistical picture of the large-scale structure of the solar wind, as well as plenty of in situ magnetospheric measurements to elucidate the magnetospheric response to external driving.

1. Solar wind properties

The interplanetary magnetic field (IMF) evolves with increasing radial distance from the Sun. One manifestation of this is the increasing winding of the field associated with the Parker Spiral (e.g. Forsyth *et al.* 1996). The measured azimuthal angles at Jupiter and Saturn are broadly in line with theory, with some small solar cycle-linked deviations which may be traced to displacements of the heliospheric current sheet (HCS) out of

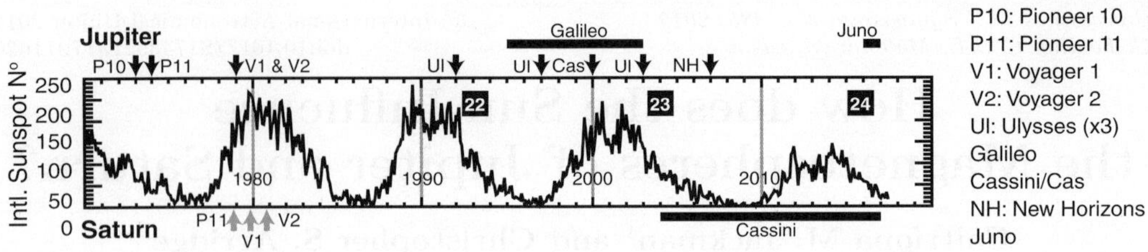

Figure 1. Schematic showing sunspot number with the timings of spacecraft close encounters (arrows) and orbital tours (horizontal bars) at Jupiter (top) and Saturn (bottom). Solar cycle numbers are in squares near each solar maximum. After Jackman & Arridge (2011). Sunspot data source: WDC-SILSO, Royal Observatory of Belgium, Brussels.

the ecliptic (e.g. Jackman *et al.* 2008; Jackman & Arridge 2011). In addition, patterns and sector structure in the IMF can develop significantly between Earth and 5-9 AU. For example, Cassini found the IMF upstream of Saturn during the declining phase of solar cycle 23 to be highly structured by compressions and rarefactions associated with Corotating Interaction Regions (CIRs) (e.g. Jackman *et al.* 2004). This structure is due to the tilt of the Sun's dipole and the interaction between fast and slow streams of solar wind. Hanlon *et al.* (2004) showed that this regular phasing can be disrupted somewhat by the merging of compression regions as the solar wind develops. However, when the phasing of compressions and rarefactions is regular, this property can be used to infer solar wind conditions even when a spacecraft is inside the magnetosphere (e.g. Bunce *et al.* 2005). Characterisation of the IMF upstream of Jupiter, during the declining phase of solar cycle 24, will be important for the interpretation of Juno data (Ebert *et al.* 2014).

Solar wind plasma parameters also change over the solar cycle, and between solar cycles, and also with heliocentric distance. Jackman & Arridge (2011) examined the distributions of dynamic pressure, plasma beta and Mach number from spacecraft upstream of both Jupiter and Saturn (Figure 2, with Saturn plasma data based on Pioneer and Voyager at solar maximum only). The Mach number distributions at both planets show highly supersonic and super-Alfvénic flow. At Jupiter the solar wind was more super-magnetosonic and super-Alfvenic at solar minimum. McComas *et al.* (2013) reported on a long-term trend for reduction in solar wind dynamic pressure, mirrored in the trends for solar wind speed, temperature and thermal pressure. If this trend continues it may have significant implications for solar wind-magnetosphere interactions in the coming years.

2. Solar wind-magnetosphere coupling

Some of the greatest debates in the outer planets community have arisen when considering the solar wind-magnetosphere interaction. One element of this debate is the role of viscous interactions between the magnetosheath flow and the magnetosphere (e.g. Axford & Hines 1961). In particular, at the giant planets this viscous interaction may lead to small-scale reconnection on the flanks and mixed boundary layers (Delamere & Bagenal 2013). Evidence that the viscous interaction can result in momentum transfer from the magnetosphere to the magnetosheath, has been presented for the case of Saturn (Burkholder *et al.* 2017). Furthermore, Delamere & Bagenal (2013) argue that the best analogy for the solar wind-gas giant magnetosphere interaction may be one of an Alfvénic solar wind-comet interaction.

In the magnetic reconnection model, the role of dayside reconnection is a topic of intense debate at the outer planets, with limited observations and numerous theoretical arguments about its relative importance. Estimates of the rate of dayside reconnection

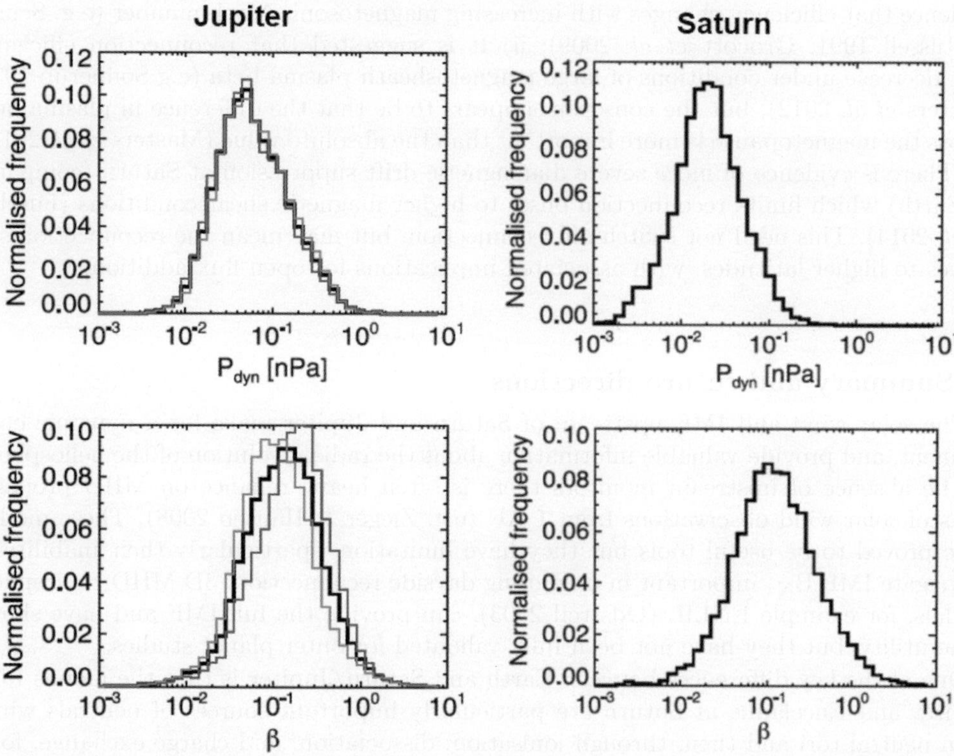

Figure 2. Distributions of solar wind dynamic pressure (top left) and solar wind plasma beta (bottom left) upstream of Jupiter with black indicating all points, blue solar minimum, and red solar maximum. The distributions have been normalised to the total number of points. The right-hand plots represent the same parameters upstream of Saturn for solar maximum only (black). Based on the databases outlined in Jackman & Arridge (2011) and following figures 4 and 5 from that paper.

rate (given as a voltage) have been obtained by adapting empirical relationships from Earth (e.g. Milan *et al.* 2004). One such relationship is shown in Eq. 2.1.

$$\Phi = V_{\mathrm{SW}} B_{\perp} L_0 \cos^4(\theta/2) \tag{2.1}$$

where Φ is the reconnection voltage, V_{SW} is the solar wind velocity, $B_{\perp}$ is the strength of the IMF in the T-N plane perpendicular to the velocity vector, θ is the 'clock angle' between the IMF vector and the planet's magnetic axis projected onto a plane perpendicular to the Sun-planet direction, and L_0 is the width of the solar wind channel in the T-N plane (perpendicular to the $B_{\perp}$ vector) in which the IMF reconnects with closed planetary field lines (Jackman *et al.* 2004; Nichols *et al.* 2006). The L_0 value is assumed to be approximately half the subsolar standoff distance (Milan *et al.* 2004). For Saturn, this gives voltages of $\sim$ 100 kV in compression regions, and $\sim$ 10 kV in rarefaction regions, compared to a long-term average voltage of $\sim$ 40 kV (Masters 2015).

Case studies show the evidence of dayside reconnection at Saturn and Jupiter, via in situ observations of accelerated flows and flux transfer events (e.g. McAndrews *et al.* 2008; Jasinski *et al.* 2016b; Ebert *et al.* 2017), and indirect observations of solar wind plasma in the cusp (e.g. Arridge *et al.* 2016; Jasinski *et al.* 2016a). However, the debate centres around the efficiency of dayside reconnection and its global effects. There are three factors which are debated as to their effects on the efficiency; i) There is conflicting evidence that efficiency changes with increasing magnetosonic Mach number (e.g. Scurry & Russell 1991; Grocott *et al.* 2009); ii) It is suggested that reconnection efficiency may decrease under conditions of large magnetosheath plasma beta (e.g Sonnerup 1974; Masters *et al.* 2012), but the consensus appears to be that the difference in plasma beta across the magnetopause is more important than the absolute value (Masters *et al.* 2012); iii) There is evidence of more severe diamagnetic drift suppression at Saturn (compared to Earth) which limits reconnection onset to higher magnetic shear conditions (Fuselier *et al.* 2014). This need not switch off reconnection, but may mean the reconnection site moves to higher latitudes, with associated implications for open flux addition.

3. Summary and future directions

The solar wind and IMF upstream of Saturn and Jupiter make for a dynamic environment, and provide valuable information about the radial evolution of the heliosphere. In the absence of upstream monitors there is often heavy reliance on MHD propagations of solar wind observations from 1 AU (e.g. Zieger & Hansen 2008). These models have proved to be useful tools but they have limitations, particularly their inability to propagate IMF B_N, important in predicting dayside reconnection. 3D MHD Heliosphere models, for example ENLIL (Odstrcil 2003), can provide the full IMF and have shown some utility, but they have not been fully validated for outer planet studies.

One of the key differences between Earth and Saturn/Jupiter is that the moons Io at Jupiter and Enceladus at Saturn are particularly important sources of neutrals which form neutral tori and then, through ionisation, dissociation, and charge exchange, form plasma tori. Photoionisation and photodissociation of material from Enceladus is an important reaction, susceptible to modulations in solar EUV over the solar cycle, and over a solar rotation. Fleshman *et al.* (2012) have shown that at solar maximum, neutral abundances due to photodissociation double, although this ultimately has a minor impact due to the importance of charge exchange. Another form of solar driving that is yet to be explored is the effect of changes in ionospheric conductivity (due to EUV or cosmic rays) over the solar cycle. The conductivity affects magnetosphere-ionosphere coupling and so changes can lead to effects on field-aligned currents, accelerating potential drops, and precipitating electron energies. These may have observable consequences for auroral emissions. This may also affect the ionospheres of natural satellites, which may affect the mass-loading of the magnetosphere in a non-linear fashion.

The Cassini dataset provides the unique opportunity to explore a full solar cycle at an outer planet with a single spacecraft. Future work may employ statistical techniques, such as quantile-quantile plotting, to examine how distributions of upstream parameters have changed across the solar cycle (e.g. Tindale & Chapman 2016). The flattened shape of Saturn's magnetopause also plays a role in modulating the magnetosheath conditions at the magnetopause, an aspect only beginning to be addressed, and will no doubt also be examined by Juno at Jupiter (e.g. Pilkington *et al.* 2014; Sulaiman *et al.* 2014).

The magnetospheres at Jupiter and Saturn are very large volumes of space relative to the terrestrial magnetospheres. As such, inferring their time-dependent response to internal and external drivers is a significant challenge. Similar problems are also being addressed for the magnetospheres of Uranus and Neptune (e.g. Arridge 2015). Rising to this challenge will require the integration of upstream parameters, in situ data, remote sensing data (e.g., energetic neutral atoms, auroral and molecular/atomic torus emissions), and models. Data assimilation may also be important in using these hetrogeneous data sets to establish a system-level picture of giant planet space weather.

References

Arridge, C. S. 2015, *Magnetotails of Uranus and Neptune (John Wiley & Sons, Inc)*, 119–133

Arridge, C. S., Jasinski, J. M., Achilleos, N., *et al.* 2016, *JGR: Space Physics*, 121, 3006

Axford, W. I. & Hines, C. O. 1961, *Can. J. Phys.*, 39, 1433

Bunce, E. J., Cowley, S. W. H., Wright, D. M., *et al.* 2005, *GRL*, 32, L20S04

Burkholder, B., Delamere, P. A., Ma, X., *et al.* 2017, *GRL*, 44, 5877

Delamere, P. A. & Bagenal, F. 2013, *JGR: Space Physics*, 118, 7045

Ebert, R., Bagenal, F., McComas, D. & Fowler, C. 2014, *Front. Astron. Space Sci.*, 1, 4

Ebert, R. W., Allegrini, F., Bagenal, F., *et al.* 2017, GRL, 44, 4401

Fleshman, B. L., Delamere, P. A., Bagenal, F. & Cassidy, T. 2012, *JGR: Planets*, 117, E05007

Forsyth, R. J., Balogh, A., Smith, E. J., Erdös, G. & McComas, D. J. 1996, *JGR*, 101, 395

Fuselier, S. A., Frahm, R., Lewis, W. S., *et al.* 2014, *JGR: Space Physics*, 119, 2563

Grocott, A., Badman, S. V., Cowley, S. W. H., *et al.* 2009, *JGR: Space Physics*, 114, a07219

Hanlon, P. G., Dougherty, M. K., Forsyth, R. J., *et al.* 2004, *JGR: Space Physics*, 109, A09S03

Hill, T. W. 2017, *Magnetosphere-Ionosphere Coupling at Jupiter and Saturn*, ed. C. R. Chappell, R. Shunk, P. Banks, J. Burch & R. Thorne

Jackman, C. M., Achilleos, N., Bunce, E. J., *et al.* 2004, *JGR: Space Physics*, 109, a11203

Jackman, C. M. & Arridge, C. S. 2011, *So. Phys.*, 274, 481

Jackman, C. M., Forsyth, R. J. & Dougherty, M. K. 2008, *JGR: Space Physics*, 113, a08114

Jasinski, J. M., Arridge, C. S., Coates, A. J., *et al.* 2016a, *JGR: Space Physics*, 121, 12

Jasinski, J. M., Slavin, J. A., Arridge, C. S., *et al.* 2016b, *GRL*, 43, 6713

Masters, A. 2015, *GRL*, 42, 2577

Masters, A., Eastwood, J. P., Swisdak, M., *et al.* 2012, *GRL*, 39, L08103

McAndrews, H. J., Owen, C. J., Thomsen, M. F., *et al.* 2008, *JGR: Space Physics*, 113, A04210

McComas, D. J., Angold, N., Elliott, H. A., *et al.* 2013, *ApJ*, 779, 2

Milan, S. E., Cowley, S. W. H., Lester, M., *et al.* 2004, *JGR: Space Physics*, 109, A04220

Nichols, J. D., Cowley, S. W. H. & McComas, D. J. 2006, *Annales Geophysicae*, 24, 393

Odstrcil, D. 2003, *Adv. Sp. Res.*, 32, 497

Pilkington, N. M., Achilleos, N., Arridge, C. S., *et al.* 2014, *JGR: Space Physics*, 119, 2858

Scurry, L. & Russell, C. T. 1991, *JGR: Space Physics*, 96, 9541

Sonnerup, B. U. Ã. 1974, *JGR*, 79, 1546

Sulaiman, A. H., Masters, A., Dougherty, M. K. & Jia, X. 2014, *JGR: Space Physics*, 119, 5651

Tindale, E. & Chapman, S. C. 2016, *GRL*, 43, 5563

Vasyliunas, V. M. 1983, *Plasma distribution and flow*, ed. A. J. Dessler, 395–453

Zieger, B. & Hansen, K. C. 2008, *JGR: Space Physics*, 113, a08107

Space Weather of the Heliosphere: Processes and Forecasts
Proceedings IAU Symposium No. 335, 2017
C. Foullon & O.E. Malandraki, eds.

doi:10.1017/S1743921317010924

The Response of the Martian Atmosphere to Space Weather

David A. Brain

Laboratory for Atmospheric and Space Physics, University of Colorado,
3665 Discovery Drive, Boulder, Colorado, USA
email: david.brain@lasp.colorado.edu

Abstract. Mars lacks a global dynamo magnetic field to shield it from the solar wind and solar storms, so may be especially sensitive to changing space weather compared to Earth. Inputs from the Sun and solar wind have been measured continuously at Mars for 20 years, and intermittently for more than 50 years. Observations of the influence of the variable space weather at Mars include compression and reconfiguration of the magnetosphere in response to solar storms, increased likelihood of aurora and increased auroral electron energies, increased particle precipitation and ionospheric densities during flare and energetic particle events, and increased ion escape during coronal mass ejection events. Continuing measurements at Mars provide a useful vantage point for studying space weather propagation into the heliosphere, and are providing insight into the evolution of the Martian atmosphere and the role that planetary magnetic fields play in helping planets to retain habitable conditions near their surface.

Keywords. planets and satellites: individual

1. Context

Broadly defined, ‘space weather’ refers to short term variability in the solar particle, field, and photon conditions throughout the heliosphere. By many measures the Martian response to space weather is the best studied in the heliosphere outside of Earth.

Charged particles and heliospheric magnetic fields encounter a much different environment at Mars than they do at Earth (see overview in Brain *et al.* 2017). Unlike Earth, Mars lacks a global dipole magnetic field and solar wind plasma is deflected around the planet in an induced magnetosphere formed through interaction of the solar wind with the conducting ionosphere (Figure 1). This obstacle to the solar wind is weak compared to a global dipole, leading to a more compact magnetosphere relative to the size of the planet. Mars is about half the size of Earth so that the entire interaction region is much smaller than Earth's magnetosphere, and the exosphere of Mars extends into the unperturbed solar wind. The Martian crust contains regions of locally magnetized crust which rotate with the planet and are sufficiently strong that they perturb the canonical picture of an induced magnetosphere, creating an obstacle to the solar wind with elements of the interactions at Venus, comets, and magnetized planets such as Earth.

One can imagine a number of possible space weather effects at Mars. Geophysically, these include compression and reconfiguration of the induced magnetosphere, stripping of atmospheric particles to space, atmospheric heating and chemistry caused by incident solar energetic particles, and aurora. From the perspective of human and robotic exploraton these include disruption of the electrical systems of orbiting and landed spacecraft and instruments (e.g. the infamous and unfortunate failure of the Mars Odyssey radiation detector due to the Halloween solar storm) and hazards to human presence at the surface. Finally, space weather in the form of ionizing radiation could frustrate the development of life at the Martian surface.

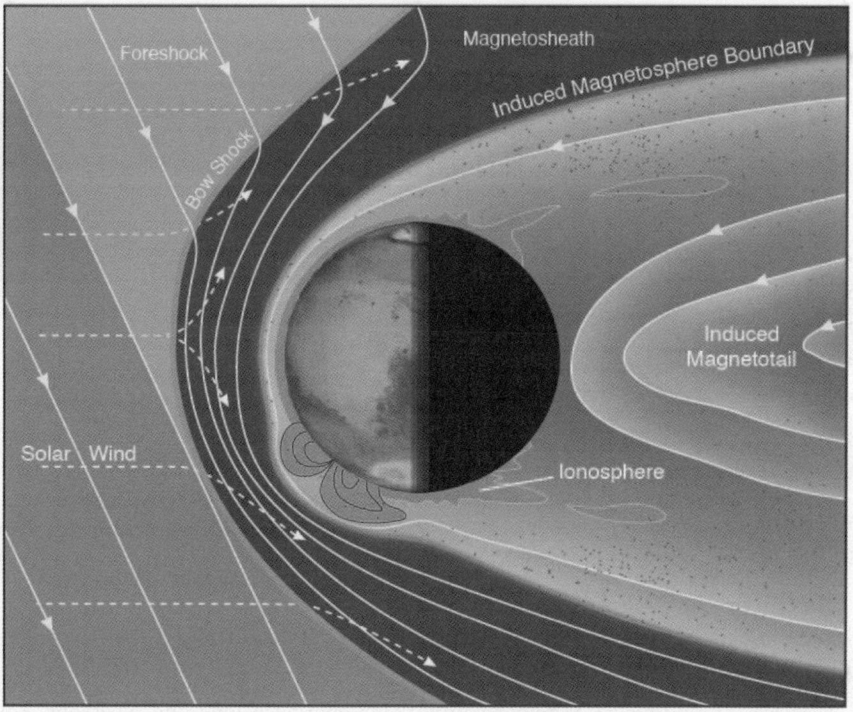

Figure 1. Diagram of the interaction of the Martian upper atmosphere with the solar wind, from Brain *et al.* (2017).

2. Space Weather Inputs

More than a dozen spacecraft missions since 1965 have made observations at Mars relevant to space weather, the magnetosphere, and the upper atmosphere. The first of these (Mariner 4, 6, 7, 9, Mars 2, 3, 5, Viking, and Phobos 2) sampled the environment near Mars in situ, and established Mars as likely to be an induced magnetosphere with at most a weak global dipole field. The Phobos 2 mission also determined that Martian atmosphere was escaping to space in respose to the interaction with the solar wind (Lundin *et al.* 1989).

Following these early missions, there has been continuous measurement of space weather at Mars since 1997, with observations supplied by Mars Global Surveyor (MGS), Mars Express (MEX), and the Mars Atmosphere and Volatile EvolutioN (MAVEN) mission. Observations from these three spacecraft comprise the longest continuous record of space weather observations at a solar system object other than Earth.

Though there is a continuous measurement record at Mars extending back two decades, the observations from MGS and MEX are less complete than those of MAVEN. For example, MGS directly measured only magnetic field and suprathermal electrons, and indirectly measured energetic ($\sim 20 - 50$ MeV) ions via its instrument backgrounds. Additionally, most of its nine year mission was spent in orbit close to the planet, so that undisturbed solar wind was not sampled directly. Mars Express has an elliptical orbit that enters the solar wind, but directly measures suprathermal electrons and ions, and indirectly measures energetic ions via instrument backgrounds. MAVEN, by contrast, has both an elliptical orbit that enters the solar wind and directly measures all of the relevant space weather drivers (Figure 2).

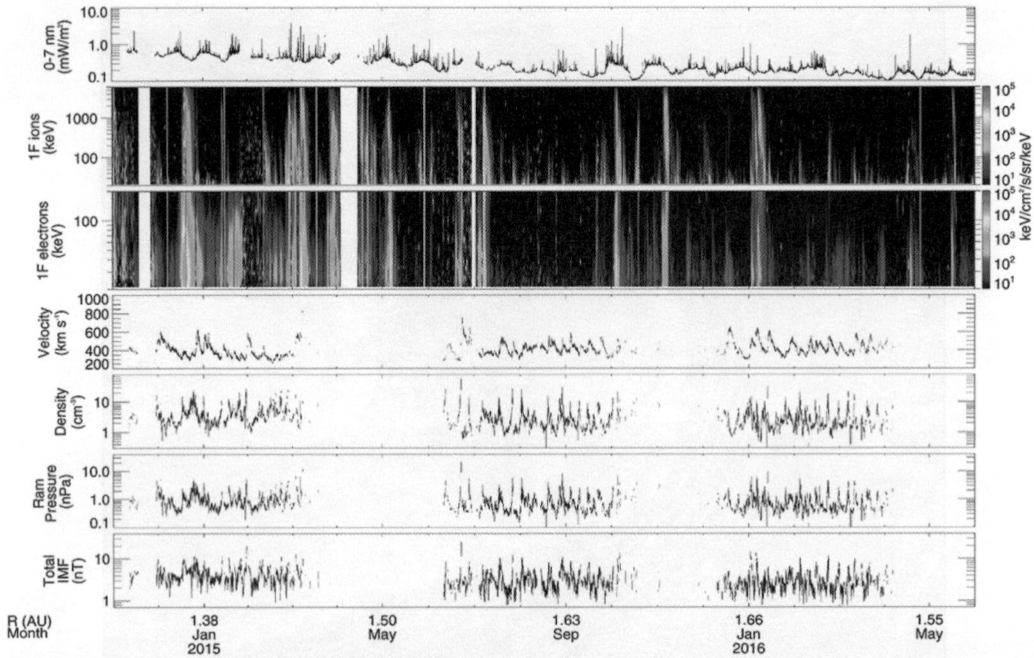

Figure 2. Space weather inputs to the Martian magnetosphere and upper atmosphere as measured by MAVEN, including EUV photon flux, solar energetic ions and electrons, solar wind velocity, density, and pressure, and interplanetary magentic field strength (Lee *et al.* 2017).

3. Martian Consequences of Space Weather

The long baseline of relevant observations at Mars has allowed the community to associate changing space environment conditions with the response of the Martian magnetosphere and upper atmosphere. Additionally, the measurements of the Martian system have allowed the community to validate models that can be used to predict the response of Mars to space weather. Here we broadly describe four categories of response.

Magnetospheric Reconfiguration - The Martian magnetosphere is comprised of several distinct plasma regions separated by plasma boundaries roughly analogous to those in Earth's magnetosphere (see Figure 1). Boundaries such as the collisionless bow shock and the Induced Magnetosphere Boundary (IMB, with some similarities to Earth's magnetopause) can become compressed in response to changing conditions in the solar wind. For example, as a Coronal Mass Ejection (CME) passes Mars these boundaries are pushed closer to the planet on the dayside, and flare away from the planet much less than during undisturbed periods (e.g. Crider *et al.* 2005; Jakosky *et al.* 2015). Figure 3 shows MHD simulation results during a CME that passed Mars in March of 2015 (Jakosky *et al.* 2015). The model agrees well with MAVEN observations along the orbit track, giving confidence that the model results are trustworthy elsewhere, and further shows a compressed and less flared magnetsphere after the arrival of the CME at Mars. Additional simulations show a complex magnetic field topology close to Mars that changes significantly during different phases of the CME passage (Luhmann *et al.* 2017). MEX observations show similiar evidence of compression of the plasma regions near Mars, with rapid compression during and after the CME shock in 2008, followed by a more gradual 'relaxation' of the plasma boundaries during the fast stream that followed (Sanchez-Cano *et al.* 2017).

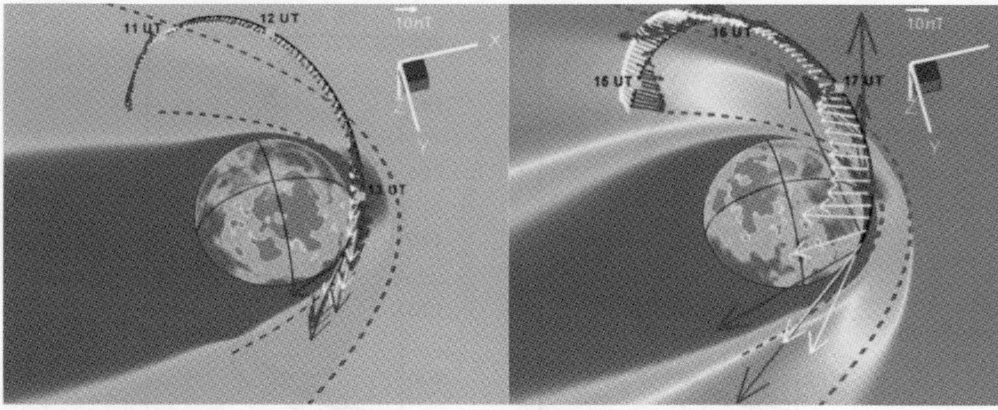

Figure 3. MHD simulations of the Martian solar wind interaction before and immediately following the passage of a CME at Mars. Crustal magnetic field strength is colored on the planet, and the simulated plasma flow speed is shown in the Mars equatorial plane, as viewed from above the southern hemisphere. Nominal bow shock and IMB locations are shown as dashed lines. The MAVEN trajectory is shown in black, with superposed measured (black) and modeled (white) magnetic field vectors. From Jakosky *et al.* (2015).

<u>*Ionospheric Enhancements*</u> - Planetary ionospheres are formed chiefly through photoionization of atmospheric neutrals by EUV and X-ray photons from the Sun, with a secondary source of ionization from particle precipitation into the atmosphere. Solar flares generate short but intense bursts of radiation at these energies that can cause additional ionization in a planet's upper atmosphere. Radio soundings of the Martian atmosphere during flare periods by MGS have shown an increase in both the peak ionospheric electron density (Gurnett *et al.* 2005), and in a soft X-ray peak near 100 km altitudes as shown in Figure 4 (Mendillo *et al.* 2006). Ionospheric densities have been observed to increase by as much as 200% in response to a flare. In addition, solar energetic particles accelerated during flares or CMEs encounter the Martian atmosphere and can increase ionization. Radar soundings of the Martian ionosphere by MEX have shown that the total electron content of the ionosphere increases substantially during both flares and some SEP events (Morgan *et al.* 2006).

<u>*Aurora*</u> - One consequence of energetic particle precipitation at Earth is aurora. Mars similarly hosts UV auroral emission, both in localized crustal magnetic field regions (Bertaux *et al.* 2005) and throughout the entire atmosphere, including far from crustal fields (Schneider *et al.* 2015). This latter aurora, termed 'diffuse aurora', is clearly associated with the presence of populations of solar energetic electrons near Mars, and is therefore a direct response to space weather. Aurora in crustal fields are more directly analogous to terrestrial aurora, and result from locally energized electron populations accelerated into the Martian atmosphere in cusp regions (Lundin *et al.* 2006; Brain *et al.* 2006). The brightest example of UV emission from crustal field aurora occurred during a SEP event (Brain *et al.* 2006), and the most energetic electron distributions in cusps tend to occur during periods of ehnanced SEP flux. Like ionospheric enhancements, aurora indicate enhanced energy input into the Martian ionosphere and upper atmosphere during space weather events.

<u>*Atmospheric Escape*</u> - Measurements from Phobos 2, Mars Express, and MAVEN all show that Martian atmospheric ions are escaping to space (e.g. Lundin *et al.* 1989;

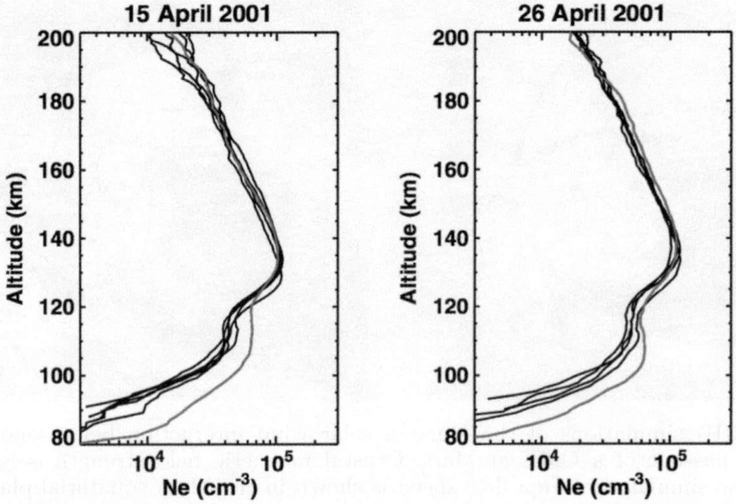

Figure 4. Electron density altitude profiles of the Martian upper atmosphere obtained via radio sounding by the MGS spacecraft for two time periods in 2001. A single profile occuring during a soft X-ray flare is colored red in each time period. From Mendillo *et al.* (2006).

Barabash *et al.* 2007; Brain *et al.* 2015). The rate of atmospheric escape varies with the driving solar and solar wind conditions - a result that is also supported by global plasma models of Mars. It is difficult for an orbiting spacecraft to measure a change in the global escape rate during a CME or SEP event due to a the limited coverage of the region over which escape occurs during the duration of the event. However, multiple studies suggest that escape rates increase during extreme events. The MEX instrument background rates increased strongly during a CME that encountered both Venus and Mars in 2006, indicating the presence of substantial SEP fluxes during the CME (Futaana *et al.* 2008). Escaping ion fluxes increased by 1-2 orders of magnitude during this time period, but this measurement is complicated by the presence of the very instrument backgrounds that indicated that a disturbed period occurred. Similarly, a moderate CME in March 2015 was observed by MAVEN, during which escaping ion fluxes on the dayside of the planet were enhanced by an order of magnitude or more (Jakosky *et al.* 2015). Continued work is currently underway to separate the influences of: the different phases of a CME, SEPs vs. increased magnetic field in the solar wind, and different geometries for CMEs encountering Mars.

4. Applications

As described above, studies of space weather at Mars reveal the response of the Martian atmosphere and near space environment to changing conditions at the Sun. But they also offer the opportunity to address several larger scientific questions.

First, the location of Mars at 1.5 AU provides an additional vantage point from which to study the propagation of the solar wind into the heliosphere. Multipoint measurements comparing conditions at Mars and Earth (in addition to Mercury, Venus, and other locations) provide useful constraints for the validation of models of CME and solar wind propagation (e.g. Falkenberg *et al.* 2011; Dewey *et al.* 2016).

Next, comparisons of the space weather response of Mars to that of Earth may teach us about the importance of a global planetary magnetic field in regulating the interaction of a planet's atmosphere with its space environment. For example, a comparison of

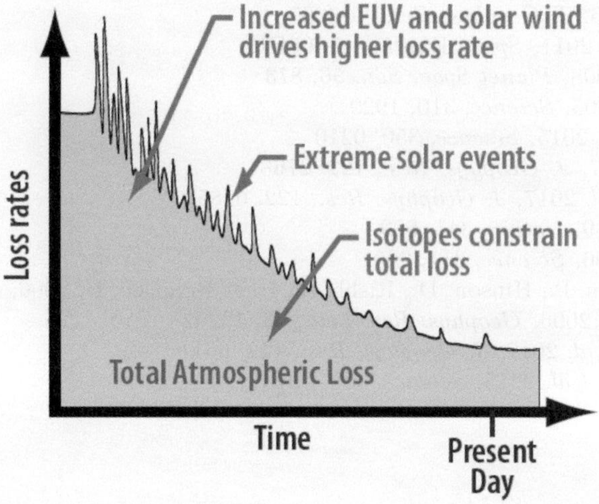

Figure 5. Cartoon showing how atmospheric loss rates from Mars have changed over time. Early in Martian history loss rates were likely to have been both greater in response to larger solar EUV flux and a more robust solar wind, and more variable in response to higher solar activity in the form of flares and CMEs. Courtesy NASA's MAVEN mission.

atmospheric escape at Earth and Mars in response to drivers from the Sun may reveal whether the presence of a global dipole planetary magnetic field reduces atmospheric loss. Of course this comparison is complicated by a number of additional differences between Earth and Mars (e.g. planetary size, heliocentric distance, etc.). Therefore, studies of Mars alone that compare escape from magnetized and unmagnetized regions of the planet present an appealing alternative to Earth-Mars comparisons.

Third, studies of atmospheric escape under different conditions at Mars can be used to understand how the different escape processes vary in response to drivers. With this understanding, it becomes possible to make simple estimates of atmospheric escape rates at earlier epochs in Martian history, when the driving conditions were different (Figure 5). This is important because abundant evidence points to an early Mars that was more habitable than that of today - where liquid water was stable on the surface for long periods of time. In short, space weather observations at Mars today help us to understand the history of the planet's habitability.

Finally, we can treat Mars as a laboratory for measuring space weather influences at unmagnetized planets in our galaxy. By gaining insight into the physical processes at work in the Martian response to space weather, we can port our understanding over to the myriad objects currently being discovered orbiting other stars. With knowledge of a star's properties we can apply the lessons learned at Mars to estimate whether a planet is likely to retain a habitable atmosphere.

References

Barabash, S., Fedorov, A., Lundin, R. & Sauvaud, J.-A. 2007, *Science*, 315, 501

Bertaux, J.-L., *et al.* 2005, *Science*, 307, 566

Brain, D. A., *et al.* 2006, *Geophys. Res. Lett.*, 33, 01201

Brain, D. A., *et al.* 2015, *J. Geophys. Res.*, 42, 9142

Brain, D., *et al.* 2017, in: R. Haberle, T. Clancy, F. Forget, M. Smith, and R. Zurek (eds.), *The Atmosphere and Climate of Mars*, (Cambridge), p. 464

Crider, D. H., *et al.* 2005, *J. Geophys. Res.*, 110, A09S21

Dewey, R., *et al.* 2016, *J. Geophys. Res.*, 7, 6207
Falkenberg, T., *et al.* 2011, *Space Weather*, 9, 00E12
Futaana, Y., *et al.* 2008, *Planet Spac. Sci.*, 56, 873
Gurnett, D., *et al.* 2005, *Science*, 310, 1929
Jakosky, B. M., *et al.* 2015, *Science*, 350, 0210
Lee, C. O., *et al.* 2017, *J. Geophys. Res.*, 122, 2768
Luhmann, J. G., *et al.* 2017, *J. Geophys. Res.*, 122, 6185
Lundin, R., *et al.* 1989, *Nature*, 341, 609
Lundin, R., *et al.* 2006, *Science*, 311, 980
Mendillo, M., Withers, P., Hinson, D., Rishbeth, H. & Reinisch, B. 2006, *Science*, 311, 1135
Morgan, D. D., *et al.* 2006, *Geophys. Res. Lett.*, 33, 13202
Sanchez-Cano, B., *et al.* 2017, *J. Geophys. Res.*, 122, 6611
Schneider, N. M. S., *et al.* 2015, *Science*, 350, 0313

Space Weather of the Heliosphere: Processes and Forecasts
Proceedings IAU Symposium No. 335, 2017
C. Foullon & O.E. Malandraki, eds.

doi:10.1017/S1743921317010420

Geomagnetopause Position and Shape Dependence on Solar Wind Plasma and IMF Parameters: Analytic Model Comparison with Observations and 3-D MHD Runs

M. I. Verigin, G. A. Kotova and V. V. Bezrukikh

Space Research Institute of Russian Academy of Sciences, Profsoyuznaya, 84/32, Moscow, 117997, Russia, email: verigin@iki.rssi.ru

Abstract. It is generally accepted to use the solar wind ram pressure ρV^2 and the IMF B_z component for empirical description of the geo-magnetopause position and shape. A specific feature of the present paper is not to use the solar wind ρV^2 but the thermal P_{th} and magnetic field P_{mag} pressures adjacent to the magnetopause for proper modelling. These pressures are deduced from the results of 3-D MHD runs and analytic solutions for post bow shock MHD flow in Lagrangian variables. The magnetopause shape variation due to B_z component changes the so called 'doubling factor' f_d, which can be analytically deduced from a Tsyganenko magnetospheric field ellipsoidal model. Including all the above effects in our analytical model leads to a good description of 'rapid' magnetopause approach to the Earth for southward IMF and to its 'stagnant' behaviour with the increase of northward IMF component.

Keywords. Magnetopause, solar wind ram pressure, magnetic and thermal pressure, doubling factor

1. Introduction

It is generally accepted that the solar wind ram pressure ρV^2 is the main factor contributing to the magnetopause stagnation pressure, which can be then approximated as $\Pi \approx k\rho V^2$ with k being a function of solar wind specific heat ratio γ and sonic Mach number M_s (Landau & Lifshits 1959):

$$k = \frac{1}{\gamma}\left(\frac{\gamma+1}{2}\right)^{(\gamma+1)/(\gamma-1)}\left(\gamma - \frac{\gamma-1}{2M_s^2}\right)^{1/(1-\gamma)} . \qquad (1)$$

Complementary factors that are influencing the magnetopause position and shape are: (i) the magnetic field tension resulting in the clock angle dependency of the magnetopause terminator cross-section (Verigin *et al.* 2009), (ii) the magnetic field pressure leading to magnetopause movement towards the Earth when the IMF cone angle approaches $\pi/2$ (Dušík *et al.* 2010), and (iii) the IMF B_z component increasing the magnetopause nose bluntness for southward B_z direction (Sibeck *et al.* 1991). It was shown (Verigin *et al.* 2014) that the contribution of the magnetic field tension to the stagnation pressure is $2\Delta/R$ times less than contribution of the magnetic pressure itself, where Δ is the magnetosheath thickness and R is the magnetic field lines curvature radius. Thus tension effects will be omitted in subsequent analysis.

The following expression presents one of the widely used relations for magnetopause shape description (see, e.g., Liu *et al.* 2015):

$$r = r_0 \cdot \left(\frac{2}{1+\cos\theta}\right)^{\alpha} \cdot \left(1 - 0.1 \cdot C\cos^2\phi\right), \qquad (2)$$

where r_o is the planetocentric distance to the magnetopause nose position, θ and φ are the polar and azimuth angles, respectively; the second multiplier is the global magnetopause shape factor, while the third multiplier describes the local cusps depression and will not be considered further. Flaring angle α and r_0 in the relation (2) are *empirical* functions incorporating physically unjustified *multiplicative* function of IMF B_z component, and power function of a *simple sum* of solar wind ram and magnetic pressures.

It seems apparent that using, not the solar wind ram pressure, but the thermal, magnetic and total pressures adjacent to the magnetopause, and better justified magnetopause shape variation with B_z are preferred for the magnetopause position and shape modeling, and it is this approach which will be used further.

2. Basic approach and relations

Empirical relations for the description of the total P_{tot}, thermal P_{th}, and magnetic field P_{mag} pressures at the magnetopause nose are based on the results of 3-D MHD modelling (Stahara 2002) and analytic solutions in Lagrangian variables (Shugaev *et al.* 2005). For the case of interplanetary magnetic field perpendicular to the solar wind velocity (Verigin *et al.* 2015):

$$P_{tot} = k\rho V^2 (1 + (\sqrt{6}/M_a^2)^{2/3}), \tag{3}$$

$$P_{th} = k\rho V^2/(1 + 25 \cdot (2 - \sqrt{6/M_s})/M_a), \tag{4}$$

$$P_{mag} = \Pi - P_{th}, \tag{5}$$

where M_a is the Alfvénic Mach number. Figure 1 shows the pressures defined in (3 - 5) as functions of $1/M_a$ and the good correspondence with results of 3-D MHD modelling (Stahara 2002).

The following important relation of our approach describes a so-called 'doubling factor' f_d, which indicates how much the internal magnetospheric field is increasing in the subsolar region due to Chapman-Ferraro currents. Using the ellipsoidal model of Tsyganenko (1989) we approximated f_d by the following relation :

$$f_d(r_0/R_0) = (4 + 5r_0/R_0)/3, \tag{6}$$

where R_o is the magnetopause nose curvature radius. Figure 2 demonstrates the good correspondence of relation (6) with the results of the model (Tsyganenko 1989). According to (6) frequently used in modeling, the doubling factor $f_d \approx 2.44$ (see, e.g., Beard 1964) corresponds to reasonable magnetopause nose shape with $R_o \approx 1.5\ r_o$ (Verigin *et al.* 2009).

Relation (6) corresponds to the theoretical case with $\boldsymbol{B}_{IMF} = 0$. The presence of non-zero $\boldsymbol{B}_{IMF}$ in the solar wind flow will lead both to r_o and R_o variations: $r_o \rightarrow r_b$ and $R_o \rightarrow R_b$. We estimated new ratio of r_b / R_b as:

$$r_b/R_b = (1 \pm \xi\sqrt{P_{mag}/P_{tot}}) \cdot r_0/R_0, \tag{7}$$

where ξ is yet an undefined parameter describing the variation of magnetic field in the magnetosphere adjacent to the magnetopause due to external non-zero magnetosheath field $\sqrt{8\pi P_{mag}}$, and +/- signs correspond to northward and southward IMF directions, respectively. The pressure balance equation for the northward IMF case can be written as:

$$B_e \cdot (R_e/r_{b+})^3 \cdot (f_d(r_0/R_0) + \xi \cdot 5r_0/3R_0 \cdot \sqrt{P_{mag}/P_{tot}}) = \sqrt{8\pi P_{tot}}, \tag{8}$$

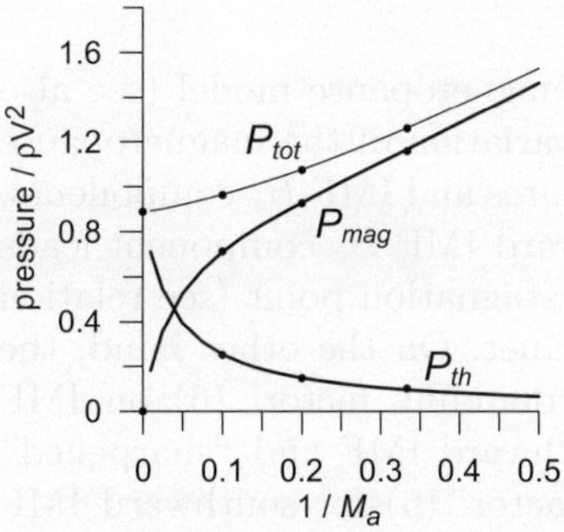

Figure 1. Comparison of 3-D MHD calculations (dots) with approximating relations (3 - 5).

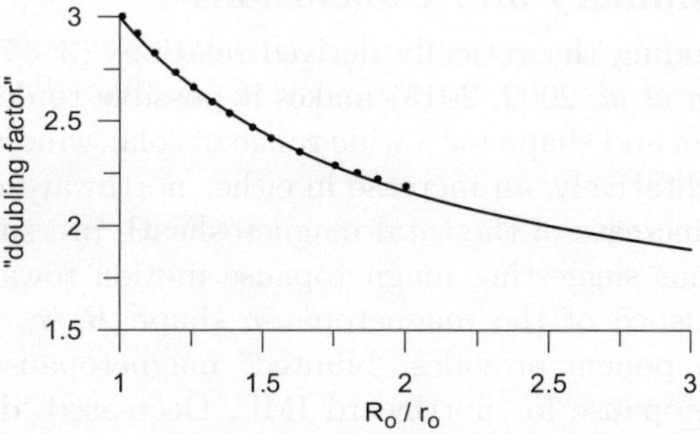

Figure 2. Comparison of Tsyganenko (1989) model results (dots) with calculations by the expression (6).

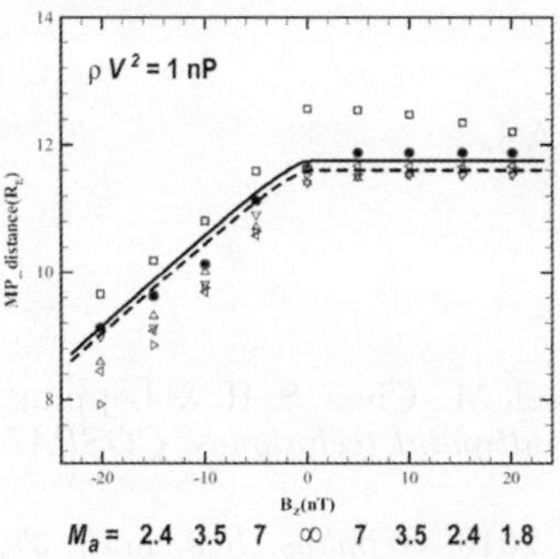

Figure 3. Comparison of the present analytic model (solid and dashed lines lines) with the modern 3-D MHD modeling results (•-, Lu *et al.* 2015) and with widely used empiric models: upward triangles (Shue *et al.* 1997), downward triangles (Shue *et al.* 1998), left triangles (Petrinec & Russell 1996), right triangles (Chao *et al.* 2002), squares (Lin *et al.* 2010).

where B_e = 31,200 nT, R_e =6,371 km. Due to the "stagnant" position of the magnetopause under northward IMF: $r_{b+} \approx r_o$. Taking into account that $B_e \cdot (R_e/r_0)^3 \cdot f_d(r_0/R_0) = \sqrt{8\pi k\rho V^2}$we can deduce the ξ factor as:

$$\xi = \sqrt{P_{tot}/P_{mag}} \cdot (\sqrt{P_{tot}/k\rho V^2} - 1) \cdot (1 + 4R_0/5r_0). \tag{9}$$

Subsequent substitution of the expression for ξ (9) into equation similar to (8), but with negative sign before ξ and $r_{b+} \to r_{b-}$, will lead to evaluation of the magnetopause nose position r_{b-} for southward IMF direction:

$$r_{b-} = r_0 \cdot \sqrt[3]{2 \cdot \sqrt{k\rho V^2/P_{tot}} - 1}. \tag{10}$$

The magnetopause nose curvature radii for northward and southward IMF direction R_{b+}, R_{b-} can now be calculated with the use of relation (7) with proper sign. It demonstrates the increase of the ratio R_b/r_b with the increase of southward IMF B_z component that provides more blunt magnetosphere nose in this case, and decrease of the ratio R_b/r_b with increase of northward IMF B_z component that provides 'sharpening' of the magnetopause nose for the northward IMF.

Figure 3 presents comparison of the present analytic model of the magnetopause nose position variation with IMF B_z component (for $M_s = 2$ – dashed line and for $M_s \to \infty$ - solid line), with the results of bulky 3-D MHD modeling by Lu *et al.* (2015). A good correspondence can be observed. Note that though condition $r_o = const$ for $B_z > 0$ was implemented in the model, coincidence of the slopes of the both modeling curves for $B_z < 0$ provides evidence of the validity of our analytic modeling approach.

3. Summary and Conclusions

Including theoretically derived relations (3 - 7) into a magnetopause model (see also Verigin *et al.* 2009, 2015) makes it possible to describe a variation of the magnetopause position and shape for a wide range of solar wind ram pressures and IMF B_z components.

Qualitatively, an increase in either northward or southward IMF B_z component leads to an increase of the total magnetosheath pressure at the stagnation point (see relation (3)) thus suggesting magnetopause motion toward the planet. On the other hand, the dependence of the magnetopause shape R_b/r_b (7) and 'doubling factor' (6) on IMF B_z component provides "blunted" magnetopause for southward IMF and "sharpened" magnetopause for northward IMF. Decreased 'doubling factor' (6) for southward IMF leads to a relatively "rapid" approach of "blunted" magnetopause toward the Earth while increased f_d for northward IMF provides "stagnant" behavior of the "sharpened" magnetopause, that qualitatively corresponds to magnetopause observations (see, e.g., Sibeck *et al.* 1991).

Acknowledgements

The paper is partially supported by the program P7 of RAS.

References

Beard, D. B., 1964, *Rev. Geophys.* 2(2), 335

Chao, J. K., Wu, D. J., Lin, C.-H., Yang, Y. H., Wang, X. Y., Kessel, M., Chen, S. H. & Lepping, R. P. 2002, in: L.-H. Lyu (eds.), *Space weather study using multipoint techniques, COSPAR Colloq. Ser.*, (Pergamon: Oxford), v.12, p.127

Dušík, Š., Granko, G., Šafránková, J., Němeček, Z. & Jelínek, K. 2010, *Geophys. Res. Lett.*, 37, L19103, doi:10.1029/2010GL044965

Landau, L. D. & Lifshitz, E. M. 1959, *Fluid Mechanics*, (Pergamon Press: London)

Lin, R. L., Zhang, X. X., Liu, S. Q., Wang, Y. L. & Gong, J. C. 2010, *J. Geophys. Res.*, 115, A04207. https://doi.org/10.1029/2009JA014235

Liu, Z.-Q., Lu J. Y., Wang C., Kabin K., Zhao J. S., Wang M., Han J. P., Wang J. Y. & Zhao M. X. 2015, *J. Geophys. Res. Space Physics*, 120, 5645, doi:10.1002/2014JA020961

Lu, J. Y., Wang, M., Kabin K., Zhao J. S., Liu Z.-Q., Zhao M. X. & Li, G. 2015, *Planetary and Space Science*, 106(2), 108

Petrinec, S. M. & Russell, C. T. 1996, *J. Geophys. Res.*, 101(A1), 137

Sibeck, D. G., Lopez, R.E & Roelof, R. C. 1991, *J. Geophys. Res.*, 96(A4), 5489

Shue, J.-H., Chao, J. K., Fu, H. C., Russell, C. T., Song, P., Khurana, K. K. & Singer, H. J. 1997, *J. Geophys. Res.*, 102(A5), 9497. https://doi.org/10.1029/97JA00196

Shue, J.-H., Song, P., Russell, C. T., Steinberg, J T., Chao, J. K., Zastenker, G., Vaisberg, O. L., Kokubun, S., Singer, H. J., Detman, T. R. & Kawano, H. 1998, *J. Geophys. Res.* 103(A8), 17691. https://doi.org/10.1029/98JA01103

Shugaev, F. V. & Kalinchenko A. P. 2005, in: V. A. Bityurin (ed.) *Proceedings of the 15 International Conference on MHD Energy Conversion and the 6 International Workshop on Magnetoplasma Aerodynamics*, (IVTAN: Moscow), v.2, p.617

Stahara, S. S. 2002, *Planet. and Space Sci.*, 50, 421

Tsyganenko, N. A. 1989, *Planet. and Space Sci.*, 37(9), 1037

Verigin, M. I., Kotova, G. A., Bezrukikh, V. V., Zastenker, G. N. & Nikolaeva, N. F. 2009, *Geomagn. and Aeronomy*, 49(8), 1176

Verigin, M., Tatrallyay, M., Erdos, G., Kotova, G., Bezrukikh, V. & Remizov, A. 2014, *The 40-th COSPAR Scientific Assembly*, 2–10 Aug. 2014, Moscow, Russia, Abstract D3.5-0014-14, 2014

Verigin, M. I., Kotova, G. A., Bezrukikh, V. V. & Remizov A. P. 2015, *EPSC Abstracts*, v. 10, EPSC2015-708

Space Weather of the Heliosphere: Processes and Forecasts
Proceedings IAU Symposium No. 335, 2017
C. Foullon & O.E. Malandraki, eds.

doi:10.1017/S1743921318000662

Analysis of Magnetic Field Variations Produced by Equatorial Electro-Jets

Ericson López[1] Franklin Aldás[2] and Akimasa Yoshikawa[3]

[1]Observatorio Astronómico de Quito, Escuela Politécnica Nacional, Quito, Ecuador
email: ericsson.lopez@epn.edu.ec

[2]Observatorio Astronómico de Quito, Escuela Politécnica Nacional, Quito, Ecuador
email: franklin.aldas@epn.edu.ec

[3]International Center for Space Weather Science and Education, Kyushu University, Japan
email: yoshi@geo.kyushu-u.ac.jp

Abstract. The Equatorial Electrojet (EEJ) is a narrow band of electrons flowing from east to west at daytime at low latitudes. The electron current produces a magnetic field variation that can be measured at different latitudes. In this work, we have used the data analysis in order to quantify the solar and lunar contributions to those variations and study the morphology of the EEJ current.

Keywords. Equatorial Electrojet, magnetic field, data analysis.

1. Introduction

Observations of the magnetic field at Huancayo in 1922 showed the existence of Equatorial Electrojet (EEJ), which is a narrow band of electrons traveling from east to west, 180 km above the surface of the Earth (Chapman 1951). Wind-driven currents, through the dynamo mechanism, produce an accumulation of positive charges at sunrise and negative at sunset (Huang and Richmond 1995), which results in considerable variations of the geomagnetic field. The variations of the magnetic field are no constant, instead, they depend on solar activity, geographical location and vary temporally with season (Rigoti *et al.* 1999).

2. Data Set

Ground magnetometers allow a deeper understanding of the terrestrial ionosphere. The Quito Astronomical Observatory, in collaboration with the University of Kyushu, has installed a Magnetic Acquisition System (MAGDAS) in Jerusalem, located 35 km northward from Quito, Ecuador. The magnetometer location has geographic coordinates GLAT: 0°0' 8.67" S, GLON: 78°21'24.87" W (López *et al.* 2014), and it is operative from October 2012 until now.

We have used the horizontal component of the magnetic field (H) recorded by the Ecuadorian MAGDAS magnetometer in combination with the data downloaded for free from the project: A Global Ground-Based Magnetometer Initiative, available at web page (`http://supermag.jhuapl.edu/mag/?`), during the period since October 2012 to December 2016. The geographic and geomagnetic locations of the stations used in this work are indicated in the Table 1.

Station	Acronym	GLON	GLAT	MLON	MLAT
Vulcan	T03	247.02	50.37	-50.40	57.60
Boulder	BOU	254.77	40.13	-38.68	48.51
Fredericksburg	FRD	282.63	38.20	-0.64	48.05
Fresno	FRN	240.30	37.10	-54.87	42.65
Bay Sait Louis	BSL	270.37	30.35	-17.88	40.69
Tucson	TUC	249.27	32.17	-43.96	39.32
San Antonio	M08	261.39	29.44	-29.35	38.74
San Juan	SJG	293.85	18.11	11.85	26.75
Jerusalem	JRS	282.64	-0.0024	-5.9800	9.67
Kourou	KOU	307.27	5.21	24.46	8.62
Huancayo	HUA	284.67	-12.05	-2.75	1.17
Villa Remedios	VRE	292.38	-17.28	3.97	-4.53
Pilar	PIL	294.47	-31.40	4.76	-18.39
Osorno	OSO	286.91	-40.34	-0.03	-26.48
Punta Arenas	PAC	289.91	-40.34	1.97	-26.64
Port Stanley	PST	302.11	-51.70	-38.96	-38.96

Table 1. Geographic and geomagnetic positions of ground stations.

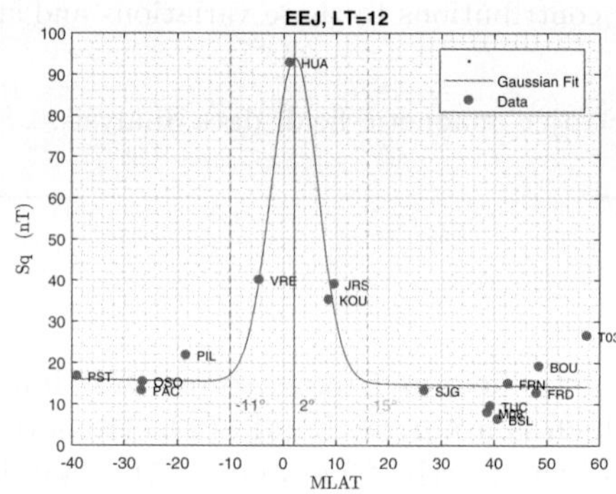

Figure 1. Profile of secular variations (S_q) as function of the magnetic latitude (MLAT).

3. Data Analysis

The magnetic field measurements provided by the ground stations is the sum of multiple contributions, among the more significant are: the Earth main magnetic field, the secular current (S_q), the field aligned currents and so on. In order to study the cyclic variations, we have removed the non-periodic variations by correction of the disturbance field (δ_H) described by Yamazaki & Maute (2017). The zero level of secular variations (S_q) has been obtained using the hourly D_{st} index which provides a quantitative measure of geomagnetic disturbance, that can be correlated with other solar and geophysical parameters. For an station with a geomagnetic latitude λ, the disturbance field δ_H over H can be related by the expression $\delta_H = D_{st} * cos(\lambda)$ (Takeda, 2002).

Fourier Series Analysis. Given the periodicity of the secular variations S_q of the terrestrial magnetic field, it is possible to expand it in Fourier series, it means to express S_q as an infinite sum of sines and cosines. In general, in order to capture the solar variability, it is enough if we use the first four solar harmonics (i.e., 24, 12, 8 and 6 hours).

The mean values of average secular variations (S_q) as function of magnetic latitude (MLAT) are plotted in the Figure 1, where we can see that the EEJ current band is symmetric, centered at 2.2° MLAT and with full-width at half-maximum of 18.12°.
Lunar Contributions to Magnetic Field Variations. We define the lunar age ν by analogy

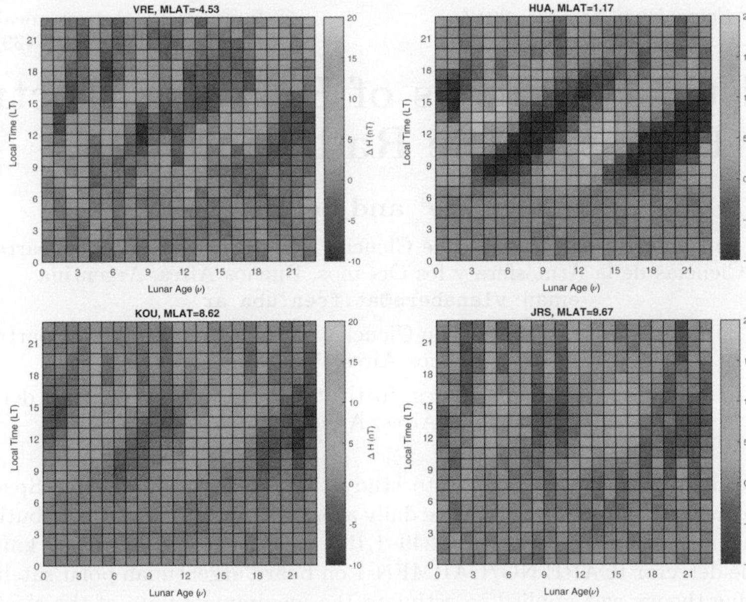

Figure 2. Lunar variations for four different stations located near to dip equator.

with the solar day. A lunar cycle (29.5306 solar days) has a duration of 24 lunar hours. This means that 0 lunar hours represents the new moon and 12 lunar hours represents the full moon. The Moon has a significant contribution during EEJ production, in order to quantify the lunar contribution we have subtracted the first harmonic calculated through the Fourier series expansion from the original data. The residual can be considered as the contribution of the Moon to the variation of the horizontal magnetic field. Then, the residuals are grouped by lunar age and local time (LT) as we show in the Figure 2.

4. Conclusions

From the Gaussian fit of the secular variations of the horizontal magnetic field component as a function of the geomagnetic latitude, we are able to derive that the full-width at half-maximum for average EEJ on America's equatorial region is 18.12° . The mean EEJ current during the years 2012 - 2016 extends from $-11°$ to $15°$ magnetic latitudes and it is centered at $MLAT = 2°$.

We found a dependence on the lunar age and the day time in the electrojet generation. Our results are similar to those presented by Yamazaki & Maute (2017), but derived by another method. Although, the Moon contribution is important in the EEJ formation, its contribution to the magnetic field is, in average, five times less important than the Sun contribution.

References

Chapman, S. 1951, *Arc. Metereol. Geophyis. Res.*, 4, 369-390
Huang, C.-M. & Richmond A. D. 1995, *J. Geophys. Res.*, 110, A05312
Gjerloev, J. W. 2012, *J. Geophys. Res.*, 117, A09213
López, E, Maeda, G. *et al.* 2014, *Sun and Geosphere*, 9, 31-34
Rigoti, A., Chamalaun, N. B. & Padilha, A. L. 1999, *Earth Planets Space*, 51, 115-128
Takeda, M. 2002, *J. Atmos. Sol.*, Terr. Phys. 64, 1617-1621
Yamazaki, Y. & Maute, A. 2017, *Space Sci. Rev.*, 2017, 206-299

Space Weather of the Heliosphere: Processes and Forecasts
Proceedings IAU Symposium No. 335, 2017
C. Foullon & O.E. Malandraki, eds.

doi:10.1017/S1743921317011000

Statistical Analysis of Extreme Electron Fluxes in the Radiation Belts

Vanina Lanabere[1] **and Sergio Dasso**[1,2,3]

[1]Universidad de Buenos Aires, Facultad de Ciencias Exactas y Naturales, Departamento de Ciencias de la Atmósfera y los Océanos, Buenos Aires, Argentina
email: vlanabere@at.fcen.uba.ar

[2]Universidad de Buenos Aires, Facultad de Ciencias Exactas y Naturales, Departamento de Física, Buenos Aires, Argentina

[3]CONICET - Universidad de Buenos Aires, Instituto de Astronomía y Física del Espacio, Buenos Aires, Argentina

Abstract. The main aim of this work is to study the frequency of extreme Space Weather events, in particular to analyse the tails of the daily averaged electron fluxes distribution function for different channels of energy between 0.249–1.192 MeV measured at $\sim$ 600 km of altitude with the particle detector ICARE-NG/CARMEN-1 on board argentinian polar satellite SAC-D. An extreme value theory was applied to estimate the maximum values of the electron flux in the outer radiation belt for different return levels. We found that the cumulative distribution function of the extreme electron fluxes presents a finite upper limit in (1) the core of the outer radiation belt for the lower energy channels and (2) in the inner edge of the outer radiation belt for energy channels larger than 0.653 keV. The results presented in this work are important to characterise Space Weather conditions.

Keywords. interplanetary medium, methods: statistical, (Sun:) solar-terrestrial relations

---

1. Introduction

The radiation belts were discovered in 1958 by van Allen. These regions in the terrestrial space environment present ions and energetic electrons (e.g., Prölss 2012) trapped by the geomagnetic field. In particular, the outer radiation belt populations present large variability in time and the fluxes can significantly increase, mainly during a geomagnetic storm, due to interplanetary plasma perturbations near Earth (e.g., Lugaz 2016). These suprathermal electrons can affect the satellite technologies in different ways depending on their energy. Electrons with energies of $\sim$ 100 keV can produce damage on the surface materials of the satellite (Koons & Fennell 2006). Larger energy electrons, a few MeV, can penetrate into the outer shield of the satellite and produce damage inside the spacecraft (Wrenn 2002). Thus, it is very important to have knowledge about extreme fluxes of energetic electrons for the development of new satellite technologies.

Koons (2001) and Meredith *et. al.* (2015) studied daily electron fluxes with energies larger than 2 MeV with GOES satellite and using the peaks over threshold (POT) method. Other analyses were done with the Maximum of Blocks method and for energies levels between some kev–MeV (e.g., O'Brien 2007 and Meredith *et al.* 2016).

In this work, we explore the extreme electron fluxes with energies in the range of 0.249 MeV to 1.192 MeV measured with the particle detector ICARE-NG on board the argentinian polar orbit satellite SAC-D. We applied the POT method in two regions of the outer radiation belt: the inner edge (L = 3.5–3.75) and in the core (L = 4.75–5), with L the McIlwain parameter. In the "Method" section, the extreme value analysis used to study the extreme electron fluxes in the outer radiation belt is described. In the

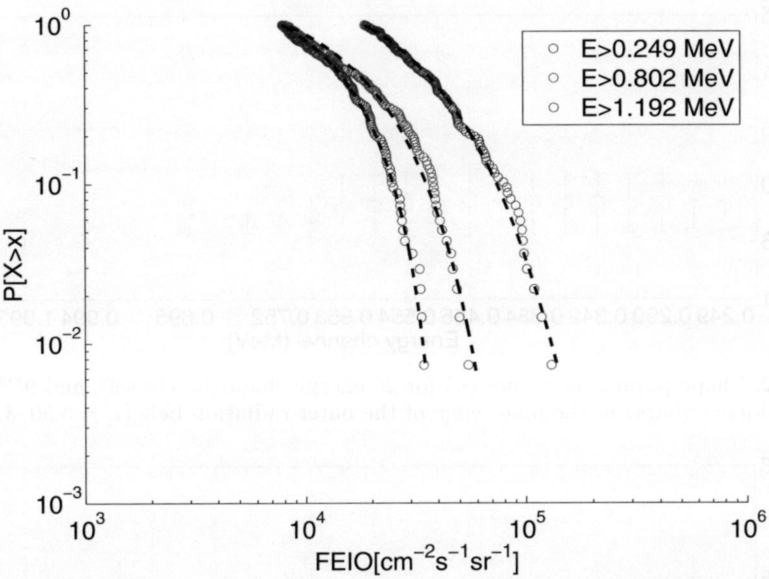

Figure 1. Extreme value analysis for three energy channels. Cumulative distribution from observations (circles) and associated fitted GP functions (dashed)

"Results" section we present the shape parameters that describe the distribution tails behaviour and the return levels for both radiation belts regions. Finally, we present the conclusions of this work.

2. Method

The extreme value theory is a statistical technique used for modelling and estimation of the tail distribution behaviour. In this work, the POT method is applied to daily averaged electron fluxes, measured by the particle detector ICARE-NG on board the argentinean satellite SAC-D during the whole mission (i.e., August/2011 to June/2015). Since the electron fluxes data have been contaminated by protons fluxes during solar proton events (SPE), we removed the data corresponding to SPE using the SPE list documented by NOAA (ftp://ftp.swpc.noaa.gov/pub/indices/SPE.txt). In particular, to study the extreme fluxes in the outer radiation belt, the threshold was defined for each energy channel and for each L value by the percentile 90% of the daily averaged electron flux. The extreme values series are reconstructed for 20 energy channels in the range of 0.249 MeV to 1.192 MeV and for the two regions in the outer radiation belt. In the case of the POT method, the appropriate function to fit the cumulative PDF of extreme events is the Generalised Pareto (GP) distribution (Picklands, 1975). We applied the maximum likelihood method to find the free parameters of the generalised Pareto cumulative distribution function, defined such as:

$$G_{(\kappa,\mu,\sigma)}(X) = \begin{cases} 1 - \left(1 + \frac{\kappa(X-\mu)}{\sigma}\right)^{-\frac{1}{\kappa}}, & \text{for} \quad \kappa \neq 0 \\ 1 - \exp\left(-\frac{X-\mu}{\sigma}\right), & \text{for} \quad \kappa = 0 \end{cases}$$

where X is the random variable associated with the electron flux (FEIO), μ and σ are the location and scale parameters, respectively. The shape parameter, κ, describes the

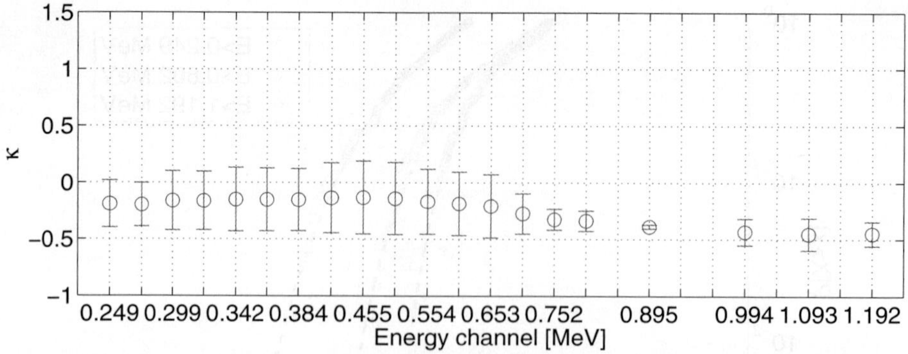

Figure 2. Shape parameter values (κ) for 20 energy channels (circles), and 95% interval confidence (bars) in the inner edge of the outer radiation belt (L = 3.50–3.75).

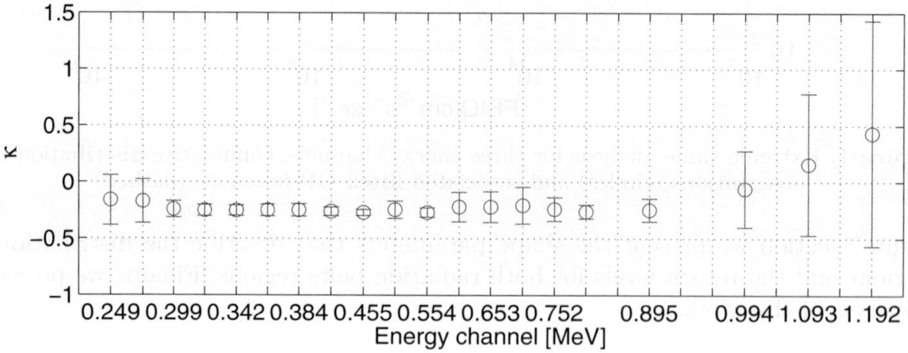

Figure 3. Same as Fig. 2 but in the core of the outer radiation belt (L = 4.75–5.00).

extreme behaviour of the distribution. The GP distribution has three basic forms depending on the value of the shape parameter: i) distributions whose tails decrease exponentially, such as the normal, lead to a GP shape parameter of zero, ii) distributions whose tails decrease as a polynomial, such as Student's t, lead to a positive shape parameter and iii) distributions whose tails are finite, such as the beta functions, as such as functions decaying as a power law, lead to a negative shape parameter (Coles, 2001).

The cumulative frequencies for three different energy channels in the inner edge of the outer radiation belt are shown in Figure 1.

3. Results and Conclusions

The shape parameter (κ) values and the 95% interval confidence for the 20 energy channels in the inner edge and in the core of the outer radiation belt are shown in Fig. 2 and Figure 3. It is observed in Figure 2 that for the higher energy channels (E >0.653 MeV to E >1.192 MeV) the shape parameter is negative with 95% confidence. While in the lower energy channels, it is not possible to determine the sign of κ. A different behaviour is observed in the core of the outer radiation belt (Figure 3). In this case the channels of lower energy (0.3 MeV< E <0.994 MeV) have values of κ significantly negative.

A statistical study of the extreme values of the daily averaged electron flux in the range of energies E = 0.249 MeV to E = 1.192 MeV in two regions of the outer radiation belt was carried out, using data from SAC-D between August/2011 and June/2015. Our results show a different behaviour in the interior and in the core of the outer radiation belt. For energies of 0.653 MeV$< E <$ 1.192 MeV, an accumulated distribution of flows with finite upper limit was found at the inner edge of the outer belt. On the other hand, in the core of the outer radiation belt, κ presents significantly negative values in channels of lower energy ($E = 0.249$ to $E = 0.895$) MeV. The expected extreme value distribution for different energy ranges, as the ones studied here, will be useful for decisions makers. In particular, it will be very important during the design of satellites and new satellite technologies, according to the orbit location and to the potential damages on the devices on board produced by energetic particles, and considering different energy ranges.

References

Koons, H. C. 2001, *J. Geophys. Res.*, A106, 2156

Koons, H. C. & Fennell J. F. 2006, *Radio Sci. Bul. Int. Union Radio SCi. (URSI)*, 316, 27–41

Lugaz, N., Farrugia, C. J., Huang C. L., Winslow R. M., Spence H. E. & Schwadron, N. A. 2016, *Nature Communications*, 7

Meredith, N. P., Horne R. B, Isles, J. D. & Rodriguez, J. V. 2015, *Space Weather*, 13, 170

Meredith, N. P., Horne R. B, Isles, J. D. & Green, J. C. 2016, *Space Weather*, 14, 136

O'Brien, T. P., Fennell, J. F., Roeder, J. L. & Reeves, G. D. 2007, *Space Weather*, 5

Picklands, J. 1975, *Ann. Stat.*, 3

Prölss, J. 2012, *Springer Science & Business Media*

Wrenn, G. L., Rodgers, D. J. & Ryden, K. A. 2002, *Ann. Geophys.*, 20, 953

Coles, S. 2001, *Springer Science & Business Media*

Space Weather of the Heliosphere: Processes and Forecasts
Proceedings IAU Symposium No. 335, 2017
C. Foullon & O.E. Malandraki, eds.

doi:10.1017/S1743921318000522

Four-Spacecraft Magnetic Curvature Analysis on Kelvin-Helmholtz Waves in MHD Simulations

R. Kieokaew[1], C. Foullon[1] and B. Lavraud[2]

[1]CGAFD, Mathematics, CEMPS, University of Exeter, Exeter, the United Kingdom

[2]Institut de Recherche en Astrophysique et Planétologie, Université de Toulouse, Toulous, France

Abstract. Four-spacecraft missions are probing the Earth's magnetospheric environment with high potential for revealing spatial and temporal scales of a variety of in-situ phenomena. Magnetic curvature is intrinsic to curved magnetic fields where the magnetic energy is stored in the form of magnetic tension. In-situ magnetic curvature has been resolved by the four-spacecraft technique called "magnetic curvature analysis" (MCA). We test the MCA on 2.5D MHD simulations of curved magnetic structures induced by Kelvin-Helmholtz (KH) waves, with increasing (regular) tetrahedron sizes of virtual spacecraft. We have found variations of the curvature vectors both in radii and orientations depending on the sizes of the tetrahedron. This is helpful to better understand the MCA measures when the technique is applied to in-situ data without knowing the scale sizes of plasma structures under consideration. This study lends support for cross-scale observations to better understand the nature of curvature and its role in plasma phenomena.

---

1. Introduction

The Earth's magnetosphere outer boundary, the magnetopause, is the site of fundamental plasma processes that allow the entry of solar wind plasma to the magnetosphere. Entry of plasma under southward IMF conditions is dominated by magnetic reconnection. Kelvin-Helmholtz (KH) instabilities, arising from a shear flow at the magnetopause boundary, have been proposed as a candidate mechanism for the penetration of solar wind plasma, via magnetic reconnection and turbulence inside rolled-up KH vortices. Entry via KH instabilities is believed to occur under northward IMF conditions when magnetic reconnection at the dayside magnetopause is less dominant. Numerous observations of KH waves at the magnetopause have been studied by the four-spacecraft *Cluster* (e.g., Foullon *et al.* 2008) and recently the Magnetospheric Multiscale (MMS) missions (e.g., Eriksson *et al.* 2016). Unlike single spacecraft, four spacecraft which form tetrahedron configurations can resolve 3D plasma geometries at their barycentre. A plasma characteristic such as 'magnetic curvature' can influence properties of plasma waves and instabilities, and motion types of charged particles. In-situ magnetic curvature has been resolved in current sheets by the four-spacecraft technique called "magnetic curvature analysis" (MCA) (Shen *et al.* 2003) by expanding a curvature $\mathbf{C} = \partial\mathbf{b}/\partial S = \mathbf{b}\cdot\nabla\mathbf{b}$, where $\mathbf{b} = \mathbf{B}/|\mathbf{B}|$ is the unit magnetic field and S is the arc-length, into $C_j = B^{-2}\Sigma_i B_i \nabla_i B_j - B^{-4} B_j \Sigma_{i,k} B_i B_k \nabla_i B_k$, where i, j, k label Cartesian coordinates $\{x, y, z\}$. A scale of the structure can be estimated from a curvature radius $R_c = 1/|\mathbf{C}|$. Truncation errors of the technique is of order $(a/D)^2$, where a is the spacecraft separation and D is the scale size of the structure (R_c). Motivated by the

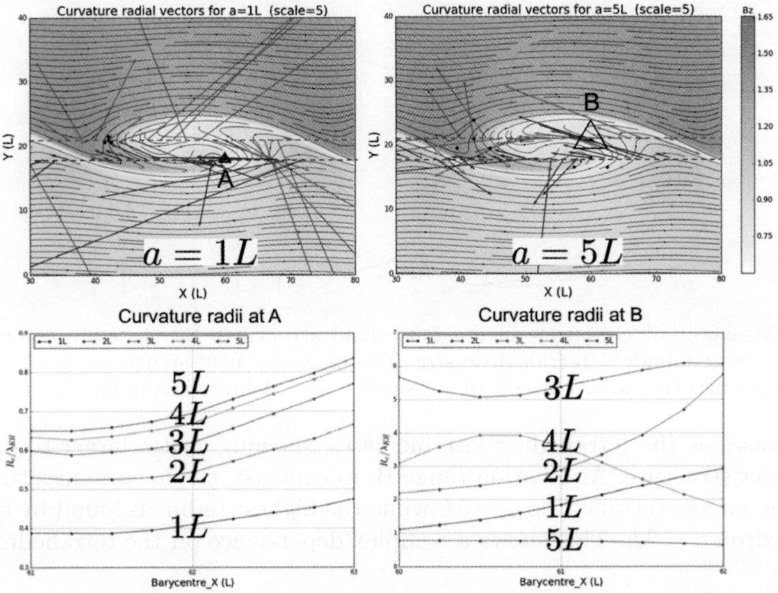

Figure 1. Measures of magnetic curvature on the KH wave by virtual spacecraft. (top) Curvature radial vectors on magnetic contours at two Y-locations from the small (left) and large (right) tetrahedron configurations. (bottom) Curvature radii from the varying tetrahedron sizes with the barycentre at A and B labeled in top panels.

curved, vortical structures of KH waves, we aim to apply the MCA on simulated KH waves to consider applications in real data.

2. Numerical Simulation and Analysis

We simulate the KH instability using Lare2d which is a resistive MHD code in 2.5D (Arber *et al.* 2001). The simulation reproduces KH waves for typical conditions along the Earth's flank magnetopause on the duskside in GSM coordinates. Magnetic field, ion density, ion temperature, and ion speed are modelled with hyperbolic tangent profiles with asymptotic values of the magnetosheath and magnetospheric sides as in Otto and Fairfield, (2000). The KH wavelength is $\lambda_{KH} = 40L$, where $L = 600$ km. We set up virtual probes, to sample across the simulation plane, in a regular tetrahedron configuration and vary the separation at typical *Cluster* scale-sizes $a = [1L, 5L]$.

Figure 1 shows measures of magnetic curvature along the X-direction of a time snapshot of the non-linear KH waves at two Y-locations. The top panels show a comparison between small ($a = 1L$, left) and large ($a = 5L$, right) tetrahedron results. Curvature radial vector projections $\mathbf{R_c} = R_c\mathbf{c_x} + R_c\mathbf{c_y}$ are plotted on magnetic field maps, with components B_z in grey and B_x, B_y as contours. Each curvature radial vector points from the tetrahedron barycentre in the direction of magnetic tension force. The small tetrahedron results show that the curvatures roughly point against the flow direction in the KH vortex (towards -X for the top cut and towards +X for the lower cut), consistent with distortion of the magnetic fields induced by the vortical flow. The large tetrahedron results, in contrast, show different curvatures, in both magnitudes and directions. In the lower panels, we further examine curvature radius for the varying tetrahedron sizes at chosen location ranges, labeled as A and B in top panels. The barycentres at A and B are in range X=[61, 63] and X=[60, 62] respectively. At location range A, the curvature

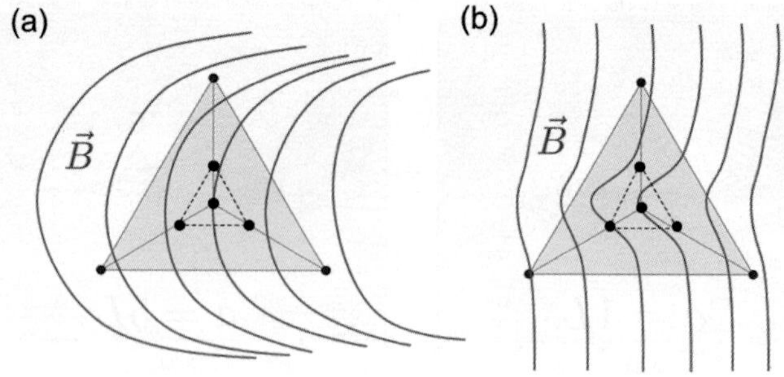

Figure 2. Measures of curvature in magnetic field structures by nested spacecraft of small (dashed) and large (shaded) tetrahedron size. The magnetic field structures have spatial variations/non-zero curvature at (a) large and (b) small scales. Adapted from Kieokaew *et al.* (2018).

radius increases as the tetrahedron size increases, showing that a larger tetrahedron resolves a larger structure. At location range B, in contrast, the lowest curvature radius is found by the largest tetrahedron $a = 5L$ while the highest radius is found by the medium size tetrahedron $a = 3L$. This shows a complex dependence on the tetrahedron size.

3. Discussions and Summary

The variations of the MCA results regarding the tetrahedron size may result from spatial variations in the magnetic structures or from limitations of the technique. Figure 2 illustrates two scenarios of magnetic field structures that lead to the results in Figure 1: those with high spatial variations at (a) large scale and (b) small scale. As the tetrahedron size increases, the scenarios would lead respectively to (a) decreasing and (b) increasing radius of curvature. The scenario (b) provides a plausible explanation for the results at location range A in Figure 1, where the curvature radius is increasing with the tetrahedron size. Combinations of both scenarios would result in the complex dependence on the tetrahedron size in location range B: the magnetic structure is consistent with the scenario (b) in the small to medium scales (1L to 3L) while it is consistent with the scenario (a) in the large scales (4L to 5L).

The MCA technique has been applied in a simulation of KH waves using varying tetrahedron sizes. The technique has revealed the magnetic distortion, useful for future analyses with real data, i.e., for identifying characteristic regions of KH waves. Magnetic curvature radius and direction vary depending on the sizes of the tetrahedron, possibly due to spatial variations in the magnetic structures. Such plasma structures with non-linear spatial variations are common in space and would be best observed by nested, cross-scale four-spacecraft. Further details of this study on magnetic curvature and complementary analysis on vorticity can be found in Kieokaew *et al.* (2018).

References

Arber, T., Longbottom, A. W., Gerrard, C. L. & Milne, A. M. 2001, *J. Comput. Phys.*, 171, 151

Chanteur, 1998, *ISSI Scientific Report*, SR-001, 349

Eriksson, S., *et al.* 2016, *Geophys. Res. Lett.*, 11, 5606

Foullon, C., *et al.* 2008, *J. Geophys. Res.*, 113, A11203

Otto, A. & Fairfield, D. H. 2000, *J. Geophys. Res.*, 105(A9), 21175

Kieokaew, R., Foullon, C. & Lavraud, B. 2018, *J. Geophys. Res.*, 123, 1-17

Shen, C., *et al.* 2003, *J. Geophys. Res.*, 108(A5), 1168

Space Weather of the Heliosphere: Processes and Forecasts
Proceedings IAU Symposium No. 335, 2017
C. Foullon & O.E. Malandraki, eds.

doi:10.1017/S1743921317007190

Space Observations to Determine the Location of Locally Vertical Geomagnetic Field

Stefania Lepidi, Domenico Di Mauro, Roberta Tozzi, Lili Cafarella, Paola De Michelis and Martina Marzocchetti

Istituto Nazionale di Geofisica e Vulcanologia, Roma, Italy

Abstract. The points where the horizontal component of the geomagnetic field vanishes are located in polar areas, far away from the geomagnetic (analytic) poles and the poles of rotation of the Earth and, differently from the geomagnetic poles, can be found experimentally with a magnetic survey to determine where the field is vertical. The experimental determination of the area where the total field is perfectly vertical, commonly known as dip pole, is not simple, due to the remoteness and harsh climatic conditions; another difficulty is related to the short term geomagnetic field variations, due to the interaction with the external solar wind, which causes the magnetospheric dynamics, particularly evident at high latitude, and as a consequence a displacement of the dip pole. Actually, the study of the dip pole displacements over short time scales can be an important tool for monitoring the magnetospheric dynamics at high latitude. In this study we present the updated location of the the dip poles, using data from the Swarm ESA's constellation of satellites along their almost polar orbits. We also analyse the spatial shift of these areas during different seasons and interplanetary magnetic field orientations.

Keywords. Solar-terrestrial relations, Earth magnetic field.

1. Introduction

The systematic observations of the Earth's magnetic field, older than those of meteorology and seismology, allowed Gauss since mid of 19th century to recognize that the main contribution to the field comes from its dipolar component. The quality of the analytic model of the Earth's magnetic field has been improved along the decades with the growing number of contributing observatories around the world. The most important contribution comes from observations collected from remote and inaccessible areas of the planet, as polar areas where the field is stronger than in any other place and its geometric dipolar pattern shows high space gradients. The points where the horizontal component of the Earth's magnetic field (H) vanishes are known as vertical or dip magnetic poles. We take the advantage of Swarm ESA's constellation with its polar orbit for determining the current position of the North Magnetic Pole (NMP) and of the South Magnetic Pole (SMP). In particular we analyse the dataset from Swarm A and B in the time interval January 1, 2014-April 30, 2016. We also analyse the NMP and SMP spatial shift under different interplanetary magnetic field (IMF) orientations and compare our results with the dip poles determination from analytic models based on ground observatory dataset interpolation.

2. Overview

The dip poles are the only magnetic poles that can be readily measured. Their shifts, non-antipodality and non-simultaneous motion (Mandea and Dormy 2003) are a clear

Table 1. Mean values of the position of the NMP and SMP for different cases.

	NMP		SMP	
	Lon(°)	**Lat(°)**	**Lon(°)**	**Lat(°)**
All	222.0	86.5	136.1	−65.6
Summer	219.1	86.6	136.2	−65.7
Equinoxes	221.7	86.5	135.1	−65.5
Winter	222.6	86.5	136.3	−65.5
Positive B_y	221.4	86.5	135.8	−65.6
Negative B_y	219.4	86.6	136.4	−65.6
Positive B_z	220.6	86.6	136.1	−65.7
Negative B_z	220.7	86.5	136.1	−65.5

proof of the important changes in the non-dipolar components of the internal field, while the geomagnetic poles, derived by the first three terms of the spherical harmonic analysis of the geomagnetic field, do not present these features. The determination of dip poles from ground spot measurements is difficult because of crustal local magnetic anomalies, limited number of data due to the harsh environment in polar areas, and random geomagnetic conditions at the moment of the measurement. Conversely, Swarm ESA's satellites, which orbit 15 times in a day covering about 8 km in 1 s, continuously collect a huge amount of data at different geomagnetic conditions; moreover, at the satellite altitude (approximately 460 km for Swarm A and 530 km for Swarm B), crustal small scale wavelength structures give very weak contributions.

3. Experimental results

We defined the experimental position of the dip pole as the point where H is minimum and lower than a fixed threshold value. The geographic coordinate system is used in this study. The number of selected points is much higher in the Northern hemisphere, so we chose a different H threshold value in the two hemispheres. In each figure we show the experimental positions for the NMP and SMP (red points), their mean value (blue point) and the position of the magnetic poles as determined from IGRF model (green star); note the different latitudinal limits for the two hemispheres. We also considered separately the three Lloyd seasons: may, jun, jul, aug (summer and winter in the northern and southern hemisphere respectively); nov, dec, jan, feb (winter and summer respectively); mar, apr, sep, oct (equinoxes). From Figure 1 it is evident that the dip pole position is more scattered for the NMP with respect to the SMP (top) and also in summer with respect to winter (bottom).

We also considered different IMF orientations, in particular the North-South (B_z) and East-West (B_y) IMF components (data from *http://cdaweb.gsfc.nasa.gov/*). As shown in Figure 2, for negative B_z the pole position has a broader displacement (top) and for positive B_y the mean SMP position shifts toward West, while the NMP shifts toward East (bottom). In Table 1 the mean values of the pole positions for all the examined cases are summarized.

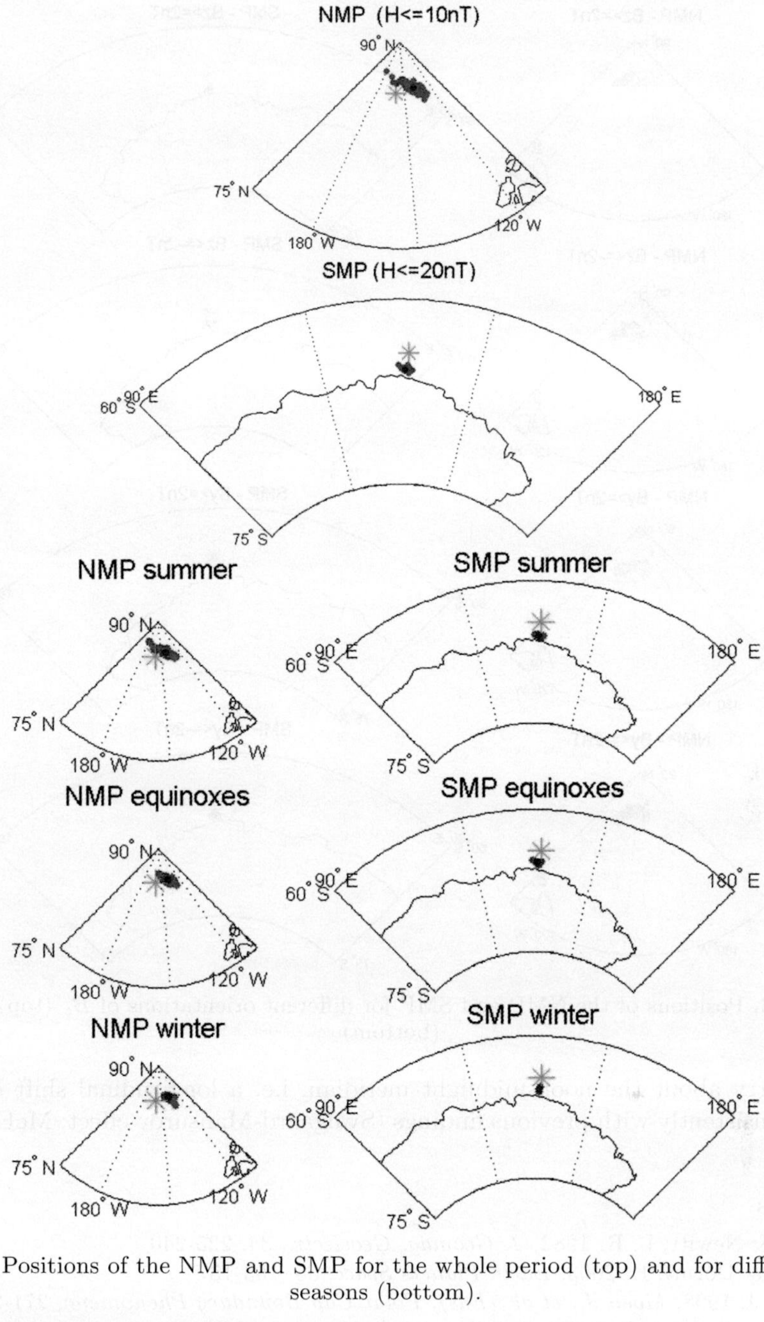

Figure 1. Positions of the NMP and SMP for the whole period (top) and for different Lloyd seasons (bottom).

4. Conclusions

Swarm satellites provide a unique dataset to study the position of the dip poles. Our results show that the NMP position has a larger variation with respect to the SMP, probably due to its faster secular motion (Dawson and Newitt 1982; Newitt *et al.* 2009). The pole position is more stable during local winter, when the ionospheric electric conductivity is reduced due to the absence of solar direct radiation, and for positive IMF B_z, i.e. closed magnetospheric conditions (Stasiewicz 1991). We also found that B_y introduces

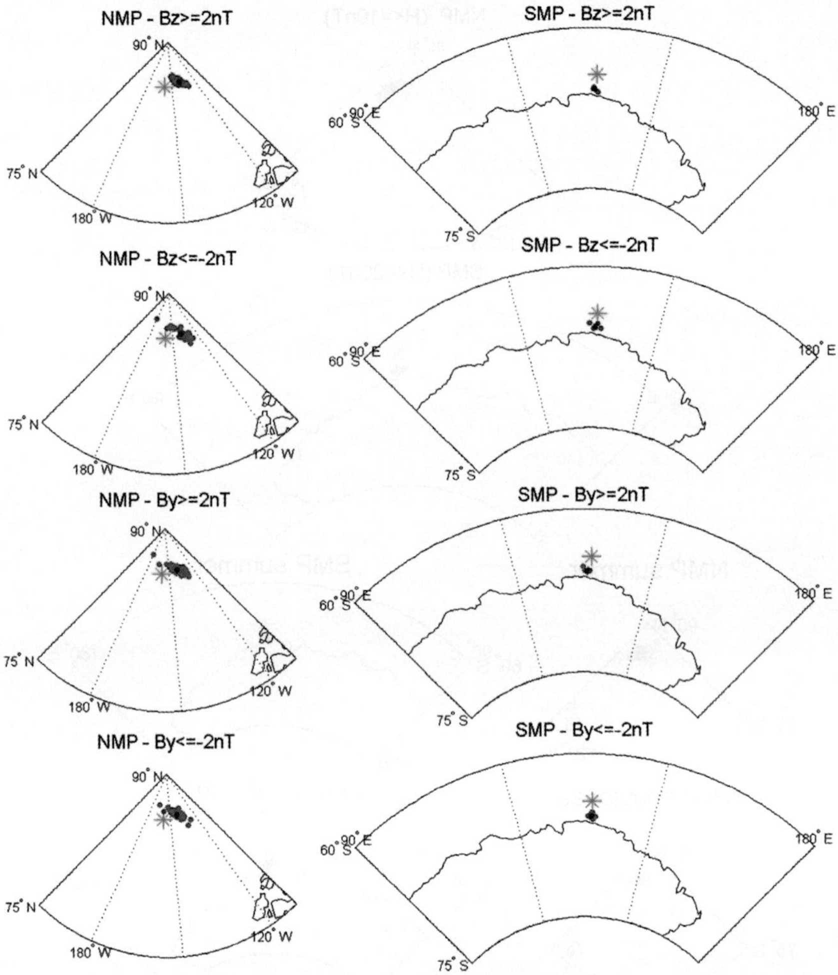

Figure 2. Positions of the NMP and SMP for different orientations of B_z (top) and B_y (bottom).

an asymmetry about the noon-midnight meridian, i.e. a longitudinal shift of the pole position, consistently with previous findings (Svalgaard-Mansurov effect; McEwen 1998).

References

Dawson, E. & Newitt, L. R. 1982, *J. Geomag. Geoelectr.*, 34, 225-240

Mandea, M. & Dormy, E. 2003, *Earth Planets Space*, 55, 153-157

McEwen, D. J. 1998, *Moen J., et al. (Eds), Polar Cap Boundary Phenomena*, 271-280

Newitt, L. R., Chulliat, A. & Orgeval, J.-J. 2009, Location of the North magnetic pole in April 2007, *Earth Planets Space*, 61, 703-710

Stasiewicz, K. 1991, *J. Geophy. Res.*, 15789-15800

Space Weather of the Heliosphere: Processes and Forecasts
Proceedings IAU Symposium No. 335, 2017
C. Foullon & O.E. Malandraki, eds.

doi:10.1017/S174392131701002X

Determining the Polar Cusp Longitudinal Location from Pc5 Geomagnetic Field Measurements at a Pair of High Latitude Stations

Stefania Lepidi[1], Patrizia Francia[2], Lili Cafarella[1], Domenico Di Mauro[1] and Martina Marzocchetti[1]

[1]Istituto Nazionale di Geofisica e Vulcanologia, Roma, Italy
[2]Università degli Studi dell'Aquila, L'Aquila, Italy

Abstract. We use low frequency geomagnetic field measurements at two Antarctic stations to statistically investigate the longitudinal location of the polar cusp. The two stations are both located in the polar cap at a geomagnetic latitude close to the cusp latitude; they are separated by one hour in magnetic local time. At each station the Pc5 power maximizes when the station approaches the cusp, i.e. around magnetic local noon. The comparison between the Pc5 power at the two stations allows to determine the longitudinal location of the cusp. Our analysis is conducted considering separately different orientation of the interplanetary magnetic field. The results, which indicate longitudinal shifts of the polar cusp depending on the selected conditions, are discussed in relation to previous studies of the polar cusp location based on polar magnetospheric satellite data.

Keywords. Solar-terrestrial relations, Earth magnetic field.

1. Introduction

The Pc5 (1.7 − 7 mHz) power at high geomagnetic latitudes can provide an indication of the polar cusp position which can depend on the season and the Interplanetary Magnetic Field (IMF) North-South (B_z) and East-West (B_y) components. In a previous study we found, indeed, that in the noon sector the power reaches a maximum at $\sim 75°$ geomagnetic latitude and it shifts from latitudes slightly lower than 75° during local winter (Nov-Dec-Jan-Feb, in the northern hemisphere) and for a southward GSM IMF ($B_z < 0$) to latitudes of $81 - 82°$ during summer and for $B_z > 0$, i.e. closed magnetospheric conditions (Lepidi and Francia 2002). These findings are consistent with the dependence of the cusp position on season and IMF obtained from optical and satellite observations. The seasonal dependence is related to the dipole tilt angle variations, in that when the dipole tilts more toward the Sun, the cusp moves poleward, to higher latitudes (Zhou *et al.* 1999). The IMF B_z and B_y components have been found to control respectively the latitudinal and longitudinal dynamics of the cusp, which moves from 80° to below 70° for strong positive and negative B_z respectively and, in the southern hemisphere, dawnward for B_y negative and duskward for B_y positive (Newell *et al.* 1989; Moen *et al.* 1999; Zhou *et al.* 2000; Pitout *et al.* 2006; McEwen *et al.* 2016). An IMF B_y component, linearly superposed with the magnetospheric field, can shift the magnetic local time (MLT) of the cusp by a maximum of 0.07-0.15 hour/nT (Stasiewicz 1991). In this study we use the Pc5 power at two Antarctic stations, Mario Zucchelli (MZS) and Scott Base (SBA), located at the same Corrected Geo-Magnetic

Table 1. Geographic and CGM coordinates and MLT noon for the two stations.

Station	Geog. Lat.	Geog. Long.	CGM Lat.	CGM Lon.	MLT noon
MZS	74.7° S	164.1° E	79.9° S	307.2° E	20 : 10 UT
SBA	77.8° S	166.8° E	79.8° S	326.5° E	19 : 00 UT

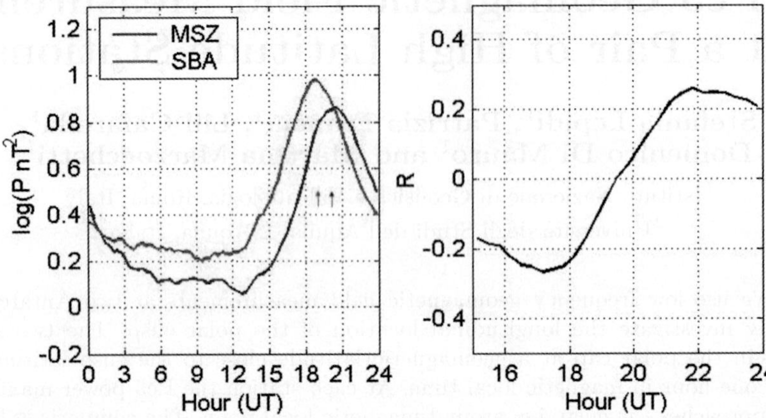

Figure 1. The UT dependence of the average Pc5 power at the two stations (left panel). The UT dependence of the ratio between Pc5 power at MZS and SBA (right panel).

(CGM, `https://omniweb.gsfc.nasa.gov/vitmo/cgm_vitmo.html`) latitude but separated by 1 hr in MLT (Table 1), to examine the longitudinal shift of the cusp for different conditions.

2. Experimental results

In this study we used 1-min values of the total horizontal component H for a year of maximum solar activity (2001). Data were differenced before computing the power spectra in order to reduce lower frequency variations. In Figure 1 (left) we show the UT dependence of the average Pc5 integrated power; it can be seen that at each station a clear power enhancement emerges around MLT noon (indicated by the arrow).

We found interesting to show the logarithmic average ratio R between the Pc5 power at MZS and SBA (Figure 1, right). The average value of R has been subtracted to eliminate any systematic difference in power level at the two stations and make more evident the different time dependence of the power. R is zero (i.e. logarithm of one) at the time T_0 ($\sim 19:30$ UT) when the power at the two stations is the same. From the power time dependence, we expect R to be equal to zero at a time between MLT noon at SBA and MZS, negative before MLT noon at SBA and positive after MLT noon at MZS.

Then, we considered different IMF conditions, restricting our analysis to a few hours containing the MLT noon, i.e. the 15-24 UT time interval (Figure 2). As can be seen, T_0 does not change significantly for different IMF B_z conditions, while it shifts to earlier times ($\sim 19:15$ UT) for positive B_y with respect to negative B_y ($\sim 19:45$ UT).

3. Conclusions

The use of the Pc5 power ratio between two stations longitudinally spaced at the same polar CGM latitude is shown to be useful to provide the polar cusp longitudinal position.

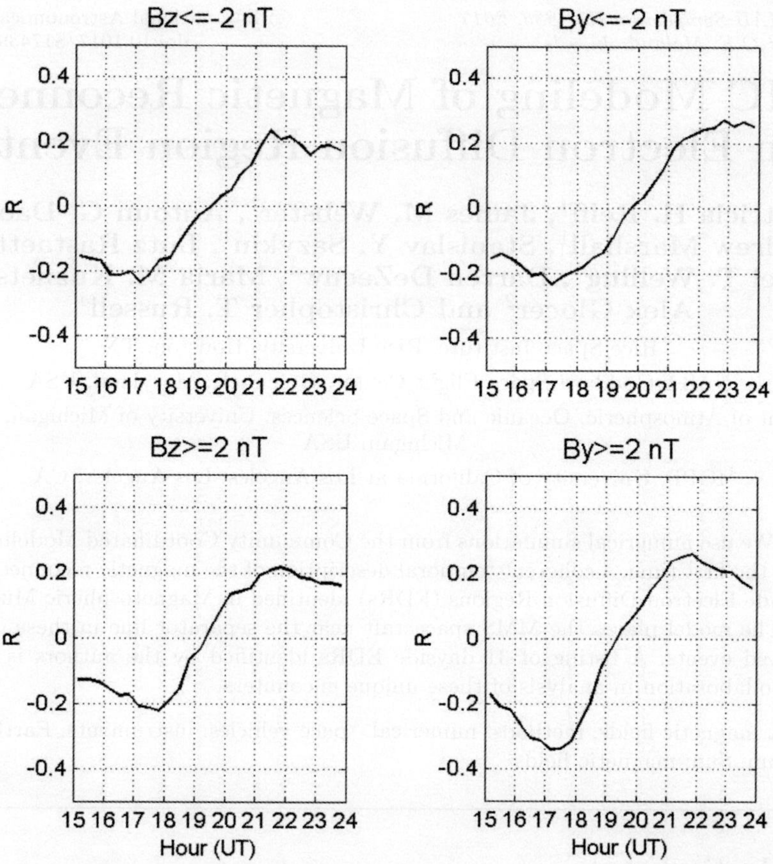

Figure 2. The ratio R during different B_z (left panels) and B_y (right panels) conditions.

It is found that the cusp position depends on the IMF B_y component, shifting to earlier times for positive B_y. The observed shift (of about 0.5 hr) is consistent with the estimate found by Stasiewicz (1991).

References

Lepidi, S. & Francia, P. 2002, *Proceedings "SOLSPA: The Second Solar Cycle and Space Weather Euroconference"*, Vico Equense (Italy), Sept. 24-29, 2001, ESA SP-477, 447-450

McEwen, *et al.* 2016, *Geophysical Mon. 215, Y. Zhang and L. J. Paxton eds.*, AGU

Moen, J., Carlson, H. C. & Sandholt, P. E. 1999, *Geophys. Res. Lett.*, 26, 1243-1246

Newell, P. T., Meng, C.-I., Sibeck, D. & Lepping, R. 1989, *J. Geophys. Res.*, 94, 8921-8927

Pitout, F., Escoubet, C. P., Klecker, B. & Reme, H. 2006, *Ann. Geophysicae*, 24, 3011-3026

Stasiewicz, K. 1991, *J. Geophy. Res.*, 96, doi: 10.1029/91JA01420

Zhou, X. W., Russell, C. T., Le, G., Fuselier, S. A. & Scudder, J. D. 1999, *Geophys. Res. Lett.*, 26, 429-432

Zhou, X. W., Russell, C. T., Le, G., Fuselier, S. A. & Scudder, J. D. 2000, *J. Geophys. Res.*, 105, 245-252

Space Weather of the Heliosphere: Processes and Forecasts
Proceedings IAU Symposium No. 335, 2017
C. Foullon & O.E. Malandraki, eds.

doi:10.1017/S1743921317010845

CCMC Modeling of Magnetic Reconnection in Electron Diffusion Region Events

Patricia H. Reiff[1], James M. Webster[1], Antoun G. Daou[1], Andrew Marshall[1], Stanislav Y. Sazykin[1], Lutz Rastaetter[2], Daniel T. Welling[3], Darren DeZeeuw[3], Maria M. Kuznetsova[2], Alex Glocer[2] and Christopher T. Russell[4]

[1]Rice Space Institute, Rice University, Houston, TX

[2]NASA Goddard Space Flight Center, Greenbelt, Maryland, USA

[3]Department of Atmospheric, Oceanic and Space Sciences, University of Michigan, Ann Arbor, Michigan, USA

[4]IGPP, University of California at Los Angeles, Los Angeles, CA

Abstract. We use numerical simulations from the Community Coordinated Modeling Center to provide, for the first time, a coherent temporal description of the magnetic reconnection process of two dayside Electron Diffusion Regions (EDRs) identified in Magnetospheric Multiscale Mission data. The model places the MMS spacecraft near the separator line in these most intense and long-lived events. A listing of 31 dayside EDRs identified by the authors is provided to encourage collaboration in analysis of these unique encounters.

Keywords. magnetic fields, methods: numerical, space vehicles: instruments, Earth, interplanetary medium, Sun: magnetic fields

---

1. Introduction

The Magnetospheric Multiscale Spacecraft (MMS) (Burch *et al.* 2015, 2016) provides unprecedented insights on magnetic reconnection, and its connection to microphysical processes occurring at Electron (EDR) and Ion (IDR) diffusion region scales. Through the Community Coordinated Modeling Center (CCMC), we ran the Space Weather Modeling Framework (SWMF) model (Powell *et al.* 1999; Tóth *et al.* 2005, 2012) including the Rice Convection Model (RCM) in highest resolution available (9.6 million cells) on several identified dayside EDR candidate events. This model provides additional insights on the global aspects of the reconnection process. For each case, we used measured solar wind and IMF data from OMNI data [http://omniweb.gsfc.nasa.gov/], propagated to the bow shock, and a steady Bx representing its average value for the 15-20 minutes before the time of the event. The SWMF at CCMC requires a steady Bx, and since we are modeling dayside events, the most recent IMF values were used; changing to a 30 or 45-minute Bx average would not have changed the results substantially. We then used the RECONX post processing tool (Glocer *et al.* 2016) to find the separator line and the positions of the magnetic nulls.

2. Modeling the EDR Events

We present field lines traced from CCMC models of two of the dayside EDR events analyzed in detail in Webster *et al.* (2017): January 10 and November 23, 2016. The full list of the 31 candidate EDR events, based on the presence of electron crescents and significant agyrotropy (Swisdak 2016), is given in Table 1. The two time periods

Table 1. 31 Dayside EDR Candidate Events.

Event	Date & Time (UTC)	X_{GSM} (R_E)	Y_{GSM} (R_E)	Z_{GSM} (R_E)	Separation (km)	$(j \cdot E')_{max}$ $nW.m^{-3}$	$(\sqrt{Q_e})_{max}$ index
A01	09/19/2015 07:43:30	6.346	5.399	-2.982	71.57	6.05	0.060
A02	10/16/2015 10:33:30	9.231	6.092	-4.403	13.87	2.42	0.052
A03	10/16/2015 13:07:02	8.310	7.078	-4.800	13.78	22.57	0.090
A04	10/22/2015 06:05:22	9.637	3.481	-1.961	16.93	7.47	0.069
A05	11/01/2015 15:08:06	7.814	6.202	-3.470	14.58	4.15	0.072
A06	12/06/2015 23:38:31	8.516	-3.916	-0.810	19.23	10.13	0.066
A07	12/08/2015 11:20:44	10.233	1.288	-1.364	15.30	8.31	0.084
A08	12/09/2015 01:06:11	9.922	-3.671	-0.928	17.34	1.07	0.051
A09	12/14/2015 01:17:40	10.131	-4.163	-1.191	16.97	7.13	0.095
A10	01/07/2016 09:36:15	8.888	-1.968	-0.733	41.75	6.78	0.047
A11	01/10/2016 09:13:37	8.808	-2.395	-0.775	40.84	13.98	0.066
A12	02/07/2016 20:23:35	3.874	-9.325	-5.720	15.99	0.38	0.057
B13	10/22/2016 12:58:41	6.406	7.700	-4.706	8.87	11.92	0.055
B14	11/02/2016 14:46:18	7.241	8.812	-3.543	8.18	1.09	0.036
B15	11/06/2016 08:40:58	7.943	4.113	-2.826	11.76	8.12	0.075
B16	11/12/2016 17:48:47	6.624	9.165	-1.104	7.35	5.12	0.048
B17	11/13/2016 09:10:41	8.958	4.563	-2.625	11.38	18.28	0.050
B18	11/18/2016 12:08:11	9.596	6.460	-2.509	4.88	1.02	0.050
B19	11/23/2016 07:49:33	9.613	3.232	-1.604	6.43	7.35	0.121
B20	11/23/2016 07:49:52	9.613	3.232	-1.604	6.43	32.32	0.089
B21	11/23/2016 07:50:30	9.620	3.245	-1.608	6.42	8.57	0.078
B22	11/28/2016 15:47:00	8.884	7.184	-0.440	6.32	1.63	0.036
B23	12/11/2016 04:41:50	9.489	-0.056	-0.448	6.89	2.07	0.026
B24	12/19/2016 14:15:02	10.204	4.170	0.934	8.42	2.33	0.050
B25	01/02/2017 02:58:13	9.647	-3.007	-0.649	9.96	2.37	0.026
B26	01/11/2017 04:22:43	10.809	-3.713	-0.154	8.17	8.53	0.055
B27	01/20/2017 12:32:07	9.634	-0.461	1.967	6.47	8.31	0.051
B28	01/22/2017 10:15:46	10.744	-2.138	1.766	5.75	4.13	0.043
B29	01/22/2017 10:15:58	10.750	-2.148	1.764	5.74	3.39	0.044
B30	01/22/2017 10:47:33	10.519	-1.790	1.837	5.86	1.98	0.036
B31	01/27/2017 12:05:43	9.270	-1.370	1.964	6.05	7.51	0.046

(corresponding to four events) modeled here were chosen because of their high values of electric fields and agyrotropy.

In the 9.6M cell adaptive grid used here, the effective grid spacing near the dayside magnetopause is 0.125 R_E (Tóth *et al.* 2005). The color contour shows the particle pressure at X_{SM}=10 R_E (near the location of MMS in both events for Fig 1 (A-E) and Fig 2 (F-J); for Fig 2 A-E, the pressure is shown at a location of Y_{SM} = 3.5 R_E, again near the MMS location. Fig 1 shows the field line tracing for January 10, using

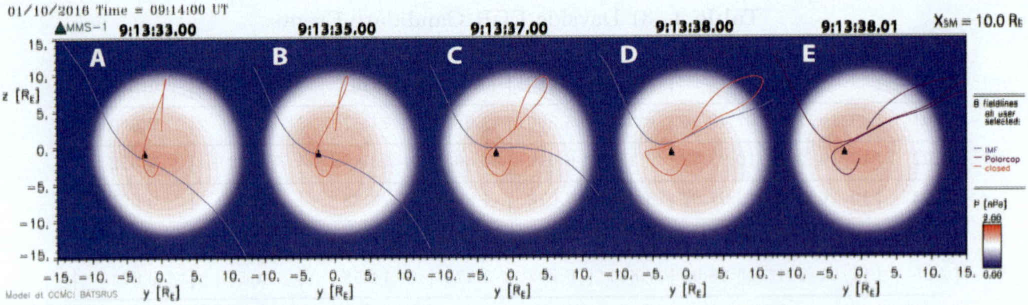

Figure 1. Time sequence for January 10, 2016, showing the plane at X_{SM} =10 R_E. The color indicates particle pressure in that plane. An .mp4 movie including these frames and more can be found in the supplemental material accessible at Cambridge's CJO site: journals.cambridge.org and also at http://mms.rice.edu/mms/Publications/IAU_335/.

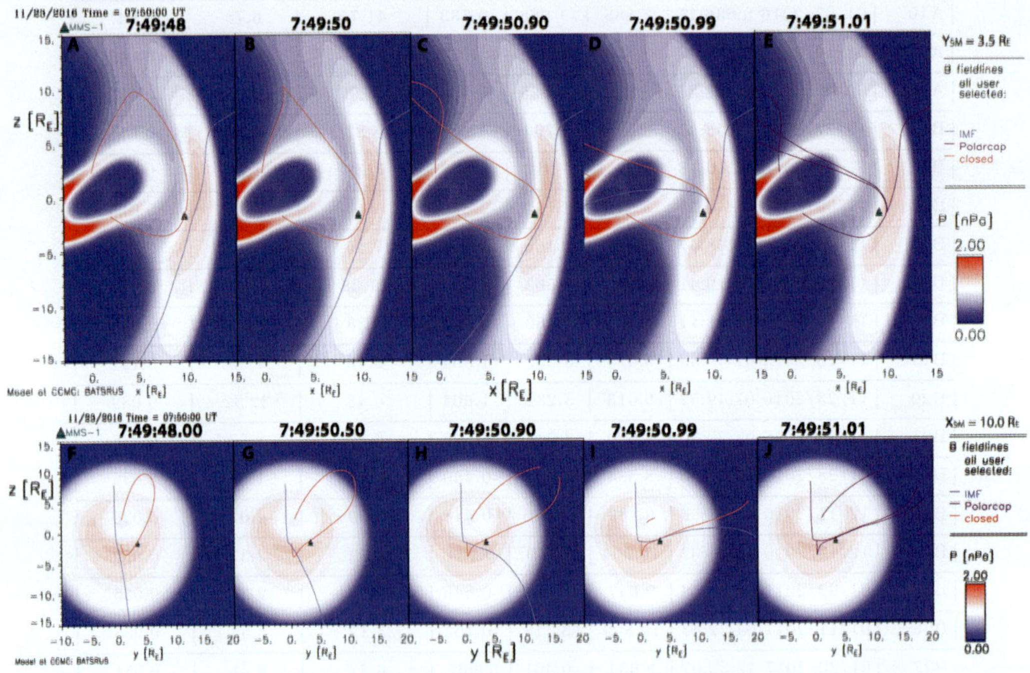

Figure 2. Field lines and particle pressures shown in Y_{SM} =3.5 R_E plane (A-E) and in the X_{SM} =10 R_E plane (F-J) for the EDR events on November 23, 2016. The color contours indicate the particle pressure in that plane. CCMC Run: James Webster_032117_1. Modeling time step for all of the panels is 07:50:00 UT. Two mp4 movies, one each for the two plane views, including these frames and more can be found in the supplemental material accessible at Cambridge's CJO site: journals.cambridge.org and also at http://mms.rice.edu/mms/Publications/IAU_335/

the model run (James_Webster_090216_1). All of the tracings shown here are done for a single time step of the model (09:14). The newly reconnected field lines closest to MMS were found (E), and then, by using the modeled velocity of the northward extension of that open field line, we map its position backwards in time in the previous panels (the blue IMF line from top left to bottom right in panels A-D). Thus the times in each panel are accurate relative times. Note that the IMF does have a substantial By component at the beginning (A), but both the solar wind field lines (blue) and closed magnetospheric field lines (red) exhibit a strongly enhanced By component just before reconnection (D) and just after (E), consistent with the MMS observations. In only 3 seconds the field line undergoes profound stretching as it approaches the reconnection site, and requires only 0.01 second to reconnect. Movies showing these frames and additional frames can be found at http://mms.rice.edu/MMS/publications/IAU_335/ . Fig 2 shows field tracings from the November 23 event, which was actually three encounters with EDRs in Table 1. Looking from the side at the location of MMS (y = 3.5 R_E), the reconnecting fields look nearly two-dimensional (panels A-C), but in D, the field mapping from the northward sheath takes a sharp turn into the y-direction as the IMF field line (blue) nears the dayside null. The lower set of panels are viewed parallel to the YZ plane at a cut near the MMS spacecraft location.

The IMF field line makes nearly a 90° turn at the null point (panel I) as the southern end of the field line continues downstream. The fact that MMS is located very near the separator is likely the reason that it observed three very intense EDR candidates in the space of a single minute (Webster *et al.* 2017). The relative timings on the plot are calculated from the flow velocities in the solar wind, so are accurate on solar wind and open field lines but approximate for the closed field lines. The MMS flow data suggest that the MMS location switches back and forth from below to above the X-line, consistent with remaining in the location of the separator for an extended time.

One striking result seen these two examples is the extreme distortion of both the closed and solar wind field lines as they near the reconnection point. Both field lines take on an induced guide field. This induced guide field is consistent with the solar wind field line being "hung up" on the dayside nose and being stretched as both ends continue to propagate downstream. The surprising result (at least to us) is that the interior field line also stretches to meet the exterior field line so that, at the exact time of reconnection, the interior closed field line and the exterior solar wind field line are actually parallel, allowing flow across the open/closed boundary. Numerical diffusion in the model allows a parallel electric field to break the frozen field condition along that line. Thus, multiple encounters seen on November 23 could be associated with temporally variable reconnection or several reconnection locations along the separator line, or motion of the separator in and out of the spacecraft suite.

Although we found the reconnection lines by interrogating the model near the position of MMS, we also ran the "RECONX" model (not shown) to postprocess the model results to find the null points and the separator line for each of these events. Not surprisingly, RECONX found the separator line to be along the stretched portion of the reconnecting field lines, with one null point on the dayside near MMS and the other null point typically in the magnetotail. In these events (January 10 and November 23, 2016), MMS was not only very near the separator, but also only a few R_E away from the dayside null point. This may explain why these two events have some of the largest electric fields, currents, and agyrotropies of the entire EDR data set.

3. Conclusions

This paper presents the first estimates of the field line travel times through realistic reconnection simulations. The simulation shows clearly how an induced guide field makes the reconnecting field lines parallel to the separator line. This distortion occurs in the last one to three seconds before reconnection. The length of time a given flux tube remains in the reconnection region is extremely small (hundredths to thousandths of seconds). The models predict severe configuration changes on the scale size of an electron gyroradius near the separator line. This paper also presents the importance of using fully 3D reconnection, including separator lines, on the dayside magnetopause.

4. Acknowledgements

The authors thank the MMS team for an amazing suite of instruments. This work was carried out using the SWMF tools developed at The University of Michigan's Center for Space Environment Modeling (CSEM), and RECONX developed by NASA. These tools are made available through the NASA Community Coordinated Modeling Center (CCMC). This work also used plasma and IMF data through CDAWEB. This study was supported by NASA under grant NNX14AN55G.

References

Burch, J. L., Moore, T. E., Torbert, R. B. & Giles, B. L. 2015, *Space Sci. Rev.*, doi:10.1007/s11214-015-0164-9

Burch, J. L. Torbert, R. B., *et al.* 2016, *Science*, 352, doi: 10.1126/science.aaf2939

Glocer, A., Dorelli, J., Toth, G., Komar, C. M. & Cassak, P. A. 2016, *J. Geophys. Res.*, Vol: 121, Pages: 140-156; doi:10.1002/2015JA021417

Powell, K. G., Roe, P. L., Linde, T. J., Gombosi, T. I. & De Zeeuw, D. L. 1999, *J. Comput. Phys.*, 154 (2), 284, doi:10.1006/jcph.1999.6299

Swisdak, M. 2016, *Geophys. Res. Lett.*, 43, 43-49, doi:10.1002/2015GL066980

Tóth, G. Sokolov, I. V., *et al.* 2005, *J. Geophys. Res.*, 110, A12226. doi: 10.1029/2005JA011126

Tóth, G., van der Holst, B., Sokolov, I. V., *et al.* 2012, *J. Comput. Phys.*, 231, 870, doi:10.1016/j.jcp.2011.02.006

Webster, J. Burch, J., *et al.* 2017, *J. Geophys. Res.*, submitted

Wednesday afternoon excursion to Sidmouth with a visit to the Norman Lockyer Observatory to enjoy a rich program of fun talk demonstrations and shows (bottom left: D. Webb and S. Strawbridge) and the scenic walk down the coastal path.

Wednesday afternoon excursion in Sidmouth with a visit to the Norman Lockyer Observatory to enjoy a rich program of fun-fair demonstrations and shows (bottom left: D. Webb and S. Strawbridge) and the scenic walk down the coastal path.

Part 4. Impact of Solar wind, Structures and Radiation on Ionospheres, Atmospheres

Space Weather of the Heliosphere: Processes and Forecasts
Proceedings IAU Symposium No. 335, 2017
C. Foullon & O.E. Malandraki, eds.

doi:10.1017/S174392131700919X

Thermospheric Dynamics in Quiet and Disturbed Conditions

Aziza Bounhir[1,2], Zouhair Benkhaldoun[1], Jonathan J. Makela[3], Mohamed Kaab[1], Brian Harding[3], Daniel J. Fisher[3], Amine Lagheryeb[1], Malki Khalifa[1], Khaoula Elbouyahyaoui[1], Mohamed Lazrek[1] and Ahmed Daassou[1]

[1]High Energy Physics and Astrophysics Laboratory, Oukaimeden Observatory, Cadi Ayyad University, Marrakech, Morocco

[2]High Energy Physics and Astrophysics Laboratory, Faculty of Sciences and Techniques, Marrakech, Morocco

[3]Department of Electrical and Computer Engineering, University of Illinois at Urbana-Champaign, Illinois 61801, USA, Urbana, IL, United States

Abstract. This paper presents the thermospheric winds and temperature properties measured with a Fabry-Pérot interferometer (FPI) over Oukaimeden observatory (31.2° N, 7.8° W, 22.8° N magnetic) in Morocco. After Three years of successful functioning from 2014 to 2017, we can address the seasonal behavior of the temperature and the winds (vertical, zonal and meridional). The dependence of the thermospheric winds and temperature on the solar cycle is also presented. The day-to-day variations of the quiet time wind pattern exhibits the importance of other type of waves superposed to the main diurnal tides. The storm time wind and temperature exhibits also a variety of ways to react to the storm. However, there is seasonal effect to the storm that will be illustrated in this paper. The signature of the MTM phenomenon is also present in the winds and temperature in geomagnetically quiet and disturbed nights. The occurrence of this phenomenon over the studied area is also addressed.

Keywords. Ionosphere, Thermosphere, Geomagnetic storm, Fabry-Perrot Interferometer.

1. Introduction

Space weather is defined as the conditions on the Sun and the near-Earth space that can influence the performances of space based and terrestrial technological instruments and endanger human health. The thermosphere/ionosphere system plays a dramatic role as a vehicle of space weather events. After a solar flare or a geomagnetic storm, the ionospheric density changes and affects the ground-to-ground and ground-to-satellite signals. Navigation and communication satellites links can be degraded or lost. Currents can be induced in power grids causing failure and satellite assets can be damaged by increased energetic particle radiation during geomagnetic storms.

One of the most challenging problems in space weather is the forecasting of the ionosphere as it is related to most of space weather events. A crucial step towards the forecasting of the ionosphere is the thorough understanding of the coupling between the charged and neutral particles in the presence of a magnetic field. Drag forces between the charged and neutral gases produce complex electrodynamic interactions. Variations in the neutral wind can drive a complex system of ionospheric currents and electric fields, which profoundly influence the structure and composition of the ionosphere.

In this paper, we will present the thermospheric measurements of a Fabry-Perrot interferometer (FPI) over the Oukaimeden observatory (31.206° N, 7.866° W, 22.84° N magnetic). The FPI (Makela & Miller 2011, Harding *et al.* 2014) measures the thermospheric wind and temperature by observing the 630-nm airglow generated by the dissociative recombination of O_2^+ which takes place at an average altitude of 250 km approximately (Link & Cogger 1988). The data from The FPI over Oukaimeden can be used to test and improve the existing thermospheric models and help improve our understanding of actual physics behind the atmosphere. The seasonal variability of the zonal and meridional winds and the temperature are presented along with the dependence on solar cycle. We will address the variability for geomagnetically disturbed and quiet conditions. During a geomagnetic storm, the thermospheric winds and temperature depart from their quiet time behavior and exhibit multiple ways to react to the storm. Finally, the seasonal effect of the storm will be illustrated.

2. Climatology of the winds and quiet time patterns

The basic climatology of the winds is well documented in Kaab *et al.* (2017). To build the climatology they adopted the method developed by Fisher *et al.* (2015). The data were over-averaged to compare with existing thermospheric models that has low spatial resolutions. In this case the climatology was compared with HWM14 empirical model (Drob *et al.* 2015). Figure 1 shows a sample of very quiet time wind data with sum-Kp index less than 6. The south (green) and north (red) look of the meridional winds are illustrated as well as the east (yellow) and west (purple) components of the zonal winds. We can clearly notice the superposition of many waves to the main diurnal tides. Small harmonics of tides and gravity waves contribute to the habitual flow of the winds. In Figure 1, the zonal components of the winds have most of the time the same magnitude. This is not the case of the equatorward flow of the meridional winds. Between 00 and 04 UT, the northern component has a larger magnitude than the southern one; of about 50 $m.s^{-1}$ for Figure 1 (a) and 90 $m.s^{-1}$ for Figure 1 (c). Given the geographical position of Oukaimeden Observatory at about 31° North, the northern component is certainly more sensitive to equatorward flows and has a larger magnitude. The southern component on the other hand is more sensitive to northward flows originating from regions of lower latitudes or from the equator. The northward surge of the Midnight Temperature Maximum (MTM) phenomenon is present in the wind flow between 02 and 04 UT.

The day-night differential solar heating drives the thermospheric wind dynamics. The eastward zonal wind combined with upward propagating tidal meridional winds and in situ tidal create a large-scale structure (Akmaev *et al.* 2010) converging onto the geographic equator. The convergence of the thermospheric tides creates an equatorial pressure bulge which bifurcates and propagates to higher latitudes in both hemispheres. The MTM peak formed by hydrostatic expansion migrates toward both polar regions and reverses the direction of the meridional component of the thermospheric wind from its normally equatorward flow to poleward flow. This northward surge transfers the F region plasma along magnetic field lines to lower altitudes, thus generating an increase in the 630.0 nm nightglow emission which creates the brightness wave and the descent of the F-region plasma, termed 'midnight collapse'.

The multiple forcing of the sun upon the Earth make it a gigantic electrical machine with many currents flowing in multiple places and make it also a gigantic non-linear resonator because Earth atmosphere displays a wave-like structure on a global scale. Multiple forms of planetary waves, tides and gravity waves superpose on the Earth's atmosphere. Tidal waves (Chapman & Lindzen 1970, Forbes & Garrett 1979) are primarily formed by the regular differential heating of the atmosphere such as the day-night and seasonal differential heating. We can mainly distinguish migrating and

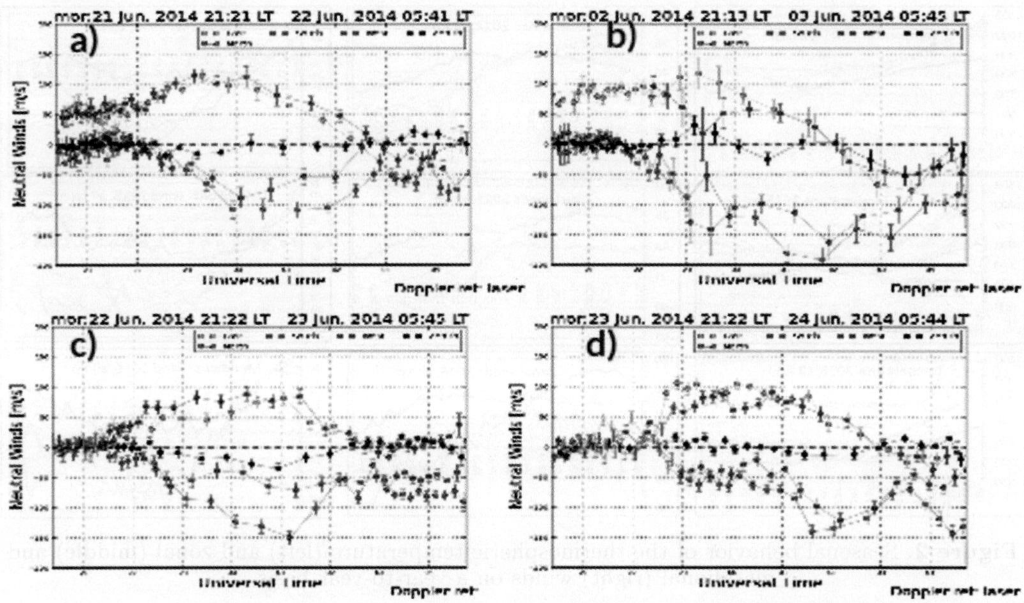

Figure 1. Very quiet time thermospheric winds dynamics. Data from Fabry-Perrot interferometer using Doppler shift and broadening of the 630 nm airglow.

non-migrating tides and stationary and upward propagating tides. Westward-migrating tidal modes dominate the upper-atmosphere response. Upward propagating waves in general can transport a huge amount of energy from low to high altitudes. As their wind amplitude grow with increasing altitude, they reach a point where they break and strong dissipation of energy occurs in the thermosphere. The change in thermospheric dynamics and properties affects immediately the composition, density and electrodynamics of the ionosphere. Therefore, modeling and observations of these type of waves are needed in order to monitor and predict changes in the ionosphere (Hagan & Forbes 2002).

In Figure 1 (d), we can notice an additional equatorward surge felt only by the northern component of the wind between 03 and 04 UT. In Figure 1 (b) between 23 and 00 UT superimposed to the wind general trend, a surge that have pushed the northern component towards the north, the southern component towards the south, the eastern component toward the east and the western component towards the west. This is a very interesting phenomenon occurring above the area under study and could be a signature of an upward propagating tides or a breaking gravity waves. This type of forcing mechanisms are mostly ignored in the thermospheric models. In is important to see that the vertical component of the wind is not negligible and can reach 30 to 50 $m.s^{-1}$ in very quiet time nights.

3. Seasonal variability

To address the seasonal variability, we have adopted the same method as for the climatology (Fisher *et al.* 2015) but averaged the data in 15 mn interval. The north and south look contribute to the meridional component of the wind and the east and west look contribute to the zonal one. Figure 2 illustrates the seasonal variability of the temperature (left), zonal winds (middle) and meridional winds, for the years 2014 (top), 2015 (middle) and 2016 (bottom). All the data have been considered regardless

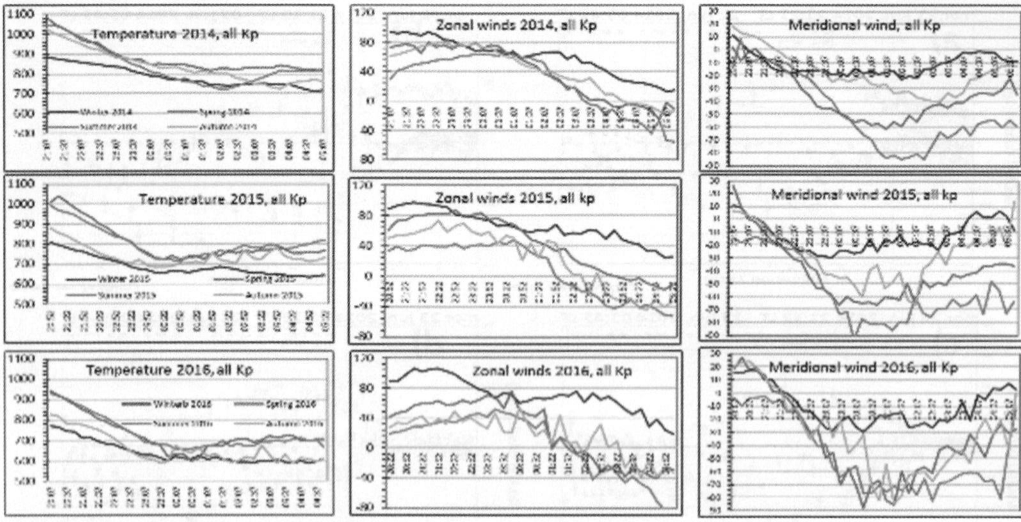

Figure 2. Seasonal behavior of the thermospheric temperature (left) and zonal (middle) and meridional (right) winds on a year-to-year basis.

of geomagnetic conditions. The contribution of all heating mechanisms is present. The temperature cools down during the night. In the early evening hours, the temperature decreases almost linearly with a slope depending on the season and the year. At around 23 to 00 UT, there is an inflexion point towards another regime where the temperature undergoes a series of increases and decreases around a constant value. These oscillations are mostly pronounced in Autumn months and in declining solar cycle. We can see in all years that the winter is the coolest and the spring the hottest along with the summer. 2014 summer temperature is particularly cold between 00 and 03 UT, even though 2014 is the maximum solar cycle. This could be due to the surprisingly geomagnetically very quiet summer.

The post-midnight oscillations are mainly caused by the Midnight Temperature Maximum (MTM) phenomenon. The heating caused by geomagnetic storms is also present in the temperature variations. Zonal winds are the strongest during winter time for all years. They are also eastward during the entire night. They abate in other seasons and reverse to westward direction in summer months. The time of reversal depend on the year. We can also notice small time scale variations especially in Autumn and spring 2016. Concerning the meridional winds, winter equatorward flow is the weakest and summer is the strongest for all years. The maximum equatorward flow shifts to later hours from spring to summer to autumn especially in 2014. Autumn and spring components are very dynamic in 2015 and 2016.

Vertical winds are illustrated in Figure 3 where we compare 2014 and 2015 seasons. We can clearly notice seasonal effect of the vertical winds. In winter time, the vertical winds are upward in the early evening hours with an average of 5 $m.s^{-1}$ and abates and reverse to downward direction before sunrise. In other seasons, the vertical winds are flowing downward in the early evening hours with an average of 5 $m.s^{-1}$ and upward before sunrise for summer, autumn and spring. We can clearly notice average seasonal values of, for example, 20 $m.s^{-1}$ in sometimes.

Large vertical wind observations have been difficult to reconcile with theory because they have an enormous impact on the dynamics, chemistry, and electrodynamics of the ionosphere/thermosphere system. Vertical gradients of composition, density, and

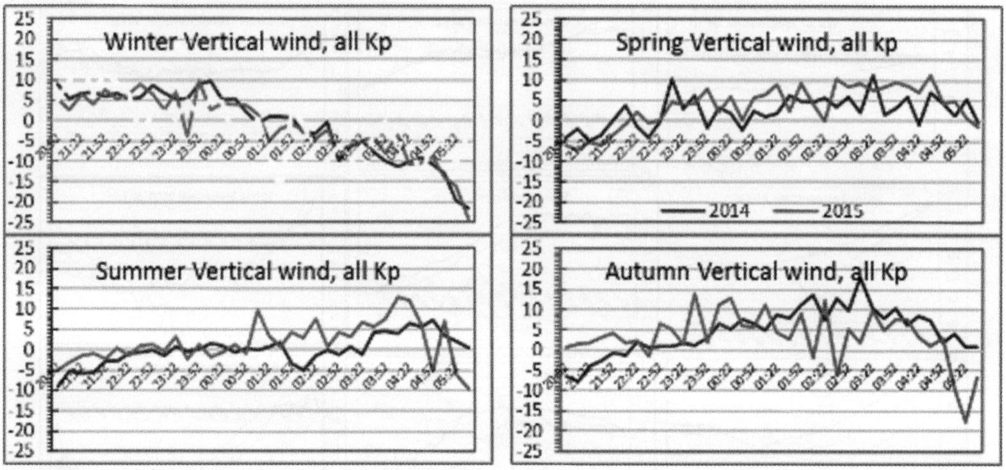

Figure 3. Seasonal behavior of the vertical winds related to 2014 and 2015 years.

temperature are larger than horizontal gradients, so even a small vertical wind causes large advective transport and adiabatic temperature changes. Vertical winds are supposed to be low of 1 $m.s^{-1}$ under hydrostatic balance of an atmosphere subject to diurnal heating (Smith *et al.* 1998), but localized source of heating and divergent horizontal flows can drive the divergent component of the vertical wind. However very high vertical winds lasting for long time seem not reasonable. (Makela & Miller 2011) reported large vertical winds lasting for several hours in many midlatitude places and concluded that these large apparent vertical winds must be an artifact of some type of contamination of non thermospheric origin. Harding (2017) developed a radiative transfer model which demonstrates that these measurements can be explained as an artifact due to the scattering of airglow light by tropospheric aerosol layer.

4. Variability with the solar cycle

To address the seasonal variability with the solar cycle, Figure 4 shows the evolution of 2014, 2015 and 2016 seasons related to thermospheric temperature, the zonal and meridional winds. We can notice that the zonal winter wind does not depend on solar cycle. In the other seasons, the magnitude of the zonal wind decreases with declining solar cycle. In summer months, the reversal time from eastward to westward direction seems to shift to earlier hours with the declining solar cycle. For autumn and spring months, zonal winds abate before sunrise in 2014 and reverse to westward direction in 2015 and 2016 months. The time of reversal depends on solar cycle. Concerning the meridional winds, we can notice very slight differences between the winter seasons. The equatorward component of the winter months gains magnitude with declining solar cycle whereas the northward component loses magnitude. For the other seasons, the magnitude of the meridional wind increases with declining solar cycle especially for spring and autumn months. We can clearly notice the presence of northward surges related to the MTM phenomenon present in all seasons but in a more pronounced way for autumn and spring months and in declining solar cycle. The thermosphere is in all seasons hotter in solar maximum and cooler in declining solar cycle. We should mention that the data have been averaged regardless of geomagnetic conditions. In the early evening hours, the seasonal slope of the temperature decrease seems to be similar for all years except for

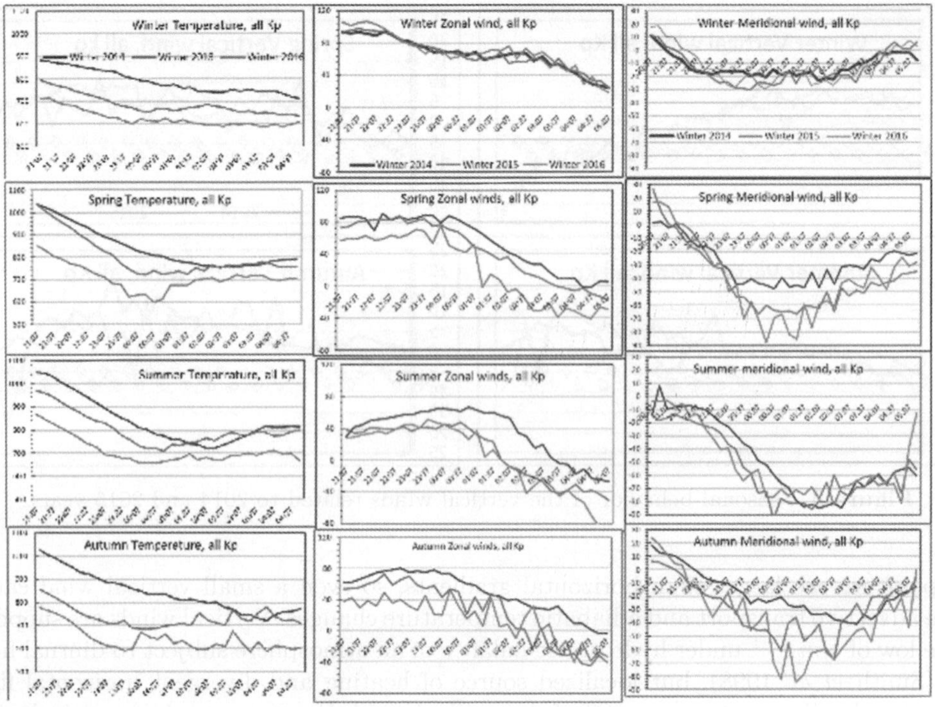

Figure 4. Variability with the solar cycle for the temperature data (left), the zonal winds (middle) and the meridional winds (right).

spring 2015. There seem to be important correlations between 2015 and 2016 seasonal temperature shape. The post-midnight heating of the thermosphere mostly related to the MTM phenomenon is more pronounced with declining solar cycle and in autumn months. The MTM phenomenon seem to occur predominantly in post-midnight hours. This is consistent with MTM observed phenomenon over many places (Meriwether *et al.* 2008, 2011).

5. Quiet and storm time variability

During geomagnetic storms, joule heating created by an increase of convective electric fields and auroral precipitation drives a global atmospheric change. The energy input drives wave surges that interact globally with the background wave mechanism (Buonsanto 1990).

The highly variable forcing sources cause disturbances to thermospheric winds which will further affect the global ionospheric dynamo (Blanc & Richmond, A. 1980). In disturbed geomagnetic conditions, thermospheric neutral winds exhibit large deviations from their quiet time climatological behavior resulting in large corresponding changes in the ionospheric plasma density, composition, temperature, and electrodynamics (Richmond & Matsushita 1975; Fuller-Rowell *et al.* 1994; Buonsanto 1990; Mendillo 2006; Meriwether *et al.* 2008; Emmert *et al.* 2004).

Through the FPI data we have noticed that in storm time, thermospheric winds depart from their habitual quiet time climatology in a variety of ways depending on the season, the magnitude of the storm, the dynamics of the storm energy release and the time of the storm. When the energy release is impulsive Traveling Atmospheric Disturbances (TADs)

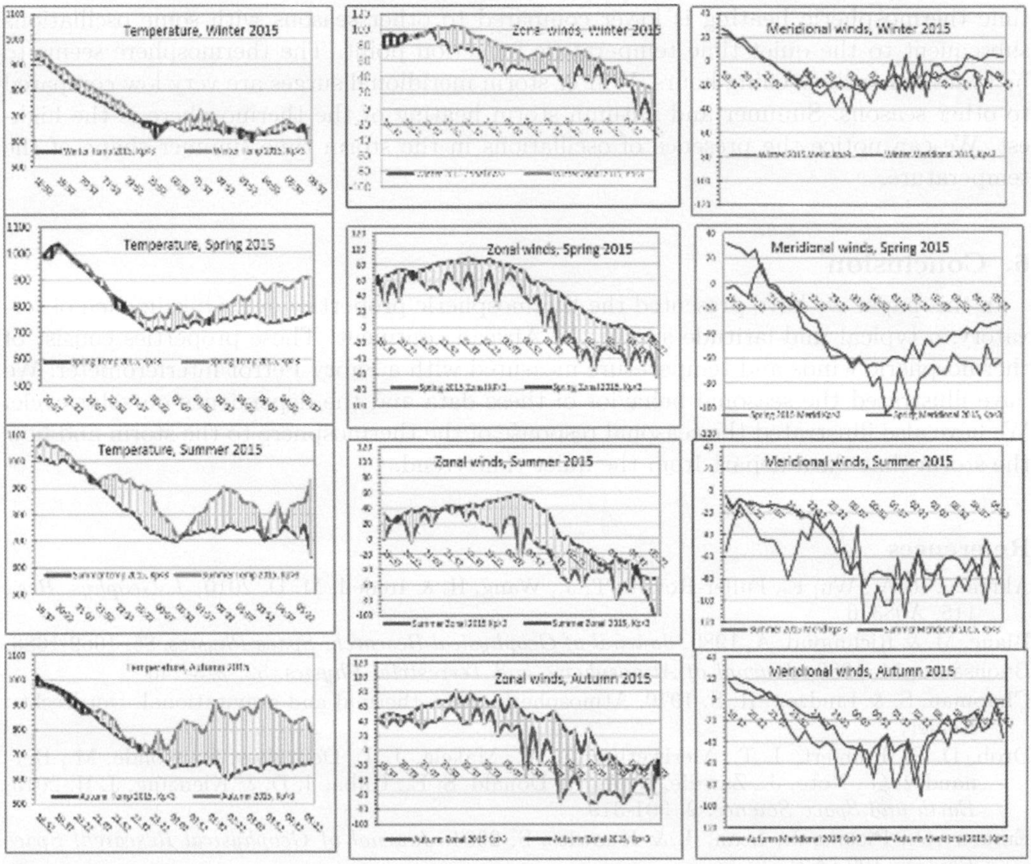

Figure 5. Seasonal effect of the storm. Quiet time data refer to data with 3 hours $K_p < 3$ and disturbed time data with 3 hours $K_p > 3$.

propagate from high to low latitudes and to the opposite hemisphere. A consequence of the TAD is the raise and fall of the ionospheric plasma along magnetic field lines creating Traveling Ionospheric Disturbances (TIDs). A storm induced Hadley cell is created flowing in opposite direction from quiet time one, with equatorward and westward winds. The variety of ways the thermospheric wind response to the storm depends on the extent reached by the storm induced Hadley cell.

Figure 5 illustrates the seasonal effect of the storm. The data have been separated into quiet-time data and geomagnetically disturbed-time data. The quiet time refer to data with 3-hour $K_p < 3$. The thermospheric data are related to the 2015 year. We must mention that the other years have the same trend as that of 2015. The averaged data reveals the general trends during storm time, but relevant information are exhibited from the study of each storm case separately. We can see in all seasons the equatorward storm surges and trans-equatorward surges in the meridional winds. The disturbed zonal winds, consisting of the storm winds minus the average quiet time winds, are westward. This result of westward and equatorward winds during the magnetic storm is consistent with previous results obtained with ground based Fabry-Perot interferometers (Makela & Miller 2011). Xiong *et al.* (2015) report that the disturbance zonal wind is mainly westward, and increases with magnetic activity and latitude. Emmert *et al.* (2004) reported westward and equator-ward night-time disturbed winds at midlatitude. Winter time thermospheric heating is lower compared to other seasons with some oscillations subsequent to the quiet time temperature inflection point. The thermosphere seems to be heated in early evening hours. Winter storm meridional surges are very low compared to other seasons. Summer and autumn storm heating of the thermosphere is the highest. We can notice the presence of oscillations in the storm time summer shape of the temperature.

6. Conclusion

In this paper we have presented the thermospheric properties over Oukaimeden observatory, a typical mid-latitude site in the African continent. These properties consist of thermospheric winds and temperature measured with a Fabry-Perrot interferometer. We have illustrated the seasonal behavior of these data and the dependence on solar cycle. We have also illustrated the seasonal response of the thermosphere to the storm and how the storm time data depart from the quiet time trend.

References

Akmaev, R. A., Wu, F., Fuller-Rowell, T. J., Wang, H. & Iredell, M. D. 2010, *J. Geophys. Res.*, 115, A08326

Blanc, M. & Richmond, A. 1980, *Journal of Geophysical Research: Space Physics*, 85, 1669-1686.

Buonsanto, M. 1990, *Journal of Atmospheric and Terrestrial Physics*, 52, 223-240.

Chapman, S. & Lindzen, R. S. 1970, Atmospheric tides: thermal and gravitational. Dordrecht : Reidel

Drob, D. P., Emmert, J. T., Meriwether, J. W., Makela, J. J., Doornbos, E., Conde, M., Hernandez, G., Noto, J., Zawdie, K. A., McDonald, S. E., Huba, J. D. & Klenzing, J. H. 2015, *Earth and Space Science*, 2, 301-319

Emmert, J., Picone, J., Lean, J. & Knowles, S. 2004, *Journal of Geophysical Research: Space Physics*, 109 (A2)

Fisher, D. J., Makela, J. J., Meriwether, J. W., Buriti, R. A., Benkhaldoun, Z., Kaab, M. & Lagheryeb, A. 2015, *Journal of Geophysical Research A: Space Physics*, 120, 6679-6693

Forbes, J. M. & Garrett, H. B. 1979, *Rev. Geophys.*, 17, 1951-1981

Fuller-Rowell, T., Codrescu, M., Moffett, R. & Quegan, S. 1994 *Journal of Geophysical Research: Space Physics*, 99, 3893-3914

Hagan, M. E & Forbes, J. M. 2002, *Journal of Geophysical Research*, 107, D24, 4754

Harding, B. J., Gehrels, T. W. & Makela, J. J. 2014, *Appl. Opt.*, 53, 666-673

Harding, B. J, Midlatitude thermospheric wind and temperature: Networked Fabry-Perot interferometer observations and radiative transfer modeling, (2017), Thesis Ph.D degree.

Kaab, M., Benkhaldoun, Z., Fisher, D. J., Harding, B., Bounhir, A., Makela, J. J., Laghriyeb, A., Malki, K., Daassou, A. & Lazrek. M. 2017, *Annales Geophysicae*, 35, 161-170

Link, R. & Cogger, L. 1988, *J. Geophys. Res.-Space*, 93, 9883-9892

Makela, J. J. & Miller, E. S. 2011, *Aeronomy of the Earth's Atmosphere and Ionosphere* , vol 2, 239-249

Meriwether, J., Faivre, M., Fesen, C., Sherwood, P. & Veliz,O. 2008, *Annales Geophysicae*, 26, 447-466

Meriwether, J. W., Makela, J. J., Huang, Y., Fisher, D. J., Buriti, R. A., Medeiros, A. F. & Takahashi, H. 2011, *Journal of Geophysical Research*, 116 , A04322

Mendillo, M. 2006, *Reviews of Geophysics*, 44, 2006

Richmond, A. & Matsushita, S. 1975, *Journal of Geophysical Research*, 80, 2839-2850

Smith, C. W., LHeureux, J., Ness, N. F., Acuña, M. H., Burlaga, L. F. & Scheifele, J. 1998, *ISpace Science Reviews*, 86, 613-632

Xiong, C., Lühr, H. & Fejer, B. G. 2015, *Journal of Geophysical Research (Space Physics)*, 120, 5137-5150

Space Weather of the Heliosphere: Processes and Forecasts
Proceedings IAU Symposium No. 335, 2017
C. Foullon & O.E. Malandraki, eds.

doi:10.1017/S1743921318000078

Variability of the Vertical Total Electron Content, from GPS data, during 2 to 8 November 2015, Using Oukaimeden and Rabat Stations in Morocco

Khaoula El bouyahyaoui[1], Aziza Bounhir[1], Zouhair Benkhaldoun[1], Rolland FLEURY[2], Christine Amory-Mazaudier[3] and Mohamed Kaab[1]

[1]Oukaimeden Observatory, LPHEA, FSSM, Cadi Ayyad University
[2]IMT Atlantique Bretagne-Pays de la Loire, campus de Brest, technopole Brest-Iroise, CS83818, 29238 Brest cedex 3.
[3]University Paris Sorbone UPMC, LPP Polytechnique, UPMC Jussieu 75005, France., Francen

Abstract. In this work we present a method for detecting the activity of the ionosphere (TEC) and we illustrate the signature of the solar activity on the vertical total electron content VTEC, during 02 to 08 November 2015, using GPS measurements obtained from two stations in Morocco, the first one in Marrakech at Observatory of Oukaimeden OUCA (31°12′23.3" N 7°51′58.8" W), the second in Rabat, Rabt ($33.9981°N; 353.1457°E, geographic$).

Keywords. Ionosphere, TEC, GPS

1. Introduction

Space weather today is a new field in science and has very interesting effects on humans, environment and technology in general. Scientists are now studying space weather with a wide range of tools to try to learn more about the physical and chemical processes taking place in the upper atmosphere and beyond. One of these tools is Global Positioning System (GPS). The ionosphere has practical importance in GPS applications because it influences the transionospheric radio wave propagation. The parameter of ionosphere that produces most of the effects on radio signals is the total electron content (TEC).

2. Total Electron Content in Ionosphere

The TEC is defined as the total number of electrons integrated along the path from the receiver to each GPS. The TEC as an indicator of ionospheric variability. It is derived by the modified GPS signal flowing through the ionosphere. TEC is measured in units of 10^{16} electrons per square meter area, where 10^{16} electrons/m^2 = 1 TEC unit (TECU) (Ouattara *et al.* 2012).

The TEC has been determined by modeling the ionosphere as a single electric layer. We show that the ionospheric TEC can be expressed as a function of the difference between two pseudo ranges or two carrier phases carried out by using the two typical GPS frequencies.

In this work, the TEC is observed at the F layer because this region has the highest variability of free electrons, causing the greatest effect on GPS received signal compared to other layers. More than two-thirds of electron concentration are located at F2 layer. This method is conducted by going through several processes (Zoundi *et al.* 2012). Figure 1 shows the flow chart of work progress to achieve the objective of the project.

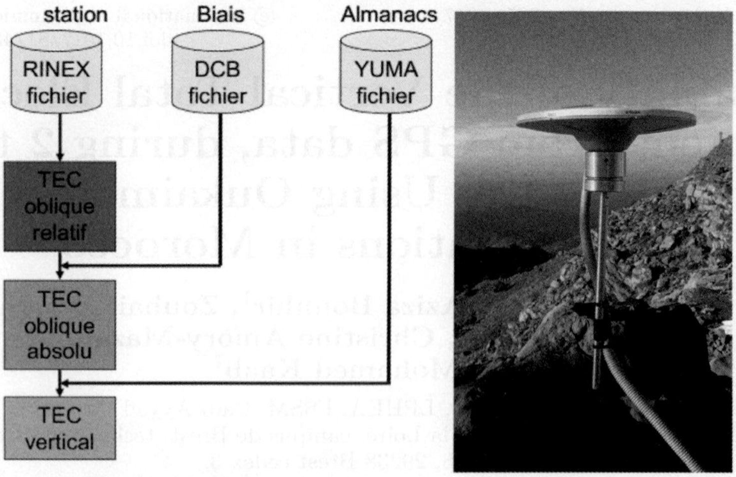

Figure 1. Left: Flowchart of TEC processing by Fleury model. Right: GPS at OUKAIMEDEN

The fundamental GPS frequency is $f_0 = 10.23MHz$ from which f_1 and f_2 are derived ($f_1 = 154f_0$ and $f_2 = 120f_0$) , defined the group delay as:

$$P_1 - P_2 = 40.3\,\mathrm{TEC}\left(\frac{1}{f_1^2} - \frac{1}{f_2^2}\right) \tag{2.1}$$

P_1 and P_2 are the pseudo ranges for f_1 and f_2 respectively. By rewriting the equation of group delay, we obtain:

$$TEC = \frac{(P_1 - P_2)}{40.3}\frac{f_1^2 f_2^2}{f_2^2 - f_1^2} \tag{2.2}$$

The TEC between the satellite and the user depends on the satellite elevation angle, this measurement is called Slant TEC (STEC). The TEC varies temporally and spatially, and depends on the solar activity, user location, and the satellite elevation angle. This measurement is named vertical TEC (VTEC). To convert STEC to VTEC, we used the following equation:

$$VTEC = (STEC - B_s - B_u)\sqrt{1 - \left(\frac{R_e \cos(\epsilon)}{R_e + h_s}\right)^2} \tag{2.3}$$

where B_s and B_u are the instrumental biases of satellites and receivers respectively, $R_e = 6371km$ is the mean radius of the Earth, ϵ is the elevation angle of the satellite (Titheridge 1972).

The process of extracting data from RINEX (Receiver Independent Exchange) file was done by using Matlab programming language whereby the RINEX file was obtained from the GPS receiver as shown in figure 1. The program will analyze and extract the information needed for calculating the TEC from the observation and navigation RINEX file. The result will show the graph of vertical TEC (VTEC) versus time. Data of VTEC were used since its value is independent on the location of satellite receiver compared to STEC (Shimeis *et al.* 2002).

3. Results

Figure 2 shows the hourly variation of VTEC for seven selected days of November 2015 based on the data availability.

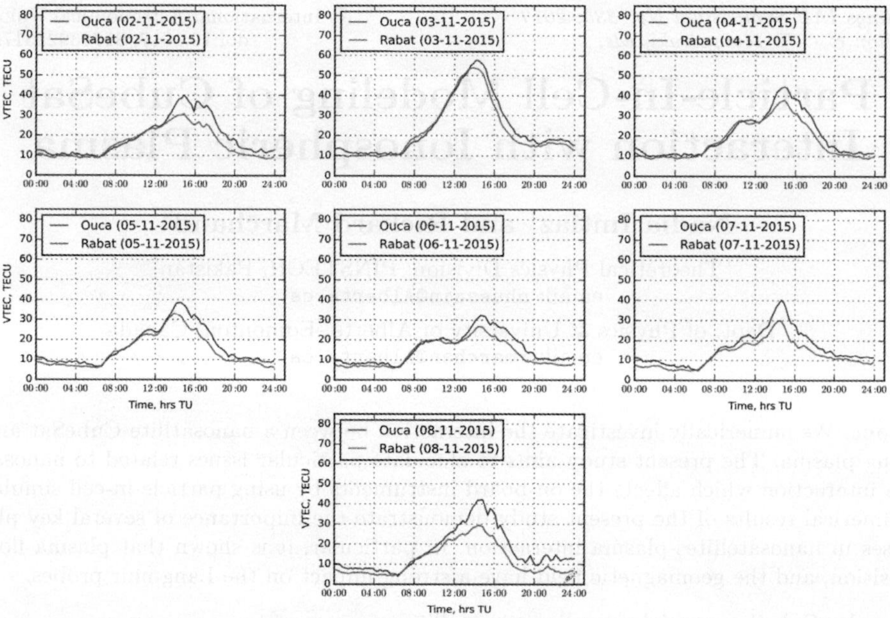

Figure 2. Local time variation of Rinex TEC for: 02-08 November 2015 from GPS at OUKAIMEDEN and Rabt.

According to ACE satellite, a co-rotating interaction region followed by higher speed stream from an equatorial coronal hole arrived at earth on 3 November 2015. The solar wind speed jumped from around 320 km/s to about 700 km/s. The high speed stream caused on 3 November a ionospheric event, as a result, the TEC increases by about 20 TECU between 12UT and 16 UT.

We can notice from Figure 2 that the TEC over Oukaimeden and the TEC over Rabat correlate, but with an enhancement of the TEC over Oukaimeden in afternoon hours. This result is expected as Oukaimeden is closer to the maximum equatorial ionization anomaly crest than Rabat.

4. Conclusion

The TEC is measured by its effect on the travel time of the signal between the satellite and the receiver. We have given the expression of the TEC and presented clearly the algorithm for calculating it. For the station of Oukaimeden, we showed for the first time the variability of the VTEC in the region. The present work illustrates the variability of the TEC in two stations located in Morroco from 02 to 08 November 2015 by using the Fleury model (Ouattara *et al.* 2012).

References

Ouattara, F., Zoundi, C. & Fleury, R. 2012, *Indian Journal of Radio and Space Physics*, 41 (6), pp 617–623

Zoundi, C. & Ouattara, F. 2012, Sciences et Médecine. Rev.CAMES-Série A, 13(2):139–143

Shimeis, A., Amory-Mazaudier, C., Fleury, R., Mahrous, A. M. & Hassan, A.F. 2014, *Advances in Space Research*, 54, 2159–2171

Titheridge, J. E. 1972, *Journal of Geophysical Research* 77 (10), pp 1978–1981

Space Weather of the Heliosphere: Processes and Forecasts
Proceedings IAU Symposium No. 335, 2017
C. Foullon & O.E. Malandraki, eds.

doi:10.1017/S1743921317007098

Particle-In-Cell Modeling of CubeSat Interaction with Ionospheric Plasma

Nadia Imtiaz[1] **and Richard Marchand**[2]

[1]Theoretical Physics Division, PINSTECH, Pakistan
email: nhussain@Alberta.ca

[2]Dept. of Physics & University of Alberta, Edmonton, Canada
email: rmarchan@Alberta.ca

Abstract. We numerically investigate the interaction between a nanosatllite CubeSat and surrounding plasma. The present study aims to elucidate particular issues related to nanosatllite-plasma interaction which affects the on-board instruments by using particle-in-cell simulations. The numerical results of the present study demonstrate the importance of several key physical processes in nanosatellites-plasma interaction. In particular, it is shown that plasma flow, ion composition, and the geomagnetic field have a strong impact on the Langmuir probes.

Keywords. CubeSat, particle-in-cell, nanosatellites

1. Introduction

The Dynamics Ionosphere CubeSat Experiment (DICE) mission funded by the National Science Foundation (NSF) and NASA Educational Launch of Nano-satellites (ELaNa) programs was launched on October 27, 2011. It consists of the two identical nanosatellites "CubeSats" which were released into an eccentric low Earth orbit. Each payload carries two identical Langmuir probes to measure in-situ ionospheric plasma parameters (Fish *et al.* 2014). For in situ measurements of space plasma, Langmuir probes have been extensively used on sounding rockets and satellites (Gurnett *et al.* 2004; Lebreton *et al.* 2006). Some numerical models have been presented to investigate the essential physical factors that are at play in the interpretation of the Langmuir probe data carried by the satellites (Imtiaz & Marchand 2015; Marchand 2016). For nanosatellites, Albarran (2015) employed the SPIS simulations to understand the interaction between the CubeSat and surrounding plasma. In this study the focus is on the importance of accurately accounting for the attitude and boom size in order to derive the ambient plasma density from spin-modulated Langmuir probe measurements on CubeSats. However, it does not account for the effect of the geomagnetic field and the ionic composition. The goal of the present study is to improve earlier approaches and enable a better interpretation of Langmuir probe measurements on CubeSats. This is accomplished by carrying out fully kinetic simulations of the interaction between a CubeSat and space environment, while accounting for the effects not included in previous studies, that is including the local geomagnetic field and ion composition. For this purpose we use the particle-in-cell simulation code 'PTetra' which is capable of simulating the time dependent interaction between a LEO satellites and plasma (Marchand 2012). PTetra computes the electrostatic sheath potential around the probes and the floating potential of the payload and booms under specified plasma conditions. In next section, we briefly explain the numerical approach along with the results and general discussion of our findings.

Table 1. Summary of the simulation parameters in reference case.

Physical parameter	Value
Flow velocity($\vec{v_d}$)	$(0, 7469, 0)m/s$
Plasma density(n_o)	$10^{10}\ m^{-3}$
Plasma temperature($T_e = T_i$)	$0.25eV$
Magnetic field ($\vec{B_o}$)	$(-4.66, -40.8, -8.72)\mu T$
Ionic composition	$24\%H^+$, $76\%O^+$
Debye Length(λ_D)	$0.02m$
ion thermal gyro radius(ρ_i)	$6.95m$
electron thermal gyro radius (ρ_e)	$0.037m$
Tetrahedral mesh resolution	$0.008m$
Number of tetrahedra	$1,739,343$
Macro-particles	$8,000,000$

Table 2. Summary of simulation cases to study CubeSat-plasma interaction.

Case	Physical conditions
a	Reference Case
b	$100\%H^+$
c	$100\%O^+$
d	$B_o = 0$
e	Stationary Plasma

2. Numerical Technique and Simulation Results

The idealized geometry of the $1.5U$ CubeSat consisting of the payload, booms and two spherical Langmuir probes is constructed with an open source mesh generator GMSH (Geuzaine & Remacle 2009) as illustrated in Figure 1. The payload/booms are at the floating potential (collecting zero net current) and the bias voltage of the two Langmuir probes ranges between $-4V$ to $+4V$. For each bias voltage considered, PTetra simulations are carried out forward in time until a steady state is reached. The current characteristics are obtained by applying the same potential to the dual Langmuir probes and calculating the current collected by each probe individually. There are number of physical processes which affect charging of the CubeSat in the Low Earth orbit ionosphere. For this purpose we considered the five cases as summarized in Table 2. The numerical results obtained in different simulation cases are analyzed to assess the effect of different physical conditions on CubeSat-plasma interaction both qualitatively and quantitatively. The interaction between CubeSat and surrounding plasma disturbs the local plasma environment. This in turn leads to the formation of the electrostatic plasma sheath and wake structures around the nanosatllite. Figure 2 illustrate the electrostatic sheaths formed around the CubeSat payload and probes due to charging of their surfaces inside the plasma for probe relative bias of $+2V$, obtained in different simulation cases. It is found that the electrostatic sheath profiles vary significantly with the physical conditions accounted for in the simulations.

The values of the floating potential computed for the CubeSat payload and Langmuir probes in different plasma conditions are given in Table 3. The variation in the floating potential depends on the charged particle fluxes to the spacecraft and hence on the charge collected by the spacecraft surface. In the reference case, the plasma consists of 24% lighter H^+ and 76% heavy O^+ ions. The thermal gyro-radii of these ions are much larger than the size of the payload and instrument. The ions are therefore practically

Table 3. Floating potentials (in *volts*) of Payload (V_{pl}), wake(V_{Lpw}) and ram(V_{Lpr}) probes.

Cases	V_{pl}	V_{Lpw} and V_{Lpr}
Reference Case	−0.49	−0.55, −0.48
Pure Hydrogen plasma	−0.37	−0.41, −0.40
Pure Oxygen plasma	−0.59	−0.59, −0.50
$B_o = 0$	−0.67	−0.61, −0.55
Stationary plasma	−0.56	−0.55, −0.55

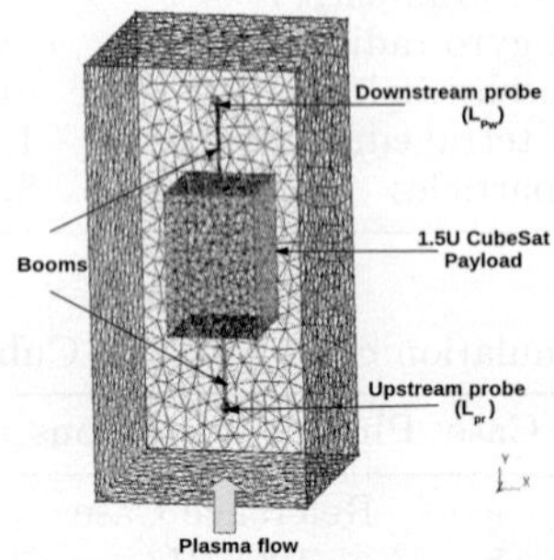

Figure 1. Simulation domain of 1.5U CubeSat payload.

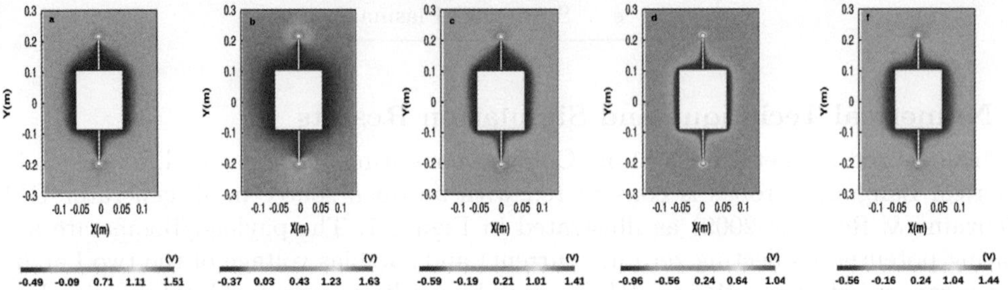

Figure 2. Sheath potential profiles around the CubeSat's payload/Langmuir probes.

unmagnetized and their current is not affected by the ambient magnetic field. On the other hand, the gyro-radius of the electron is comparable of the scale lengths of the system and therefore, electrons are strongly magnetized. The strong magnetization of electrons makes their dynamics restricted and only electrons contained in the magnetic flux tube of radius of the order of two thermal gyro-radii contribute to the collected electron current. Therefore, in order to balance the electron and ion currents such that the net collected current becomes zero, the floating potential has to be less negative when a magnetic field is accounted for in the simulations. In a plasma with 100%O^+ ions, the floating potential is strongly negative compared to the reference case. This behavior arises due to the higher mobility of electrons compared to the heavier O^+ ions. In order to balance the electron and ion fluxes, the potential barrier opposing the electron current should be larger than the reference case. However, in a 100% H^+ ionic plasma, the thermal speed of H^+ ion is 4 times larger than the O^+ thermal speed. The relatively high mobility of a lighter H^+ ion leads to the higher ion flux incident on the spacecraft surface compared to the reference and heavy ion cases. In order to achieve the balance between the ion and electron fluxes, a weaker electron repulsion is required. This in turn leads to the less negative floating potential. It can be seen that a strongly negative floating potential is obtained in case without magnetic field. This behavior arises due to unrestricted motion

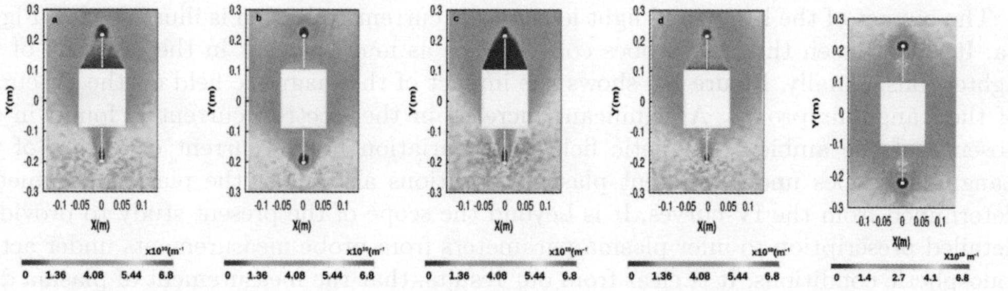

Figure 3. Wake structures behind the CubeSat's payload obtained in different cases.

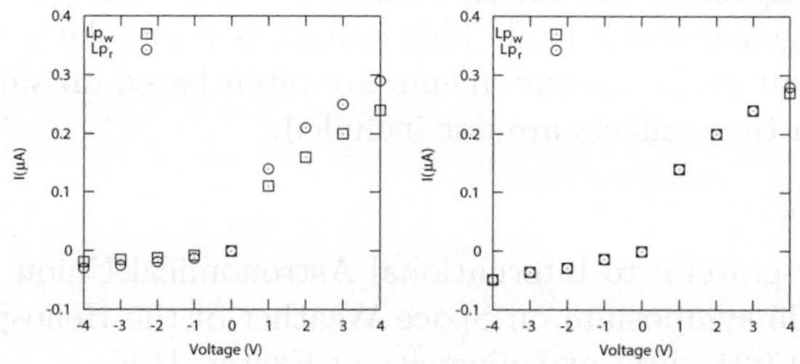

Figure 4. IV-curves of the Langmuir probes in flowing (left) and stationary (right) plasmas.

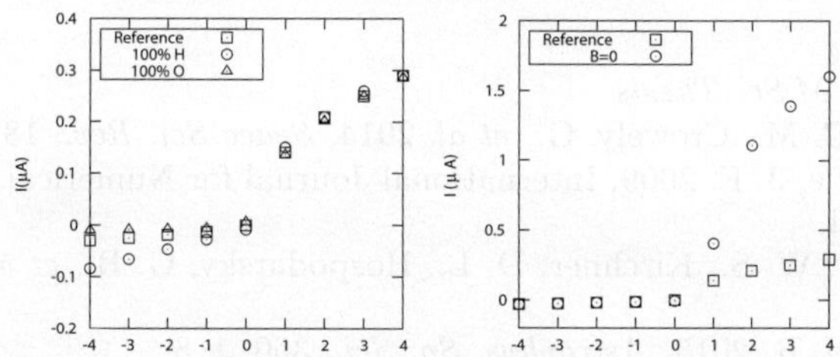

Figure 5. IV-curves of upstream L_{pr} for ionic composition (left), and magnetic field (right).

of the electrons which leads to the higher electron flux to the spacecraft. In order to ensure the net collected current to be zero, a strong potential barrier is required to repel the electrons. Therefore, the value of the floating potential is slightly more negative than its value in the reference case.

The supersonic motion of the CubeSat through a plasma modifies the plasma density and velocity distributions around it. This is manifest from a significant depletion in the ion density in the wake region. The properties of the wake structure depends sensitively on the plasma flow and ionic composition as illustrated in Figure 3. It is found that an enhanced ion density depletion wake is formed in a high Mach plasma ($M = 4$); i.e., with 100% O^+ compared to that in the a low Mach number plasma ($M = 1$); i.e., with 100% H^+, as shown in Figure 3. Of course, there is no wake formation in the stationary plasma case (with $M = 0$).

The current characteristics of the upstream/downstream spherical Langmuir probes in the range bias voltages between $-4V$ to$+4V$ have been computed for different plasma conditions. As expected, in all cases with plasma flow, the probe in the upstream(ram) region collects large current as compared to that in the downstream (wake) region as shown in Figure 4. This assymetry is a direct consequence of the plasma density depletion in the wake, and the resulting reduction in particle fluxes to the probe located in that region.

The impact of the heavy and light ions on the current collection is illustrated in Figure 5a. It can be seen that the probes collect twice as much current in the presence of the lighter ions. Finally, Figure 5b, shows the impact of the magnetic field on the IV-curves of the Langmuir probes. A significant increase in the electron current is found in the absence of the ambient magnetic field. The variation in the current collection of the Langmuir probes under different plasma conditions also affect the plasma parameters determined from the IV-curves. It is beyond the scope of the present study to provide a detailed prescription to infer plasma parameters from probe measurements under actual ionospheric conditions. It is clear from our results, that the measurement of plasma density and temperature must account for several effects such as the geometry and location of the probes with respect to the satellite body, the ion composition, and the strength and direction of the geomagnetic field. This remark is particularly important in view of the fact that Langmuir probe measurements are often based on simple analytic models (e.g., OML) in which these effects are not included.

Acknowledgements

The author is very grateful to International Astronomical Union for financial support to attend the IAU 335 symposium on Space Weather of the Heliosphere: Processes and Forecasts from July 17-21, 2017 at University of Exeter, U.K.

References

Albarran, R. M. 2015, *M.Sc. Thesis*

Fish, C. S., Swenson, C. M., Crowely, G., *et al.* 2014, *Space Sci. Rev.*, 181, 61–120

Geuzaine, C. & Remacle, J. F. 2009, International Journal for Numerical Methods in Engineering, 79, 1309–1331

Gurnett, D. A., Kurth, W. S., Kirchner, D. L., Hospodarsky, G. B., *et al.* 2004, *Sp. Sci. Rev.*, 114, 395–463

Imtiaz, N. & Marchand, R. 2015, *Astrophys. Sp. Sci.*, 360, 1–8

Lebreton , J. P., Stverak, S., Travnicek, P., Maksimovic, M., Klinge, D., Merikallio, S., Lagoutte, D., Poirier, B., Blelly, P. L., Kozacek, Z. & Salaquarda, M. 2006, *Planet. Space Sci.*, 54, 472–486

Marchand, R. 2012, *IEEE Trans. Plasma Sci.*, 40, 217–229

Marchand, R. 2017, *IEEE Trans. Plasma Sci.*, 45, 1923–1926

Space Weather of the Heliosphere: Processes and Forecasts
Proceedings IAU Symposium No. 335, 2017
C. Foullon & O.E. Malandraki, eds.

doi:10.1017/S1743921317007219

Possible Influence of the Solar Eclipse on the Global Geomagnetic Field

Jung-Hee Kim[1,2] and Heon-Young Chang[1,2]

[1]Department of Astronomy and Atmospheric Sciences, Kyungpook National University, Daegu 702-701, Korea
email: hyc@knu.ac.kr

[2]Research and Training Team for Future Creative Astrophysicists and Cosmologists (BK21 Plus Program)

Abstract. We investigate the geomagnetic field variations recorded by INTERMAGNET geomagnetic observatories. We confirm that the effect of solar eclipse can be seen over an interval of 180 minutes centered at the time of maximum eclipse on a site of a geomagnetic observatory. It is found that the effect of the solar eclipse on the geomagnetic field becomes conspicuous as the magnitude of a solar eclipse becomes larger. The effect of solar eclipses is more evident in the second half of the path of Moon's shadow. We also find that the effect can be overwhelmed, more sensitively by geomagnetic disturbances than by solar activity of solar cycle.

Keywords. Solar Eclipse, geomagnetic Field, data Analysis

1. Introduction

A solar eclipse is such an uncommon event that one has to be regarded as serendipitous for observing it at a given location on the Earth (e.g., Ahn and Lee 2004). During the solar eclipse, solar radiation to the Earth is obscured by the Moon. The solar eclipse subsequently causes intricate changes in the atmosphere of the Earth, which occur at all heights from the surface layer to the upper ionosphere, or even up to the plasmasphere (Anderson 1999, Baran *et al.* 2003). In this contribution, we investigate an ensemble average of the 207 geomagnetic field variations observed by INTERMAGNET geomagnetic observatories when the Moon's umbra or penumbra passed over them on a day of the solar eclipse. Unlike previous studies in which efforts are mainly focused on a certain solar eclipse as a case study, we analyze as a whole results acquired by 100 geomagnetic observatories. We carry out examinations by dividing the data set into 2 or 3 subsets on the basis of parameters of solar eclipses, the geomagnetic field and solar activity to see whether the effect of the solar eclipse shows any dependence on circumstances under which solar eclipses occur.

2. Data

Procedures we have collected geomagnetic field data are as follows. Firstly, we complied the geographic longitude and latitude of the geomagnetic observatories from the 150 INTERMAGNET website†. Secondly, with information on solar eclipses occurring during the period from January in 1991 to September in 2016, which is obtained from the NASA eclipse website‡, we computed local circumstances for every solar eclipse visible

† http : //www.intermagnet.org/
‡ https : //eclipse.gsfc.nasa.gov/solar.html

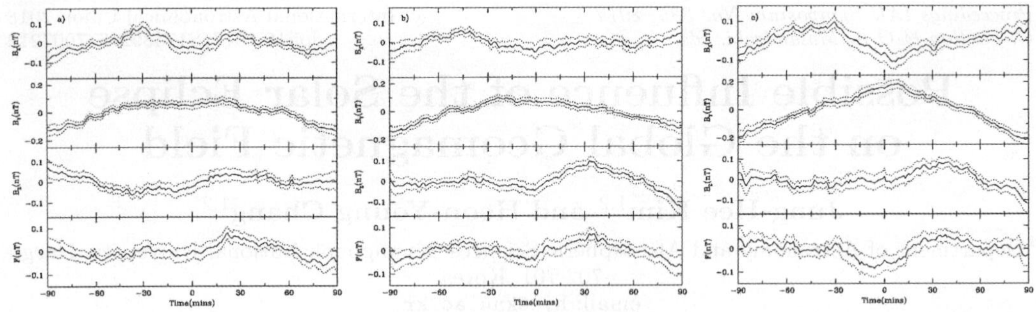

Figure 1. Variations of the three components of the geomagnetic field B_X, B_Y, B_Z and the total strength F are shown from top to bottom, respectively. Continuous and dotted curves represent the average and 1-σ envelopes, respectively. Plots result from data sets: (a) $0.7 \leqslant$ magnitude < 0.8, (b) $0.8 \leqslant$ magnitude < 0.9, and (c) $0.9 \leqslant$ magnitude.

from 150 geomagnetic observatories using the JavaScript provided by the NASA website. Thirdly, we identified only the geomagnetic observatories where the solar eclipse with its magnitude greater than 0.7 can be observed. Finally, for each geomagnetic observatory we obtained from the INTERMAGNET website minute mean values of the 4 geomagnetic components (B_X, B_Y, B_Z, F) from -90 minutes to +90 minutes with respect to the local time of maximum eclipse at the site of the geomagnetic observatory. We have thus obtained 207 geomagnetic manifestations observed by 100 INTERMAGNET geomagnetic observatories at events of the 39 solar eclipses spanning from the maximum of the solar cycle 22 to the end of the solar cycle 24.

3. Dependence on Circumstances of Solar Eclipse

To examine *patterns* of the geomagnetic field variations induced by solar eclipses, we have firstly detrended slowly varying parts of geomagnetic field variations with the first order polynomial fit before taking an average of time series of the geomagnetic field data. When averaging, we have further normalized the detrended geomagnetic field variations so that the height of the curve from the minimum value to the maximum value is equal to unity. By doing so, in some sense we average a shape of the geomagnetic field variation possibly produced by solar eclipses.

Magnitude of solar eclipse.

In Figure 1, we show three data sets divided according to the magnitude of a solar eclipse, respectively. That is, from Figure 1a to Figure 1c results are shown from data sets where $0.7 \leqslant$ magnitude < 0.8, $0.8 \leqslant$ magnitude < 0.9, and $0.9 \leqslant$ magnitude, respectively. Obviously, the effect of the solar eclipse on the geomagnetic field, such as, an increase in B_Y and decreases in B_X, B_Z and F, becomes conspicuous as the magnitude becomes larger.

Location in the path.

In Figure 2, we show results from two data sets where observing site is located in the first half of the path, and observing site is located in the second half of the path, respectively. An increase in B_Y is clear in Figure 2a, but other possible effects seem undetectable. On the other hand, decreases in B_X and F are also clear in Figure 2b. Hence, we conclude that the effect of solar eclipses depends on observing site. That is, the effect of solar eclipses is more evident in the second half of the path of Moon's shadow.

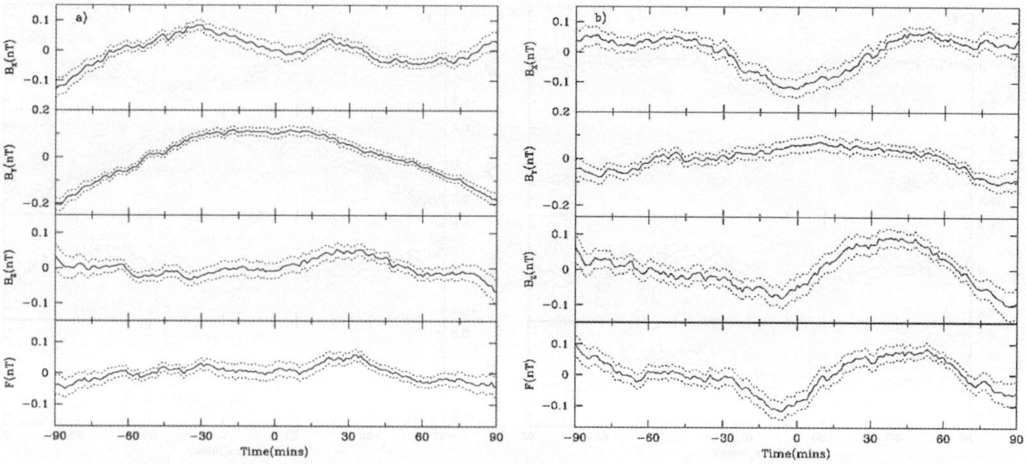

Figure 2. Similar plots as in Figure 1. Plots result from data sets where: (a) observing site is located in the first half of the path, (b) observing site is located in the second half of the path.

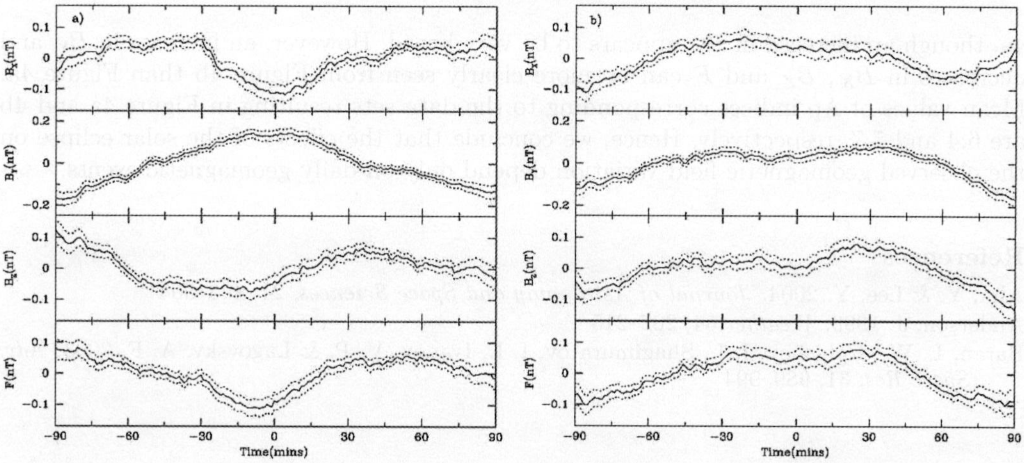

Figure 3. Similar plots as in Figure 1. Plots result from data sets where: (a) the Ap index is less than or equals to 7 representing magnetically quiet days, (b) the Ap index is greater than 7.

Geomagnetic activity.

In Figure 3, we show results from two data sets where the Ap index is less than or equals to 7 and where the Ap index is greater than 7, respectively. An increase in B_Y and decreases in B_X, B_Z and F can be clearly seen from Figure 3a. On the contrary, any signatures of solar eclipse effects are absent in the patterns of the geomagnetic field variation data shown in Figure 3b. Our finding here, therefore, implies that when the day of the solar eclipse is geomagnetically quiet an evidence of the solar eclipse effects is likely to be revealed and that when the geomagnetic field is disturbed on the day effects of the solar eclipse on the observed geomagnetic field variation seems to be overwhelmed by other geomagnetic disturbances.

Solar activity.

In Figure 4, we show results from two data sets due to solar eclipses occurring during separate time intervals, respectively, which correspond to periods around the solar minimum and the solar maximum. Decreases in B_X and F can be noticed from Figure

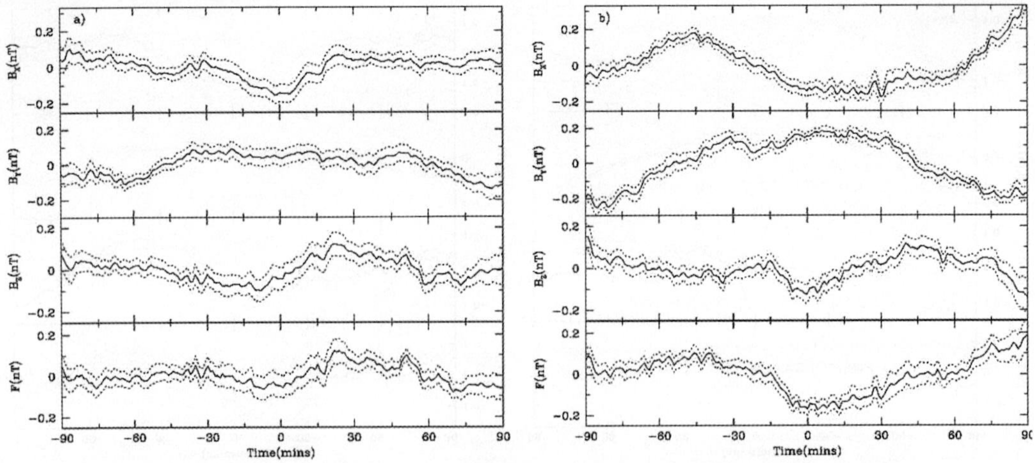

Figure 4. Similar plots as in Figure 1. Plots result from solar eclipses occurring during two separate time intervals, respectively, corresponding to periods around: (a) the solar minimum, and (b) the solar maximum.

4a, though an increase in B_Y appears to be very broad. However, an increase in B_Y and decreases in B_X, B_Z and F can be more clearly seen from Figure 4b than Figure 4a. Mean values of Ap indices corresponding to the data sets resulting in Figure 4a and 4b are 6.4 and 5.7, respectively. Hence, we conclude that the effects of the solar eclipse on the observed geomagnetic field variation depend only on daily geomagnetic events.

References

Ahn, Y. & Lee, Y. 2004, *Journal of Astronomy and Space Sciences*, 21, 493–504

Anderson, J. 1999, *Weather* 54, 207–215

Baran, L. W., Ephishov, I. I., Shagimuratov, I. I., Ivanov, V. P. & Lagovsky, A. F. 2003, *Adv. Space Res.* 31, 989–994

Space Weather of the Heliosphere: Processes and Forecasts
Proceedings IAU Symposium No. 335, 2017
C. Foullon & O.E. Malandraki, eds.

doi:10.1017/S1743921317007220

About Factors of Solar Radiation Affecting the Ionosphere

Aleksander V. Rakhlin,[1] Olga A. Sheiner,[1] Feodor I. Vybornov[1] and Aleksander V. Pershin[1]

[1]Radiophysical Research Institute, National Lobachevsky State University,
25/12A Bol'shaya Pecherskaya str., 603950, Nizhny Novgorod, Russian Federation
email: rakhlin.alex@nirfi.unn.ru

Abstract. In this report we present an attempt to find a characteristic set of the space weather parameters allowed to identify the dominant physical connections. This study is based on the data of vertical and oblique sounding of the ionosphere in 2015-2016.

Keywords. (Sun:) solar-terrestrial relations, methods: data analysis

1. Introduction

Within the framework of space weather studies, research were conducted for a long time to determine the geo-effectiveness of various processes occurring on the Sun and near-Earth space. In most cases, the relationship between changes in near-space parameters measured by artificial satellites and parameters of the upper atmosphere of the Earth is quite clearly expressed. In particular, measurements of ionospheric plasma parameters can serve as a criterion that unequivocally testifies to the degree of geo-efficiency of processes in near-Earth space. The most common methods of ground-based measurements are methods of vertical and oblique sounding of the ionosphere. As one of the parameters of the upper atmosphere of the Earth, it is proposed to consider the time behavior of the critical frequency (f_0F2) of the ionospheric $F2$ layer, measured by vertical sounding of the ionosphere. The behavior of the maximum observed frequency for oblique sounding trajectories (MOF), that is determined in time using ionospheric oblique sounding stations, can be used for analysis, as well.

2. Overview

To increase the sensitivity, to eliminate stationary dependencies (for example, exclude diurnal and seasonal behaviour of f_0F2) we use method to study the disturbances in the data of vertical ionospheric sounding, based on the second order deviation of critical frequency of ionospheric layer $F2$ (Δf_0F2) (Sheiner *et al.* 2002). The data of critical frequency f_0F2 have been determined from uniform ionograms obtained with the modern digital ionosonde CADI. This ionosonde is installed at the NIRFI landfill "Vasilsursk" (near Nizhny Novgorod), and working program of regular observations allowed to obtain ionograms at least once in 15 minutes. The accuracy of determining the critical frequency was less than 50 kHz. The data of maximum applicable frequency for oblique sounding trajectories MOF have been determined using an ionosonde-direction finder with chirp signal on the Cyprus–Nizhny Novgorod chirp-sounding path. This device measures the key characteristics of the ionospheric channel simultaneously in real time over the entire range of transmission frequencies of HF radio signals in all propagation modes (Uryadov *et al.* 2016).

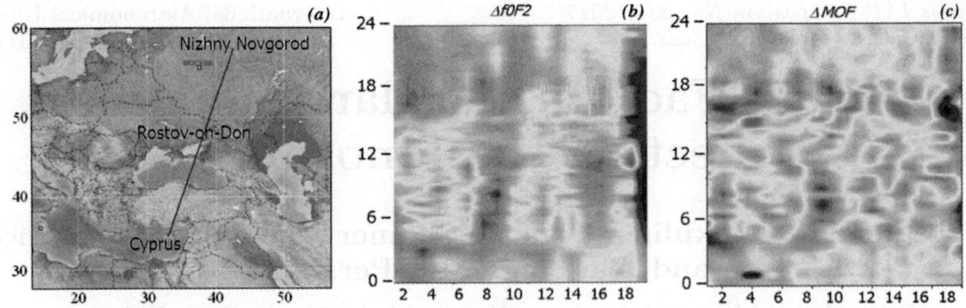

Figure 1. *a)* – The map with the chirp-sounding pass Cyprus–Nizhny Novgorod; *b)* and *c)* – behaviour of differential parameters Δf_0F2 (measured in "Vasilsursk") and ΔMOF, vertical axis – time of the day (UT), horizontal – days of the month.

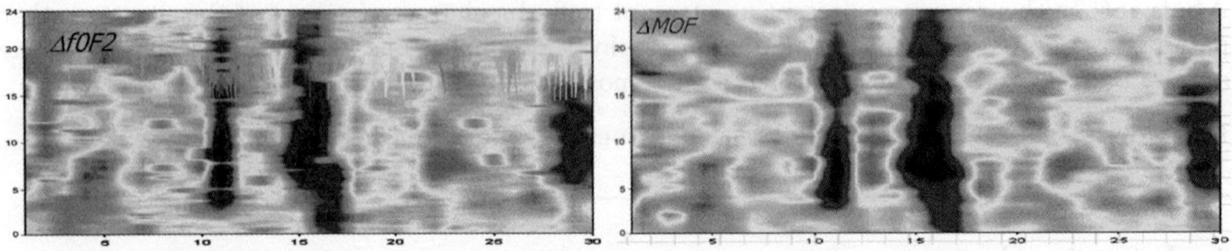

Figure 2. Behaviour of differential parameters Δf_0F2 (measured in Rostov-on-Don) and ΔMOF; vertical axis – time of the day (UT), horizontal – days of the month.

For example, Figure 1 illustrates the chirp-sounding pass Cyprus–Nizhny Novgorod (a) and the behaviour of differential parameters Δf_0F2 (b) and ΔMOF (c) for March 1-18, 2015. We have to add, comparison of two pictures (ΔMOF received using Cyprus-Nizhny Novgorod pass and Δf_0F2 measured in Rostov-on-Don, city located just in the middle of the pass, using vertical sounding of the ionosphere) shows good agreement in behavior (observations in April, 2015, see Figure 2).

To find the most significant parameters of the influence of solar activity on the ionosphere we conducted joint analysis of the data of the GOES satellites and the data of the oblique sounding of the ionosphere on the Cyprus–Nizhny Novgorod pass. Figure 3 presents the results of such analysis. In the bottom panels are graphs of the deviations of the maximum observed frequency (ΔMOF). A significant decrease by a few MHz (blue) of the instantaneous ΔMOF in comparison with the monthly average ΔMOF is clearly correlated with the behavior of the planetary Kp index (Figure 3a). At the same time, we have to mention that the correlation of the ΔMOF behavior with the density of the proton flux and the intensity of X-ray radiation is reliably traced only in the summer observation cycle.

Comparison of ΔMOF behavior with the measured parameters of the near-Earth space obtained on the GOES satellites (Electron Flux, Proton Flux, X-ray) shows a good cross-correlation in most cases when the increase (decrease) of ΔMOF relative to the mean value is caused by the corresponding changes in the parameters measured on the satellites. But sometimes one can see that the changes in the ΔMOF behavior are only partially correlated with the behavior of the characteristics of the near-Earth space measured on the GOES satellites (see Figure 3b). At the same time, joint analysis of the ΔMOF with the data of solar wind proton speed from the ACE RTSW satellite demonstrates the role of high-speed solar wind. Decrease by a few MHz (blue) of the instantaneous ΔMOF may be due to the after-effect of high-speed solar wind passage (Figure 4).

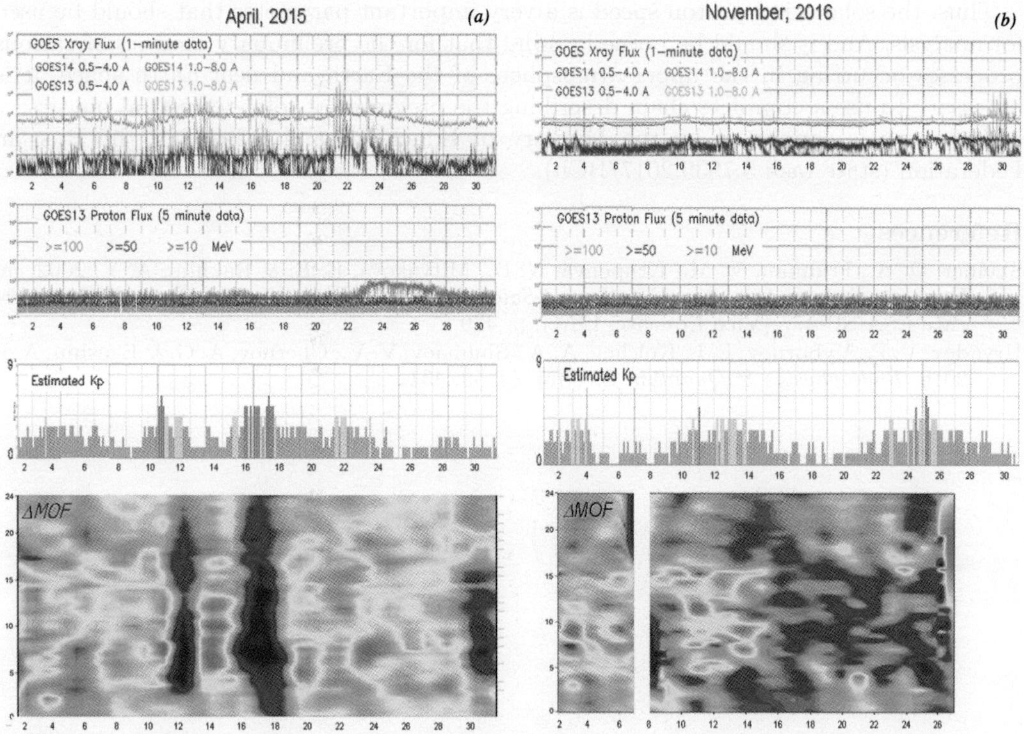

Figure 3. Results of joint analysis of the data of the GOES satellites and the data of the oblique sounding of the ionosphere on the Cyprus-Nizhny Novgorod pass for April 2015 *a)* and November 2016 *b)*, horizontal axis – days of the month; vertical axis for ΔMOF – time of the day (UT).

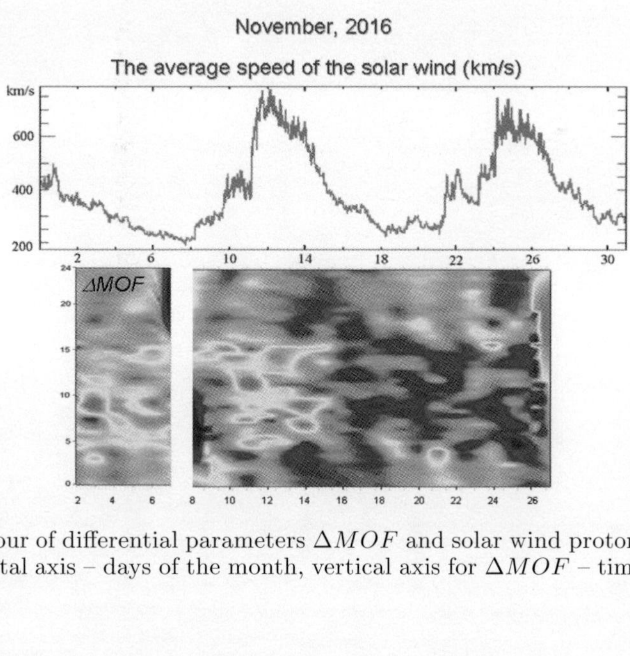

Figure 4. Behaviour of differential parameters ΔMOF and solar wind proton speed on November, 2016. Horizontal axis – days of the month, vertical axis for ΔMOF – time of the day (UT).

Thus, the solar wind proton speed is a very important paremeter that should be used for analysis. And it should be borne in mind that for the preliminary analysis of various processes occurring in the upper atmosphere of the Earth and near-Earth space, it is necessary to use a complete chain describing the processes of solar-terrestrial physics.

The work is supported by the Ministry of Education and Science of the Russian Federation (state task 3.7939.2017/BCh).

References

Sheiner, O. A., Fridman, V. M., Krupenya, N. D., Mityakova, E. E. & Rakhlin, A. V. 2002, in: Huguette Sawaya-Lacoste (ed.), *Second Solar Cycle and Space Weather*, Proc. Euroconference ESA SP-477 (Vico Equense: ESA), p. 479

Uryadov, V. P., Vybornov, F. I., Kolchev, A. A., Shumaev, V. V., Chernov, A. G. & Egoshin, A. I. 2016, *Radiophysics & Quantum Electronics*, 59, 431

Part 5. Long-term Trends and Predictions for Space Weather

Space Weather of the Heliosphere: Processes and Forecasts
Proceedings IAU Symposium No. 335, 2017
C. Foullon & O.E. Malandraki, eds.

doi:10.1017/S1743921317007967

Predicting a Solar Cycle Before its Onset Using a Flux Transport Dynamo Model

Arnab Rai Choudhuri
Department of Physics, Indian Institute of Science,
Bangalore – 560012, India
email: arnab@iisc.ac.in

Abstract. We begin with a review of the predictions for cycle 24 before its onset. After summarizing the basics of the flux transport dynamo model, we discuss how this model had been used to make a successful prediction of cycle 24, on the assumption that the irregularities of the solar cycle arise due to the fluctuations in the Babcock–Leighton mechanism. We point out that fluctuations in the meridional circulation can be another cause of irregularities in the cycle.

Keywords. Sun: activity, Sun: magnetic fields, MHD.

1. Introduction

Let us begin with a disclaimer. This review will focus on the physics of predicting solar cycles from dynamo models and will refrain from presenting any detailed prediction for the upcoming cycle 25, which is nowadays becoming a hot topic of research.

Now that we know what the present cycle 24 has been like, let us take a look at the many predictions of cycle 24 before its onset. Pesnell (2008) produced a plot combining all the different predictions of the peak sunspot number of cycle 24. Figure 1 is adopted from this plot, indicating the two predictions based on theoretical dynamo models. The first theoretical prediction by Dikpati *et al.* (2006) was that cycle 24 would be a very strong cycle, whereas the other prediction by Choudhuri *et al.* (2007) was that it would be a fairly weak cycle. All the other predictions shown in Figure 1 were based on various precursors and empirical projections. We can see that the predictions covered almost the entire range of possible values of the peak sunspot number from $\approx$40 to $\approx$190. The horizontal line indicates the actual peak sunspot number of cycle 24 and was added by us while preparing this presentation. It is clear that Choudhuri *et al.* (2007) predicted the cycle 24 peak almost correctly.

If several people make several predictions covering the entire possible range, then somebody's prediction has got to come out right! Were Choudhuri *et al.* (2007) simply the lucky persons whose prediction accidentally turned out to be correct? Or did they get it correct because they figured out the correct physics for making such predictions? We would like to argue that they figured out the correct physics partially, but not fully. Their success in predicting cycle 24 was due to a combination of intuition and luck.

In a classic work, Parker (1955) envisaged that the solar cycle is produced by an oscillation between the toroidal and poloidal magnetic fields of the Sun. Sunspots form out of the toroidal field due to magnetic buoyancy and provide an indication of the strength of the toroidal field. On the other hand, the magnetic fields in the polar regions of the Sun are a manifestation of the poloidal field. We now know that there is truly an oscillation between these two field components. The polar (i.e. poloidal) field becomes strongest around the time when the sunspot number (i.e. toroidal field) has its lowest value and vice versa. Svalgaard *et al.* (2005) and Schatten (2005), whose predictions for

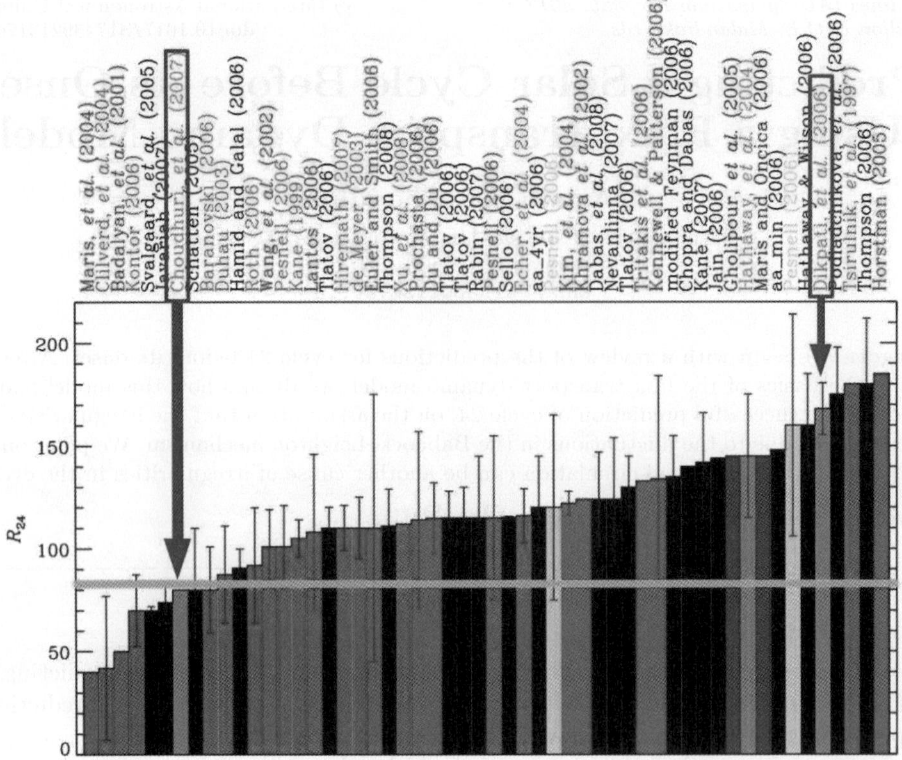

Figure 1. Different predictions of the strength of cycle 24, adopted from Pesnell (2008). The two predictions based on theoretical dynamo models (Dikpati *et al.* 2006; Choudhuri et al. 2007) are indicated by arrows. The horizontal line added by us indicates the actual peak strength of cycle 24 reached around April 2014.

cycle 24 are included in Figure 1, suggested that the polar field at the beginning of a cycle is a good precursor for the strength of the cycle and used the weak polar field at the beginning of cycle 24 to predict essentially the same low value of the cycle peak that was predicted from the theoretical dynamo calculations of Choudhuri *et al.* (2007). While discussing the physics of cycle prediction, we need to address the question of why the polar field at the beginning of a cycle acts as such a good precursor of the cycle strength.

2. Basics of flux transport dynamo model

The flux transport dynamo model, which started being developed in the 1990s (Wang et al. 1991; Choudhuri *et al.* 1995; Durney 1995; Dikpati & Charbonneau 1999; Nandy & Choudhuri 2002) and has been recently reviewed by several authors (Choudhuri 2011, 2014; Charbonneau 2014; Karak *et al.* 2014a), has emerged as an attractive theoretical model of the solar cycle. We describe the basics of this model in this Section before getting into the question of cycle prediction in the next Section.

In order to have an oscillation between the toroidal and poloidal magnetic components, we need to have mechanisms to generate each component from the other. The generation of the toroidal component from the poloidal component due to stretching by differential rotation is a straightforward process. To complete the loop, we need a mechanism to generate the poloidal component back from the toroidal component. The flux transport dynamo model assumes that this is achieved by the Babcock–Leighton (BL) mechanism

(Babcock 1961; Leighton 1969), in which the toroidal field first gives rise to bipolar sunspots which emerge with a tilt due to the action of the Coriolis force (D'Silva & Choudhuri 1993) and then the poloidal field arises from the decay of these tilted bipolar sunspots (Hazra *et al.* 2017).

A dynamo model based just on the idea of toroidal field generation by differential rotation and poloidal field generation by the BL mechanism gives a poleward propagating dynamo wave—in accordance with what is called the Parker–Yoshimura sign rule (Parker 1955; Yoshimura 1975). In order to explain the observation that sunspots appear increasingly at lower latitudes with the progress of the solar cycle, we need an extra mechanism to reverse the direction of the dynamo wave and make it propagate equatorward. Choudhuri *et al.* (1995) realized that this extra mechanism is provided by the meridional circulation (MC) of the Sun, which is observed to be poleward near the surface of the Sun. In order to avoid mass piling up in the polar region, there has to be a subsurface reverse flow towards the equator. The majority of flux transport dynamo models assume an one-cell MC with the equatorward flow at the bottom of the convection zone. However, Hazra *et al.* (2014) have shown that the flux transport model can work even with a multi-cell MC structure as long as there is an equatorward flow at the bottom of the convection zone. The toroidal field produced there due to the strong differential rotation discovered by helioseismology is advected by this flow equatorward, making sure that sunspots appear at lower latitudes with the progress of the cycle. The poloidal field generated by the BL mechanism near the surface is advected poleward by the poleward MC there—in accordance with observational data.

Perhaps, at this stage, it is a good idea to state clearly what we mean by flux transport dynamo. We would refer to a solar dynamo model as a flux transport dynamo model if it satisfies the following characteristics: (i) the Babcock–Leighton (BL) mechanism for generating the poloidal field at the solar surface is included in the model; and (ii) the meridional circulation (MC) plays an important role by transporting the poloidal field poleward near the surface and the toroidal field equatorward at the bottom of the convection zone. While different authors in the past might have meant slightly different things by flux transport dynamo, we believe that the definition we give here has come to be regarded as the universally accepted definition of flux transport dynamo at the present time.

Detailed theoretical calculations based on the flux transport dynamo model are broadly in agreement with observational data pertaining to the solar cycle. A flux transport dynamo code based on mean field equations tends to give a periodic solution, unless something special is done to make the solution irregular. After our brief summary of the flux transport dynamo model, we come to the question as to what causes irregularities of the cycle and whether an understanding of that will help us in predicting future cycles.

3. Fluctuations in the Babcock–Leighton (BL) mechanism

It has been known for some time that fluctuations in the poloidal field generation mechanism can cause irregularities in the cycle (Choudhuri 1992). Choudhuri *et al.* (2007) suggested that the fluctuations in the BL mechanism would be the main source of irregularities in the dynamo process. It is not difficult to understand how the fluctuations in the BL mechanism may arise. This mechanism depends on the tilts of bipolar sunspot pairs—a pair with a larger tilt making a more significant contribution to the poloidal field. The average tilt of sunspot pairs is given by Joy's law. However, one finds a distribution of tilt angles around this average (Stenflo & Kosovichev 2012)—presumably

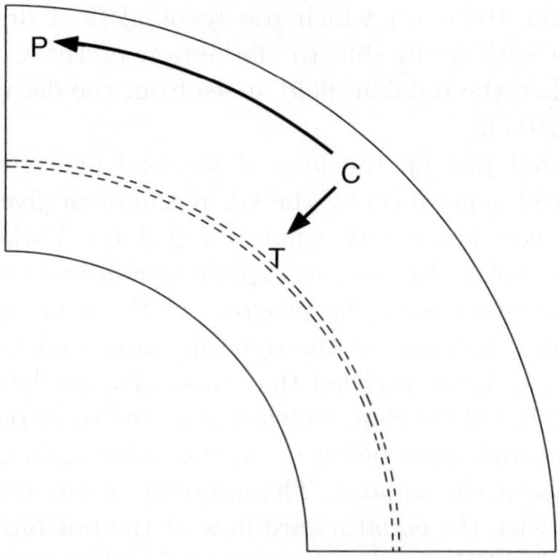

Figure 2. A sketch explaining how the correlation between the polar field at the end of a cycle and the strength of the next cycle arises. From Jiang *et al.* (2007).

caused by the effect of the turbulence in the solar convection zone through which flux tubes rise to form bipolar sunspot pairs (Longcope & Choudhuri 2002).

Let us first come to the question how the observed correlation between the polar field at the beginning of a solar cycle and the strength of the cycle arises. Refer to Figure 2 taken from Jiang *et al.* (2007), who provided an explanation of this. The BL mechanism produces poloidal field at C in mid-latitudes near the surface. This poloidal field will simultaneously be advected poleward by the meridional circulation (MC) to produce the polar field at P at the beginning of the next cycle and will also diffuse to T at the bottom of the convection zone, where it will act as the seed of the next cycle. If the fluctuations in the BL mechanism make the poloidal field produced at C in a cycle stronger than the usual, then both the polar field at P at the beginning of the next cycle and the seed of the next cycle at T will be stronger than usual. On the other hand, if the poloidal field produced at C is weaker than the usual, then the opposite of this will happen. We believe that this is how the correlation between the polar field at the beginning of a cycle and the strength of the cycle arises. It may be pointed out that the turbulent diffusion has to be sufficiently high to ensure that the poloidal field diffuses from C to T in time of the order of 5–10 yr, in order to produce the required correlation. Choudhuri et al. (2007) used a value of turbulent diffusivity on the basis of mixing length arguments, which made the correlation come out beautifully. However, Dikpati *et al.* (2006) used a rather unrealistically low value of turbulent diffusion and the diffusion time across the convection zone in their model is of the order of 200 yr. Their model would not give a correlation between the polar field at the beginning of a cycle and the strength of the cycle. It may be mentioned that solar dynamo models with high and low diffusivities have very different characteristics (Jiang *et al.* 2007; Yeates et al. 2008). There is enough evidence now that models with higher diffusivity are closer to reality. Higher diffusivity helps in explaining such observational features as the dipolar parity (Chatterjee *et al.* 2004; Hotta & Yokoyama 2010) and the lack of hemispheric asymmetry (Chatterjee & Choudhuri 2006; Goel & Choudhuri 2009).

In order to model actual cycles, one needs to incorporate the actual fluctuations in the BL mechanism into the code. Choudhuri *et al.* (2007) devised a scheme of figuring out actual fluctuations of the BL mechanism from the observational data of the poloidal fields and then incorporating these in the dynamo code. Since such data are available only from the 1970s, actual cycles could be modelled only from that time. Choudhuri *et al.* (2007) succeeded in modelling cycles 21–23 reasonably well and cycle 24 was predicted to be a weak cycle. Their prediction of cycle 24 was a robust prediction, since they had incorporated the weakness of the polar field at the beginning of cycle 24 in their model and the high diffusivity of their model would make this correlated with the strength of cycle 24. As we have already pointed out, this prediction has been borne out triumphantly—making this the first successful prediction of a solar cycle based on a theoretical dynamo model in the history of this subject.

4. Fluctuations in the meridional circulation (MC)

When Choudhuri *et al.* (2007) made their prediction, it was not realized that the meridional circulation (MC) probably has occasional large fluctuations and these also may produce irregularities in the solar cycle. These fluctuations are different from the periodic variation of MC with the solar cycle, presumably due to the Lorentz force of the dynamo-generated magnetic field (Hazra & Choudhuri 2017). It has been known for some time that the strength of MC determines the period of the dynamo—faster MC making cycles shorter and vice versa (Dikpati & Charbonneau 1999). Although we have MC data only for about two decades, we have data pertaining to durations of cycles for about two centuries and these data indicate that there have fluctuations in MC with correlation time of the order of a few decades (Karak & Choudhuri 2011). What kinds of irregularities will the fluctuating MC introduce? Suppose MC has slowed down, making cycles longer. Then diffusion will have more time to act, making cycles weaker. We may thus expect longer cycles to be weaker and shorter cycles stronger. Since a shorter cycle would imply a faster rise time of the cycle, we may expect a correlation between the rise time and the strength of a cycle. Such a correlation is found in the observational data and is known as the Waldmeier effect. Karak & Choudhuri (2011) gave an explanation of this effect by introducing fluctuations in MC in their model.

On the basis of such studies, we conclude that there are two important sources of irregularities in the solar cycle—fluctuations in the BL mechanism and fluctuations in MC. Choudhuri *et al.* (2007) made their prediction of cycle 24 on the basis of the assumption that the irregularities are produced by fluctuations in the BL mechanism alone. Presumably their prediction was so successful because there had not been large fluctuations in MC in the last few years. Now that we know the fluctuations of MC to be another factor introducing irregularities in solar cycles, we need to develop cycle prediction methods taking this into account. We are presently working on this problem.

It may be mentioned that one big challenge in this field is to develop a theory of grand minima like the Maunder minimum in the 17th century. While it has been shown that grand minima can be induced by fluctuations in the BL mechanism alone (Choudhuri & Karak 2009) or by fluctuations in MC alone (Karak 2010), we need both of these to develop a comprehensive theory of grand minima (Choudhuri & Karak 2012; Karak & Choudhuri 2013).

5. Conclusion

Within the last few years, the flux transport dynamo model has emerged as an attractive model for explaining the solar cycle and there is increasing evidence that other solar-like stars also may have similar dynamos working inside them (Karak *et al.* 2014b; Choudhuri 2017). It is important that we understand how the irregularities in the cycle arise, since such an understanding may enable us to predict a future cycle before its onset.

It appears that fluctuations in the BL mechanism and fluctuations in MC are the two main sources of irregularities in the solar cycle. Before the beginning of cycle 24, the role of MC fluctuations was not generally appreciated. The successful theoretical prediction of Choudhuri *et al.* (2007) was based on the assumption that irregularities in the solar cycle are caused only by fluctuations in the BL mechanism. With the realization that MC fluctuations also can introduce additional irregularities, it is necessary to develop prediction methods taking this into account.

Acknowledgements

I thank Gopal Hazra for help in preparing the manuscript. My research is supported by DST through a J.C. Bose Fellowship.

References

Babcock, H. W. 1961, *ApJ*, 133, 572
Charbonneau 2014, *ARAA*, 52, 251
Chatterjee, P. & Choudhuri, A. R. 2006, *Solar Phys.*, 239, 29
Chatterjee, P., Nandy, D. & Choudhuri, A. R. 2004, *A&A*, 427, 1019
Choudhuri, A. R. 1992, *A&A*, 253, 277
Choudhuri, A. R. 2011, *Pramana*, 77, 77
Choudhuri, A. R. 2014, *Indian J. Phys.*, 88, 877
Choudhuri, A. R. 2017, *Science China Phys. Mech. Astron.*, 60, 019601
Choudhuri, A. R., Chatterjee, P. & Jiang, J. 2007, *Phys. Rev. Lett.*, 98, 131103
Choudhuri, A. R. & Karak, B. B. 2009, *Res. Asron. Astrophys.*, 9, 953
Choudhuri, A. R. & Karak, B. B. 2012, *Phys. Rev. Lett.*, 109, 171103
Choudhuri, A. R., Schüssler, M. & Dikpati, M. 1995, *A&A*, 303, L29
D'Silva, S. & Choudhuri, A. R. 1993, *A&A*, 272, 621
Dikpati, M. & Charbonneau, P. 1999, *ApJ*, 518, 508
Dikpati, M., de Toma, G. & Gilman, P. A. 2006, *Geophys. Res. Lett.*, 33, 5102
Durney, B. R. 1995, *Solar Phys.*, 160, 213
Goel, A. & Choudhuri, A. R. 2009, *Res. Asron. Astrophys.*, 9, 115
Hazra, G. & Choudhuri, A. R. 2017, *MNRAS*, in press
Hazra, G., Choudhuri, A. R. & Miesch, M. S. 2017, *ApJ*, 835, 39
Hazra, G., Karak, B. B. & Choudhuri, A. R. 2014, *ApJ*, 782, 93
Hotta, H. & Yokoyama, T. 2010, *ApJ*, 714, L308
Jiang, J., Chatterjee, P. & Choudhuri, A. R. 2007, *MNRAS*, 381, 1527
Karak, B. B. 2010, *ApJ*, 724, 1021
Karak, B. B. & Choudhuri, A. R. 2011, *MNRAS*, 410, 1503
Karak, B. B. & Choudhuri, A. R. 2013, *Res. Asron. Astrophys.*, 13, 1339
Karak, B. B., Jiang, J., Miesch, M. S., Charbonneau, P. & Choudhuri, A. R. 2014a, *Space Sc. Revs*, 186, 561
Karak, B. B., Kitchatinov, L. L. & Choudhuri, A. R. 2014b, *ApJ*, 791, 59
Leighton, R. B. 1969, *ApJ*, 156, 1
Longcope, D. & Choudhuri, A. R. 2002, *Solar Phys.*, 205, 63
Nandy, D. & Choudhuri, A. R. 2002, *Science*, 296, 1671
Parker, E. N. 1955, *ApJ*, 122, 293
Pesnell, W. D. 2008, *Solar Phys.*, 252, 209
Schatten, K. 2005, *Geophys. Res. Lett.*, 32, L21106
Stenflo, J. O. & Kosovichev, A. G. 2012, *ApJ*, 745, 129
Svalgaard, L., Cliver, E. W. & Kamide, Y. 2005, *Geophys. Res. Lett.*, 32, L01104
Wang, Y.-M., Sheeley, N. R. & Nash, A. G. 1991, *ApJ*, 383, 431
Yeates, A. R., Nandy, D. & Mackay, D. H. 2008, *ApJ*, 673, 544
Yoshimura, H. 1975, *ApJ*, 201, 740

Space Weather of the Heliosphere: Processes and Forecasts
Proceedings IAU Symposium No. 335, 2017
C. Foullon & O.E. Malandraki, eds.

doi:10.1017/S1743921317010754

Towards Estimating the Solar Meridional Flow and Predicting the 11-yr Cycle Using Advanced Variational Data Assimilation Techniques

Ching Pui Hung[1,2], Allan Sacha Brun[2], Alexandre Fournier[1], Laurène Jouve[3,2], Olivier Talagrand[4] and Mustapha Zakari[1]

[1]Institut de Physique du Globe de Paris, Sorbonne Paris Cité, Université Paris Diderot UMR 7154 CNRS, F-75005 Paris, France
email: hung@ipgp.fr

[2]Laboratoire AIM Paris-Saclay, CEA/IRFU Université Paris-Diderot CNRS/INSU, 91191 Gif-Sur-Yvette, France

[3]Université de Toulouse, UPS-OMP, Institut de Recherche en Astrophysique et Planétologie, 31028 Toulouse Cedex 4, France

[4]Laboratoire de météorologie dynamique, UMR 8539, Ecole Normale Supérieure, Paris Cedex 05, France

Abstract. We present in this work the development of a solar data assimilation method based on an axisymmetric mean field dynamo model and magnetic surface data. Our mid-term goal is to predict the solar quasi cyclic activity. We focus on the ability of our variational data assimilation algorithm to constrain the deep meridional circulation of the Sun based on solar magnetic observations. Within a given assimilation window, the assimilation procedure minimizes the differences between data and the forecast from the model, by finding an optimal meridional circulation in the convection zone, and an optimal initial magnetic field, via a quasi-Newton algorithm. We demonstrate the capability of the technique to estimate the meridional flow by a closed-loop experiment involving 40 years of synthetic, solar-like data. We show that the method is robust in estimating a (stochastic) time-varying flow fluctuating 30% about the average, and that the horizon of predictability of the method is ~ 1 cycle length.

Keywords. Dynamo, meridional circulation, data assimilation, numerical

1. Introduction

Solar activity impacts space-weather and influences our modern technology-based society significantly, so it is important to obtain good solar predictions. The most common index to quantify solar activity is the sunspot number (SSN) (Clette *et al.* 2014; Vaquero *et al.* 2016), and Wolf reconstructed the past SSN from 1749. In addition, the surface magnetic field of the Sun is also an important observable. Observations of solar magnetic field can at least be traced back to 1908 (Hale 1908). Solar activity is quasiperiodic with sunspot cycle of 11 ± 3 years (Clette & Lefèvre 2012). Those cycles, however, vary in both their period and amplitude (Svalgaard, Cagnotti & Cortesi 2015), which raises questions regarding their predictability (Ossendrijver *et al.* 2002; Tobias *et al.* 1998). Data assimilation is an emerging technique in solar prediction, which incorporates observations in numerical models (Brun 2007). For examples of application in studying solar activity, see Dikpati, Anderson & Mitra (2014), Dikpati, Anderson & Mitra (2016) and Jouve, Brun & Talagrand (2011). In this work, we develop a variational data assimilation technique to analyze magnetic observations of the Sun. By controlling selected parameters of the physical model, an optimal fit of data is obtained over the time window when observations are available. We adopt the 4D-Var method, which minimizes the misfit or objective function, by using the adjoint model (Fournier *et al.* 2010; Talagrand 2010). This is based on the Babcock-Leighton flux transport mean field dynamo model (Babcock 1961; Leighton 1969; Jouve & Brun 2007) and its adjoint. The initial magnetic field and the meridional flow are the control parameters, as they determine the phase and the cycle length of the magnetic field. While recent estimate of meridional flow from helioseismology below the solar surface suggests the possibility of more complex structures other than a unicellular one (Zhao *et al.* 2013; Schad, Timmer & Roth 2013; Haber *et al.* 2002; Kholikov, Serebryanskiy & Jackiewicz 2014), there is no unique conclusion on the meridional structure, which raises the interest of estimating the meridional flow with a dynamo model. We show that we get a satisfactory estimate of the true time varying profile of the meridional circulation, and reconstruct the magnetic trajectory.

2. Data Assimilation Framework

We develop a data assimilation technique to estimate the initial magnetic field and the flow of the Sun on the meridional plane. We aim to minimize an objective function $\mathcal{J}$, defined in terms of the differences between the observations and the model trajectory,

$$\mathcal{J} = \sum_{\alpha} \sum_{i=1}^{N^o_{\alpha,t}} \sum_{j=1}^{N^o_{\alpha,\theta}} \frac{[y_\alpha(\theta_j, t_i) - y^o_\alpha((\theta_j, t_i)]^2}{\sigma^2_\alpha(\theta_j, t_i)}, \tag{2.1}$$

where α denotes the type of magnetic proxy y to be compared. The proxies with the superscript ^{o} stand for observations, and without for the forecast values, and σ_α stands for the errors. For each type α we sum over the observation times and latitudes, $N^o_{\alpha,t}$ and $N^o_{\alpha,\theta}$, respectively. In the assimilation setup, the magnetic observations are compared with the forecast values from the flux transport dynamo model. The latter is controlled by the state vector $\mathbf{x}$, which consists of the initial magnetic field and the flow of the model on the meridional plane. We adopt a quasi-Newton algorithm in the minimization of the objective function (Gilbert & Lemaréchal 1989), and the gradient of the function with respect to $\mathbf{x}$ is obtained by the adjoint model (Hung *et al.* 2015, 2017).

3. Closed-loop experiment and results

We verify and study our assimilation framework with a closed-loop numerical experiment. We generate synthetic observations from the same model used in the assimilation procedure with a known set of control parameters, contribute to the truth state $\mathbf{x}^t$. Then, by ingesting the synthetic observations, we show that we are able to estimate the truth state. In our example shown, we create an meridional circulation as a linear combination of 2 time varying expansion coefficients of the basis functions on the meridional plane. These 2 basis functions are of opposite parity about the equator, so that the flow and and corresponding field can be equatorially asymmetric. The scale of the stochastic time variability ($\sim$ 3 years) is chosen to be of the same order as the observed flow (Ulrich 2010). Furthermore, the amplitude of the fluctuation (30% of the average) is adjusted so that the corresponding fluctuation of the cycle length in the model is close to the observed 11 ± 3 years. We show the time-latitude plot of the surface flow in Figure 1. The synthetic magnetic proxy we used in this experiment is the surface magnetic field at the line of sight (B_{los}), and the toroidal field energy at the tachocline in both hemispheres (which mimic the hemispheric SSNs), and the data is noised artificially with 10% of its root mean square. We show the magnetic butterfly diagrams of the B_{los} and the toroidal field at the tachocline ($r = 0.7R_\odot$) in Figure 2. We assimilate the synthetic data yearly

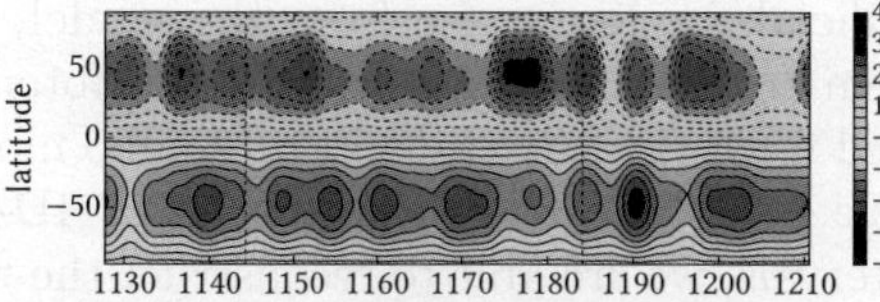

Figure 1. Latitudinal component of the flow at the surface as a function on arbitrary time unit. The assimilation period in the numerical experiment which follows is indicated by the dashed vertical lines. The sign convention is positive for a flow due south.

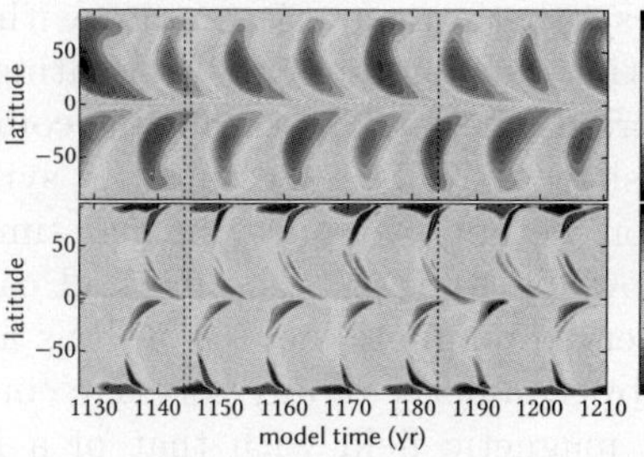

Figure 2. Top: time-latitude plot of the toroidal field at the tachocline. Bottom: Same for B_{los} at the surface.

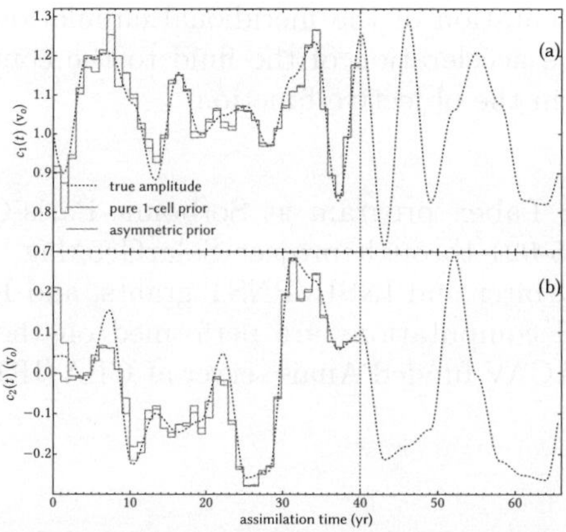

Figure 3. (a) Time series of the coeff. of the unicellular component of the meridional circulation. The reference time series is in broken line. The piecewise constant red (blue) curve is the result of assimilating synthetic observations from a unicellular (asymmetric) prior information. (b) Same for the antisymmetric component.

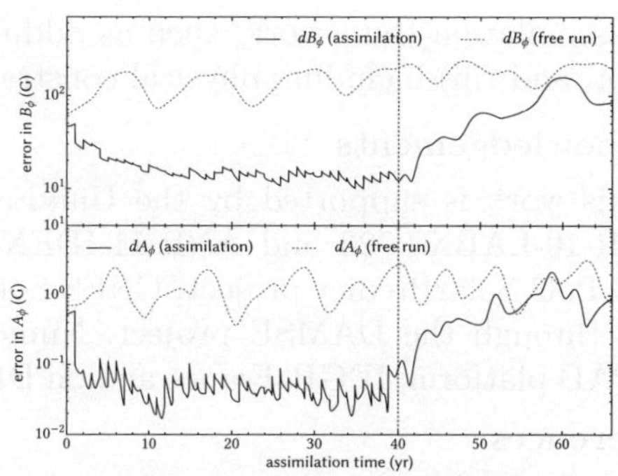

Figure 4. Top: Absolute difference between various estimates of the toroidal magnetic field and the true magnetic field versus time. Blue broken: free run of the dynamo model (unconstrained by data). Black: data assimilation estimate, with data consisting of B_{los} and pseudo sunspot number (toroidal field energy at the tachocline). Bottom: Same for the poloidal magnetic field.

for 40 consecutive years. Figure 3 shows the result of the estimate of the expansion coefficients of the meridional circulation. We approximate the true, smooth, time varying profiles with piecewise constant functions, by ingesting the synthetic observations yearly and assuming the flow is constant within each year. This estimate reasonably captures the time variability of the flow. Figure 4 shows the total error of the estimated poloidal field potential (A_ϕ) and the toroidal field (B_ϕ) (black curves), defined by integration of the differences between the truth and the estimate over the meridional plane. The errors are compared with the counterparts of a free dynamo run without assimilation (blue broken curves). When the synthetic observations are available, the errors in the estimate are 5% or less than the counterparts of the free run. After year 40, the errors start to grow and reach the free run case after $\sim$ 15 years without input of the data, which indicates the predicative capability in this example.

4. Summary

Our numerical experiment shows the capability of the data assimilation framework in estimating the deep meridional circulation of the Sun using magnetic proxies. We construct synthetic magnetic proxies with artificial random noise from the model, like B_{los} and the sunspot number, by relating them to the surface poloidal field potential and the toroidal field in the tachocline computed from the flux transport dynamo model. We then show that, by ingesting the synthetic monthly observations using a 4D-Var assimilation method over one-year long time intervals, we are able to reconstruct the time varying flow for 40 years. The method can also account for the equatorial asymmetry of the observed magnetic field as well as an asymmetric meridional flow, hence it is not impaired by symmetry of any sort. By comparing the time evolution of the error of the estimated magnetic field with that of a free dynamo run where no data assimilation is done, we conclude that in this experiment, the predictive capability of the method is about 15 years. More details and analysis can be found in Hung *et al.* (2017). Our future development include (i) analyzing magnetic proxies of the real Sun with the data assimilation method, (ii) improving the representation of the meridional circulation in the assimilation framework, such as adding the acceleration of the fluid to the control vector, and (iii) including physical constraints in the objective function.

Acknowledgements

This work is supported by the UnivEarthS Labex program at Sorbonne-Paris-Cité (ANR-10-LABX-0023 and ANR-11-IDEX-0005-02) through project SolarGeoMag, the ERC PoC SolarPredict project, CNES Solar Orbiter and INSU/PNST grants, and Idex SPC through the DAMSE project. Numerical computations are performed on the S-CAPAD platform, IPGP, France and on DIM-ACAV funded Anais server at CEA/IRFU.

References

Babcock, H. W. 1961, *ApJ*, 133, 572
Brun, A. S. 2007, *AN*, 328, 329
Clette, F. & Lefèvre, L. 2012, *SWSC*, 2, A06
Clette, F., Svalgaard, L., Vaquero, J. M. & Cliver, E. W. 2014, *Space Sci. Revs*, 186, 35
Dikpati, M., Anderson, J. L. & Mitra, D. 2014, *GRL*, 41, 5361
Dikpati, M., Anderson, J. L. & Mitra, D. 2016, *ApJ*, 828, 91
Fournier, A., Hulot, G., Jault, D., Kuang, W., Tangborn, A., Gillet, N., Canet, E., Aubert, J. & Lhuillier, F. 2010, *Space Sci. Revs*, 155, 247
Gilbert, J. C. & Lemaréchal, C. 1989, *Mathematical programming*, 45, 407
Haber, D. A., Hindman, B. W., Toomre, J., Bogart, R. S., Larsen, R. M. & Hill, F. 2002, *ApJ*, 570, 855
Hale, G. E. 1908, *ApJ*, 28, 315
Hung, C. P., Jouve, L., Brun, A. S., Fournier, A. & Talagrand, O. 2015, *ApJ*, 814, 151
Hung, C. P., Brun, A. S., Fournier, A., Jouve, L., Talagrand, O. & Zakari, M. 2017, *ApJ*, 849, 160
Jouve, L., Brun, A. S. & Talagrand, O. 2011, *ApJ*, 735, 31
Jouve, L. & Brun, A. S. 2007, *A&A*, 470, 239
Kholikov, S., Serebryanskiy, A. & Jackiewicz, J. 2014, *ApJ*, 784, 145
Leighton, R. B. 1969, *ApJ*, 156, 1
Ossendrijver, M., Stix, M., Brandenburg, A. & Rüdiger, G. 2002, *A&A*, 394, 735
Schad, A., Timmer, J. & Roth, M. 2013, *ApJ*, (Letters) 778, L38
Svalgaard, L., Cagnotti, M. & Cortesi, S. 2015, *Solar Phys.*, 292, 34
Talagrand, O. 2010, *Data Assimilation: Making Sense Of Observations* (Springer), 41-67
Tobias, S. M., Brummell, N. H., Clune, T. L. & Toomre, J. 1998, *ApJ*, 502, L177
Ulrich, R. K. 2010, *ApJ*, 725, 658
Vaquero, J. M., Svalgaard, L., Carrasco, V. M. S., Clette, F., Lefèvre, L., Gallego, M. C., Arlt, R., Aparicio, A. J. P., Richard, J. G. & Howe, R. 2016, *Solar Phys.*, 291, 3061
Zhao, J., Bogart, R. S., Kosovichev, A. G., Duvall, T. L. & Hartlep, T. 2013, *AAS/Solar Physics Division Meeting*, 44, 204.02

Space Weather of the Heliosphere: Processes and Forecasts
Proceedings IAU Symposium No. 335, 2017
C. Foullon & O.E. Malandraki, eds.

doi:10.1017/S1743921317008195

Reconstruction of the Filament Properties, Based on Centenarian Daily Observations of the Sun in H_α Line

Kseniya A. Tlatova[1], Valeriya Vasil'eva[1] and Andrey Tlatov[1,2]

[1]Kislovodsk mountain astronomical station of the Pulkovo observatory,
357700, Gagarina str. 100 Russia
email: k.tlatova@mail.ru

[2]Kalmyk State University, Elista, Russia

Abstract. Preliminary results of distinguishing solar filaments on daily observation data at the H_α spectral line of the Kodaikanal Solar Observatory (1912-2002) are presented.

To distinguish the boundaries of solar filaments, methods have been developed, based on automatic procedures for distinguishing low-contrast objects on the solar disk as well as on editing the boundaries of selected structures in semi-automatic mode. An analysis of solar filaments' characteristics has been performed. We are considered variation of the average tilt-angle and the radius of curvature of the filaments in 15-23 cycles of activity.

Keywords. Sun, solar filaments, long-term variations

1. Introduction

The appearance of filaments on the Solar disk is associated with the distribution of active region magnetic fields as well as with large-scale magnetic fields. Filaments are observed in almost every latitude of the solar atmosphere.

Systematic H_α observations began around 1915 at the Kodaikanal Observatory (India), when J. Evershed used a spectroheliograph with a large variance spectral grid. In 1919, Meudon Observatory (France) started its observations using a spectrogeliograph, and since 1959 the Kislovodsk Observatory (Russia) started using Lio filter.

Long-term solar filament data can provide significant information about the properties of the solar periodicity. The solar filaments are located on the lines where the large-scale magnetic field changes. Thus, the analysis of filament data allows for the restoration of a large-scale magnetic field topology over a long period of time (McIntosh 1972, Makarov & Tlatov 2000, Mouradian & Soru-Escaut 1994, Tlatov 2009).

Previously, filaments at the synoptic maps for the period from 1919 to 2014 were digitized (Tlatov *et al.* 2016, Li 2010). This data included the digitization of intermediate lines of filaments, which were drawn manually on the synoptic maps of Meudon and Kislovodsk Observatories.

The purpose of this work was to create a series of filament features due to the daily observation data for the period of about 100 years, to create a data bank for the vector contours of filaments' boundaries, stored in the daily images of various observatories, and to provide a preliminary analysis of the data received.

2. Source data and processing method

This work focuses mainly on presenting the results of processing the daily data of Kodaikanal Observatory (India) for the period from 1912 to 2002. We used images of

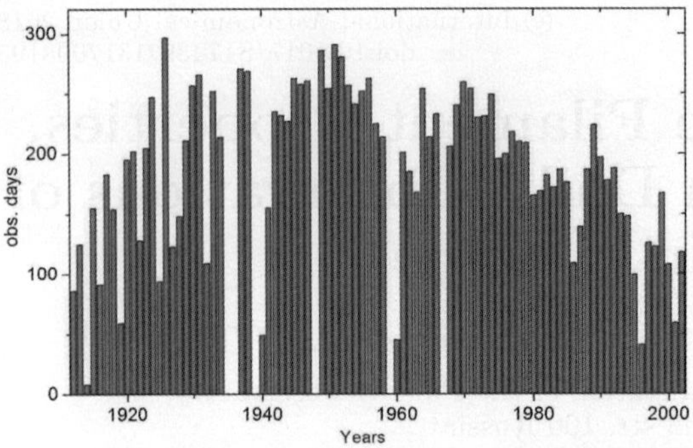

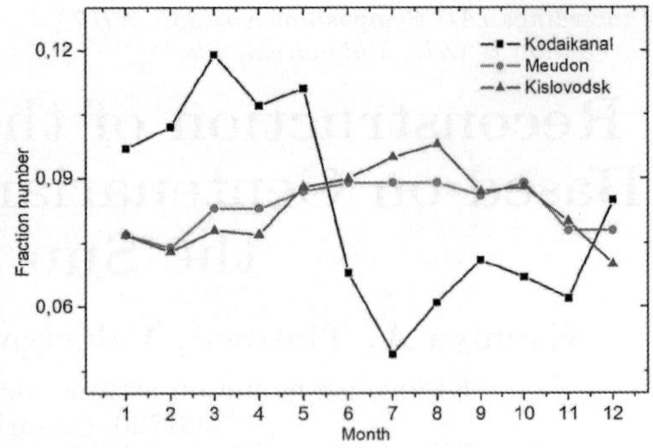

Figure 1. (On the left) The number of processed photographic plates of the Kodaikanal Observatory per year.

Figure 2. (On the right) A change in the number of filaments in observations of the Kodaikanal, Meudon and Kislovodsk Observatories depending on the month of a year.

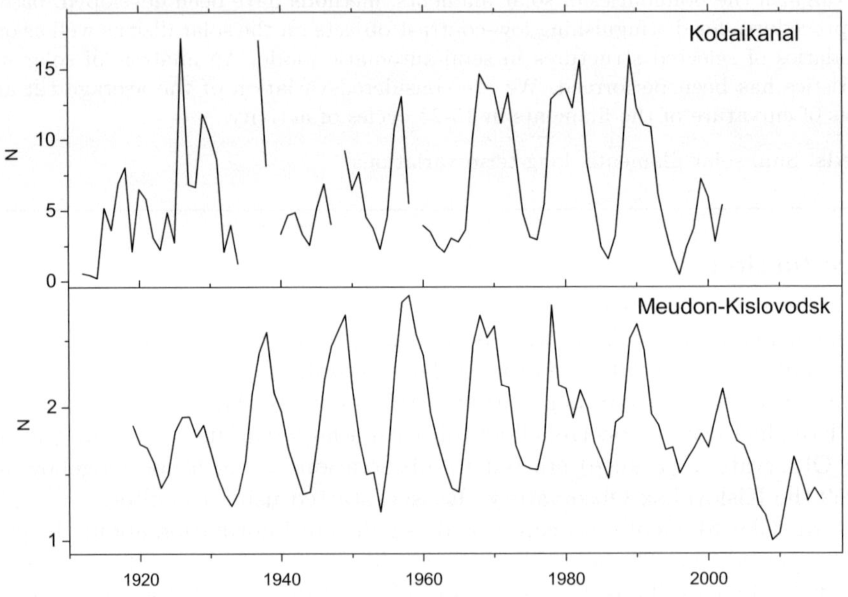

Figure 3. The average number of filaments per day, shown on the daily plates of the Kodaikanal observatory and synoptic maps of Meudon and Kislovodsk observatories.

800x800 pixels available on the web-site (http://kso.iiap.res.in:8080/KodaiArchiveData/) as a source data of Kodaikanal Observatory, as well as, in part, the digitized photographic plates, performed by A.G. Tlatov in 2000 at Kodaikanal Observatory.

Special software was developed to distinguish solar filaments. The algorithm includes several steps: distinguishing the solar limb and grid overlapping, dimness adjustment to the edge, finding the P-angle marker, distinguishing filaments above the level of the quiet Sun, and editing the selected objects by the operator. Our software provides both a fully automatic selection mode and a manual and semi-automatic editing mode, which distinguishes this software from other methods (Fuller *et al.* 2005, Hao *et al.* 2015, Li 2010).

In total we have allocated over 326 thousand filaments on 24308 images of Kodaikanal Observatory for the period from 1912 to 2002. Figure 1 shows the number of observation

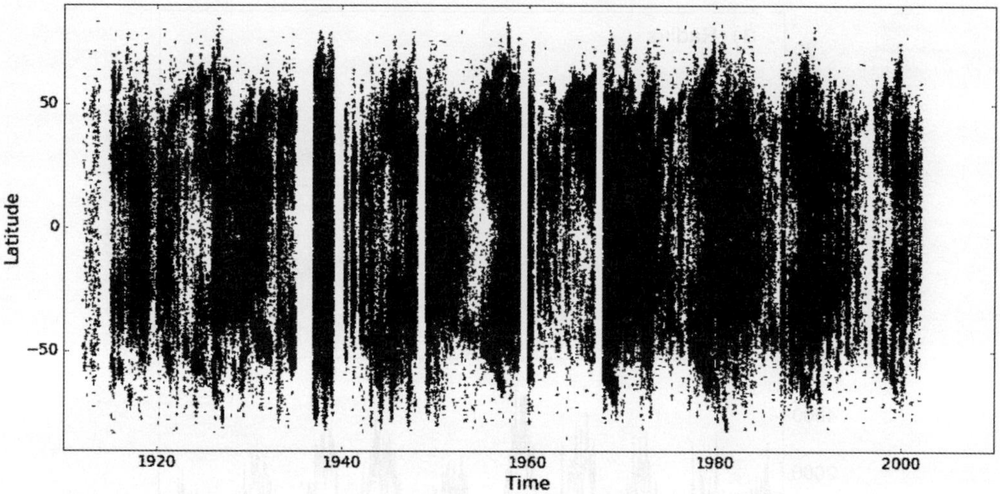

Figure 4. A latitude-time diagram of the position of filament centers.

days for the year that we have processed. We skipped a number of days, for example, in 1935-1936 and in 1959 due to lack of data on the web-site. We were processing all the photographic plates and sometimes even more than one per day. The number of filaments for recordings of no more than one plate per day made over 209 thousand pieces. The database includes both vector data of the filament shapes and text bases of measurements regarding filament parameters.

The processing of daily data from different Observatories is an important requirement for the creation of a unified filament database. Figure 2 shows a change in the relative number of filaments for Kodaikanal, Meudon and Kislovodsk Observatories. Data for Meudon Observatory was collected from the digitized synoptic maps (Tlatov *et al.* 2016). Data for Kislovodsk Observatory are presented at the website of the Observatory: http://solarstation.ru. We would like to mention that maximum observations in Kodaikanal Observatory were made during the first half of a year, while Kislovodsk and Meudon Observatories focused on the second half of a year.

The daily observation data provide much more information than the data presented on the synoptic maps. Figure 3 shows the average number of filaments per day received after processing daily plates of Kodaikanal Observatory and the number of filaments per day received from synoptic maps from Meudon Observatory (1919-2002) and Kislovodsk Observatory (2003-2014). The average number of filaments per day received from the daily plates of Kodaikanal Observatory made ~ 7, and from the synoptic maps ~ 1.8.

The new database regarding to the features of solar filaments provides a comprehensive analysis of solar activity at various latitudes of the solar atmosphere. Figure 4 provides a latitude-time diagram of the position of filament centers according to the data of Kodaikanal Observatory. Data for a number of years are not available, and this is due to an incomplete catalog available on the Internet. It is also possible to note the annual variation in the number of filaments, apparently related to weather conditions and the quality of the original photographic plates. The diagram shows filament drifting towards the equator, associated with distribution of the solar sunspots activity, and drifting to the poles associated with the process of pole-ward transport of magnetic flux in a cycle of activity.

Data of digitization of Kodaikanal filaments, supplemented with data of other Observatories, allows studying the topological properties of filaments for about 100 years.

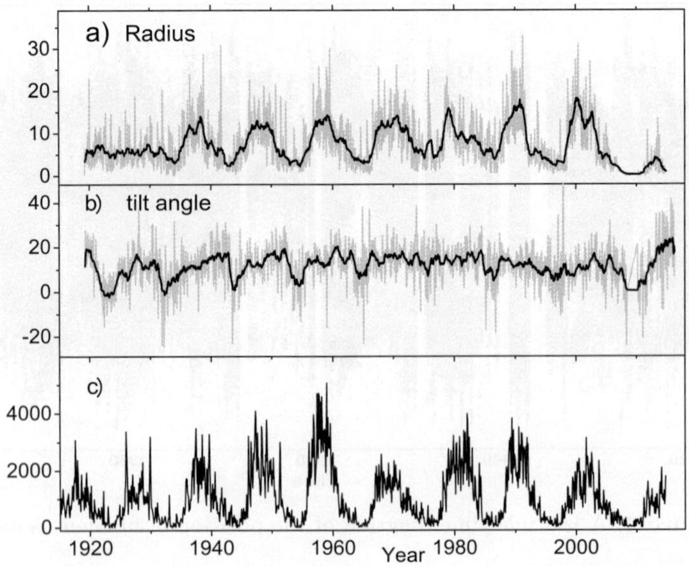

Figure 5. The variation in time a) the radius of curvature of the filaments in heliographic degrees; b) tilt-angle solar filaments c) the sunspots area for comparison.

Figure 5 shows the variation in time of the radius of curvature of the filaments in heliographic degrees, tilt-angle of solar filaments and the sunspots area for comparison.

3. Conclusion

A number of research groups are currently working on the reconstruction of long-term solar activity indices through the processing of long-term observational archives. The H_α observations are an important part of this work. Preliminary results of the processing of observation data from Kodaikanal Observatory for the years between 1912 and 2002 are presented in this paper. The daily data made it possible to distinguish filaments and analyse the activity in the H_α chromospheric line for over a long period of time. The resulting line showed greater stability than other available data lines. The data obtained have a great potential for analysing the properties of solar filaments and a large-scale magnetic field.

The works were performed with partial support of the RFBR Project N 15-02-03900 and RSF Project N 15-12-20001.

References

Fuller N., Aboudarham, J. & Bentley, R. D. 2005, *Solar Phys.*, 227, 61

Li, K. J. 2010, *MNRAS*, 405, 1040

Hao, Q., Fang, C., Cao, W. & Chen, P. F. 2015, *Astrophysical Journal Supplement Series* 221, 33

Makarov, V. I. & Tlatov, A. G. 2000, *Astronomy Reports*, 44, 759

McIntosh, P. S. 1972, *Reviews of Geophysics and Space Physics*, 10, 837

Mouradian, Z. & Soru-Escaut, I. 1994, *Astrop.and Astroph.*, 290, 279

Tlatov, A. G. 2009 *Solar Phys.*, 260, 465

Tlatov, A. G., Kuzanyan, K. M. & Vasil'yeva V. V. 2016, *Solar Phys.*, 291, 1115

Tlatov, A. G., Vasilieva, V, V., Makarova, V. V. & Otkidychev, P. A. 2014 *Solar Phys.*, 289, 1403

Space Weather of the Heliosphere: Processes and Forecasts
Proceedings IAU Symposium No. 335, 2017
C. Foullon & O.E. Malandraki, eds.

doi:10.1017/S1743921317007104

Space Weather in the Heliosphere

C. T. Russell

Earth, Planetary, and Space Sciences
University of California
Los Angeles, CA 90095–1567, USA
email: ctrussell@igpp.ucla.edu

Abstract. We live on a very special planet in a very special solar system. Our planet has a benign climate. Our star has several habitable planets and is not so active as to inhibit the exploration and future colonization of these planets. In this short paper we review how the solar wind interacts with the planets, what factors matter in this interaction, and how active is our star.

Keywords. Space weather, solar wind, magnetospheres, reconnection, solar out bursts

1. Introduction

The star we call 'the Sun' burns hydrogen producing helium and excess heat. This energy leaves the fusion zone in the interior of the Sun and is conducted, and convected to the photosphere where it is radiated into space. In the process of moving through the convection zone it generates a magnetic field that has an approximate 22-year cycle (Russell *et al.* 2010) in which the magnetic field forms flux ropes that are unstable to eruptions that produce energetic charged particles. At the current epoch this solar activity produces some hazards to the industrial activities on the surface of the populated, third planet that we call Earth. We refer to this solar energetic particle variability and its associated phenomena as space weather. In this brief treatise we discuss how space weather varies across the solar system and remark how space weather may also occur within the magnetic envelopes of some of the planets. We close with a discussion of the frequency of occurrences of extreme space weather events and their intensity.

2. Magnetic Reconnection in and with the Solar Wind

The most important process in a magnetized plasma is arguably magnetic reconnection. It can change magnetic energy into both thermal energy and directed flows. It can link moving plasmas and thereby transfer momentum. Space weather can be affected by reconnection in the solar wind and between the solar wind and planetary magnetospheres. An example of reconnection in the solar wind is the interaction regions of oppositely directed magnetic fields both rooted in the Sun and extending into space. An example of reconnection of a magnetosphere with the solar wind occurs when the flow of magnetized plasma from the Sun arrives at Earth and generates a magnetic storm. The former is affected by the beta value of the plasma (the ratio of the thermal energy density to the magnetic energy density) and the latter is affected by the magnetosonic Mach number of the solar wind flow relative to the planetary obstacle to that flow. The Mach number in turn controls the beta of the post-shock plasma that reaches the obstacle and therefore affects the reconnection rate at the boundary we call the magnetopause. High Mach numbers produce high beta in turn weakening reconnection (Scurry *et al.* 1994).

Figure 1 shows the expected heliocentric variation of the beta value (right hand scale) and the magnetosonic Mach number (M_{ms}) on the left-hand scale. The changes in B and

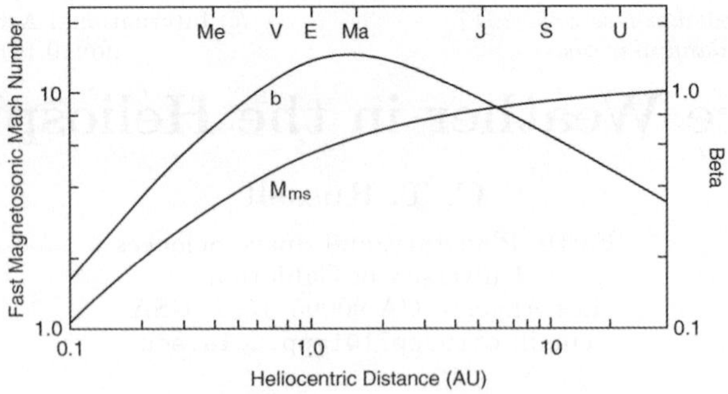

Figure 1. Expected values of plasma beta and solar wind Mach number versus heliocentric distance (Russell et al, 2016).

M_{ms} here reflect expected changes in the solar wind due to expansion and do not include any effects of the interaction with the local interstellar medium. Reconnection in the solar wind (governed by beta) should happen most rapidly near the Sun and far from the Sun. Reconnection between the interplanetary magnetic field and any planetary magnetic fields (governed by Mach number) should happen most readily at Mercury, Venus and Earth and less so at Jupiter, Saturn and Uranus. Flux transfer events (Russell and Elphic 1978) are small and occur rapidly at Mercury (Russell and Walker 1985), and they produce large infrequent events at Earth. They are seldom seen at Jupiter (Walker and Russell 1985) and almost never at Saturn (Lai *et al.* 2012). While no study has attempted to study sector boundary reconnection rates versus heliocentric distance, solar eruptions show that reconnection is very efficient and fast near the Sun.

3. The Role of Size

Size matters in the space weather of magnetospheres in two ways. Figure 2 illustrates the relative sizes of magnetospheres with each other and with their host planets. If the host planet (like Mercury) occupies a major fraction of the magnetosphere then a robust radiation belt will not form. Mercury has no radiation belt. Earth and Saturn have modest radiation belts and Jupiter has a particularly intense radiation belt. Part of energization of the jovian magnetosphere is the agitation of the inner magnetosphere by the material produced by Io's volcanoes. Uranus and Neptune appear to have very quiet magnetospheres and low trapped particle fluxes.

Size can also be judged relative to plasma scale lengths. Figure 3 uses a hybrid simulation to show the change in the disturbance in a flowing magnetized plasma in terms of the size of the obstacle as a function of the ion inertial length (the speed of light divided by the natural frequency of the ions) that is determined by the plasma density. The smallest sized obstacle produces a whistler wave (top pair of panels). A larger obstacle produces a standing compressional wave and a wake. A larger obstacle size produces a quasi-magnetosphere with a bow wave. At largest scales (bottom) the interaction has a bow wave, magnetosheath, and fully developed shock, a magnetopause and plasmas sheet with reconnection on the forward magnetopause and in the tail.

4. Momentum and Energy Transfer

J-W. Dungey (1961, 1963) was the first to understand how reconnection led to momentum transfer from the solar wind to the magnetosphere so that the plasma in the magnetosphere circulated. A moving magnetized plasma has an electric field in the frame of the

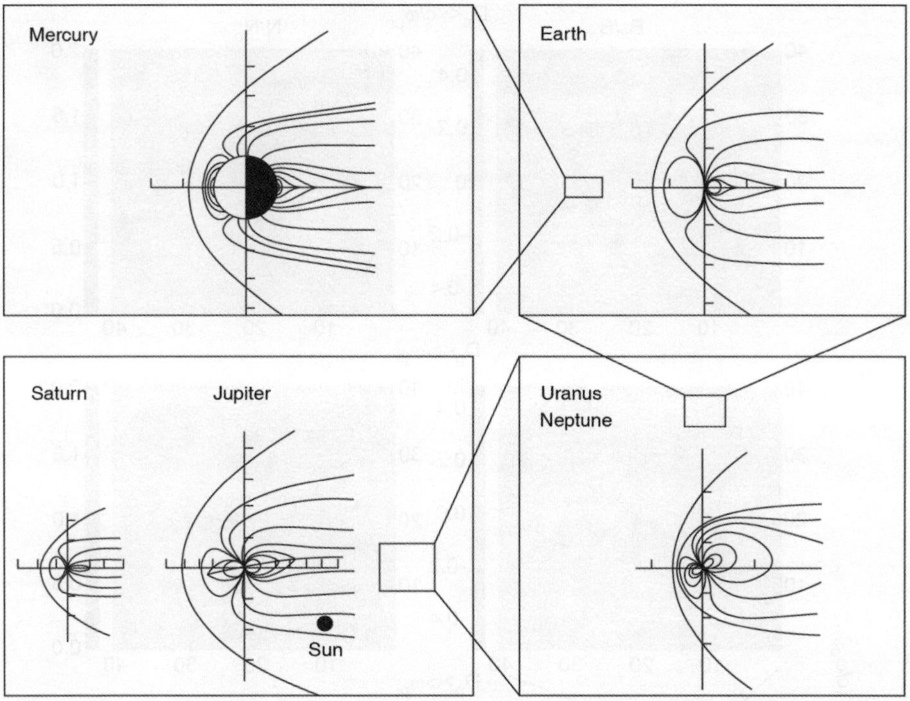

Figure 2. The relative size of the magnetospheres of the magnetized planets.

planet. This electric field affects the magnetosphere differently depending on the orientation of the planetary dipole field. Figure 4 shows the two key diagrams from Dungey's work. The top panel shows the situation for a northward magnetic field convected to the Earth. Antiparallel fields at the boundary allow for interconnection downstream from the magnetosphere. This can produce flows in the magnetosphere that move sunward at high latitudes and tailward at low latitudes.

If the magnetic field in the solar wind is southward, the reconnection can occur at noon and cause a circulation of plasma over the poles away from the Sun and back toward the Sun at low latitudes. Unsteadiness in the process leads to substorms. A strong steady flow will create a magnetic storm.

5. Space Weather in the Magnetospheres of Jupiter and Saturn

The volcanic moon Io, inside the jovian magnetosphere adds mass to that magnetosphere which becomes ionized by Jupiter's radiation belt particles. The rapid rotation of Jupiter spins up that plasma and it stretches the magnetized plasma until it "breaks" in the jovian magnetotail. Most of the plasma is then ejected down Jupiter's tail and the emptied flux tubes "float" inward against the outward flow to reach Io, be mass loaded again, and repeat their journey (Russell *et al.* 2000)). Enceladus, a moon at Saturn has a water plume that massloads Saturn's magnetosphere and reconnection in Saturn's tail allows the flux tube to return to Enceladus and be refilled (Lai *et al.* 2016)

6. Space Weather Events

The first extensively studied space weather event is the Carrington event of September 1, 1859 whose magnetogram is shown in Figure 5 (top panel). It is characterized by a very

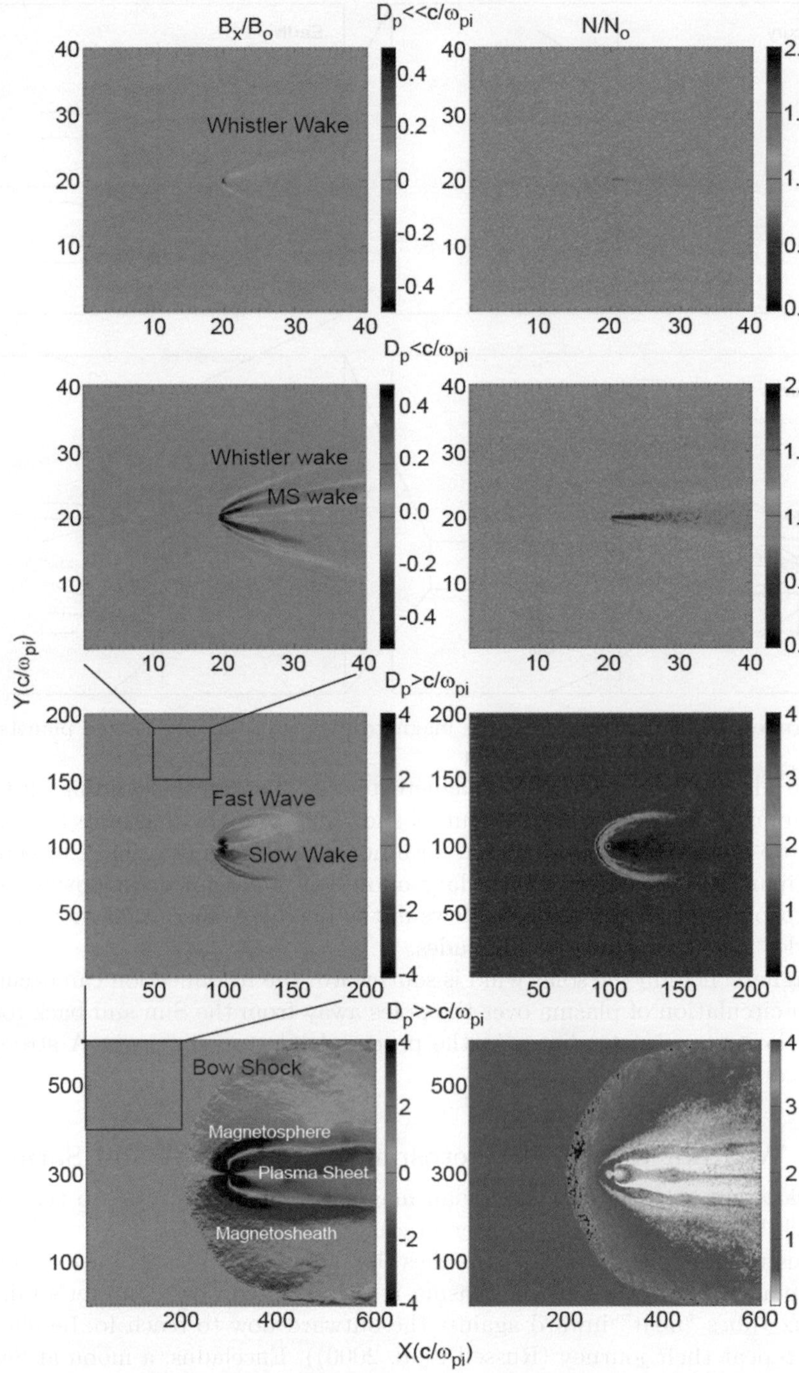

Figure 3. Magnetic field along the flow and the plasma density for four different hybrid simulations of increasing ratio of obstacle size to the ion inertial length. A hybrid simulation uses a particle code for ions and a massless fluid for electrons. The smallest obstacle produces only whistler mode waves when the largest obstacle deflects the flow and produces magnetosonic and Alfven waves (Blanco-Cano et al. 2004).

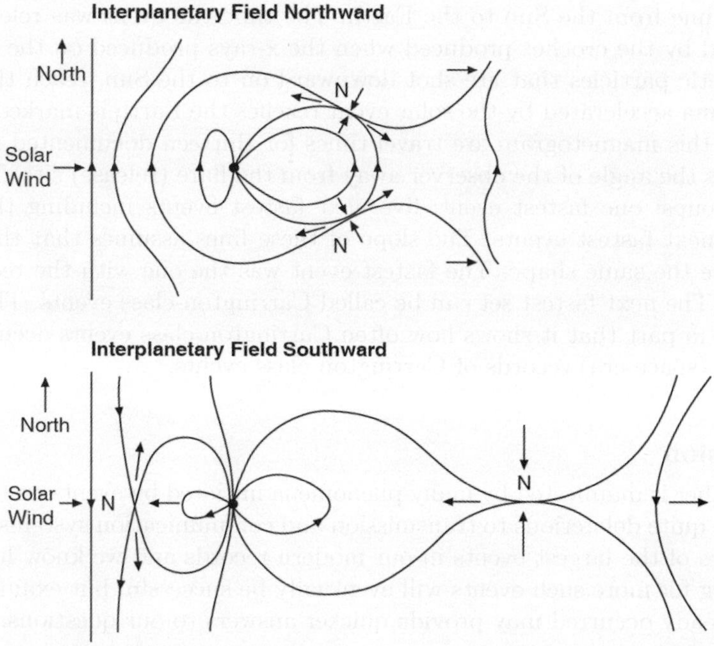

Figure 4. A magnetic reconnection of a due northward interplanetary magnetic field and a due southward magnetic field in the noon-midnight meridian (Dungey 1961; 1963).

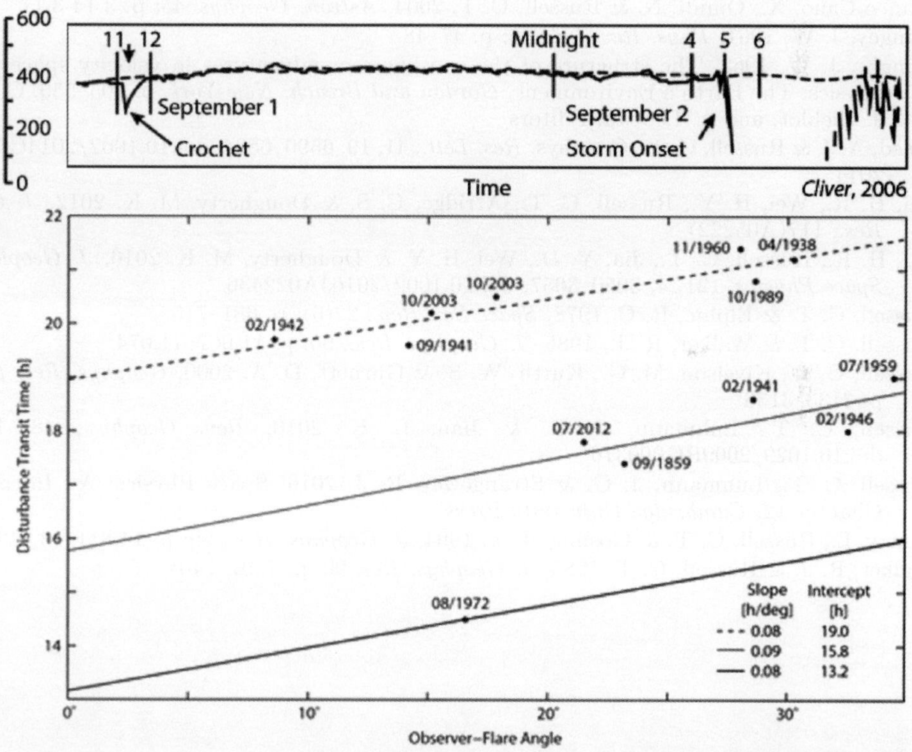

Figure 5. Magnetogram for the Sept 1, 1859 magnetic storm seen by Carrington together with the transition for 13 fastest solar outbursts (Freed and Russell 2014).

short travel time from the Sun to the Earth. The time the event was released from the Sun is marked by the crochet produced when the x-rays produced on the Sun's surface, by the energetic particles that are shot downward on to the Sun, reach the Earth. The time the plasma accelerated by the solar event reaches the Earth is marked by the storm onset. Below this magnetogram are travel times for thirteen documented fastest events, plotted versus the angle of the observer away from the flare (release) site. These self-sort into three groups: one fastest event; five next fastest events including the Carrington event; seven next fastest events. The slope of these lines assumes that the disturbance fronts all have the same shape. The fastest event was the one with the record energetic particle flux. The next fastest set can be called Carrington-class events. The importance of this line is in part that it shows how often Carrington class events occur and that we have modern (space era) records of Carrington class events.

7. Conclusion

Space weather is manifested by many phenomena initiated by events on the Sun. These events can be quite deleterious to transmission and communication systems on Earth. We have examples of the largest events in our modern records and we know how to identify them. Waiting for more such events will eventually be successful but examining the ones that have already occurred may provide quicker answers to our questions.

References

Blanco-Cano, X., Omidi, N. & Russell, C. T. 2004, *Astron. Geophys.* 45: p. 3.14-3.17

Dungey, J. W. 1961, *Phys. Rev. Lett.*, 6: p. 47–48

Dungey, J. W. 1963, The structure of the exosphere or adventures in velocity space, in Geophysics: The Earth's Environment, *Gordon and Breach: New York*. p. 505–550. C. Dewitt, J. Hieblot, and A. Lebeau, Editors

Freed, A. J. & Russell, C. T., Geophys. *Res. Lett.*, 41, 19, 6590–6594, doi:10.1002/2014GL061353, 2014

Lai, H. R., Wei, H. Y., Russell, C. T., Arridge, C. S. & Dougherty, M. K. 2012, *J. Geophys. Res.*, 117(A05222)

Lai, H. R., Russell, C. T., Jia, Y. D., Wei, H. Y. & Dougherty, M. K. 2016, *J. Geophys. Res. Space Physics*, 121, 4, 3050–3057, doi:10.1002/2016JA022436

Russell, C. T. & Elphic, R. C. 1978, *Space Sci. Rev.*, 22(6): p. 681–715

Russell, C. T. & Walker, R. J., 1985, *J. Geophys. Res.*, 90: p. 11,067–11,074

Russell, C. T., Kivelson, M. G., Kurth, W. S. & Gurnett, D. A. 2000, *Geophys. Res. Let.t*, 27: p. 3133–3136

Russell, C. T., Luhmann, J. G. & Jian, L. K. 2010, *Rev. Geophys.*, 48, RG2004, doi:10.1029/2009RG000316

Russell, C. T., Luhmann, J. G. & Strangeway, R. J. 2016, Space Physics: An Introduction, Chapter 12. *Cambridge University Press*

Scurry, L., Russell, C. T. & Gosling, J. T. 1994, *J. Geophys. Res.*, 99: p. 14,811–14,814

Walker, R. J. & Russell, C. T. 1985, *J. Geophys. Res.* 90: p. 7397–7404

Space Weather of the Heliosphere: Processes and Forecasts
Proceedings IAU Symposium No. 335, 2017
C. Foullon & O.E. Malandraki, eds.

doi:10.1017/S1743921317010146

Principal Component Analysis of Geomagnetic Activity: New Information on Solar Wind

Kalevi Mursula and Lauri Holappa

ReSoLVE Centre of Excellence, Space Climate Research Unit, University of Oulu, Finland
email: `kalevi.mursula@oulu.fi`

Abstract. We use the principal component analysis to study geomagnetic activity at annual resolution using a network of 26 magnetic stations in 1966-2015, and an extended network of 40 stations in 1980-2015. The first principal component describes the long-term evolution of global geomagnetic activity, and has an excellent correlation with indices like the Kp/Ap index. The two networks give identical results for the first principal component. The second principal component is highly correlated with the annual percentage of high-speed streams (HSS). The extended network has a slightly higher sensitivity to HSSs. We verify the non-trivial latitudinal distribution of the second empirical orthogonal function. We find that the amplitude of the 22-year variation of geomagnetic activity has a closely similar latitudinal distribution as the second empirical orthogonal function. This verifies that the 22-year variation of geomagnetic activity is related to HSSs. The most likely cause is the Russell-McPherron mechanism.

Keywords. Space climate, Geomagnetic activity, High-speed solar wind, Principal component analysis

1. Local geomagnetic activity in a global network of stations

Geomagnetic activity is a measure of the effect of the solar wind and the accompanied heliospheric magnetic field to the Earth's magnetic field. This interaction drives various electric current systems in the magnetosphere and ionosphere that are monitored by ground-based magnetometers as geomagnetic activity. The two main solar wind structures driving geomagnetic activity are the coronal mass ejections (CME) and high speed solar wind streams together with the related corotating interaction regions (CIRs). CMEs and HSS/CIRs are responsible for the most severe types of geomagnetic activity, the substorms and storms.

We extend our earlier analysis (Holappa *et al.* (2014)), where we used data from a network of 26 ground-based magnetic stations from 1966 to 2009, both temporally, including data from 1966 until 2015 and spatially by using an extended network of 40 stations from 1980 to 2015. Both networks include stations from a large range of latitudes and longitudes in the northern hemisphere. The criteria for choosing the observatories were the continuity, length and quality of data, and the global coverage of stations. The list of the original 26 stations is given in Table 1 of Holappa *et al.* (2014). The additional stations of the extended network are Guam, Honolulu, Hurbanovo, Furstenfeldbruck, Boulder, Novosibirsk, Saint Johns, Newport, Ottawa, Yellowknife, Barrow, Hornsund, Cambridge Bay, and Resolute Bay. The data for these magnetic stations were retrieved from the World Data Center of Edinburgh, UK.

Figure 1 shows the annual A_h indices (Mursula and Martini, 2007) for the 26 stations from 1966 to 2015 and the additional 14 stations from 1980 to 2015. A_h index is a measure of local geomagnetic activity, which follows the K-index method but uses hourly digital

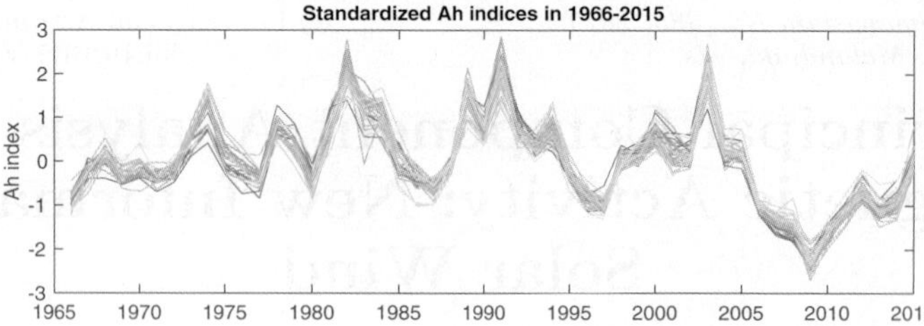

Figure 1. Annual means of A_h indices of local geomagnetic activity for 26 stations in 1966-2015 and for 14 stations in 1980-2015.

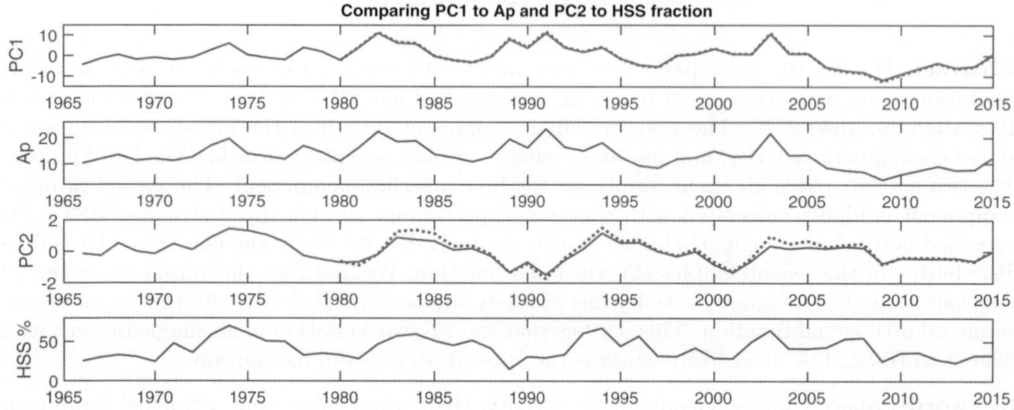

Figure 2. [top] PC1 for 26 stations in 1966-2015 (solid line) and for 40 stations in 1980-2015 (dotted line) [second] Global Ap index of global geomagnetic activity [third] PC2 for 26 stations in 1966-2015 (solid line) and for 40 stations in 1980-2015 (dotted line) [bottom] Annual percentages of HSSs in solar wind.

magnetic data, and therefore can be calculated straightforwardly for all stations whose magnetic field measurements are given as digital hourly values. We have standardized (remove mean; normalize by standard deviation) the annual A_h means for each station in order to remove the latitudinal variation of geomagnetic activity and to have all stations on an equal footing. Figure 1 shows a uniform temporal evolution of geomagnetic activity, with roughly equal relative solar cycle variation in all stations. One can also see the recent decadal decline of geomagnetic activity, with annual means from 2006 until 2014 remaining below zero. Only in 2015 geomagnetic activity reached the long-term mean (zero). This shows the exceptional quietness of the near-Earth space in recent years.

2. Principal component analysis (PCA)

We have made the PCA for 26 stations in 1966-2015 and, separately, for 40 (26+14) stations in 1980-2015. Figure 2 shows the first two principal components (PC1 and PC2) of these two networks in panels 1 and 3, respectively. In both cases PC1, which gives a measure of global geomagnetic activity, takes some 96% of the total variance in annual means, and the two PC1s can be hardly distinguished from each other in Figure 2. PC1 is highly correlated with the Ap index, which can, based on this comparison, be verified as a reliable measure of global geomagnetic activity at annual scale.

PC2 explains only some 2% of total variance in annual means in both cases. The large fraction of PC1 should, however, not mask the important information included in PC2. The two PC2s depict small differences between each other around 1984, 1994 and 2003. These are the times of most intense HSSs, as shown in Figure 2 (bottom

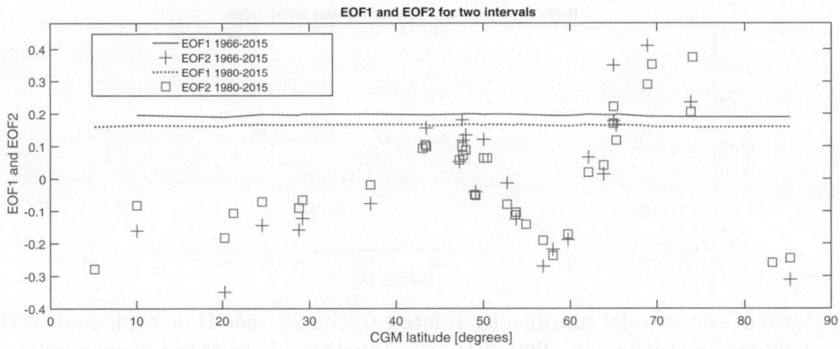

Figure 3. EOF1 (lines) and EOF2 (pluses and squares) for 26 stations in 1966-2015 (solid line) and for 40 stations in 1980-2015 (dotted line) as a function of corrected geomagnetic latitude.

panel). Correlation coefficient between the extended network and HSS percentage is 0.86, while it is 0.83 for the 26 stations during the same time interval in 1980-2015 (and 0.81 over the whole time 1966-2015). This difference is the virtue of using a larger station network, which takes a better account of the HSS effect to geomagnetic activity in different regions of the world. Anyway, in both cases, the HSSs explain a very large fraction of the variability of PC2 (73% and 70%, respectively), demonstrating their crucial importance for PC2.

The empirical orthogonal functions (EOFs, or scalings) of the first two principal components (EOF1 and EOF2) are depicted in Figure 3. EOF1s are almost straight lines, indicating that the fraction of PC1 is roughly the same for all stations of a network. (EOF1 of the extended network is smaller because of the larger number stations). The latitudinal distribution of EOF2 has a maximum at the auroral oval, a local minimum at subauroral latitudes at about $55 - 60°$ and another maximum at mid-latitudes of about 40°-50°. This distribution is due to the difference in the average location and intensity of substorms related to CMEs and HSSs (Holappa *et al.* 2014). At auroral latitudes the effect of HSSs to geomagnetic activity is largest, leading to the peak in EOF2. Towards lower latitudes, the effect of HSSs decreases but the decrease is not uniform with latitude. Rather, the relative effect of CMEs is largest at sub-auroral latitudes, leading to the local minimum of EOF2 at these latitudes.

3. Solar magnetic cycle at different latitudes

Two successive 11-year solar activity cycles affect to the Earth's magnetic field slightly differently, producing a 22-year variation of geomagnetic activity. This is related to the variation of solar polarity during the 22-year magnetic cycle, also called the Hale cycle. A mechanism which naturally produces a polarity dependent response in alternating cycles is the Russell-McPherron (R-M) mechanism (Russell and McPherron, 1973). The horizontal component of the heliospheric magnetic field produces, because the tilt of the Earth's rotation axis with respect to solar equator, a component which can enhance or suppress geomagnetic activity depending on solar polarity. During positive polarity times the R-M effect enhances geomagnetic activity (in spring and fall), and reduces that in negative polarity times (Cliver *et al.* 1996).

We have calculated the wavelet spectrograms of daily A_h indices for the 26 stations using Morlet wavelet. Figure 4 depicts the ratio between the wavelet amplitudes of 22-year and 11-year variations as a function of corrected geomagnetic latitude. The latitudinal variation of this ratio is strikingly similar to that of the EOF2. Accordingly, the Hale cycle in geomagnetic activity is much stronger at auroral and even at mid-latitude stations

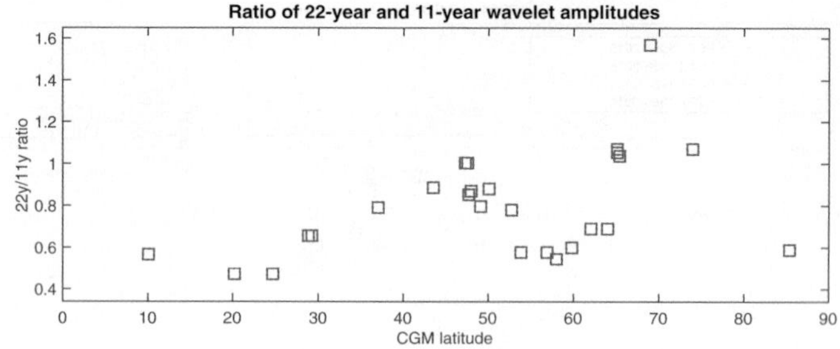

Figure 4. Ratio of the wavelet amplitudes related to the 22-year Hale cycle and to the 11-year Schwabe cycle for 26 stations in 1966-2015 as a function of corrected geomagnetic latitude.

than in sub-auroral stations. As explained above, PC2 describes the fraction of HSSs in solar wind, and the latitudinal variation of EOF2 shows where the HSS effect is relatively larger or smaller compared to CMEs. Therefore, the similarity of the latitudinal variation of the 22-year amplitude and EOF2 strongly suggests that the Hale cycle of geomagnetic activity is due to effects related to high-speed streams. A natural explanation is the R-M mechanism since the polarity of HSSs faithfully follows the dominant polarity in the source hemisphere. We have also verified that the Hale ratio maximizes during the minimum times of solar cycle, when the solar polarity is most clearly ordered with heliographic latitude (not shown here). At these times the R-M mechanism enhances (reduces) geomagnetic activity during positive (negative, resp.) polarity minima.

4. Conclusions

We have used here the principal component analysis to study long-term geomagnetic activity. PC1 describes the global geomagnetic activity, and is quite insensitive to the number of stations used in PCA. It correlates well with the traditional Kp/Ap index, thus verifying the validity of Kp/Ap as a measure of global geomagnetic activity. PC2 is highly correlated with the annual percentage of high-speed streams. We found very similar results for PC2 from the two networks, although the extended network has a slightly better sensitivity to high-speed streams. We verified the interesting, non-trivial latitudinal distribution of the EOF2. We also found that the amplitude of the 22-year (Hale cycle) variation of geomagnetic activity has a closely similar latitudinal distribution as EOF2. This demonstrates that the Hale cycle of geomagnetic activity is due to the effect of HSSs. The most likely cause is the Russell-McPherron mechanism.

Acknowledgements

We acknowledge the financial support by the Academy of Finland to the ReSoLVE Centre of Excellence (project no. 272157). Magnetometer data were obtained from the World Data Center for Geomagnetism, Edinburgh (`http://www.wdc.bgs.ac.uk/`), and solar wind data from the OMNI database (`http://omniweb.gsfc.nasa.gov/`). We appreciate the continuous work made at the observatories around the world.

References

Cliver, E. W., Boriakoff, V. & Boumar, K. H. 1996, *J. Geophys. Res.*, 101, 27,091
Holappa, L., Mursula, K., Asikainen, T. & Richardson, I. G. 2014, *J. Geophys. Res.*, 119, 4544
Mursula, K. & Martini, D. 2007, *Geophys. Res. Lett.*, 34, L22, 107
Russell, C. T. & McPherron, R. L. 1973, *J. Geophys. Res.*, 78, 92

Space Weather of the Heliosphere: Processes and Forecasts
Proceedings IAU Symposium No. 335, 2017
C. Foullon & O.E. Malandraki, eds.

doi:10.1017/S1743921317008006

Predicting the Loci of Solar Eruptions

N. Gyenge[1,2,3] and R. Erdélyi[1,3]

[1]Solar Physics and Space Plasmas Research Centre (SP^{2}RC),
School of Mathematics and Statistics, University of Sheffield
email: `n.g.gyenge@sheffield.ac.uk`

[2]Debrecen Heliophysical Observatory (DHO), Konkoly Observatory,
Research Centre for Astronomy and Earth Sciences
Hungarian Academy of Sciences, Debrecen, P.O.Box 30, H-4010, Hungary

[3]Dept. of Astronomy, Eötvös L. University, Pázmány P. sétány 1/A
Budapest, H-1117, Hungary

Abstract. The longitudinal distribution of solar active regions shows non-homogeneous spatial behaviour, which is often referred to as Active Longitude (AL). Evidence for a significant statistical relationships between the AL and the longitudinal distribution of flare and coronal mass ejections (CME) occurrences is found in Gyenge *et al.* 2017 (*ApJ*, 838, 18). The present work forecasts the spatial position of AL, hence the most flare/CME capable active regions are also predictable. Our forecast method applies Autoregressive Integrated Moving Average model for the next 2 years time period. We estimated the dates when the solar flare/CME-capable longitudinal belts face towards Earth.

Keywords. Sun: flares, Sun: coronal mass ejections (CMEs), Sun: sunspots, Sun: rotation

1. Motivation

The existing flare and CME forecasting tools are usually based on the local dynamics of active regions, such as the magnetic topology (e.g., Korsós *et al.* 2014, Korsós & Erdélyi 2016). However, certain global phenomena are also able to provide opportunities for predicting the locus of solar eruptive events, such as the longitudinal inhomogeneous properties of active regions (e.g., Bumba *et al.* 1965; Bogart 1982; Berdyugina *et al.* 2006; Balthasar 2007; Zhang *et al.* 2011). In our previous study, Gyenge *et al.* 2017 (hereinafter GY17), we found that the most complex active regions appear near the Active Longitude (AL). We concluded that the source of the most probably flare/CME-capable active regions is at the AL. In this paper, we combine our recently developed AL tracking method by Autoregressive Integrated Moving Average (ARIMA) model for forecasting the longitudinal position of the AL numerous Carrington Rotations (CRs) in advance.

2. Forecasting the Active Longitude

We employ the Debrecen Photoheliographic Data (DPD) sunspot catalogue for determining the position of the AL. The DPD covers 540 Carrington Rotations, which equals to about five solar cycles. Panel *A* of Figure 1 shows the migration of the AL. The horizontal axis represents time. The vertical axis shows the Carrington Phase ($CP = L/360$), which is the longitudinal position L of the most significant sunspot group clusters in a certain CR. The clustering method (DBSCAN) groups together points that are relatively closely packed together in a high-density region and it marks outlier points that stand alone in low-density regions (Ester *et al.* 1996). The longitudinal location of the clusters

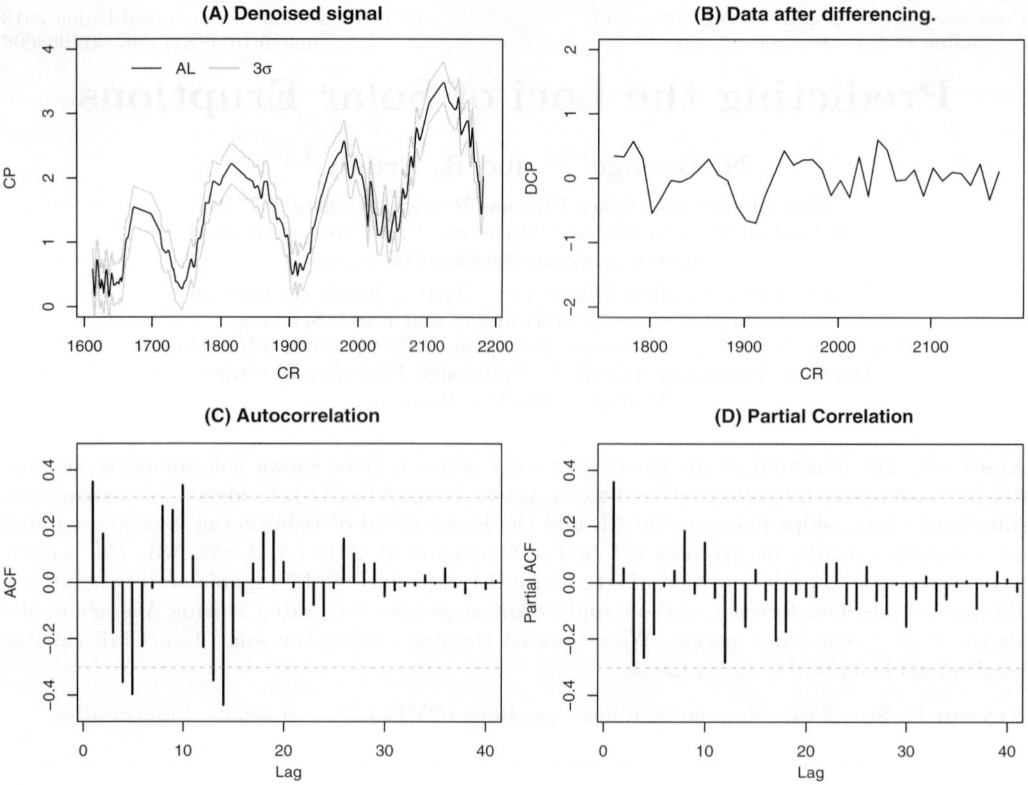

Figure 1. Migration of the AL (Panel *A*), the normalised denoised signal (Panel *B*), the autocorrelation (Panel *C*) and partial correlation (Panel *D*) of the differentiated data for the northern hemisphere are shown. One lag represents around a year.

represent the position of the AL. The AL identification method developed based on an empirical approach, described in GY17. The data is distinguished by the hemispheres.

Seasonal ARIMA model (e.g. Ho & Xie. 1998) is used for projecting the future values of time series, derived from the parameters ARIMA(p,d,q) $\times$ (P,D,Q). The parameters p and P are the number of non-seasonal and seasonal autoregressive (AR and SAR) terms. The parameters q and Q are the non-seasonal and seasonal moving average (MA and SMA) terms. Finally, the parameters d and D are the non-seasonal and seasonal differences (Box & Jenkins. 1968).

The application of the ARIMA model requires stationarity data; if y_t is considered as a stationary time sample, then the distribution of any subsample $(y_t, ..., y_s)$ is independent on t for all s (Chatfield 1975). The signal is denoised by applying a moving average with 3 CRs window (Panel *A* of Figure 1). The black line the marks actual position of the AL clusters and the grey belt within the migration demonstrates the mean squared error. The data is non-stationary if a pronounced trend is present. Dickey-Fuller (ADF) tests are used to study the trend-stationarity. The null hypothesis of the test is that the unit root is present in a time series. The alternative hypothesis could mean trend-stationarity. The results of the test also suggest that the data is non-stationary (northern hemisphere data: p-value $= 0.6351$ and southern hemisphere: p-value $= 0.3054$).

The trend-stationarity data can be archived by differencing. Now, the first difference of the time sample means the change between consecutive data points, and it is written as

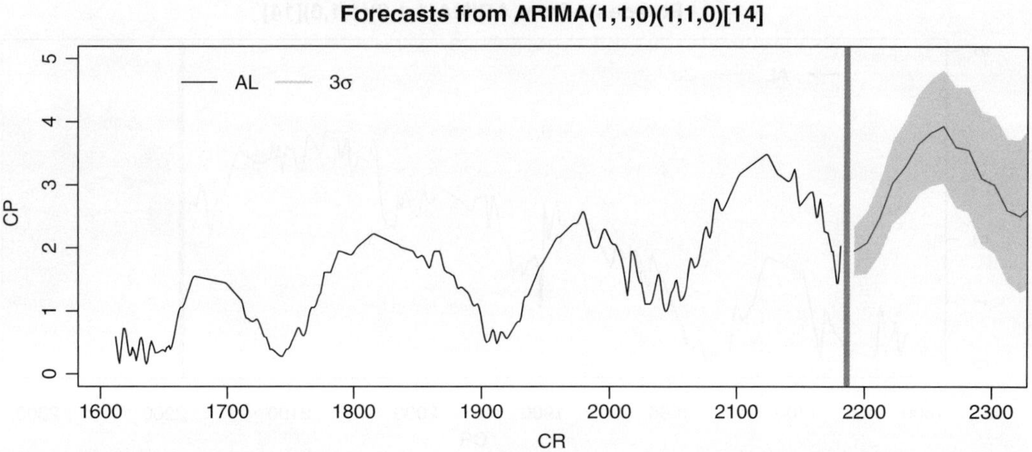

Figure 2. Forecasting the AL for the norther hemisphere. The vertical axis is the Carrington Phase (CP) and the horizontal axis stands for the Carrington Rotation (CR). The 3σ standard deviation is displayed by the grey band.

$y'_t = y_t - y_{t-1}$. Seasonality makes it so that the mean of the observations is not constant, but instead it evolves according to a cyclical pattern. The seasonally differenced series has a similar definition but the difference here means the difference between an observation and the corresponding observation from the previous "AL-cycle": $y'_t = y_t - y_{t-n}$, where, n is the "AL-cycle". In GY17, we found that the pattern of the AL migration does not correspond well with the 11-year solar cycle. The lengths of the cycloid AL pattern iterates between 8 years and 14 years. However, there is an 11-year cyclic behaviour on average. Panel B of Figure 1 shows the seasonal differenced and a first differenced data to obtain stationary time sample. The ADF tests confirms the results obtained with differencing and seasonal differencing (northern hemisphere: p-value = 0.09696 and southern hemisphere: p-value = 0.04897). We found that seasonal differencing $D = 1$ and differencing $d = 1$ are useful terms for the ARIMA model estimation.

The next step of the parametric estimation of the ARIMA model is the examination of the autocorrelation (ACF) and partial autocorrelation (Partial ACF) of the differenced data. Applying ACF and Partial ACF provides some initial guess value about the appropriate ARIMA model parameters (Box *et al.* 1994). Due the down-sampled data, lag 10 corresponds to 140 CR, which equals to the 11-year solar cycle. Panel C of Figure 1 (ACF) shows a harmonic wave decay pattern and the Partial ACF (Panel D of Figure 1) cuts off quickly. The ACF also reveals the existence of the fluctuation at positive lag around lag 10 (also 20 and 30). If the Partial ACF displays a sharp cut-off while the ACF decays more slowly, this behaviour clearly suggests an AR signature $p = 1$ (Box *et al.* 1994). The seasonal spikes in the ACF seem to be clear, they fade away in the ACF and cut off after lag 1 in the Partial ACF. This is the sign of a seasonal SAR model $P = 1$ (Box *et al.* 1994). So, it seems ARIMA(1,1,0) $\times$ (1,1,0) is a good model for estimating the future values of the time series to the northern hemisphere (Figure 2). Similar methodology is applied for the southern hemisphere data (Panel B of Figure 3), however, the statistical error is an order of magnitude larger than in the other hemisphere. Due the large error the souther hemisphere data does not seem to provide reliable forecast.

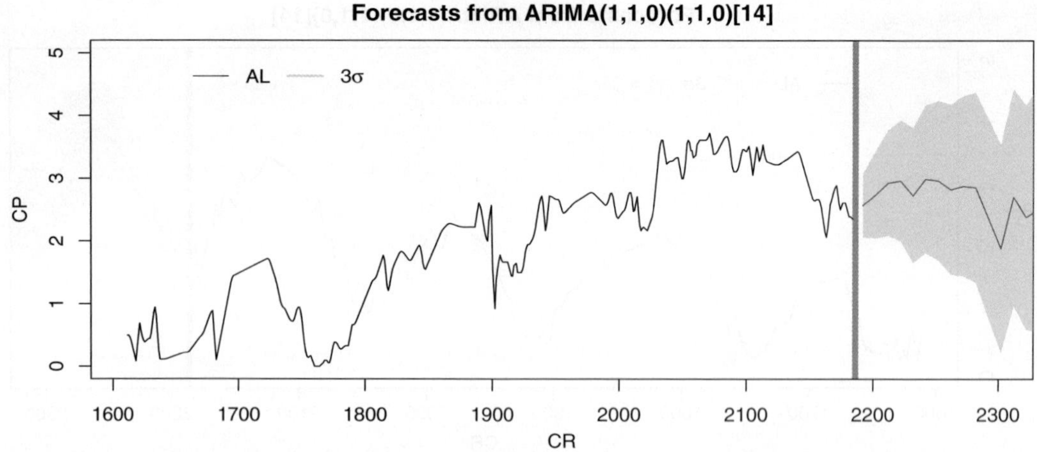

Figure 3. Forecasting the AL for the southern hemisphere. The vertical axis is the Carrington Phase (CP) and the horizontal axis stands for the Carrington Rotation (CR). The 3σ standard deviation is displayed by the red band.

3. Conclusion

By applying our method outlined in this short report, we are able to forecast the potential flare and/or CME sources several CRs in advance. We predict that the enhanced CME and flare active longitudinal belt will face towards Earth in weeks 41, 42, 46, 50 in 2017 and weeks 2, 5, 9, 13, 17, 20, 24, 27, 35, 39 in 2018. These estimates indicate 60% of flare and CME activity and the highest probability of fast CME occurrence (see GY17).

Acknowledgement

RE and NG are grateful to STFC and The Royal Society (UK) for the support received.

References

Balthasar, H., 2007, *A&A*, 471, 281
Berdyugina, S. V., Moss, D., Sokoloff, D. & Usoskin, I. G., 2006, *A&A*, 445, 703
Bumba, V., *et al.* 1965, *ApJ*, 141, 1492-1517
Bogart, R. S., 1982, *Sol. Phys*, 76, 155
Box G E P, Jenkins G M, 1968, *Applied Statistics*, 17(2), 91-109
Box, G E P, Jenkins, G M, Reinsel G C, 1994, *Time Series Analysis*, Forecasting and Control
Chatfield C., 1975, *Chapman and Hall*, London (see also, 6th ed., 2003)
Ester, M., Kriegel, H. P., Sander, J. & Xu, X. 1996, pp. 226231
Gyenge, N., Singh, T., Kiss, T. S., Srivastava, A. K. & Erdélyi, R. 2017, *ApJ*, 838, 1, 10 pp
Ho, S. L. & Xie, M. 1998, *Computers & industrial engineering*, 35, 213-216
Korsós, M. B. & Erdélyi, R. 2016, *ApJ*, 823, 2, 11 pp
Korsós, M. B., Baranyi, T. & Ludmány, A. 2016, *ApJ*, 789, 2, 7 pp
Zhang, L. Y., Mursula, K., Usoskin, I. G. & Wang, H. N. 2011, *JASTP*, 73, 2-3, p. 258-263

Space Weather of the Heliosphere: Processes and Forecasts
Proceedings IAU Symposium No. 335, 2017
C. Foullon & O.E. Malandraki, eds.

doi:10.1017/S1743921317007207

Sunspots Areas and Heliographic Positions on the Drawings Made by Galileo Galilei in 1612

Mikhail V. Vokhmyanin and Nadezhda V. Zolotova

St. Petersburg State University,
7/9 Universitetskaya nab., St. Petersburg, 199034 Russia
email: m.vokhmyanin@spbu.ru; ned@geo.phys.spbu.ru

Abstract. In 1612, Galileo Galilei made very accurate drawings of the solar disk. Currently, 47 of them are in the open access: 9 in May 3 – 11, 35 in June and July, and 3 in late August. Unfortunately, reports have not provided the clock time, which results in uncertainty of sunspots heliographic coordinates. In the present study, we determine the exact time of the drawings by comparing the positions of the same spots from day to day. The time of the observations, which varies from 12 to 16 UT, gives us the direction of the solar rotation axis and the position of the helioequator. Unlike the spots drawn by Christopher Scheiner in 1611 – 1612, none of the analyzed spots lies within the helioequator. This confirms the quality of the Galileo's drawings.

Keywords. Sun: activity, (Sun:) sunspots

1. Introduction

Historical sunspot observations are highly potential data source on the long-term behavior of the solar activity. One of the long-standing problem is the reconstruction of homogeneous time-series of the solar indexes in the past. Detailed revision of historical astronomical reports should help to resolve whether the Sun is a star with a regular periodicity or there are periods without the Schwabe cycle.

One of the first telescopic monitoring of sunspots was carried out in 1612 by Galileo Galilei. His pupil, a monk of Cassino named Benedetto Castelli, invented the projection technique to observe sunspots and draw sunspot positions on the solar disk in details. Continuous observations from June 2 to July 8 and August 19, 20, and 21 were published in the book by Galilei (1613, *Istoria e dimostrazioni intorno alle macchie solari e loro accidenti comprese in 3 lettere*). Note that these observations are highly accurate in comparison with sunspot drawings of other observers in the beginning of the seventeen century (Zolotova & Ponyavin 2016). Thin structure of a spot, umbra and penumbra, were depicted. Every day, a new image of the solar disk was made.

June and July drawings were processed by Casas *et al.* (2006). To define the differential rotation speed, they provided positions of sunspots which appear on the next day, besides spots near the limb were also excluded from analysis. Casas *et al.* (2006) roughly suggested that Galilei observed at local noon.

In this paper, we present the reconstruction of the heliographic coordinates of all sunspot groups, observed by Galilei in June – July 1612, their area and number. Basing on our technique which minimizes the discrepancy between the sunspots latitudes from day to day, we find out that sunspots were drawn after the solar noon.

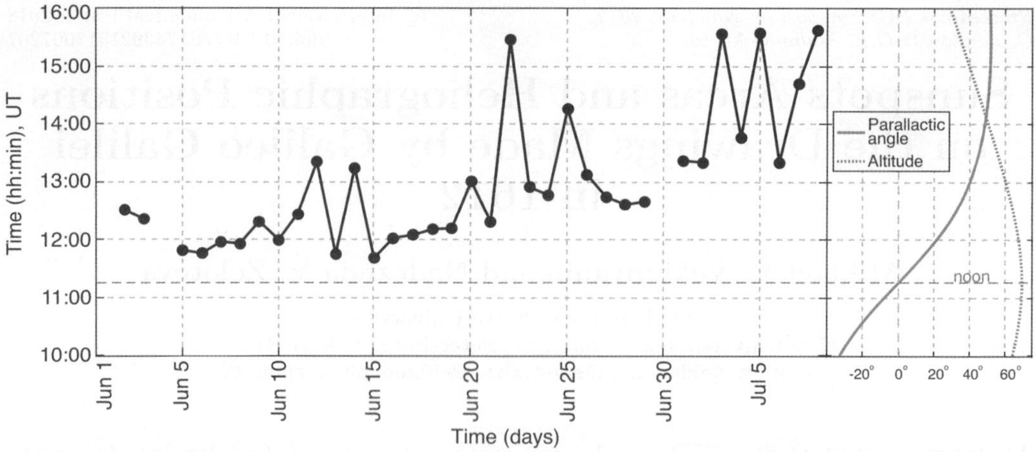

Figure 1. Black points joined by black line shows the time when spots were drawn. On the left, blue denotes the parallactic angle and red the solar altitude for mid-June. Horizontal dashed green line marks the solar noon in Villa delle Selve.

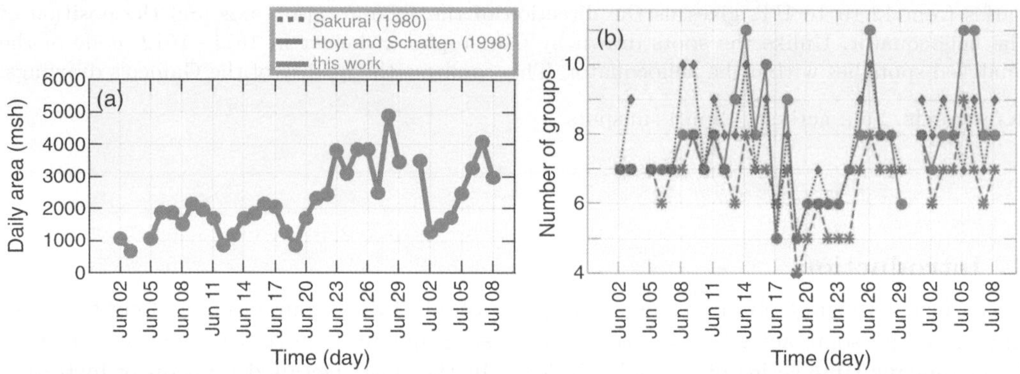

Figure 2. (a) Daily sunspot area. (b) Green denotes the number of groups according to Sakurai (1980), red, those by Hoyt & Schatten (1998), and blue, this work.

2. Results

Time on clock and geo-location are required to define the heliographic sunspot coordinates. Unfortunately, sunspot drawings by Galilei (1613) do not provide time of observations that results in uncertainty of ephemeris of the Sun. To deduce time of observation, we developed the iterative method which minimizes the difference between the latitudes of sunspot groups or subgroups in a given day and average latitudes of these groups over the range of preceding or successive days. Although, the uncertainty of the spots position increases with distance from the center of the disk, we find out that the near limb spots are valuable for adjustment of the heliographic grid.

Figure 1 shows time of observation in UT. Green dashed line defines the local solar noon, when altitude of the Sun is maximal, but parallactic angle equals zero. The farther from noon the observations were made, the greater an error in determination of time of observation.

One may see that sunspots were drawn after the noon. Here, we can speculate that the upper culmination could be engaged for astrometric measurements. The other reason of not fixed time of drawing could be clouds.

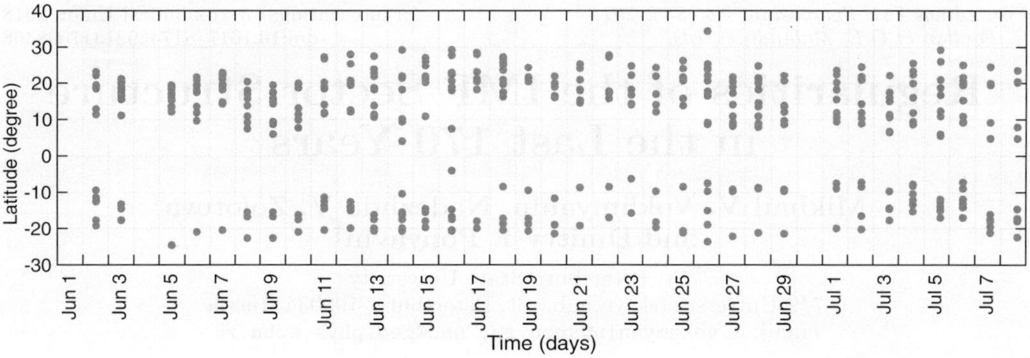

Figure 3. Time-latitude distribution of sunspots in June – July 1612.

Figure 2a depicts the total daily sunspot area in the millions of the solar hemisphere. Figure 2b compares number of sunspot groups defined by Sakurai (1980) shown by green dotted line, by Hoyt & Schatten (1998), red dashed line, and by us, blue solid line. Note that group counts defined by Sakurai (1980) and by us systematically exceed those by Hoyt & Schatten (1998).

Figure 3 shows the butterfly diagram. Sunspots sit on high- and mid-latitudes, that corresponds to the ascending phase or maximum of a solar cycle. Note that the near-equator region is spotless. On the contrary, according to early observations by Scheiner (Arlt *et al.* 2016) spots in 1612 emerged on the equator. The fact is that Scheiner had not been interested in accurate sunspot mapping, but demonstrated that spots are not an optical artifact.

3. Conclusions

We analyzed continuous set of drawings of sunspots on the solar disk published by Galilei (1613). For June – July 1612, we provide the heliographic coordinates of sunspot groups, their areas and number. We define the time of observations by minimizing the discrepancy between the latitudes of sunspot groups (or subgroups) in a given day and average latitudes of these groups over the range of preceding and successive days. In the further work, we will process observations by Galilei of May 3 – 11 and August 19 – 21 1612.

Acknowledgements

We use data from the Galileo project by Al Van Helden and Owen Gingerich. The reported study was funded by RFBR according to the research projects No. 15-02-06959-a and No. 16-02-00300-a.

References

Arlt, R., Senthamizh Pavai, V., Schmiel, C. & Spada, F. 2016, *A&A*, 595, A104

Casas, R., Vaquero, J. M. & Vazquez, M. 2006, *Solar Phys.*, 234, 379

Galilei, G. 1613, *Istoria E dimostrazioni intorno alle macchie solari E loro accidenti comprese in tre lettere scritte all'illvstrissimo signor Marco Velseri ...*, G. Mascadi, Roma

Hoyt, D. V. & Schatten, K. H. 1998, *Solar Phys.*, 179, 189

Sakurai, K. 1980, *Journal for the History of Astronomy*, 11, 164

Zolotova, N. V. & Ponyavin, D. I. 2016, *Solar Phys.*, 291, 2869

Space Weather of the Heliosphere: Processes and Forecasts
Proceedings IAU Symposium No. 335, 2017
C. Foullon & O.E. Malandraki, eds.

doi:10.1017/S1743921317008468

Regularities of the IMF Sector Structure in the Last 170 Years

Mikhail V. Vokhmyanin, Nadezhda V. Zolotova and Dmitry I. Ponyavin†

St. Petersburg State University,
7/9 Universitetskaya nab., St. Petersburg, 199034 Russia
email: m.vokhmyanin@spbu.ru; ned@geo.phys.spbu.ru

Abstract. The interplanetary magnetic field (IMF) controls magnetospheric currents which cause variations of the ground-based magnetic field. Regular magnetic observations made in the 19th century allow us to infer daily IMF polarities back to 1844. The results coincide with satellite data in about 79% days. Moreover, for the most part of the 19th and 20th centuries, proxies obtained from various geomagnetic data (Helsinki, Saint-Petersburg, Potsdam, and Ekaterinburg) show the same patterns. This suggests that the reliability of the proxies is sufficient to study the IMF in the past. The large-scale organization of the IMF polarities, the so-called sector structure, reveals semi-centennial north-south displacements of the heliospheric current sheet (HCS).

Keywords. Sun: magnetic fields, solar wind, evolution

1. Introduction

Continuously extending solar corona forms the interplanetary magnetic field (IMF). This field is stretched out along the Parker spiral and is directed either away (positive polarity) or toward (negative polarity) the Sun depending upon the source field. The IMF of one polarity is usually observed for several days reflecting the sectoral organization of the IMF at the ecliptic plane. Heliospheric current sheet (HCS) divides areas of opposite IMF. It has been found that one of the IMF polarity may dominate over the other for long periods. On a seasonal scale, this happens due to higher heliographic latitudes of the Earth in Fall and Spring. Thus, when solar dipole is axisymmetric, we observe more IMF with the polarity from the corresponding hemisphere (Rosenberg & Coleman 1969). Observations made on board of Ulysses spacecraft in 1994–1995 revealed southward inclination of the HCS (Crooker *et al.* 1997; Smith *et al.* 2000). At the ecliptic plane, this causes the overall excess of the IMF from the northern solar hemisphere. According to the IMF polarities derived in Echer & Svalgaard (2004), the north-south asymmetry of the HCS exists at least from 1926 (Hiltula & Mursula 2006, 2007). This result is based on a combination of three polarity proxies obtained in Svalgaard (1972), Mansurov (1969), and Vennerstroem *et al.* (2001).

In the present work, we analyze reconstructed IMF polarities from Vokhmyanin & Ponyavin (2016). This data set covers additional 80 years allowing to investigate long-term changes of the N-S asymmetry.

2. Results and discussion

The IMF polarity is generally defined by the sign of the IMF By component in GSM coordinate system. Depending upon the IMF By sign, the ionospheric field-aligned current

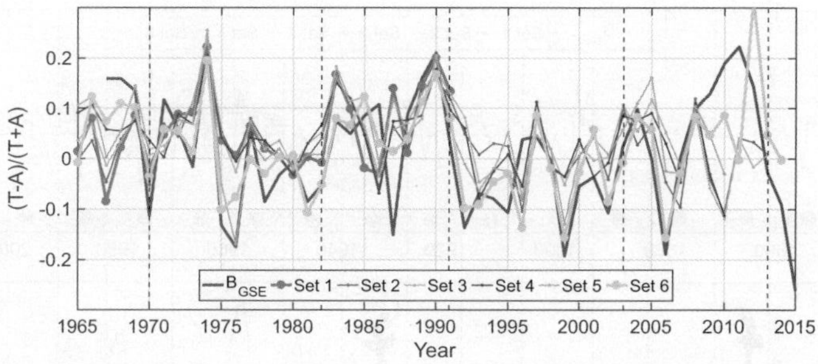

Figure 1. The annual ratio of days with negative (T) and positive(A) IMF polarities according to satellite observations (red) and from geomagnetic data sets 1–6.

Table 1. Reliability of the inferred IMF polarities.

	Set 1	**Set 2**	**Set 3**	**Set 4**	**Set 5**	**Set 6**
Success rate, %	79.5	80.3	82.4	82.5	85.4	87.9
$< \lvert\Delta TA\rvert >$	0.10	0.10	0.10	0.09	0.07	0.06

system rotates to dawn or dusk causing a mid-latitude geomagnetic effect (Vennerstrom *et al.* 2007). So the IMF polarity can be inferred in the pre-satellite era. Using the prolonged observations at the first European magnetic stations, we inferred the IMF sector structure back to 1844 (Vokhmyanin & Ponyavin 2013, 2016). Here, we divide the IMF polarity proxies into six groups to estimate their reliability in the past. Data Set 1 is based on geomagnetic observations from Saint-Petersburg and Helsinki (available since 1844). In Set 2, we add results from Ekaterinburg and Potsdam (1896), in Set 3 from De Bilt and Sitka (1902), in Set 4 from Sodankyla and Eskdalemuir (1911), in Set 5 from Godhavn and Lerwick (1926), and in Set 6 from polar station Thule (1947). The success rate over 1965 – 2010 is presented in Table 1. It is defined as the relative number of coincidences with actual IMF polarity — the sign of the (By-Bx).

In order to estimate the HCS state, we calculate the $(T - A)/(T + A)$ ratio (hereinafter referred to as TA), where T and A are the annual numbers of days with negative and positive IMF polarities, respectively. In Figure 1, the ratio calculated for the IMF polarities from satellites is shown by the red curve, and for the reconstructed polarities by orange, yellow, violet, green, and blue colors. Near the solar magnetic field reversals (dashed vertical lines) the average TA and, thus, the dominant polarity change sign. Regularity of such alterations means the direction of the HCS shift stays the same. The largest difference between the actual and inferred TA is observed within the periods of low geomagnetic activity, when the success of the reconstruction decreases (Vokhmyanin & Ponyavin 2016). The average of the absolute difference is presented in Table 1. Although, it is about 0.10 for the proxies based on data sets 1–4, we still can discern the wavy evolution of the TA ratio.

In Figure 2, we show TA from 1844 to 2017. The years of solar magnetic field reversal are arbitrary indicated with vertical dashed lines. For more clarity, we also show prospective wave of TA (purple curve) drawn by hand. The maxima of this wave lie within the solar activity minima, when solar dipole is presumably axisymmetric. Significant deviations from the purple curve in 1856 – 1858, in 1876 – 1879, and in 1897 – 1901 coincide with extremely low geomagnetic activity. As we mentioned, this could be the cause of low success rate and hence unreliable TA estimation. Thus, the purple curve was drawn according to the TA values at the declining and growing phases of solar cycles. The

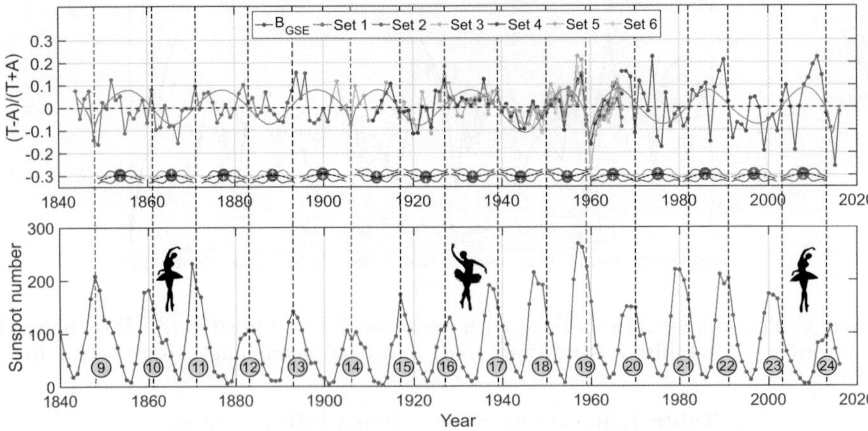

Figure 2. *Top*: TA ratios similar to Figure 1; purple wave – approximate TA wave; *bottom*: the annual sunspot numbers from the WDC-SILSO; the HCS asymmetry is illustrated by the ballerina's skirt.

assumed wave stutters around maxima of solar cycles 14 and 19. This means that before solar cycle 19 the HCS was shifted northward, while before solar cycle 14 — southward (see schematic illustration under the top plot). Therefore, the polar field was stronger in the northern hemisphere during solar cycles 15 – 19, and in southern during cycles 10 – 14 and 20 – 24. Our result on the HCS displacements is supported by other studies of the north-south asymmetries found in differential rotation (Pulkkinen & Tuominen 1998; Zhang *et al.* 2013) and solar activity (Verma 1993). If this scenario consistently repeats and the N-S asymmetry indeed has periodic behavior, we might expect northward solar field strengthening in the upcoming solar cycle 25.

Acknowledgements

The reported study was funded by RFBR according to the research projects No. 15-02-06959-a and No. 16-02-00300-a.

References

Rosenberg, R. L. & Coleman, P. J. 1969, *J. Geophys. Res.*, 74, 5611

Crooker, N. U., Lazarus, A. J., Phillips, J. L., Steinberg, J. T., Szabo, A., Lepping, R. P. & Smith, E. J. 1997, *J. Geophys. Res.*, 102, 4673

Smith, E. J., Jokipii, J. R., Kota, J., Lepping, R. P. & Szabo, A. 2000, *ApJ*, 533, 1084

Echer, E. & Svalgaard, L. 2004, *Geoph. Res. Lett.*, 31, L12808

Hiltula, T. & Mursula, K. 2006, *Geophys. Res. Lett.*, 33, L03105

Hiltula, T. & Mursula, K. 2007, *Adv. Sp. Res.*, 40, 1054

Svalgaard, L. 1972, *J. Geophys. Res.*, 77, 4027

Mansurov, S. M. 1969, *Geomagn. Aeron.*, 9, 622

Vennerstroem, S., Zieger, B. & Friis-Christensen, E. 2001, *J. Geophys. Res.*, 106, 16011

Vokhmyanin, M. V. & Ponyavin, D. I. 2016, *J. Geophys. Res.*, 121, 11943

Vennerstrom, S., Friis-Christensen, E., Olsen, N. & Moretto, T. 2007, *Geophys. Res. Lett.*, 34, L16101

Vokhmyanin, M. V. & Ponyavin, D. I. 2013, *Geophys. Res. Lett.*, 40, 3512

Verma, V. K. 1993, *ApJ*, 403, 797

Zhang, L., Mursula, K. & Usoskin, I. 2013, *A&A*, 552, A84

Pulkkinen, P. & Tuominen, I. 1998, *A&A*, 760, 755

Space Weather of the Heliosphere: Processes and Forecasts
Proceedings IAU Symposium No. 335, 2017
C. Foullon & O.E. Malandraki, eds.

doi:10.1017/S1743921317010912

Reinforcing a Double Dynamo Model with Solar-Terrestrial Activity in the Past Three Millennia

V. V. Zharkova[1], S. J. Shepherd[2] E. Popova[3] and S. I. Zharkov[4]

[1]Northumbria University, Faculty of Engineering and Environment, Newcastle upon Tyne, UK
email: valentina.zharkova@northumbria.ac.uk

[2]School of Engineering, Bradford University, Bradford, UK
email: s.j.shepherd@bradford.ac.uk

[3]Moscow University, Skobeltsyn Institute of Nuclear Physics, Moscow 119991, Russia
email: popovaelp@phys.msu.ru

[4]Hull University, Department of Physics and Mathematics, Kingston upon Hull, UK
email: s.zharkov@hull.ac.uk

Abstract. By applying Principal Components Analysis (PCA) to solar magnetic synoptic maps in cycle 21-23 obtained with Wilcox Solar Observatory we derived analytical expressions for two principal components and their summary curve of solar magnetic field oscillations defined by dipole magnetic sources. In this paper we extrapolate backwards three millennia the summary curve describing solar activity and compare it with the relevant historic data. The extrapolated summary curve shows a remarkable resemblance to the sunspot and terrestrial activity reported in the past millennia: the Maunder Minimum (1645-1715), Wolf minimum (1200), Oort minimum (1010-1050), Homer minimum (800-900 BC), the medieval warm period (900-1200), the Roman warm period (400-10BC). We note that Sporer minimum (1460-1550) derived from the increased abundance of isotope Δ^{14}C is likely produced by a strong increase of galactic cosmic rays caused by a supernova Vela Junior occurred in the Southern hemisphere.

Keywords. solar activity, dynamo, geoactivity

1. Introduction

Our understanding of solar activity is tested by the accuracy of its prediction. The latter became very difficult to derive from the observed sunspot numbers and to fit sufficiently close into a few future 11 year cycles or even into a single cycle until it is well progressed (Pesnell 2008). The fact that a large number of model predictions for cycle 24 disagreed with the measured sunspot numbers suggests that the appearance of sunspots on the surface during a solar cycle is governed by the action of some physical processes of the solar dynamo that are not yet fully considered in the models.

In order to reduce dimensionality of these processes in the observational data, Zharkova *et al.* (2012) applied Principal Component Analysis (PCA) to the low-resolution full disk magnetograms captured by the Wilcox Solar Observatory in cycles 21-23. This approach revealed a set of 8+ eigen vectors of temporal variations of solar background magnetic field (SBMF), which seem to appear in pairs. These two principal components (PCs) reflecting the strongest waves of solar magnetic oscillations, are indicated by Zharkova *et al.* (2015) having the highest eigen values covering about 39% of the data variance, or a one $\sigma = 0.67$ interval.

The main pair of PCs is shown to be associated with the two magnetic waves attributed to the poloidal magnetic field produced by dipole magnetic sources (see Popova *et al.* 2013; Zharkova *et al.* 2015) describing a double dynamo action in two different layers

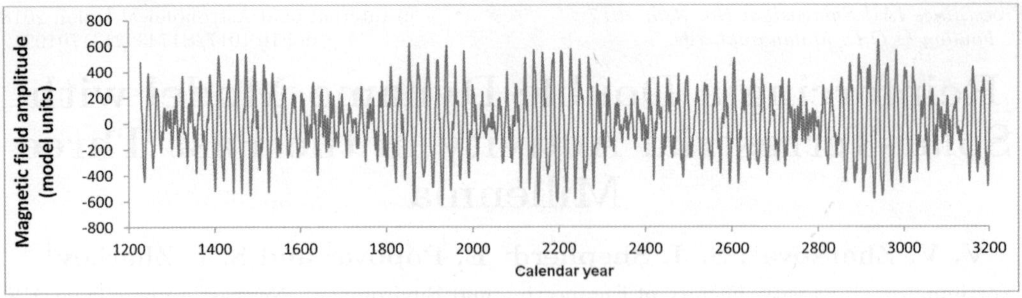

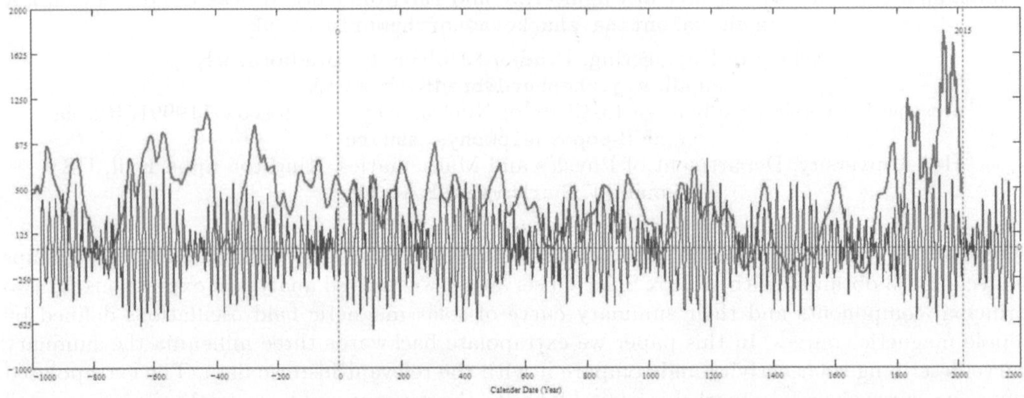

Figure 1. Top plot: the summary curve constructed using two Principal Components in arbitrary units (axis Y) for 1200AD to 3000AD, similar to Fig. 3 in Zharkova *et al.* (2015), with the removed horizontal pointers. The bottom plot: solar activity prediction backward 3000 years with a summary curve (blue line) of the two principal components (PCs) of solar background magnetic field (SBMF) (Zharkova *et al.* 2015, Zharkova *et al.* 2015) versus the reconstruction by Solanki and Krivova (2011) (red line) by merging the sunspot curve (17-21 centuries) and carbon dating curve (before the 17 century).

of the solar interior. These waves are found originating in the opposite hemispheres and travelling with an increasing phase shift to the Northern hemisphere (in odd cycles) and the Southern hemisphere (in even cycles) (Zharkova *et al.* 2012, 2015). The maximum (or double maximum for the double waves with a larger phase shift) of solar activity for a given solar cycle coincides with the time when each of the waves approaches their maximal amplitudes. The hemisphere where it happens becomes the most active one. This can naturally account for the north-south asymmetry of solar activity often reported in many cycles. The resulting summary curve if these two PCs describe the resulting solar activity shown by Shepherd *et al.* (2014).

The existence of two waves in the poloidal (and toroidal) magnetic fields instead of a single one for each of them, used in most prediction models, and the presence of a variable phase difference between the waves was shown by Zharkova *et al.* (2015) to naturally explain the difficulties in predicting sunspot activity with a single dynamo wave on a scale longer than one solar cycle as noted by by Karak and Nandy (2012). In support of this statement Shepherd *et al.* (2014) showed that the sunspot activity is associated with the modulus summary curve, which is a derivative from the two waves generated in two layers of the solar interior and not from a single layer accepted by most models.

In the current paper we expand the summary curve to the past three thousand years using findings from Zharkova *et al.* (2015) and compare this curve with the recent long-term reconstruction of stochastic solar activity suggested by Solanki and Krivova (2011) and Usoskin (2013).

2. Solar activity in the past three millennia with the summary curve

Findings by Zharkova *et al.* (2015) reveal very stable magnitudes of the two principal eigen values for each and every solar cycle considered, whether they derived from a single cycle, or a pair of cycles in any combination, or all three of them (21-23) (see Figure 1, top plot). This proves that the parameters of own oscillations of the Sun are maintained the same over a large period of time re-assuring a very good health of the Sun's heartbeat, or the solar dynamo. We have shown that visible variations of the wave parameters occur owing to different conditions in these two layers, so that the waves are generated with close but not equal frequencies. These variations are shown to cause a beating effect of these two magnetic waves, which produces an envelope grand cycle with the grand minima regularly occurring every 350-400 years.

Now we extend this extrapolation backwards to 3000 years (to 1000 BC) as shown in Figure 1 (bottom plot, blue curve). For comparison, we have overplotted the curve of the supposed solar activity derived by Usoskin (2013); Solanki and Krivova (2011) (red curve), which used sunspot numbers until 17 century and the isotope Δ^{14}C abundances prior to the 17 century.

It can be noted that our summary curve in Figure 1 reveals a remarkable resemblance to the sunspot and terrestrial activity reported in the past centuries showing accurately the recent grand minimum (Maunder Minimum) (1645-1715), the other grand minima: Wolf minimum (1300-1350), Oort minimum (1000-1050), Homer minimum (800-900 BC); also the Medieval Warm Period (900-1200), the Roman Warm Period (400-150 BC) and so on. These minima and maxima reveal the presence of a grand cycle in solar activity with a duration of about 350-400 years, with the next grand cycle minimum (Modern minimum) approaching in 2020-2055 as reported by Zharkova *et al.* (2015).

It turns out that longer grand cycles have a larger number of regular 22 year cycles inside the envelope of a grand cycle but their amplitudes are lower than in shorter grand cycles. This means that there are significant modulations of the magnetic wave frequencies generated for different grand cycles in these two layers: a deeper layer close to the bottom of the Solar Convective Zone (SCZ) and shallow layer close to the solar surface whose physical conditions derive the dynamo wave frequencies and amplitudes. The larger the difference between these frequencies the smaller the number of regular 22 years cycles and the higher their amplitudes.

In addition, to a grand cycle of 350-400 years, one can also derive a larger super-grand cycle of about 1900-2000 years by combining the curves in Figure 1 (bottom plot, blue curve) and the curve used for the prediction for the next 1000 years in Fig. 3 of Zharkova *et al.* (2015) reproduced here in Figure 1 (top plot). This super-grand cycle can be distinguished by comparing the five grand cycles in Figure 1 (top plot) (years 1001-3000 AD) with the next 5 grand cycles restored for 1000 BC-1000 AD) (bottom plot), which clearly show the repeating patterns for every 5 grand cycles.

It has to be noted that so far we only used the two PCs of the solar magnetic field oscillations assigned by Zharkova *et al.* (2015) to dynamo waves produced by dipole magnetic sources. Popova *et al.* (2017) shown that consideration of the next pair of magnetic waves produced by quadruple magnetic sources allows us to recover Dalton minimum (1790-1820) and another centennial cycles (caused by Gleissberg cycle).

However, Sporer minimum (1460-1550) is not present in our summary curve (the blue curve) plotted in Figure 1, showing instead during the same period of time a maximum

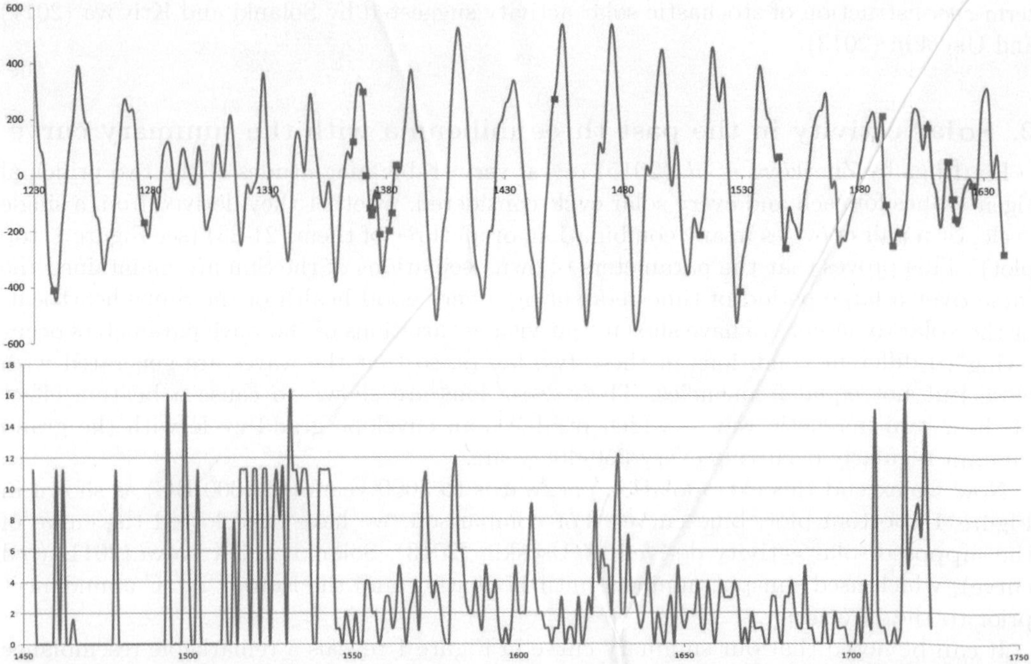

Figure 2. Top plot: verification of the summary curve (blue line) with large sunspots (red dots) placed at the time of their observations from the pre-telescope visual observations in Sporer minimum (1450-1550) from Wittman (1978); Wittman and Xu (1987). Note, that red dots for large sunspot were placed onto the activity curve at the exact times of their occurences during Sporer minimum. Bottom plot: distributions of terrestrial auroras over the period of Sporer minimum revealing the strongest auroras ever observed (from Schlamminger (1990); Schroder and Treder (1999)).

of the grand cycle previous to the modern one (17-21 centuries). We discussed this discrepancy in Zharkova *et al.* (2017) uncovering some important properties of the solar activity caused by quadruple magnetic sources. For the particular grand cycle including Sporer minimum the solar activty curve was surprisingly well verified by the naked eye observations of sunspots (see Figure 2, top plot).

Moreover, the intensities of auroras during the period of Sporer minimum are much higher than in any periods of maximal solar activity in the past. In fact, the aurora intensities in this period significantly exceed the intensities of auroras ever observed on Earth in the past 500 years (see Figure 2, bottom plot).

Keeping in mind that, normally, a) strong auroras coincide with the maxima of solar activity and not with its minima, and b) in the telescope era such strong auroras, as they were seen in the 14-16 centuries were never observed as indicated by Figure 2 (bottom plot), it is logical to assume that in that period there were some other sources, which increased a flow of relativistic particles causing these auroras. The Sun could not provide such the increase as its grand period for these centuries was not much different from the previous and the following grand cycles as shown in Figure 1.

Hence, we need to look at contributions of extra-terrestrial effects, e.g of supernovae, which were somehow overlooked despite supernova remnants are the major sources of galactic cosmic rays. There were a number of supernovae recorded in the past centuries which could affect the Earth as shown by Zharkova *et al.* (2017). Remnants of the mysterious supernova Vela Junior, which exploded in 1290 and became a neutron star, was found in 1998 located in the southern sky in the constellation Vela only 650-700 light-years away, which could cause the terrestrial effects during 13-15 centuries.

This supernova, at 46 degrees south, may have been too far south for observers in the Northern hemisphere to notice it, especially if it obtained peak brightness during the northern summer. At this declination, the supernova would be invisible above about 45 degrees north, making it invisible to the majority of Europe. This is the supernova, which could be the source of strong cosmic rays occurred in 13-15 centuries leading to large terrestrial auroras, and as consequence, to a long streak of epidemics in the southern countries on Earth, including China (see Reischauer *et al.* 1960).

3. Conclusions

Given a number of grand minima and grand maxima correctly fit by the summary curve combined with other means of the curve verification discussed above, we can confidently reinforce our previous findings (Zharkova *et al.* 2015) that the basic features of the solar activity are well reproduced by the two magnetic waves generated by dipole magnetic sources in two different layers of the solar interior. This finding also emphasizes the fact that the solar activity has a very well-maintained periodicity of these waves maintained over millennia reflecting a very good dynamo-health of the Sun.

References

Karak, B. B. & Nandy, D. 2012, *Astrophys. J. Lett.*, 761, L13
Pesnell, W. D. 2008, *Solar Phys.*, 252, 209
Popova, E., Zharkova, V. & Zharkov, S. 2013, *Annales Geophysicae*, 31, 2023
Popova, E., Zharkova, V., Shepherd, S. J. & Zharkov, S. 2017, *JASTP*, https://doi.org/10.1016/j.jastp.2017.05.006
Reischauer, O., Fairbank, J. K. & Craig, A. M. 1960, *East Asia: The great Tradition*, 64
Schlamminger, L., 1990, *MNRAS*, 247, 6.7
Shepherd, S. J., Zharkov, S. I. & Zharkova, V. V. 2014, *Astrophys. J.*, 795, 46
Schroder, W. & Treder, H. J. 1999, *Geofisica Internecional*, 39, 197
Solanki, S. K. & Krivova, N. A. 2011, *Science*, 334, 916
Usoskin, I. G. 2013, *Living Reviews in Solar Physics*, 10
Wittmann, A., 1978, Astron. Astrophys., 66, 93
Wittmann, A. D. & Xu, Z. T. 1987, *Astron. Astrophys. Suppl.*, 70, 83
Zharkova, V. V., Shepherd, S. J. & Zharkov, S. I. 2012, *MNRAS*, 424, 2943
Zharkova, V. V., Shepherd, S. J., Popova, E. & Zharkov, S. I. 2015, *Nature Scientific Reports*, 5, 15689
Zharkova, V. V., Popova, E., Shepherd, S. J. & Zharkov, S. I. 2017, *JASTP*, *https://doi.org/10.1016/j.jastp.2017.09.019*

Part 6. Challenges and Strategic Plans for Earth and the Heliosphere

Space Weather of the Heliosphere: Processes and Forecasts
Proceedings IAU Symposium No. 335, 2017
C. Foullon & O.E. Malandraki, eds.

doi:10.1017/S1743921318000571

Challenges and Strategic Research Plans for Earth and Heliosphere: Research Infrastructures, Projects and Initiatives

Ingrid Mann[1], Kirsti Kauristi[2], Ruth Bamford[3], Ian McCrea[3], Joeran Moen[4], Kjellmar Oksavik[5], Masanori Yamauchi[6], Magnar Johnsen Gullikstad[7] and Esa Turunen[8]

[1]UiT, The Arctic University of Norway, Tromsoe, Norway
[2]Finnish Meteorological Institute, Helsinki, Finland
[3]Rutherford Appleton Laboratories, United Kingdom
[4]University of Oslo, Norway
[5]University of Bergen, Norway
[6]Swedish Institute of Space Physics, Kiruna, Sweden
[7]Troms Geophysical Observatory, Tromsoe, Norway
[8]Sodankylae Geophysical Observatory, Finland

Abstract. We describe existing research infrastructures relevant for space weather and open issues of space weather research including the need for sustainable observation networks and for high-quality data products as basis for model development. The local relevance in Europe for studies of the ionosphere at high latitude is described. We propose as possible a way forward to sustain space weather research in Europe to establish a European research infrastructure project for space weather research and observations.

Keywords. Heliospheric physics, ionospheric variability, space weather

Introduction

The region around the Sun that is filled with the solar wind, the heliosphere includes almost the entire solar system. It is influenced by plasma, particles and radiation disturbances launched by the Sun, and by the physical processes that they cause. Research to study of these phenomena started in the 1970's, when observations from space provided observational data of the solar corona and the interplanetary medium inside Earth orbit and the first space missions to the outer system were launched. As a result, the International Astronomical Union (IAU) devoted a new commission to the topic of heliospheric studies (described in a recent summary of the commissions history, Mann *et al.* 2016). The first report of the new IAU commission pointed out that, in future, it would be useful to develop a capability for predicting, from solar observations, the state of the solar wind at the Earth (cf. Axford *et al.* 1976). The report also noted already that such a predictive capability would need to be based on advanced theoretical understanding. While some researchers realized very early the influence of space processes on Earth and sometimes used the term weather in this context, the term space weather is frequently used since the 1990s (cf. Cade and Chan-Park 2015). Space weather (SW) describes the immediate impact on the Earth of the time-varying cosmic environment; this cosmic environment is influenced predominantly, but not exclusively by plasma, particles, magnetic field, and radiation launched by the Sun, the disturbances and the physical processes that they cause. Hence SW is linked to a number of different areas of basic research including areas of interest represented within the IAU. Today we have a wealth of data, but the need to integrate and understand them in many ways poses a greater challenge than ever: How can we coordinate our efforts, make results accessible and sustain relevant observation initiatives? This article will address some of the open issues, it provides a short description of the top-down international efforts, local space weather challenges, existing infrastructures and the plans for a Research Infrastructure for Space Weather Research in Europe. The IAU can play an important role of providing a forum of bottom-up science discussions and initiatives related to space weather research.

1. Space Weather: International top-down coordination

During recent years much progress has been made in using the data from missions and activities to provide, develop and maintain services for Space Weather (SW) environment specification, now-casting and short-term forecasting. To improve international coordination, both in science and in operational services, COSPAR, the international Committee on Space Research, has addressed this challenge together with NASAs ILWS (International Living With a Star) programme, by publishing a Roadmap (Schrijver *et al.* 2015), which gives a review of the science questions that need to be solved in order to answer the questions posed by future forecast services. It is recognized that there is a need for reliable forecasts with < 1 day lead times and more accurate information about the onset time, duration, amplitude and spatial distribution of SW events than can be provided today.

One of the key recommendations made by the Roadmap is the development of innovative approaches for the incorporation of data into models (including data-driving, assimilation and ensemble modelling). For example, longer lead times in the forecasting of Interplanetary Magnetic Fields (IMF) north-south component, Bz would need better understanding of how a CME evolves as it leaves the Sun and propagates through the heliosphere. Similarly, resolving the key factors controlling different energy dissipation modes in the disturbed magnetosphere-ionosphere system would enable more accurate specification of the near-Earth radiation environment, particle precipitation and electric fields, informing strategies to mitigate harmful effects due to Geo-magnetically Induced Currents (GICs) in power transmission lines. This kind of prediction cannot reliably be performed at present, because our current capabilities in predicting ionospheric storm evolution are still moderate, due to the limited understanding of some crucial magnetosphere-ionosphere-thermosphere linkages and consequent deficiencies in the physics-based models. In order to improve the current situation in modelling and forecasting, the COSPAR Roadmap makes recommendations both for enhanced teaming inside the solar-terrestrial research community, and involving collaborations between agencies and with other communities (e.g. meteorology, solid-earth). These recommendations are of particular interest for European research groups who typically push forward their SW projects with sporadic funding, compiled from different national and international sources.

2. International research infrastructures and global networks

There are already a number of international efforts to integrate and coordinate the activities that are relevant to space weather. Within Europe, the participants in ESAs SSA space weather programme develop a coordinated set of Expert Service Centres in the areas of solar weather, heliospheric weather, ionospheric weather, space radiation and geomagnetic conditions. The results from these are mediated through a Space Weather

Coordination Centre, which coordinates the provision of space weather services and operates a helpdesk, dealing with queries about the precursor service provision or space weather conditions in general. An ESA SWE data centre hosts a space weather web portal, in which are embedded a number of space weather applications, provided by the Expert Service Centres. The ESA SSA programme is also working toward the development of underpinning space-based operational capabilities, notably the Lagrange missions to the L1 and L5 points, being pursued in collaboration with NASA, which will ensure the availability continuous monitoring data, from both upstream of the Earth, and from a location well away from the Sun-Earth line, allowing both in-situ observation and remote sensing of potentially Earth-impacting space weather events. All of this is already a major step forward, compared to the situation a few years ago; however there are still two areas of major weakness which ESA funding is unable to address. These relate to the sustainability of the ground-based instrument networks which underpin the provision of space weather data and the R2O development of the modelling capabilities which use these data in order to produce simulations and predictions.

The ground-based instruments are very diverse in nature, ranging from relatively small and cheap detectors operated by individual research groups to large, complex instruments which are already international research infrastructures in their own right. The EISCAT Scientific Association, for example, has effectively functioned as an international research infrastructure since 1975, providing a network of incoherent scatter radars in the European Arctic, with funding support from relevant research organisations in Norway, Sweden, Finland, UK, Japan and China; in earlier times the EISCAT funders also included France and Germany. Since 1998, EISCAT has been working within the framework of ESFRI and EU funding to plan its new EISCAT_3D radars, three large phased arrays (cf. Wannberg *et al.* 2010), to be situated in the northern parts of Norway, Sweden and Finland. This facility, which is about to enter its construction phase, is capable of continuous operation with modes which can be adapted as geophysical conditions change. A focus of the project lies in studies related to ionospheric processes and to the long-term variability in the atmosphere and global change (cf. McCrea *et al.* 2015). As well as investigating fundamental physical processes, a significant part of the case for the new radar has been based on its ability to provide continuous data for the emerging set of space weather monitoring and forecast capabilities. It should also be noted in passing that EICSCAT_3D has capabilities for the surveillance and tracking of space debris, whose dynamics are linked to variations in space weather and potentially point to a significant future user base for space weather services.

To study global phenomena also requires a world-wide coordination of observations. The EISCAT radars already run a number of experiments per year which are jointly coordinated with similar radars in Canada, the US, Puerto Rico, Peru and Greenland, with occasional participation by radars in Russia, Ukraine, China and Japan, planned on an international level but sustained on a best efforts basis according to the available funding. Another notable international initiative in radar collaboration is SuperDARN, a collaboration of nine countries, which has been continuously running a network of HF oblique sounding radars in the high-latitude regions to study plasma convection and other phenomena for more than 20 years (Greenwald *et al.* 1995) and has recently added poleward (PolarDARN) and equatorward (StormDARN) extensions of the instrument network. An example for a wide-spanning network established on national level is the Chinese meridian project that involves a chain of ground-based observatories roughly along 120 degree E longitude and 30 degree N latitude. that are equipped with multiple instruments to measure key parameters of the ionosphere at 20 to 1000 km (Wang *et al.* 2013).

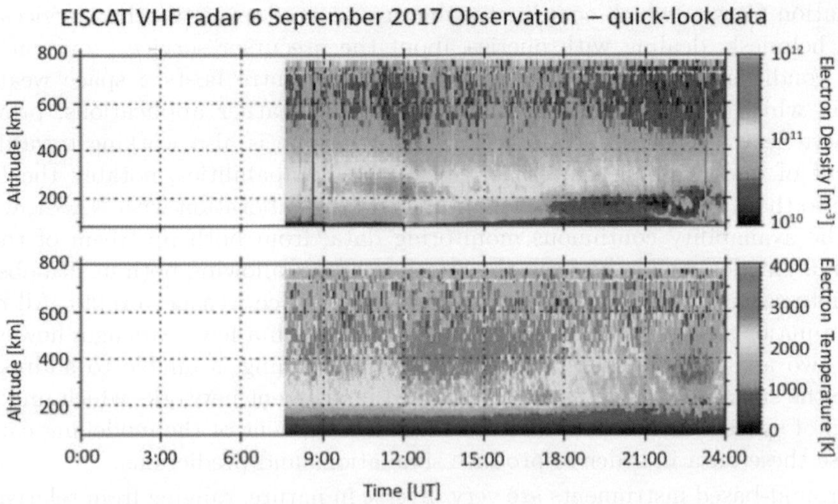

Figure 1. The ionospheric electron density and temperature measured with EISCAT on 6. September 2017 showing a clear signature in electron temperature at about 200 to 600 km altitude following the X9.3 solar flare around 12:00 UT.

3. Local infrastructures

At the other end of the spectrum, small instruments, such as magnetometers, are operated by individual research groups around the world, but their data are coordinated together by collaborations such as SuperMAG, which provides a central repository for over 300 globally distributed instruments in a standard form, with common standards of data reduction and artefact removal. Similar examples of coordinated use of multiple distributed instruments can be seen in the MIT Automated Process of GPS (MAPGPS) system, which integrates data from over 2000 globally distributed GPS receivers, and the GLORIA riometer consortium, which provides a database of instruments measuring ionospheric absorption worldwide and offers some degree of data synthesis (e.g. to create multi-instrument movies). Even at European level, these networks tend to be operated on a best efforts basis, sometimes by a single funder and occasionally by a collection of national funding agencies acting together. Successful examples are often operated by one or more national funding agencies to operate a set of instruments locally that are integrated in an international structure. For example, the Finnish Meteorological Institute leads the MIRACLE network of magnetometers, all-sky cameras and GPS measurements, distributed across Norway, Sweden and Finland, the magnetometer data from which also form part of the SuperMAG database, while the GPS data are also used in global assimilative studies. In general terms, however, the present arrangements for funding space weather instrument operations in Europe are far from satisfactory, threatening the stability of any system based on exploiting their data.

4. Model development and validation

The development of space weather models has, to a large extent, always been a pure research activity in which groups with appropriate capabilities have developed models as research tools to be applied to understanding particular physical problems. This has meant that the design, documentation and validation of such models has always varied greatly. In particular, within Europe at least, such models have tended to remain as research tools, where the understanding and ability to extend them for particular purposes, or to use them to their full potential, has been restricted to a relatively small group of developers. There are several examples of models which have become obsolete as developers have moved on and such knowledge has been lost. Up to now there has been no good mechanism for models to be translated from research to operational use, though this is beginning to change as both ESA and national space weather centres are realising the requirement for operational models. In the US, the job of upgrading models for operational use has been taken on, at least in part, by NASAs Community Coordinated Modelling Center (CCMC), which works both with developers and operational users to bridge this gap, turning research models into operational tools. This method has been very successful, because it has ensured developer involvement in the operational use of their models and also ensured that the intellectual property of developer groups is acknowledged and respected. Nothing similar currently exists in Europe and establishing a proper mechanism for model development in Europe would be important and timely.

5. Local space weather challenges

Some of the space weather challenges have a particular local relevance. Europe includes populated regions in the Arctic and with the intensification of human activity in the Polar Regions relying on satellite signals, a detailed understanding of the different mechanisms structuring the ionosphere at high latitudes is of increasing practical importance (cf. Moen *et al.* 2013). Ionospheric irregularities are known to degrade the signal quality of the Global Navigation Satellite System (GNSS), and this may reduce GNSS position accuracy (Jin *et al.* 2015). Intense GNSS phase scintillation and loss of signal lock has been found in the dayside polar cap (Oksavik *et al.* 2015). In the central polar cap the scintillation is often localized to high-density storm-time tongues of ionization or polar cap patches that are in transit. The strongest scintillation events are found in the night-side polar cap, where dense polar cap patches enter the night-side auroral oval. As the irregularities can vary both in time and space, a 3D multi-scale instrument approach is required to understand underlying mechanisms, instability modes, and growth rates causing the plasma turbulence. Characterizing the evolution of ionospheric irregularities and the transfer of energy between the different scales can be regarded as a study of fundamental plasma physics from a theoretical point of view, but it also has direct applications to developing reliable space weather forecast.

6. European initiative for a space weather research infrastructure

The need to establish a sustainable setting for regular and coordinated transition of research models to operational space weather services has triggered some informal discussions among European research groups. The Research-To-Operation (R2O) projects conducted under the Space Situational Awareness (SSA) programme of ESA are acknowledged, but on the other hand it is noticed that some research groups possessing valuable capabilities may not have easy access to the SSA programme. Therefore, it would be advisable to have a research oriented framework (with working title European Sustainable SW Network, ESWNET) to facilitate efficient networking between SW modelers and observers, based initially on the distributed network of experts in existing models and data products as the substrate for developing and validating new models. ESWNET would exploit EU investments for e-infrastructures facilitating advanced computation environments and the network would work closely with academic research groups and

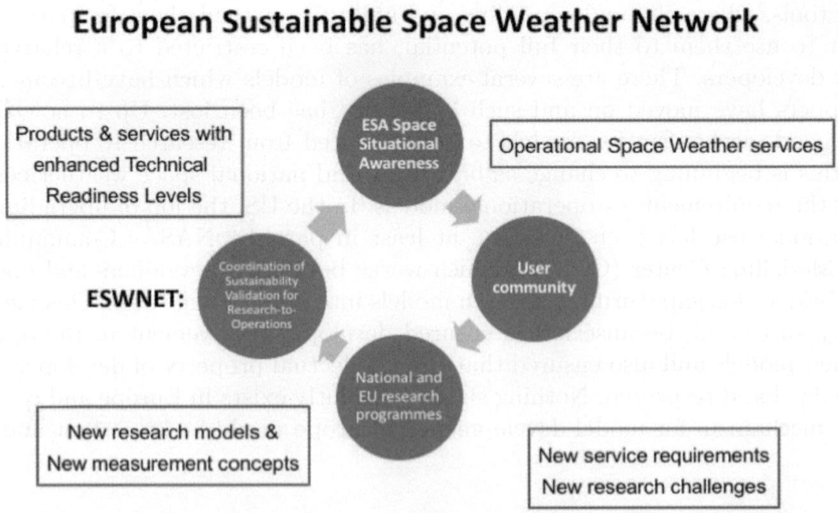

Figure 2. A new network of European solar-terrestrial research groups, ESWNET, is suggested to support ESA and EU in coordinating both operational and research activities for continuously improving Space Weather services.

other stakeholders (e.g. ESA and EUMETSAT) who maintain ground- and space-based instrumentation.

Figure 2 presents a schematic, showing how ESWNET could be located within the European environment of SW activities. The new network would facilitate communication and coordination between ESA SSA SW services and national and EU research programmes, with the goal to integrate the best expertise from distributed sources into one joint Sun-Earth observatory. ESWNET could support SSA by serving as the platform for upgrading Technical Readiness Levels and validation of new SW models and measurement concepts. In an optimal situation, this network could even influence the planning of European and national SW funding programs, so that SW service user requirements could be addressed with coordinated research missions.

Obviously ESWNET itself would also need some funding in order to be an attractive platform for research groups to collaborate. Getting a position in the Roadmap of the European Strategy Forum on Research Infrastructures (ESFRI) could be a feasible pathway for establishing ESWNET. The ESFRI framework, which operates with the backing of the European Commission, provides good models for agreeing on governance structures, data policy and IPR issues. Furthermore, as partners in an ESFRI initiative, ESWNET research groups would have a stronger position in the competition for sustainable national funding lines.

7. Summary

The discipline of space weather is becoming an area in which scientific understanding needs to inform the requirements for real-time monitoring and future prediction. This need, which will only become more urgent as society becomes increasingly dependent on space-based technologies, can only be fulfilled by using the capabilities of our globally distributed observing instruments, complemented by the best of our models, optimised to work together in a real-time operational setting. While some organisations, notably ESA in Europe and NOAA and NASA in the US, are taking steps toward this, some key capabilities are still missing. In particular, there is a worrying lack of stability and sustainability in the funding arrangements for the operation of ground-based instruments, which is notable in Europe but also applies on a global scale. Another missing piece in the jigsaw is the effort, and governance framework, needed to turn research tools such as models into real operational systems, capable of using appropriately validated multi-instrument data to make predictions with appropriately quantified accuracy. These responsibilities would seem to fall somewhere between the research community, which uses instrument data and develops models to study fundamental physical mechanisms, and operational agencies which have a requirement to use data and models in a standard way for operations. A new European Space Weather Infrastructure, as proposed here, would supply the missing pieces in this jigsaw, and deliver a stable European contribution to this burgeoning international collaborative effort.

Acknowledgements

The quick-look data presented in Figure 1 are courtesy of EISCAT and for illustration purpose only. EISCAT is an international association supported by research organizations in Norway (NFR), Sweden (VR), Finland (SA), Japan (NIPR and STEL), China (CRIPR), and the United Kingdom (NERC). EISCAT VHF and UHF data are available under http://www.eiscat.se/madrigal/. KK and IM acknowledge discussions on SWNET with Jean Lilensten, Consuelo Cid Tortuero, Iwona Stanislawska, Jens Berdermann, Johan de Keyser, Margit Haberreiter, Peter Gallagher, Stefaan Poedts, Vladimir Kalegaev, and Ari Viljanen. KO is grateful for being selected as the 2017-2018 Fulbright Arctic Chair, and his sabbatical at Virginia Tech is sponsored by the U.S.-Norway Fulbright Foundation for Educational Exchange. KO is also supported by the Research Council of Norway under contract 223252.

References

Axford, W. I., Hollweg, J. V., Suess, S. T., Blum, P. W. & Fahr, H. J. 1976, *Trans. Int. Astron. Union (Reports on Astronomy)* 16A, 175-187

Greenwald, R. A., *et al.*1995, Darn/Superdarn, *Space Sci. Rev.* 71, 761-796

Jin, Y., Moen, J. I. & Miloch, W. J. 2015, *J. Geophys. Res. Space Physics* 120, 9176-9191, doi:10.1002/2015JA021449

Mann, I., Manoharan, P. K., Gopalswamy, N., Briand, Chashei, I. V., Gibson, S. E., Lario,D. Hanaoka, Y., Malandraki,O., Kontar, E. & Richardson, J. D. 2016, Division E Commission 49, *Transactions of the IAU* 29A, 300-315

McCrea, I, *et al.* 2015, *Prog. Earth Planet. Sci.* 2, 21, doi:10.1186/s40645-015-0051-8

Moen, J., Oksavik, K., Alfonsi, L., Daabakk, Y., Romano V. & Spogli, L. 2013, *J. Space Weather Space Clim.* 3, A02, doi:10.1051/swsc/201302

Oksavik, K., van der Meeren, C., Lorentzen, D. A., Baddeley, L. J. & Moen, J. I. 2015, *J. Geophys. Res. Space Physics* 120, 9161-9175, doi:10.1002/2015JA021528

Schrijver, C. J., Kauristie, K., *et al.* 2015, *Advances in Space Research* 55, 2745-2807

Wang, M., Lou, W. Y., Li, P., Shen, X. H. & Li, Q. 2013, *J. Atmos. Sol. Terr. Phys.*, 102, 261–268

Wannberg, U. G., Eliasson, L., Johansson, J., *et al.* 2010, *URSI Radio Science Bulletin* 2010 (333), 75-88 doi: 10.23919/URSIRSB.2010.7911050

Space Weather of the Heliosphere: Processes and Forecasts
Proceedings IAU Symposium No. 335, 2017
C. Foullon & O.E. Malandraki, eds.

doi:10.1017/S1743921317008857

Defining and Characterising Heliospheric Weather and Climate

Mauro Messerotti[1,2]

[1]INAF-Astronomical Observatory of Trieste,
Loc. Basovizza 302, 34149, Trieste, Italy
email: messerotti@oats.inaf.it

[2]Dept. of Physics, Trieste University,
Via A. Valerio 2, 34127 Trieste, Italy
email: mauro.messerotti@ts.infn.it

Abstract. At large distance scales, space exploration in the last decades has significantly helped in better locating the boundaries of the Heliosphere and outlining its shape as well as in probing the various plasma domains that separate the inner heliospheric region from the interstellar one. At shorter distance scales, a fleet of spacecraft has been probing the outer and inner Solar System plasma with a high level of detail.

This monitoring, complemented by space- and ground-based observations of processes relevant to the Heliosphere, has pointed out a series both of intrinsic and extrinsic perturbations that characterise the physical state of heliospheric plasmas both at small and large spatial scales and on short and long temporal scales.

By means of concept maps that schematise the association among concepts, in this work we will present a new domain ontology for the definition and characterisation of Heliospheric Weather and Climate.

Keywords. Solar-terrestrial relations, solar system: general, Galaxy: general, galaxies: general

1. Introduction

Many definitions of Space Weather exist. Are such definitions using a terminology, which is correct both from a physical and from a semantic point of view? What can be a possible approach to correctly match the requirements for knowledge coding? In the following, we will concisely consider a holistic approach to the definition of Heliospheric Weather and Climate, entirely based on concept maps.

2. A European definition for Space Weather

A number of definitions for Space Weather has been formulated by different organisations, that are based on the respective missions and goals, often biased by the end user needs. A dedicated working group of COST Action 724 analysed many definitions and elaborated a European definition for Space Weather that is semantically and physically consistent (Lilensten *et al.* (2008)). It reads, "Space Weather is the physical and phenomenological state of natural space environments. The associated discipline aims, through observation, monitoring, analysis and modelling, at understanding and predicting the state of the Sun, the interplanetary and planetary environments, and the solar- and non-solar-driven perturbations that affect them; and also at forecasting and nowcasting the possible impacts on biological and technological systems".

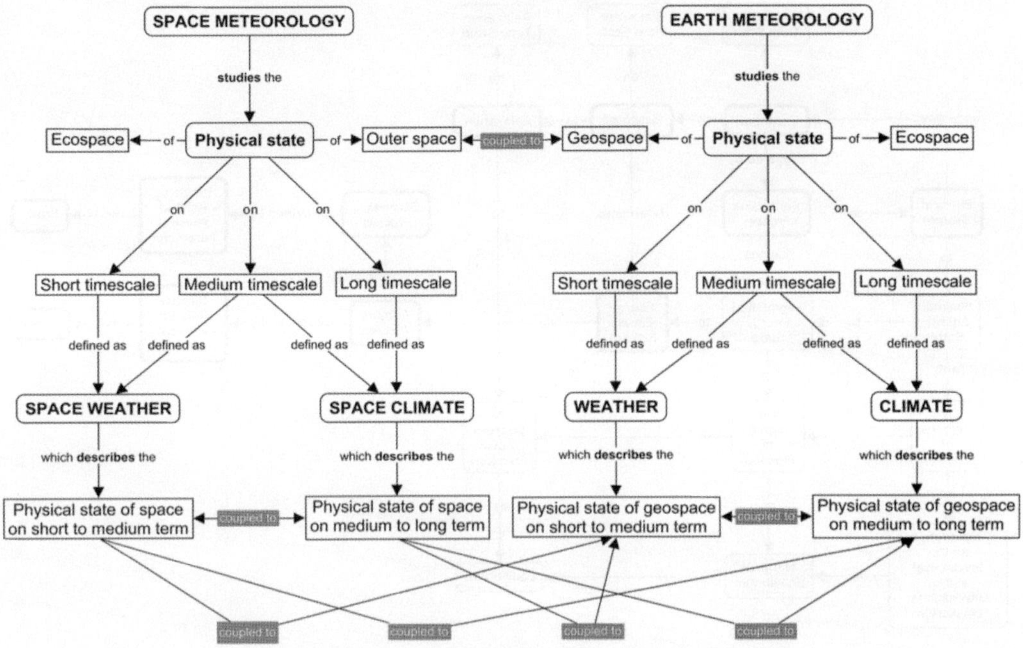

Figure 1. C-map that defines Meteorology of Space and Earth, and their relationships.

3. The ontological approach

Concept maps (c-maps) are conceptual tools that allow a flexible representation of concepts and their relationships, i.e., the schemes for knowledge coding. Hence, c-maps are suitable for building up a domain ontology, such as, e.g., a Space Weather (SWx) ontology (Messerotti (2007), Messerotti (2008), and references therein). Furthermore, c-maps can be used to embed knowledge in scientific data bases as metadata (Messerotti (2002)). In this way, one can define Meteorology of Space as the discipline that studies the physical state both of Ecospace and Outer Space on short timescale (Space Weather) and on long timescale (Space Climate), and, similarly, Meteorology of Earth, by emphasising the interrelationships as in Figure 1.

3.1. *General definition of Meteorology, Weather and Climate*

The term "Meteorology", used in Earth and planetary sciences, refers to the discipline which deals with the atmosphere and its phenomena, and studies the "Weather", i.e., the physical state of the atmosphere at a specific time and place, and the "Climate", i.e., the average state of weather on the long term.

Similarly, one can define as "Meteorology of Space" the discipline which deals with the Space Domain and its phenomena, and their impacts on Earth and planetary environments. It studies, respectively, "Space Weather", i.e., the physical state of the Space Domain at a specific time and place, and "Space Climate", i.e., the average state of Space Weather on the long term.

3.2. *Description of the physics*

The c-map approach is quite effective in describing the physics of processes from the highest level of abstraction, which depicts the commonalities in physical system triggering, response, and relaxation (e.g., Figure 2), to a higher level of detail (e.g., Figure 3).

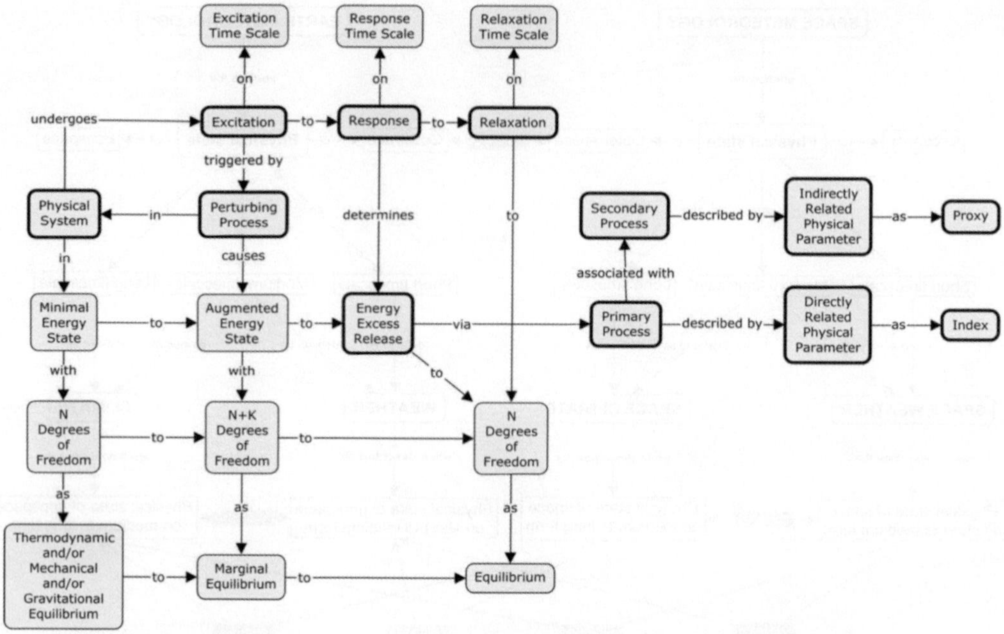

Figure 2. C-map that outlines physical system triggering, response, and relaxation.

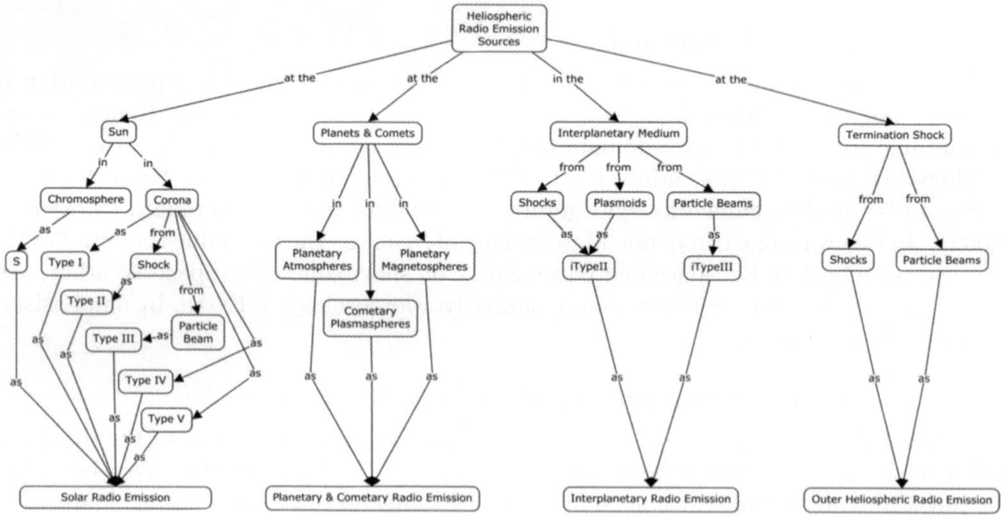

Figure 3. C-map that summarises heliospheric radio emission sources (Messerotti (2011)).

4. Semantics of Meteorology of Space

As stressed in Section 3.1, Space Weather refers to the physical state of space and planetary environments as perturbed by energetic photons and particles from both the Sun and outside the solar environment, and the discipline that studies Space Weather is Meteorology of Space.

Anyway, it must be noticed that quite often Space Weather is used to mean both the physical state of the environment and the discipline which studies it, but we stress that this is a misuse from an ontological and linguistic point of view. With regard to that, a never-ending discussion has been in progress, due to the widespread misuse and the concern for not being understood, e.g., by end-users as they are familiar with the wording Space Weather, and a change could be source of confusion. Notwithstanding, we are convinced that this is not an argument to justify a distortion of the terminology.

4.1. *The acronym for Space Weather*

Following a coherent approach, we consider the acronym for Space Weather as follows.

In particular, we emphasise that the universal acronym for "Weather" is "WX", i.e., the Morse code shorthand for the word "Weather", which is used as such in all messaging systems.

By analogy, the acronym for "Space Weather" should be "SWX" or "SWx". In fact, The World Meteorological Organisation (WMO) is organising a global SWx service for aviation safety according to the requirements by ICAO (International Civil Aviation Organization), and it will be operational in 2018.

Hence, the adoption of SWx as acronym for Space Weather in the messaging is a natural extension of the meteorological framework and unambiguous.

Other acronyms are misleading, such as, e.g., SWE (Society of Women Engineers) and SW typically used to mean "Solar Wind" in solar and heliospheric physics.

5. The Space Weather workflow: basic assumptions

SWx activities can be subdivided into three categories: (1) SWx monitoring; (2) SWx science; and (3) SWx operations.

Despite that there is a natural overlapping of the three categories, they can be clearly identified and should not be interchanged to properly focus the different roles. This consideration applies both to data sets and to observing facilities.

For instance, a non-active (closed) data set is useful for SWx science, but not for SWx operations, and a spacecraft that provides data with a latency of days to months has not been designed for SWx operations. Similarly, a science model has to be transitioned to an operational model through a long process of testing and validation.

6. The Space Domain as coupled astrophysical domains

The Space Domain is a complex physical system, which consists of a nonlinearly coupled complex subsystems. This structure extends in spacetime at increasing time, space, and energy scales. Hence, the sources of SWx have to be identified respectively in cluster of galaxies, galaxies, the Galaxy (Milky Way), the solar neighbourhood, the Heliosphere, and the Sun. In fact, high energy phenomena occur, e.g., in the Milky Way, where Magnetars, Hypernovae, and Supernovae outbursts contribute to Galactic Cosmic Rays (GCRs) and originate Gamma Ray Bursts (GRBs), which perturb the physical state of the Heliosphere. Hence, it is justified to define and consider the "extragalactic" and "galactic" climate, which are studied by "cosmoclimatology", being the relevant phenomena evolving on quite long timescales. Cosmoclimatology is still at a very preliminary stage.

7. Definition of Heliospheric and Solar Weather and Climate

Instead of using the term SWx for any domain, like, e.g., SWx in the Heliosphere, it is more appropriate to define the terms "Heliospheric Weather" (HSWx) and "Heliospheric Climate" (HSC) as a natural extension of Heliophysics (Messerotti (2011)). These definitions are reported in Figure 4, where the outer and inner forcing agents are listed.

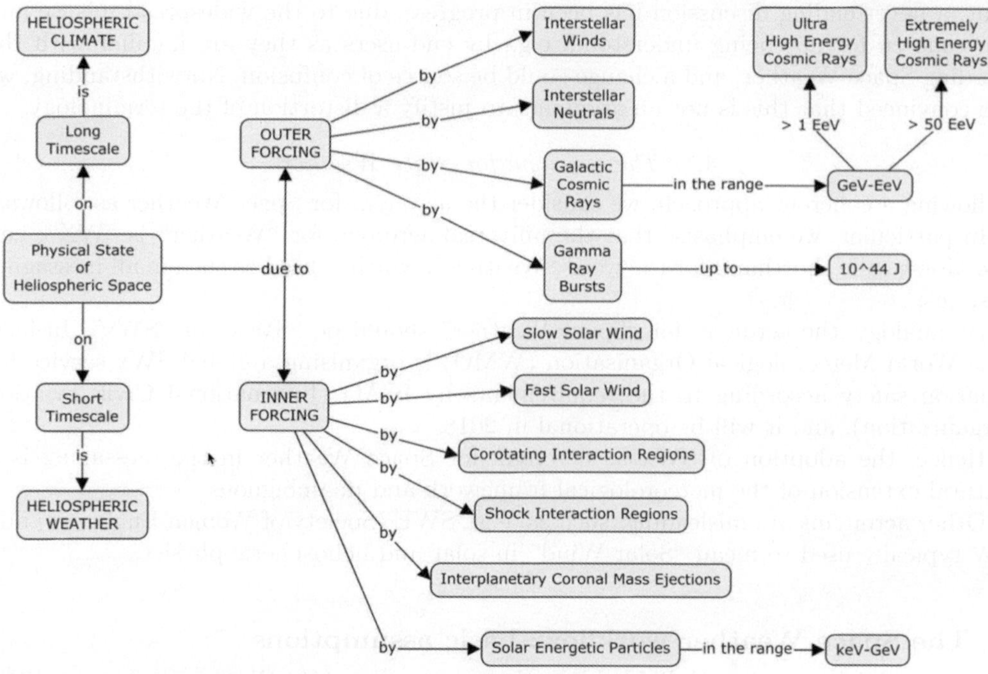

Figure 4. C-map that defines Heliospheric Weather and Climate.

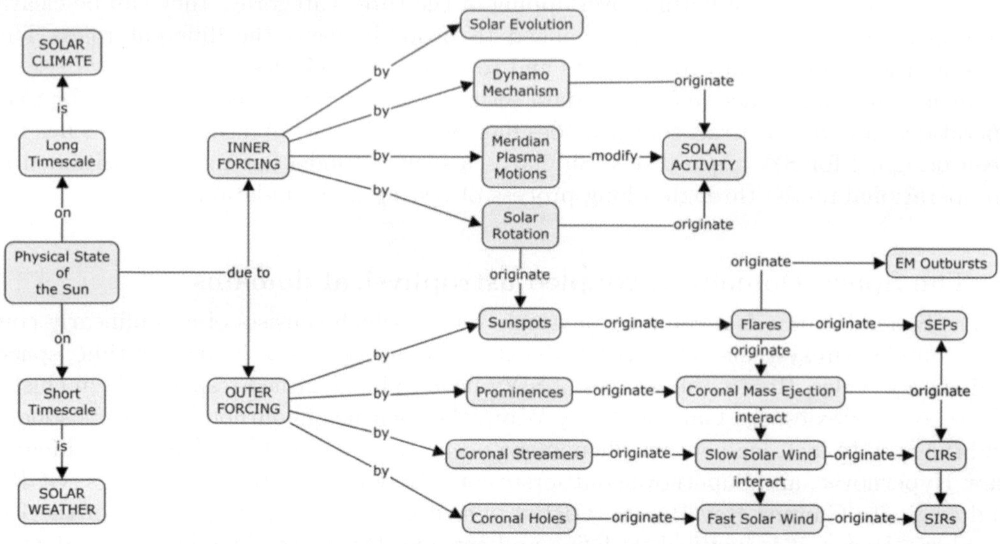

Figure 5. C-map that defines Solar Weather and Climate.

Among the former ones, GCRs and GRBs are considered, whereas the latter ones include Solar Wind structures (CIRs, SIRs), Interplanetary Coronal Mass Ejections (ICMEs), and Solar Energetic Paricles (SEPs).

Energy ranges span from GeV to EeV (CRs), up to 10^{44} J (GRBs), and keV to Gev (SPEs). The characteristic times span from milliseconds to days.

Ultra and Extremely High-Energy Cosmic Rays are quite rare, but they cannot be disregarded, as potential sources of impacts on biological and technological systems.

7.1. *Definition of Solar Weather and Climate*

Analogously, Solar Weather and Climate are the physical state of the Sun on short and long timescales, respectively (Figure 5). SWx impacts on Earth and planets are originated by perturbations that occur at the Sun as manifestations of its physical state, in turn, determined by inner and outer forcing processes (see, e.g., Messerotti *et al.* 2008, 2009). The inner forcing agents span from the Sun's evolution, dynamo mechanism, and meridional plasma motions to solar rotation, which originate and modulate solar activity. Outer forcing agents are the activity manifestations like photospheric sunspots, chromospheric and coronal prominences, coronal streamers, and Coronal Holes, which originate flares, CMEs, slow and fast Solar Wind streams, SEPs, CIRs, and SIRs.

8. Conclusions

The Heliosphere is the most relevant domain for the human beings and its perturbations affect both technological and biological systems.

Notwithstanding, it is just a physical domain nonlinearly coupled with all other domains in the Universe, whose physical state, despite of their distance, significantly biases the Heliospheric Weather and Climate.

Hence, outer high energy sources must be taken into account when assessing Heliospheric Weather and Climate, in the framework of Cosmoclimatology.

This poses the issue of a multi-disciplinary approach that spans over many astrophysical subjects, related to events that occur on cosmic timescales and for which a detailed climatology does not exist yet.

An approach based on knowledge coding by concept maps can help in defining the physical domains, the phenomenology and the involved physical processes, and their relationships, according to correct physical and semantic approaches.

Furthermore, this approach is quite promising in advanced data handling and knowledge discovery applications, as it provides correct terms and attributes for metadata in advanced data models.

References

Lilensten, J., Sanahuja, B. & Messerotti, M. 2008, in: J. Lilensten, A. Belehaki, M. Messerotti, R. Vainio, J. Watermann & S. Poedts (eds.), *Developing the scientific basis for monitoring, modelling and predicting Space Weather*, COST 724 Final Report (Luxembourg: Office for the Official Publications of the European Communities), p. 298

Messerotti, M. 2002, in: H. Sawaya-Lacoste (ed.), Proc. Second Solar Cycle and Space Weather Euroconference (Noordwijk: ESA Publications Division), p. 607

Messerotti, M. 2007, in: H. Lundstedt (ed.), *Solar Activity: Exploration, Understanding and Prediction*, Proc. Lund Workshop (Noordwijk: ESA Publications Division), p. 11

Messerotti, M. 2008, in: J. Lilensten, A. Belehaki, M. Messerotti, R. Vainio, J. Watermann & S. Poedts (eds.), *Developing the scientific basis for monitoring, modelling and predicting Space Weather*, COST 724 Final Report (Luxembourg: Office for the Official Publications of the European Communities), p. 350

Messerotti, M., Zuccarello, F. & Lundstedt, H. 2008, in: J. Lilensten, A. Belehaki, M. Messerotti, R. Vainio, J. Watermann & S. Poedts (eds.), *Developing the scientific basis for monitoring, modelling and predicting Space Weather*, COST 724 Final Report (Luxembourg: Office for the Official Publications of the European Communities), p. 20

Messerotti, M., Zuccarello, F., Guglielmino, S., Bothmer, V., Lilensten, J., Noci, G., Storini, M. & Lundstedt, H. 2009, *Space Sci. Revs*, 147, 3-4, 121

Messerotti, M. 2011, in: M. P. Miralles & J. Sánchez Almeida (eds.), *The Sun, the Solar Wind and the Heliosphere*, Proc. Sopron IAGA Conference (Berlin: Springer), p. 209

Space Weather of the Heliosphere: Processes and Forecasts
Proceedings IAU Symposium No. 335, 2017
C. Foullon & O.E. Malandraki, eds.

doi:10.1017/S1743921317010456

Monitor and Prediction of Near-Earth Radiation Environment in the Frame of Space Monitoring Data Center at Moscow State University

Irina N. Myagkova[1], **Vladimir V. Kalegaev**[1], **Mikhail I. Panasyuk**[1,2], **Yuliya S. Shugai**[1], **Sergey A. Dolenko**[1], **Sergey Yu. Bobrovnikov**[1], **Vera O. Barinova**[1], **Minh Duc Nguyen**[1], **Vladimir R. Shiroky**[1], **Valery E. Eremeev**[1], **Andrey V. Bogomolov**[1], **Oleg G. Barinov**[1], **Natalia A. Vlasova**[1] **and Nikolay V. Kusnetsov**[1]

[1]Skobeltsyn Institute of Nuclear Physics of Lomonosov Moscow State University, 1(2), Leninskie gory, Moscow, 119991, Moscow, Russia,
email: irina@srd.sinp.msu.ru

[2]Physical Department of Lomonosov Moscow State University, 1(2), Leninskie gory, 119991, Moscow, Russia

Abstract. Radiation environment of near-Earth space is one of the most important factors of space weather. Space Monitoring Data Center of Moscow State University provides operational monitor and forecast of radiation conditions both at Geostationary Orbits (GEO) and at Low Earths Orbits (LEO) of the near-Earth space using data of recent space missions (Vernov, CORONAS series) and current (Lomonosov, Meteor-M, Electro-L) ones. Internet portal of Space Monitoring Data Center of Skobeltsyn Institute of Nuclear Physics of Lomonosov Moscow State University (SINP MSU - [swx.sinp.msu.ru]) provides possibilities to monitor and analyze the space radiation conditions in the real time mode together with the geomagnetic and solar activity including hard X-ray and gamma-emission of solar flares.

Keywords. Sun: Flares, Space weather: radiation, Space weather: magnetic storm, Space weather: forecast

1. Introduction

Radiation environment is one of the main factors of near-Earth space that should be taken into account in the analysis of space weather. With the development of space technology, we are becoming much more dependent on the processes that occur in near-Earth space, and the ability to quickly predict these processes becomes very important. With increasing number of space missions, there is an issue of operational receipt, storage and real time access to the data of space measurements. Sources of this data as well as modern computing and communication permit to provide new opportunities for space conditions operational analysis and forecasting based on space weather applications developed during the last years.

Solar UV and soft X-ray fluxes, produced by solar flares, can cause short-wave communication disruption in high-latitude areas, as well as lead to malfunctions in navigation systems. According to the estimates of experts (see e.g.: Cole, 2003), more than half of failures in the operation of on board spacecraft equipment are due to the adverse impact of space weather factors. Charged particles cause degradation of semiconductor materials, malfunctions in microchips in systems of spacecrafts, electrilization of a spacecraft.

Energetic charged particles can lead to damage and failure of electronics and harm people on board spacecraft. If solar protons come to Earth during the main phase of a magnetic storm, when the boundary of their penetration into the magnetosphere shifts to lower latitudes, not only astronauts, but also pilots and passengers of airliners during transpolar flights are exposed to space radiation. So we have to create a complex set of coupled operational models from Sun to Earth that permit to predict the radiation risk for satellites in different orbits, and also the magnetic storms. The solution of this problem is to collect all relevant space-related data sets acquired by multiple satellites of various purposes and capabilities as well as by ground stations, and to develop the operational services that use space weather models and collected data to produce forecasts.

To correctly evaluate the radiation conditions, it is necessary to take into account the influence of geomagnetic activity on the acceleration, propagation and scattering of charged particles in the magnetosphere. Also, it is necessary to take into account the characteristics of solar activity: the state of the solar corona, the neutral radiation of solar flares - UV, X- and gamma-rays.

There are many centers around the world which provide such kind of service: e.g. Space Weather Prediction Center (SWPC/NOAA), integrated Space Weather Analysis System (iSWA), Space Situation Awareness (SSA) etc. The main monitoring space missions as well as models of space factors provide real-time data for analysis of space weather. Each center provides analysis based only on its own set of implemented operational models. The monitoring data necessarily contain the measurements at GEO and LEO, spacecrafts in solar wind, solar observatories to show the whole picture of space weather.

Space Weather Analysis Centre of SINP MSU (SWX) [swx.sinp.msu.ru] provides information about the current state of near-Earth's space in the real time. For data analysis, the models of the space environment, working in off-line as well as in on-line mode, have been implemented. Interactive services allow one to retrieve and analyze data in a given time moment. Space Monitoring Data Center (SMDC) of SINP MSU [smdc.sinp.msu.ru] also uses the data from these spacecrafts, and, in addition, it uses data from Russian satellites, Electro-L, Meteor-M and Lomonosov. All centers provide access to satellite data, space weather models, and forecasts online through their websites. Data are available through either an FTP server or an HTTP server. In this paper we will give a short overview of SWX, its data sources and operational services.

2. Interactive Data Analysis

The analysis of data measurement is the most important stage of the near-Earth space investigation. The study of various processes in the Sun, in the solar wind and in the Earth's magnetosphere requires different data for different time periods. Data sets often overlap, sometimes it is difficult to determine what data are required for the current scientific research. The service presented at this page allows the researcher to select the necessary data for his analysis, to preview it in graphical representation for subsequent calls to the database of online space monitoring.

Operational monitor and forecast of the Earth's radiation environment is very topical both for solving fundamental scientific problems of solar-terrestrial physics, and for providing safety of space missions and polar aviation. Therefore, data of experiments on board LEO (low-altitudes polar) spacecraft are very important. Now, a lot of data of experiments are available, including measurements of LEO spacecraft like Meteor-M No. 1 and POES NOAA series. The latest Russian satellites - RELEC, Meteor-M No. 2, Lomonosov were also launched to LEO orbit. However, data transmitted from LEO spacecraft have peculiarities connected with the features of LEO orbit: a spacecraft

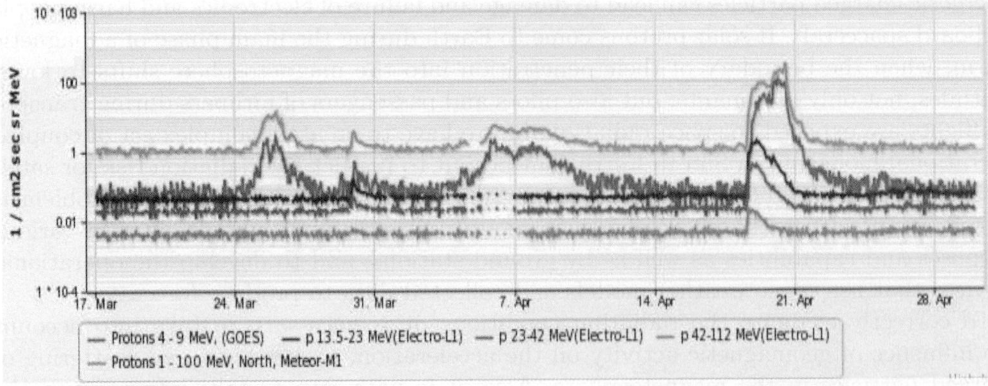

Figure 1. Time dependencies of SEP proton flux measured in experiments on board the Meteor-M No. 1, GOES and Electro-L1 satellites from March, 24 to April, 24 2014

consistently passes different areas of near-Earth space - polar caps, area of outer Earth's Radiation Belts (ERB), middle latitudes, inner ERB. No public systems intended for analysis of radiation conditions at low altitudes, which could allow quick comparison of data obtained in L1 point with those from LEO and GEO, were created until now.

The automatic system of separation of LEO-satellite data developed in SINP MSU permits to carry out comparative analysis of variations of SEP flux in polar caps and particles of the outer ERB at low altitudes with ones measured at GEO, and between different LEO experiments data also. The result of such comparison of SEP flux measured at LEO in polar caps (after automatic separation) with ones measured at GEO during one month of 2014 is shown in Figure 1.

The storage of information managed by Oracle's high-performance database provides the ability to perform intelligent processing and fast analysis of incoming information. There are the following threshold values for the key space environment factors corresponding to the values used in NOAA's Space Weather Center, the excess of which indicates the development of potentially dangerous processes and phenomena in the near-Earth space: 1) The solar energetic proton flux with energies > 10 MeV exceeds 10 particles / (cm^2-s-sr); 2) The flux of electrons of outer Earth's belt with energies > 2 MeV exceeds 10^3 particles / (cm^2-s-sr); 3) Flare soft X-ray (GOES data) index exceeds the level of M5.0; 4) Geomagnetic activity is characterized by the values of Dst <-50 nT and / or Kp>5. The warning system carries out continuous comparison of measurement data with specific threshold values and generates bulletins with reports on changes in the radiation and geomagnetic situation. The current state of the space radiation environment is presented in the bulletin [swx.sinp.msu.ru/bulletin.html].

3. Space weather forecast problem

The main sources of disturbances of the Earth's magnetosphere are the coronal mass ejections, reaching the Earth's orbit, and high-speed streams of solar wind (SW). A necessary (and perhaps a sufficient) condition for the emergence of magnetic storms is the presence of the south (negative) z-component of the interplanetary magnetic field (IMF Bz), which enables the transfer of energy from the SW into the Earth's magnetosphere. So, for short-term prediction it is especially important to have operative information about the values of SW and IMF parameters. The authors of this paper have their experience of the Dst index prediction using scientific models based on artificial neural

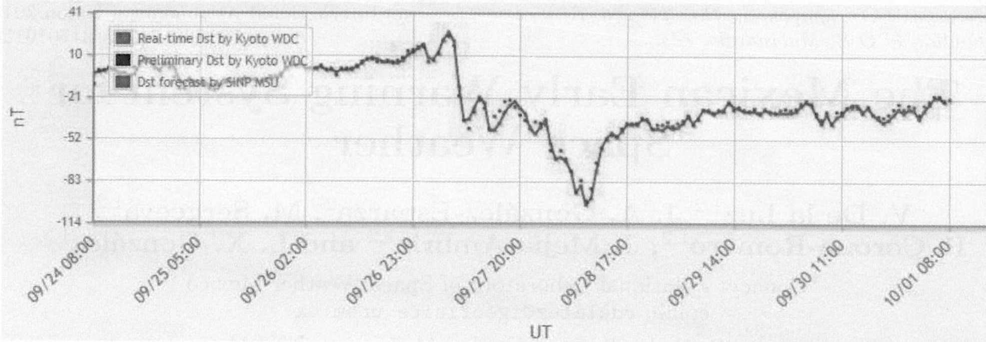

Figure 2. The short-term forecasting of Dst-variations obtained in real time during September 2017

networks (ANN) technology, as well as experience of using adaptive (including ANN) methods for prediction of electrons flux in the outer radiation belt of the Earth (Efitorov *et al.* 2016). It was shown that the best quality of the Dst-index forecast is achieved when constructing a neural network model that uses both the history of the Dst index and the parameters of the solar wind (its velocity and density) and IMF (modulus and Bz component) as input data. In Figure 2, results of the short-term forecasting of Dst variation during September 2017 in real time using ANN are presented - forecast shown in green points, experimental data - as the red curve.

The empirical model based on CORONAS-F and the "University-Tatiana" data obtained during magnetic storms in 2001 - 2006, was created. It allows one to estimate the position of the extreme value (minimum invariant latitude) of the boundary of SEP protons penetration into the Earth's magnetosphere in the evening and night-time sector (Myagkova *et al.* 2009). On this basis and using the ANN forecast of Dst and Kp index, the operation model for the realization of short-term (1 hour ahead) forecast of the extreme position of the SEP penetration boundary was created.

4. Summary

The new methodological concept was developed of automatic processing of the experimental data of spacecraft at LEO (circular polar orbit at the low altitude), which is based on separation of the experimental data in accordance with the physical areas passed by the satellite as it moves along its orbit, and thereby with the origin of energetic charged particles, which are detected at these parts of orbits. To provide short-term forecasting of the Earth's magnetosphere state - the geomagnetic indexes and the flux of relativistic electrons in the outer ERB - we used artificial neural networks. The directions of research to increase the horizon and to improve the quality of the prediction are considered.

This study was performed at the expense of the Russian Science Foundation, grant no.16-17-00098.

References

Cole, D. G. 2003, *Space Sci. Rev.* 107, 295

Efitorov, A., Myagkova, I., Sentemova, N., Shiroky, V. & Dolenko, S. 2016, *Adv. Int. Sys. Comp.* 449, 281

Myagkova, I., Bogomolov, A., Yushkov, B. & Kudela, K. 2009, *Bull. RAS* 73, 322

Space Weather of the Heliosphere: Processes and Forecasts
Proceedings IAU Symposium No. 335, 2017
C. Foullon & O.E. Malandraki, eds.

doi:10.1017/S1743921317010158

The Mexican Early Warning System for Space Weather

V. De la Luz[1,2] **J. A. González-Esparza**[2], **M. Sergeeva**[1,2]
P. Corona-Romero[1,2], **J. Mejía-Ambriz**[1,2] **and L. X. González**[1,2]

[1]Conacyt - National Laboratory of Space Weather Mexico
email: vdelaluz@igeofisica.unam.mx

[2]Instituto de Geofisica Unidad Michoacan, UNAM

Abstract. Early Warning of Space Weather phenomena is one of the most important products produced by Mexican Space Weather Service (SCiESMEX). The aim is to deliver a verified warning to Mexican National Center for Disaster Prevention (CENAPRED) as well as to general public in near-real time automatically. First, the international pubic warning (or alert) is produced by Space Weather Prediction Center of National Oceanic and Atmospheric Administration, U.S.A. Further, the alert is received by SCiESMEX system, decodified, translated in Spanish and put in the context of local current conditions. The alerts are transformed and validated in agreement to Mexican policies of civil protection and CENAPRED protocols. The automatic system segments the information and delivers the corresponding web- page-alerts, emails, twits, facebook messages and SMS. The latency of the alert is 5 minutes. The code is running since January of 2015 without major interruptions.

Keywords. Sun: activity, Sun: solar-terrestrial relations, solar system: general

1. Introduction

A complex infrastructure of radio communications and energy generation and distribution has been developed in Mexico during the last decades. The transmission system of the National Electricity Company (in Spanish CFE, Comisión Federal de Electricidad) has a capacity of 163,571.7 megavolts ampere (MVA) in 475 power substations and 57,453 km of transmission lines. Additionally, the CFE has in bailment 24,897 MVA with a length of 1,632 km in transmission lines. The CFE distribution system has a capacity of 54,624 MVA with 1,910 power substations and 500,000 km in high and medium voltage lines. According to CFE data, there were 1,380,589 operating electric transformers in 2015 (CFE, 2015). The national capacity for electricity generation is approximately 54,374.7 megawatts (MW) with 1,081 generation units divided between five regions: Northwest (15%), North (17%), West (27%), Central (16%), and Southeast (24%). At the same time, the Mexican Oil Company (in Spanish PEMEX) has 4,100 km of pipelines for gas transportation in order to move 7,000 million of cubic feets of gas per day. The Ministry of Communications and Transportation (SCT) and the Federal Institute of Telecommunications (IFT) are responsible for the quality of radio communications. The study of the use of the electromagnetic spectrum in Mexico performed by SCT (CFT, 2013) showed that there were more than 30 different users operating between 30 MHz and 300 GHz including the following: fixed and mobile services of radio communications for voice and data, maritime, aeronautic in route, mobile by satellite, broadcast, amateur radio, astronomy, Space Weather, aeronautic radio navigation, meteorology, radio localization, space operations, vehicle radio localization, media broadcast,

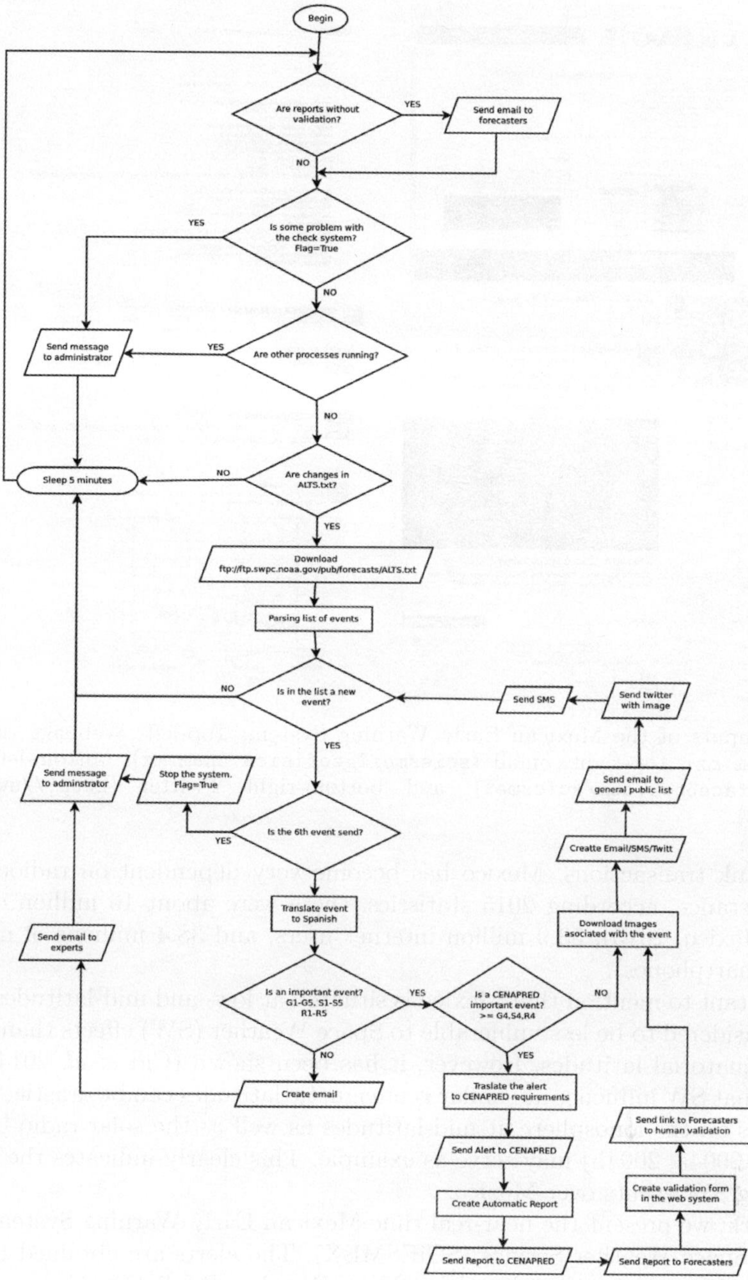

Figure 1. Flow chart of the Mexican Early Warning System developed by SCiESMEX - LANCE.

media links, AM and FM radio, mobile no geostationary, digital broadcast, GPS, horary signals and pattern frequencies for calibration, national system for public security, internet, satellite, tv satellite, point-to-point microwave, radio broadcast by satellite, governmental satellite services, and armed forces radio communications services. There are 62,996 registered dealers operating between 30 MHz and 3 GHz, mainly focused in the range of (148-173) MHz. The Bank of Mexico is the institution that supervises the

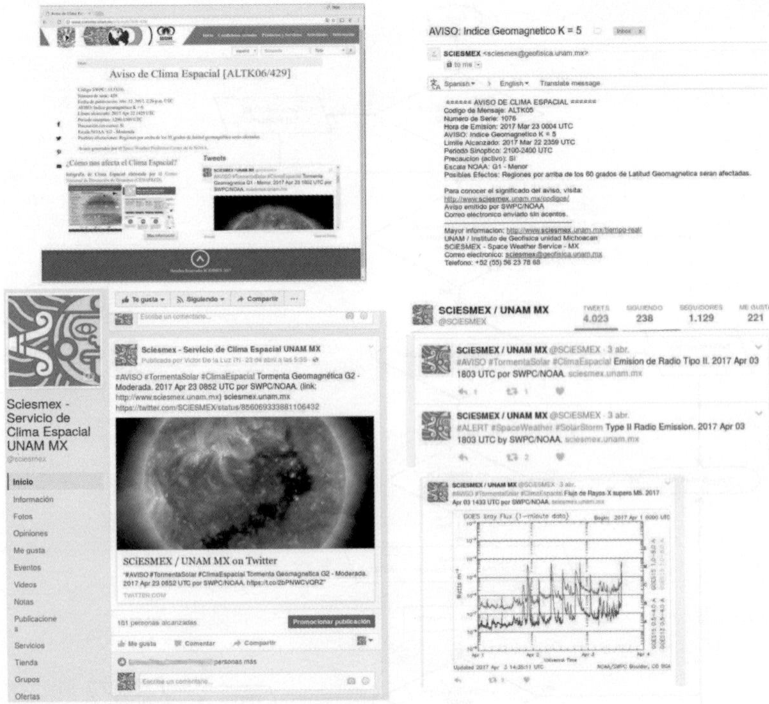

Figure 2. Ouputs of the Mexican Early Warning System: Top-left, Webpage (`http://www.sciesmex.unam.mx`); top-right, email (`sciesmex@geofisica.unam.mx`), bottom-left, Facebook (`http://www.facebook.com/sciesmex`); and bottom-right Twitter (`http://www.twitter.com/sciesmex`).

electronic bank transactions. Mexico has become very dependent on radiocommunications. For instance, according 2015 statistics, there were about 16 million credit cards (Banco de Mexico, 2015), 53.9 million internet users, and 38.4 millions of mobile users (58% uses smartphones).

It is important to mention that Mexico is situated at low- and mid-latitudes, which are generally considered to be less vulnerable to Space Weather (SW) effects than the regions of high or equatorial latitudes. However, it has been shown (Cid *et al.* 2014, Yizengaw *et al.* 2013) that SW influence at similar geomagnetic latitudes can be drastic. The impact of solar flares on the ionosphere at mid-latitudes as well as the solar radio burst effects (Balch *et al.* 2004a, 2004b) may serve as example. This clearly indicates the importance of monitoring SW events over Mexico.

In this work, we present the near-real time Mexican Early Warning System developed by Mexican Space Weather Service (SCiESMEX). The alerts are obtained from the international pubic warnings produced by Space Weather Prediction Center of National Oceanic and Atmospheric Administration, U.S.A. (SWPC, NOAA), and adapted automatically. Then the warning messages are send by email, social media (Facebook and Twitter), SCiESMEX webpage, and SMS network.

2. Flowchart

Figure 1 illustrates the flowchart of the system. The input for the code are the SW warning messages from SWPC. The system downloads the alerts each 5 minutes and parses the contents, in case of new alert was released. Further, the message is automatically traslated into Spanish. The following step is to check if the particular alert is important to general public. The NOAA scales for SW (G1-G5, S1-S5, or R1-R5) are used for reference. If negative, a warning email is generated and is sent only to the group of SW experts. Otherwise, the alert is important for National Center to Prevent Natural Hazards (CENAPRED in Spanish). In this case the information is sent directly to CENAPRED and a brief note for general public is prepared. The messages are sent via Facebook, Twitter, Email, SMS and SCiESMEX webpage. For safety purposes, if there are more than 5 messages within a 5 minute period, the automated sending system is paused until human check is performed.

3. Conclusion

Figure 2 presents an example of results produced by the warning system for four platforms: Webpage (`http://www.sciesmex.unam.mx`), email (`sciesmex@geofisica.unam.mx`), Facebook (`http://www.facebook.com/sciesmex`), and Twitter (`http://www.twitter.com/sciesmex`).

The System of Early Warning of Space Weather phenomena in Mexico has been operational without interruption since the end od 2014. The direct connection to Twitter and Facebook helps to increase significantly the number of warning recipients Warning recipients. This is an important issue as the objectives of SCiESMEX as a part of National SW Laboratory (LANCE) include the informing the general public about the current SW state over Mexico. The project is under development and we are currently working on second step human validation, automatic report generation, a double check for downloaded file from SWPC, extensive stress tests, and ISO 9001 Certification.

4. Acknowledgements

The authors thank Catedras Conacyt (Conacyt Fellow) for supporting this work. V. De la Luz acknowledges Conacyt-254497 for Ciencia Basica and CONACYT-268273 for Repositorios Institucionales projects. SCiESMEX is partial supported by CONACyT-AEM 2014-01-247722, CONACyT LN 279764 y UNAM-PAPIIT IN106916.

References

Balch, C., Biesecker, D., Combs, L., Crown, M., Doggett, K., Kunches, J., Singer, H. & Zezula, D. 2004a, Halloween space weather storms of 2003. Technical report, Tech. Memo. OAR SEC

Balch, C., Murtagh, B., Zezula, D., Combs, L., Nelson, G., Tegnell, K., Crown, M. & McGehan, B. 2004b, Intense space weather storms october 19–november 07, 2003, service assessment. *NOAA, US Dep. of Commerce, Silver Spring, MD*

Banco de Mexico (2015). Indicadores basicos de tarjeta de credito. Technical Report 1, Banco de Mexico

Cid, C., Palacios, J., Saiz, E., Guerrero, A. & Cerrato, Y. 2014, *Journal of Space Weather and Space Climate*, 4(27):A28

Comision Federal de Electricidad (2015). Informe anual 2014. Technical Report 1, Instituto Federal de Electricidad

Comision Federal de Telecomunicaciones (2013). El espectro radioelectrico en mexico. estudios y acciones. Technical Report 1, Instituto Federal de Telecomunicaciones

Yizengaw, E., Doherty, P. & Fuller-Rowell, T. 2013, *Space Weather*, 11(7):389–391

(Top) 24 of the 28 participants to the young scientist poster competition outside of Reed Hall, main venue for the poster program, with A. Hillier coordinating (center). (Bottom) 5 worthy winners were celebrated during the conference dinner at the Imperial Hotel, Torquay: from left to right, R. Chiba, N. Hussain, V. Pant, S. Mahajan and M. Korsos; with A. Hillier and C. Foullon standing behind and to the right authors of book prizes A. Choudhuri, J. Luhmann and C. Russell.

Part 7. Forecasting Models

Space Weather of the Heliosphere: Processes and Forecasts
Proceedings IAU Symposium No. 335, 2017
C. Foullon & O.E. Malandraki, eds.

doi:10.1017/S1743921318000534

Numerical Short-Term Solar Activity Forecasting

Huaning Wang[1,2], Yihua Yan[1,2], Han He[1], Xin Huang[1], Xinghua Dai[1], Xiaoshuai Zhu[1], Zhanle Du[1], Hui Zhao [1] and Yan Yan[1]

[1]Key Laboratory of Solar Activity, National Astronomical Observatories, Chinese Academy of Sciences,
A20 Datun Road Chaoyang District, Beijing 100012, China
email: hnwang@bao.ac.cn
[2]University of Chinese Academy of Sciences

Abstract. It is well known that the energy for solar eruptions comes from magnetic fields in solar active regions. Magnetic energy storage and dissipation are regarded as important physical processes in the solar corona. With incomplete theoretical modeling for eruptions in the solar atmosphere, activity forecasting is mainly supported with statistical models. Solar observations with high temporal and spatial resolution continuously from space well describe the evolution of activities in the solar atmosphere, and combined with three dimensional reconstruction of solar magnetic fields, makes numerical short-term (within hours to days) solar activity forecasting possible. In the current report, we propose the erupting frequency and main attack direction of solar eruptions as new forecasts and present the prospects for numerical short-term solar activity forecasting based on the magnetic topological framework in solar active regions.

Keywords. solar activity, magnetic field, forecast

1. Introduction

Solar eruptions have been revealed to be disturbances propagating through interplanetary space, resulting in space weather events in nearby the Earth, which are dangerous to high-tech systems including aerospace, communication, power grids etc. Short-term solar activity forecasting models provide users products described with probability of solar eruptions within hours to days. Conventional forecasting models are mostly based on the statistical relation between solar eruptions and measures from observations. The most popular measures are obtained through the classification of sunspot groups (e.g. McIntosh 1990). In the era with only simple optical imaging solar telescope, the classification of sunspot groups was the best way to forecast solar activity. With the development of solar instrumentation, multi-band solar observations with high spatial and temporal resolution have been realized in the ground and the space since 1950s. The vector magnetic field measurement of the solar surface provides us with a numerical boundary to understand MHD processes in the solar atmosphere. It is reasonable that some of parameters derived from observational data have been taken as measures of solar activity forecasting models supported with statistical method or artificial intelligence technology (Falconer *et al.* 2003; Cui, *et al.* 2006; Wang *et al.* 2008; Yu *et al.* 2009; Bobra & Couvidat 2015; Liu *et al.* 2017) The current products of short-term solar activity forecasting models, however, can hardly meet the demand of our users from different fields. Wang (1998) first proposed a concept of solar magnetic synoptic meteorology, in which magnetic interfaces among independent flux loop systems are regarded as frontier surfaces where violent solar events frequently happen. With incomplete theoretical modeling for eruptions in the solar atmosphere, dynamic processes at the magnetic interface cannot be easily described with MHD equations, although many theoretical works have been done (Chen 2011; Amari *et al.* 2014; Jiang *et al.* 2016). On the other hand, the "big data" from ground and space solar observations stimulate us to begin our work on numerical short-term solar activity forecasting. For this reason, the current paper is structured as follows: a new concept of solar activity forecasting is proposed in the second section; solar magnetic topological framework is introduced in the third; the analysis of two cases is presented in the fourth section, before a final summary.

2. New Concept of Solar Activity Forecasting

Disturbances caused by solar eruptions propagate through interplanetary space, and create shock waves in the solar wind interacting with the magnetosphere. This means that there exists two problems in forecasting solar eruption: (1) When and how will it happen? (burstiness); (2) Will it have an effect on the Earth? (harmfulness). The burstiness means that the solar eruption is regarded as a phenomenon of large scale free energy release from solar magnetic field, which is always triggered by some unexpected internal or external disturbances. This is the reason why solar eruptions are hard to predict. The harmfulness means that the solar magnetic field and it's rotation make the environment of interplanetary space complicated. The current propagation model of solar disturbances suffers from uncertainties in the upper boundary at $2.5R_{\odot}$ and the observational characteristic of energetic particles is related to solar rotation, shock, IMF, and orbit motion of the Earth (Reames *et al.* 1996).

It is obvious that solar eruptions impacting on the Earth cannot be fully provided with a normal solar activity forecast presented in probability. We suggest that the erupting frequency and the main attack direction of solar eruptions should be included in forecasts. According to historical records, a group of solar eruptions is more dangerous than occasional one. In fact all major solar storms originated in super active regions, which had multiple eruptions, such as 1859's Carrington event, 1989's great solar storm, 2000's Bastille day event, 2003's Halloween event. The erupting frequency is defined as a number of solar eruptions during a specific forecasting time window, which can be considered as a parameter of solar active regions. The topology of local and global magnetic fields in the solar corona plays an important role in determination of solar eruption's behaviors and characteristics. In order to provide our users with new forecasts concerning the erupting frequency and the main attack direction of solar eruption, we need numerical reconstruction of 3D magnetic field in the corona. The domain for the numerical reconstruction is currently confined in solar active regions, because a global magnetic field reconstruction requires real time global solar magnetic field observations at different sites in deep space.

3. Solar Magnetic Topological Framework

Taking vector magnetic fields on the solar surface as boundaries, we can reconstruct solar coronal magnetic fields with different models. Here we employ a non-linear force-free field (NLFFF) model proposed by Yan and Sakurai (2000). The coronal field is reconstructed with boundary integration based on a Green's function:

$$Y(\lambda, r) = \frac{cos(\lambda r)}{4\pi r}, \tag{3.1}$$

where the parameter λ is determined with a best fitting algorithm for a specific active region, and then the 180 degree ambiguity is also removed with a reference field in linear

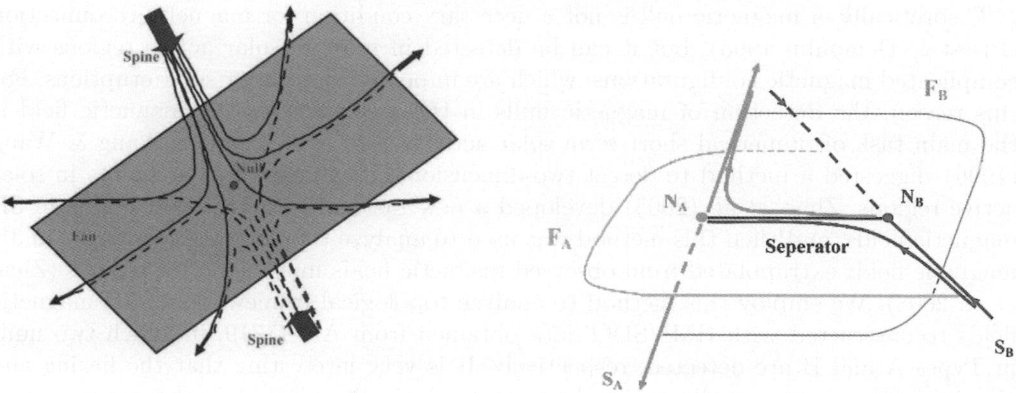

Figure 1. Topological features of Type B null (left) and topological connections between nulls in Types A and B(right)

force-free field (LFFF) model suitable for this parameter (Wang *et al.* 2001). We take a solar active region with NOAA number 11719 (AR 11719) as an example. The vector magnetic field observed with the HMI/SDO instrument at 07:16:00UT on April 11, 2013 is taken as the boundary. When the value of λ is well determined, a reasonable alignment between the reconstructed 3D field and the EUV features will be reached. The best fitting value of λ is about $-9.0 \times 10^{-7} m^{-1}$, which means that the dominant magnetic flux in this active region has negative force-free factor.

It has been widely accepted that the magnetic reconnection is regarded as an important process of magnetic energy release for solar flares since Sweet (1958) proposed a model concerning the neutral point theory of solar flares, in which the topology of solar magnetic field is inevitably changed. Up to now several magnetic configurations beneficial to the reconnection have been investigated in detail (Longcope 2005). Null points in solar magnetic fields are regarded as a kind of important configurations associated with the reconnection (Lau & Finn 1990; Priest *et al.* 1997). Near a null point located at $\mathbf{x} = 0$, the magnetic field $\mathbf{B}$ can be given with Taylor expansion,

$$\mathbf{B}(\mathbf{x}) = \delta\mathbf{B} \cdot \mathbf{x}, \tag{3.2}$$

where $\delta\mathbf{B}$ is a real matrix with 3×3 elements. The topological property of a magnetic null is determined by eigenvalues of $\delta\mathbf{B}$. Here we only consider the situation that all eigenvalues are real because our reconstructed solar corona magnetic field is derived from a NFFF model (Lau & Finn 1990). Since the trace of this matrix is zero, the sign distribution of three eigenvalues has two types: one is $(+--)$, which we will call "Type A", and the other $(-++)$, *i.e* Type B. The left panel of Figure 1 shows the topological framework of Type B, while that of Type A has inverse field line directions. The right of Figure 1 demonstrates topological connections between two types of nulls, in which topological components of nulls, such as spine and fan, are indicated. It should be note that the intersection between fans F_A and F_B, namely separator, is the most important interface where solar events might be initiated due to magnetic reconnections. Here the magnetic topological framework is defined as a set including topological features of nulls and their connections. Magnetic fields extrapolated from solar data have quite different topological frameworks, which would be useful to determine the erupting frequency and the main attack direction of solar eruptions.

Theoretically, a magnetic null is not a necessary condition for magnetic reconnection (Priest & Démoulin 1995), but it can be detected in most of solar active regions with complicated magnetic configurations, which are important for major solar eruptions. For this reason, the detection of magnetic nulls in the reconstructed 3D magnetic field is the main task of numerical short-term solar activity forecasting model. Wang & Wang (1996) suggested a method to detect two-dimensional magnetic singular points in solar active regions. Zhao *et al.* (2005) developed a new method to detect null points in 3D magnetic fields, and then this method was used to analyze topological frameworks in 3D magnetic fields extrapolated from observed magnetic fields in some active regions (Zhao *et al.* 2008). We employ this method to analyze topological frameworks of 3D magnetic fields reconstructed with HMI/SDO data obtained from AR 11719, in which two nulls in Types A and B are detected, respectively. It is very interesting that the flaring and ejecting processes are exactly confined by the topological framework in this active region. An enhanced study for this active region will be done and relevant results will be reported in a separated paper.

4. Case Analyses

It is widely believed that the free energy building in solar active regions is driven by the evolution of photospheric magnetic field, and then released by magnetic reconnection in separators of topological frameworks. Although how solar eruptions are triggered by has not been well described with modern solar physics, how topological frameworks effect solar eruptions can be understood. Separators in topological frameworks can be located with numerical computation, and some hot points related to these separators can also be found by multi-band solar observations. The information concerning the complexity of separators and the location of hot points can tell us the erupting frequency and which parts of topological frameworks will get involved in the eruptions. We show two active regions as our case analyses.

The first case is AR 11158 from February 14, 2011, when this region appeared near the center of solar disk. The data set includes SDO/HMI vector magnetograms; SDO/AIA 171Å images; SoHO/Lasco images and GOES 15 soft X-ray flux. Table 1 is obtained from observational data of GOES 15 and SoHO/Lasco during 00h-20h(UT), in which nine C or M flares and five CMEs are listed. The second CME might be originated from other locations because it's central position angle (CPA) is quite different from that of others. Sun *et al.* (2012) selected one of several homologous non-radial eruptions to study how these eruptions were modulated by the local 3D magnetic field in which a null was detected at about 9Mm above the solar surface. In the current work, we employ the model suggested by Yan and Sakurai (2000) for the coronal field reconstruction, the tool proposed by Zhao *et al.* (2005) for the topological framework analysis. There exist two nulls in the topological framework shown in Figure 2. One of them is the same as that obtained by Sun *et al.* (2012), while another one might be ignored as it has a height over 80Mm. A separator is detected within a height range of (9Mm, 90Mm). We select a small sector area ($\theta = 6$ deg, $r = 200$Mm) along the direction of outflow to construct a space-time diagram, in which 400 AIA 171Å images with 3min interval are used (see left panel of Figure 3). It is interesting that a stable point is detected, which is probably the intersection between the sector and the separator in the transition region, and variations of EUV intensity well correspond to all of the eruptions originated in AR11158 (see right panel of Figure 3). Sun *et al.* (2012) believe that homologous non-radial eruptions was only caused by the null at the lower height, but we strongly suggest that the topological framework provides a robust platform for these eruptions. Actually the flux systems involved in eruptions are

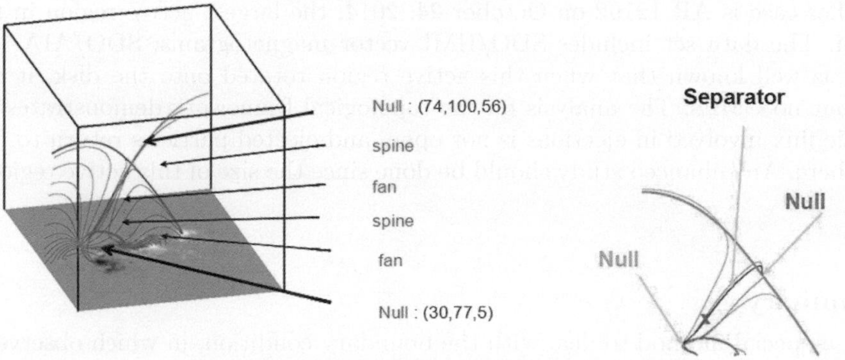

Figure 2. Topological frameworks in AR11158. (Left) All components of the topological framework, 1 unit=1.5 Mm; (Right) Separator connecting nulls in Types A and B

date	begin	end	peak	class
20110214	0235	0246	0242	C1
20110214	0429	0509	0449	C8
20110214	0651	0703	0658	C6
20110214	0839	0904	0849	C1
20110214	1151	1226	1200	C1
20110214	1241	1258	1253	C9
20110214	1347	1442	1427	C7
20110214	1720	1732	1726	M2
20110214	1923	1936	1930	C6

CME Number	Begin Time	Central PA(deg)	Angular width(deg)	Linear fit (km/s)	Accel (m/s^2)
1	08:12:06	288	057	303	06.3
2	11:00:05	051	007	217	07.4
3	14:00:07	306	022	380	-23.3
4	17:24:05	263	038	229	-08.5
5	18:24:05	Halo	360	326	04.6

Table 1. Solar flares (left) and CMEs (right) during 00h-20h(UT)of February 14,2011

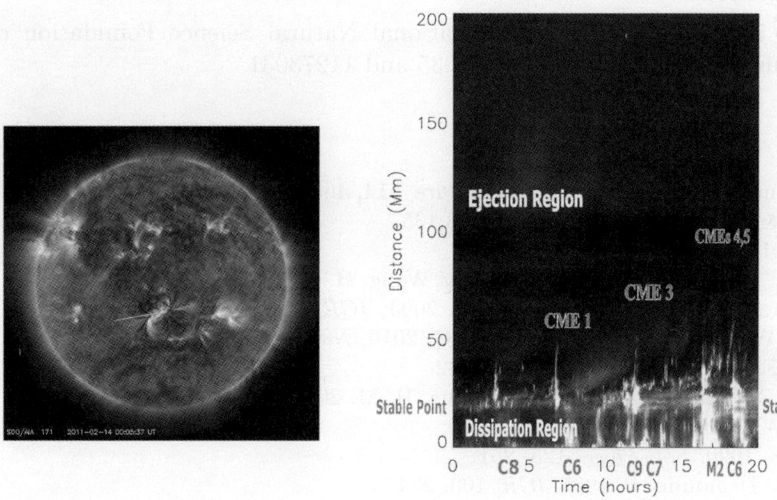

Figure 3. AIA 171Å image(left) and space-time diagram from a small sector area of the AIA images (right)

non-radial. This case analysis demonstrates that the topological framework determines the erupting frequency and main attack direction of solar eruptions, and plays a role like the frontier surface where violent solar events frequently happen.

Another case is AR 12192 on October 24, 2014, the largest active region in the solar cycle 24. The data set includes SDO/HMI vector magnetograms; SDO/AIA 171Å images. It is well known that when this active region rotated onto the disk, it only had flares, but no CMEs. The analysis on the topological framework demonstrates that the magnetic flux involved in ejections is not open, and ejected particles return to the lower atmosphere. An enhanced study should be done since the size of this active region is very large.

5. Summary

With a special method to deal with the boundary condition, in which observed vector fields are incorporated as part of the bottom boundary conditions consistently, we are able to reconstruct 3D magnetic fields in the corona as force-free fields and then determine the topological framework. The analyses on two cases demonstrate that the magnetic topological framework in solar active regions plays a role like the frontier surface where violent solar events frequently happen. This situation could be termed as solar magnetic synoptic meteorology. The erupting frequency and the main attack direction of solar eruptions can be forecasted according to properties of topological features obtained from numerical computation. The numerical short-term solar activity forecasting provides our users new products.

In order to improve the numerical short-term solar activity forecasting, the following points should be emphasized: improved models for coronal magnetic field reconstruction; reliable tools for magnetic topological analysis; improved solutions to the MHD equations.

Acknowledgement

This work is jointly supported by National Natural Science Foundation of China (NSFC) through grants 11403044, 11673035 and 11273031

References

Amari, T., Canou, A. & Aly, J. J. 2014, *Nature*, 514, 465
Bobra, M. G. & Couvidat, S. 2015, *ApJ*, 798,135
Chen, P. F. 2011 *LRSP*, 8, 1C
Cui, Y. M., Li, R., Zhang, L. Y., He, Y. L. & Wang, H. N. 2006, *Sol. Phys.*, 237, 45
Falconer, D. A., Moore, R. L. & Gary, G. A. 2003, *JGR*, 108, 1380
Jiang, C. W., Wu, S. T., Feng, X. S. & Hu, Q. 2016, *NatCo*, 7, 111522
Lau, Y. T. & Finn, J. M. 1990, *ApJ*, 350, 672
Liu, C., Deng, N., Wang, Jason T. L. & Wang, H. M. 2017, *ApJ*, 843, 104L
Longcope,D. W. 2005, *LRSP*, 7, 2C
McIntosh, P. S. 1990, *Sol. Phys.*, 125, 251
Priest, E. R. & Démoulin, P. 1995 *JGR*, 100, 23443
Priest, E. R., Bungey, T. N. & Titov, V. S. 1997, *Geophys. Astrophys. Fluid Dyn.*, 84, 127
Reames, D. V., Barbier L. M. & Ng C. K. M. 1996, *ApJ*, 466, 473
Sun, X. D., Hoeksema, J. T., Liu, Y., Chen, Q. R. & Hayashi, K. J. 2012, *ApJ*, 757, 149
Sweet, P. A., 1958, in Electromagnetic Phenomena in Cosmical Physics, (Ed.) Lehnert, B. & Cambridge University Press, Cambridge, U. K. 1958, *IAUS*, 6, 123
Wang, H. N. & Wang, J. X. 1996, *A&A*, 313, 285
Wang, H. N., Yan, Y. H. & Sakurai, T. 2001, *Sol. Phys.*, 201, 323
Wang, H. N., Cui, Y. M., Li, R., Zhang, L. & He, H. 2008, *AdSpR*, 42, 1464

Wang, J. X., in Synoptic Solar Physics, eds Balasubramaniam, K. S., Harvey, J. W. & Rabin, D. M. 1998, *ASPC*, 140, 121

Yan, Y. H. & Sakurai, T. 2000, *Sol. Phys.*, 195, 89

Yu, D., Huang, X., Wang, H. N. & Cui, Y. 2009, *Sol. Phys.*, 255, 91

Zhao, H., Wang, J. X., Zhang, J. & Xiao, C. J. 2005, *ChJAA*, 5, 443

Zhao, H., Wang, J. X., Zhang, J. & Xiao, C. J. 2008, *ChJAA*, 8, 133

Space Weather of the Heliosphere: Processes and Forecasts
Proceedings IAU Symposium No. 335, 2017
C. Foullon & O.E. Malandraki, eds.

doi:10.1017/S1743921317007244

Sandpile Models and Solar Flares: Eigenfunction Decomposition for Data Assimilation

Antoine Strugarek[1,2], Allan S. Brun[1], Paul Charbonneau[2] and Nicole Vilmer[3]

[1]Laboratoire AIM Paris-Saclay, CEA/Irfu Université Paris-Diderot CNRS/INSU, F- 91191 Gif-sur-Yvette
email: antoine.strugarek@cea.fr

[2]Département de physique, Université de Montréal, C.P. 6128 Succ. Centre-Ville, Montréal, QC H3C-3J7, Canada

[3]LESIA, Observatoire Paris, CNRS, UPMC, Universite Paris-Diderot, 5 place Jules Janssen, 92195 Meudon, France

Abstract. The largest solar flares, of class X and above, are often associated with strong energetic particle acceleration. Based on the self-similar distribution of solar flares, self-organized criticality models such as sandpiles can be used to successfully reproduce their statistics. However, predicting strong (and rare) solar flares turns out to be a significant challenge. We build here on an original idea based on the combination of minimalistic flare models (sandpiles) and modern data assimilation techniques (4DVar) to predict large solar flares. We discuss how to represent a sandpile model over a reduced set of eigenfunctions to improve the efficiency of the data assimilation technique. This improvement is model-independent and continues to pave the way towards efficient near real-time solutions for predicting solar flares.

Keywords. Sun: flares

1. Introduction

Solar eruptive phenomena have multiple impacts on the heliosphere and on the environment around our Earth (e.g. Schwenn 2006; Pulkkinen 2007). Along with Coronal Mass Ejections (CMEs), the solar flares are one of the most studied space weather phenomenon and can be accompanied by the acceleration of highly energetic particles. The energy and duration of solar flares have been shown to be distributed as power-laws (for a review, see Aschwanden *et al.* 2016), spanning more than eight orders of magnitude in energy. This striking self-organization (Charbonneau 2013) over the whole corona have led multiple authors to use basic, toy models such as sandpiles (e.g. Lu & Hamilton 1991) to model the statistical distribution of flares.

In the context of space weather, and also to better constrain the physical mechanisms behind flares, it is thus interesting to try to predict their occurence. Such a task was recently shown to be very difficult (Barnes *et al.* 2016), as all of the modern techniques developed to predict flares most often fail. Alternative methods and ideas are thus needed today to explore other paths to provide robust predictions of solar flares. We build here on an idea first developed by Bélanger *et al.* (2007); Strugarek & Charbonneau (2014) based on the combination of simple sandpile models reproducing the solar statistical distribution (Strugarek *et al.* 2014), and modern data assimilation techniques using the solar X-ray flux observed with the GOES satellites.

The aim of this proceeding is to report on an attempt to facilitate the data assimilation algorithm by reducing size of the state vector of the model. This is achieved by projecting

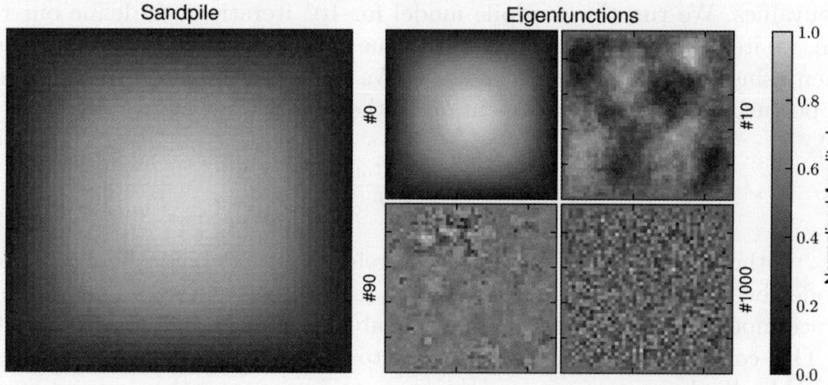

Figure 1. *Left.* 2D Sandpile model. *Right.* Four representative eigenfunctions of the sandpile.

the model itself on a basis of eigenfunctions (§ 2), which we couple to the data assimilation package (§ 3) that was originally developed in Strugarek & Charbonneau (2014). We conclude and lay out the perspective of this preliminary work in § 4.

2. Deterministically-driven sandpile and eigenvalue decomposition

We base our analysis on the deterministically-driven sandpile model developed by Strugarek *et al.* (2014). However, the procedure described here is independent of the model and could be applied to other types of modelling (e.g. Hung *et al.* 2017).

The aim of data assimilation is to adjust a set of control parameters (a 'state vector') of the model to get an optimal fit to observed (or synthetic) data. One straightforward strategy is to consider the initial condition of the model as this state vector. The advantage of this approach is that we ensure that we do not tune the model itself (*i.e.* the model retains its statistical properties no matter what we do to the set of control parameters), but we rather try to find a moment in the evolution of the model that fits the observational data.

One big disadvantage of this approach lies in the size of the set of control parameters. Indeed, the size of our set will be N^2 (N being the number of points in our 2D sandpile model), which will become quickly numerically prohibitive for most minimization techniques as N gets large. In order to mitigate this limitation, we explore here the possibility to project the model on a series of eigenfunctions to reduce the size of our state vector, e.g. by adjusting only the most relevant ones. We now describe in details our procedure.

We construct the eigenfunction basis of our model using its covariance matrix. Let us define the state of our sandpile model at time t by $\mathcal{S}(t)$, taken as a 1D vector of size N^2. We define $\mathcal{S}_0 = \mathcal{S}(t_0)$. A covariance matrix $\mathcal{C}$ of our model is a $N^2 \times N^2$ matrix that is defined as

$$\mathcal{C} = \overline{(\mathcal{S}(t) - \mathcal{S}_0)^\intercal \cdot (\mathcal{S}(t) - \mathcal{S}_0)}\,, \tag{2.1}$$

where the overline denotes a time average. We diagonalize the covariance matrix to obtain a set of N^2 eigenvectors of our model. This decomposition is illustrated in Figure 1 where we show the sandpile model itself along four representative eigenfunctions (out of 2304 in this case).

Now that we have a systematic method to derive a set of eigenfunctions of our model, we want to assess whether a subset of the eigenfunctions is enough to robustly control the large events triggered by the model in a given time window. We first choose one random state of our model $\mathcal{S}$, which we project on the eigenfunctions basis to obtain a set of eigenvalues. We run the sandpile model for 10^4 iterations to define our reference case. Then, we iteratively perturb each eigenvalue and compare the result of the sandpile model when using the perturbed initial state. We use a cost function $\mathcal{J}$ to estimate how much the perturbation affects the evolution of the sandpile, which we define as

$$\mathcal{J} = 1 - \left(\alpha \sum_{\text{match}} \frac{E_{\text{obs}}}{E_{\text{tot}}} - \beta \sum_{\text{miss}} \frac{E_{\text{obs}}}{E_{\text{tot}}} - \gamma \sum_{\text{FalseAlarm}} \frac{E_i}{E_{\text{tot}}} \right), \tag{2.2}$$

where E_{obs} is the energy of the events in the reference model, E_i the energy of events in the model being compared, $E_{\text{tot}} = \sum E_{\text{obs}}$ is the total energy of all the events in the reference model, and α, β and γ are adjustable coefficient that we set here to 1, 0.5 and 0.25. This cost function gives more weight to the large events that we most want to predict reliably. It also gives more weight to matching events, then missed events, and finally less weight to false alarms. If the model fits perfectly the reference model, $\mathcal{J} = 0$. Conversely, if none of the events fit the reference model, $\mathcal{J} \sim 1.75$ (no match, all events are considered as false alarms).

We repeat this procedure for 10 different random number sequences used in the sandpile model to ensure that the subset of eigenfunctions we are trying to isolate is robust and order the eigenfunctions in ascending order with their score $\mathcal{J}$. We now use this ordered basis of our model to perform data assimilation over synthetic data.

3. Data assimilation using the eigenvalue decomposition

The data assimilation procedure was briefly described in Strugarek & Charbonneau (2014) and can be summarized as follows.

We first construct synthetic data (upper right panel in Figure 2) by running our model over 4000 iterations and filtering events higher than 10^4 E_0 (E_0 being the energy normalization of our adimensional model, see Strugarek & Charbonneau 2014). This will be our target observation that we will try to match by finding a new initial state for our system. Then, we generate random states of our model that will be used as initial states for our data assimilation algorithm. We project these initial states on our ordered eigenfunction basis. After these initialization steps, we minimize the cost function $\mathcal{J}$ (Eq. 2.2) with a simulated annealing algorithm. The set of control parameters for this algorithm is defined by the amplitude of the first N_e eigenvalues, which we allow to evolve. The remaining $N - N_e$ eigenvalues are thus kept fixed during the assimilation procedure. We tested our algorithm for values of N_e ranging from 10 to N_{max} ($N_{\text{max}} = 2206 < N_e$ due to the boundary conditions of the sandpile forcing it to zero on its edges). The resulting final state and execution time of the data assimilation procedure are shown in Figure 2.

On the top left panel we display the final cost function after the data assimilation procedure. The blue bars denote assimilation with varying N_e, while the orange bar corresponds to a standard assimilation run for which no eigenfunction decomposition was used (thus the set of control parameters corresponds to the N^2 points in the initial sandpile, see Strugarek & Charbonneau 2014). We immediately remark that if we use too few eigenfunctions in the data assimilation procedure, the system is too constrained and we are not able to reproduce satisfactorily the reference model. When $N_e \geqslant 50$, we empirically observe that we are able to match equally or better the reference model ($\mathcal{J}$ is smaller) than the standard assimilation case, and with a much faster execution time (lower left panel). The reference model used in the data assimilation procedure is shown in the right panels in black and grey. Three realisations of the model resulting from the assimilated initial states are shown in the three other panels.

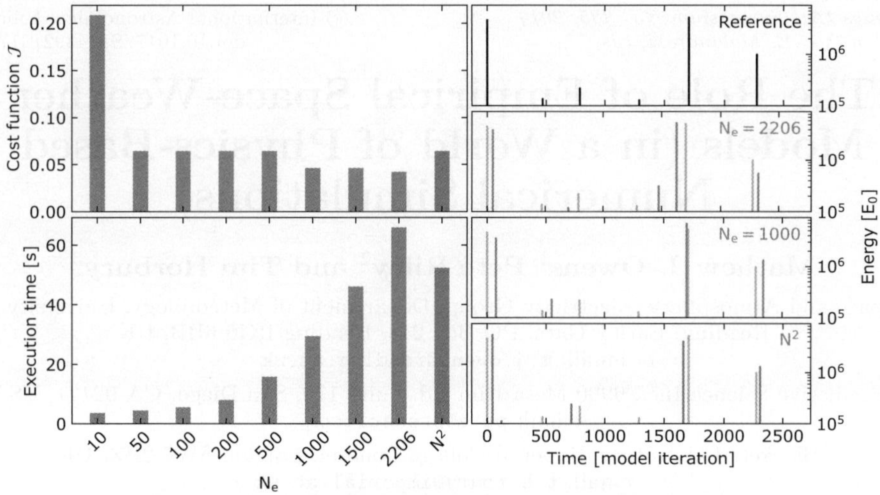

Figure 2. *Left.* Final cost function (top) and execution time (bottom) of the assimilation procedure, as a function of the number of eigenfunctions N_e involved in the assimilation. The orange bar corresponds to an asssimilation with no eigenvector projection, using all the $N^2 = 2304$ points in the sandpile. *Right.* Results from the sandpile models. The reference case being assimilated is at the top and reported in gray in all panels. The three lower panels are representative cases.

4. Conclusions

We have shown in this proceeding how to automatically decompose a sandpile model on a set of eigenfunctions. We found that the shape of the eigenfunctions described both the large-scale structure of the sandpile, as well as nested structures associated with locations in the sandpile likely to give rise to avalanches (see Figure 1).

In the context of data assimilation for predicting solar flares (Strugarek & Charbonneau 2014), this decomposition allows to reduce significantly the number of control parameters in the data assimilation procedure. Only a subset of eigenvalues can be used to control the subsequent evolution of a sandpile, keeping the other eigenvalues constant. This opens promising perspectives in designing a predictive tool based on sandpile models, even using expensive numerical minimization techniques such as simulated annealing.

We plan to make use of the eigenfunction decomposition to test our data assimilation procedure on real data in a future work. The speed up allowed by this decomposition will allow us to explore the parameter space of the minimization procedure on a large sample of flaring epochs for the Sun, and ultimately characterize the predictive capabilities of sandpiles when applied to a real system such as the flaring Sun.

References

Aschwanden, M. J., Crosby, N. B., Dimitropoulou, M., *et al.* 2016, *Space Sci Rev*, 198, 47

Barnes, G., Leka, K. D., Schrijver, C. J., *et al.* 2016, *ApJ*, 829, 89

Bélanger, E., Vincent, A. & Charbonneau, P. 2007, *Sol. Phys.*, 245, 141

Charbonneau, P. 2013, in Self-Organized Criticality Systems, ed. M. J. Aschwanden (Open Academic Press), 404

Hung, C. P., Brun, A. S., Fournier, A., *et al.* 2017, *ApJ* submitted

Lu, E. T. & Hamilton, R. J. 1991, *ApJL*, 380, L89

Pulkkinen, T. 2007, *LRSP*, 4, 1

Schwenn, R. 2006, *LRSP*, 3, 2

Strugarek, A. & Charbonneau, P. 2014, *Sol. Phys.*, 289, 4137

Strugarek, A., Charbonneau, P., Joseph, R. & Pirot, D. 2014, *Sol. Phys.*, 289, 2993

Space Weather of the Heliosphere: Processes and Forecasts
Proceedings IAU Symposium No. 335, 2017
C. Foullon & O.E. Malandraki, eds.

doi:10.1017/S1743921317007128

The Role of Empirical Space-Weather Models (in a World of Physics-Based Numerical Simulations)

Mathew J. Owens[1] **Pete Riley**[2] **and Tim Horbury**[3]

[1]Space and Atmospheric Electricity Group, Department of Meteorology, University of Reading, Earley Gate, PO Box 243, Reading RG6 6BB, UK
email: m.j.owens@reading.ac.uk

[2]Predictive Science Inc., 9990 Mesa Rim Rd, Suite 170, San Diego, CA 92121, USA
email: pete@predsci.com

[3]Blackett Laboratory, Imperial College London, London SW7 2BZ, UK
email: t.horbury@imperial.ac.uk

Abstract. Advanced forecasting of space weather requires prediction of near-Earth solar-wind conditions on the basis of remote solar observations. This is typically achieved using numerical magnetohydrodynamic models initiated by photospheric magnetic field observations. The accuracy of such forecasts is being continually improved through better numerics, better determination of the boundary conditions and better representation of the underlying physical processes. Thus it is not unreasonable to conclude that simple, empirical solar-wind forecasts have been rendered obsolete. However, empirical models arguably have more to contribute now than ever before. In addition to providing quick, cheap, independent forecasts, simple empirical models aid in numerical model validation and verification, and add value to numerical model forecasts through parameterization, uncertainty estimation and 'downscaling' of sub-grid processes.

Keywords. solar wind, solar-terrestrial relations

1. Introduction

State-of-the-art space-weather forecasting typically involves a chain of coupled numerical models with domain-specific approximations to the relevant physics (Merkin *et al.* 2006; Toth *et al.* 2005). Photospheric magnetic field observations are used to constrain a coronal model (e.g., Mikic *et al.* 1999), which then drives a magnetohydrodynamical (MHD) heliospheric model (e.g., Odstrcil, 2003; Riley *et al.* 2001), out to 1 AU. Note that even in full MHD schemes, semi-empirical methods are still necessary to relate coronal structure and solar wind speed (see section 4.1). Time-dependent phenomena, such as coronal mass ejections (CMEs), are incorporated in an ad-hoc manner (Odstrcil *et al.* 2004; Lionello *et al.* 2013). Forecast lead times are generally on the order of the solar wind Sun-Earth transit time, approximately 2-5 days. Advances in computing power and code efficiency have seen a steady increase in the spatial and temporal resolution of models. Similarly, better understanding of the physics of the corona and solar wind has led to better a representation of physical processes within models.

An alternative to the first-principles, physics-based approach is the use of statistical methods and empirical models. Despite their relative simplicity, such models can often approximate physical features and processes that the numerical simulations have difficulty with. These empirical models provide quick, cheap, independent forecasts and aid in state-of-the-art model validation and verification. Most importantly, they add value to numerical model forecasts through parameterization of missing or poorly represented physical processes, uncertainty estimation and 'downscaling' of sub-grid and stochastic processes. Using examples from near-Earth solar wind forecasting, these concepts are illustrated in the following sections.

2. Providing independent forecasts

There are a number of near-Earth solar wind prediction methods that can provide independent forecasts to complement the state-of-the-art numerical forecasts. In-ecliptic solar wind in particularly difficult to forecast due to the dominance of slow solar wind, which is poorly understood relative to the fast polar wind (Cranmer *et al.* 2017).

The simplest possible forecast, if it can really be called that, is climatology, wherein the value of a given solar wind parameter at all times in the future is simply forecast to be the mean value of that parameter over some long-term period in the past (say, 1965-2016). Thus unlike other forecasts, the predictive capability of climatology is independent of forecast lead time. For near-Earth solar wind speed [V_{SW}], the climatological mean (and therefore forecast) is 433 km s^{-1}, whereas for the out-of-ecliptic heliospheric magnetic field component [B_Z], it is close to 0 nT. Using simple point-by-point metrics, such as correlation or mean-square error, to compute forecast ability relative to observations, climatology often compares favorably with state-of-the-art numerical forecasts (e.g., Owens *et al.* 2008). In reality, this is often the result of numerical models suffering 'double penalties', whereby small timing errors in the forecast of a large event results in both a forecast miss and a forecast false alarm, whereas climatology only records a miss (see Figure 4 of Owens *et al.* 2005). Clearly, climatology is not useful as an 'actionable' forecast. It can, however, be valuable for long-term planning, such as setting design requirements for space hardware.

A slightly more useful forecast is persistence, whereby the solar wind in the future is assumed to be the same as that at the forecast time. Clearly, the performance of a persistence forecast reduces with forecast lead time, but speed of this degradation varies greatly with different solar wind parameters. For V_{SW}, persistence provides a more accurate forecast than climatology out to a lead time of around a day. For B_Z, however, persistence falls behind climatology at forecast lead times beyond an hour (see Figure 3 of Owens *et al.* 2017). But perhaps the simplest useful forecast comes from exploiting the recurrent nature of the solar wind (Bartels, 1934). All solar wind parameters display autocorrelation at a lag of around 27 days, the synodic rotation period of the Sun, as well as its harmonics with decreasing amplitude (Owens *et al.* 2013). Thus a recurrent forecast (sometimes also called 27-day persistence) predicts the solar wind in 27 days time will be the same as that today. It out performs climatology for V_{SW}, but not B_Z.

The recurrent forecast is valuable for a number of reasons. First, it is entirely independent of the state-of-the-art forecasts, providing forecast robustness when used in conjunction. Second, the observational input to recurrence is *in situ* solar wind observations, rather than photospheric magnetograms, providing forecast resilience to data availability. Third, as the forecast uses recent solar wind observations to predict future solar wind observations, there is almost zero forecast bias, such as, e.g., a systematic underestimate of the magnetic field strength. Fourth, recurrence offers forecast lead times well in excess of the 2-5 days provided by numerical simulations.

A further level of empirical forecast sophistication can be achieved using so-called 'analogue forecasts', whereby past solar wind observations are a good analogue of the future. Past analogues are defined as times in historic observations when the solar wind was similar to, say, the last 24 hours. The analogue periods can then be used to make a probabilistic forecast of the future (Riley *et al.* 2017; Owens *et al.* 2017). In addition to the benefits of recurrence forecasts, analogue forecasts are also much more accurate, outperforming climatology for forecast lead times of many tens of days for V_{SW}, and even around 10 hours for B_Z. The same analogue forecasting method can be applied to centennial and millennial reconstructions of the heliospheric magnetic field, enabling forecasting of space climate decades into the future (Barnard *et al.* 2011).

3. Model validation and verification

It is vital for both model development and model selection that forecasts are accurately verified and validated. In isolation, simple metrics, like correlation or mean-square error, are rather meaningless; they are unable to distinguish between periods where there is a genuine problem with the forecast and periods where the solar wind is simply unpredictable. By comparing the performance with a baseline forecast, however, this distinction can be made. Climatology can provide a baseline, as was implicitly done in the discussion in the previous section, but recurrence or analogue forecasts provide a more appropriate comparison, as they contain realistic solar wind structure (Siscoe *et al.* 2004).

4. Adding value to state-of-the-art numerical forecasts

Perhaps most importantly, empirical models can be used to add value to state-of-the-art numerical forecasts. There are three broad ways in which this is achieved.

4.1. *Parameterization of missing physics*

The MHD approximation, computational limits in the spatial and temporal scales, and limitations in observational boundary conditions means some physical processes are not (well) captured by numerical simulations. Of particular relevance to near-Earth solar wind forecasting, the heating of the corona and subsequent solar wind acceleration does not produce sufficient contrast between fast and slow wind regimes (Riley *et al.* 2015). To enable correct inputs to solar wind models, this process is parameterized in coronal models using empirically-derived relations to coronal magnetic field topology (Wang & Sheeley, 1990; Riley *et al.* 2015). In addition to enabling more accurate solar wind forecasts, refinement of these empirical relations also aids scientific understanding of coronal heating and thus can potentially inform a first-principles model of solar wind acceleration.

4.2. *Uncertainty estimation*

Numerical models are typically used to provide a single, deterministic forecast of solar wind conditions in near-Earth space. Use of numerical model ensembles enables a measure of forecast uncertainty, but due to computational expense ensembles are typically limited to a few tens of members. Empirical methods for propagating the solar wind from Sun to Earth are not quite as accurate as the full 3-dimensional MHD solution (Riley & Lionello, 2011), but they are much less computationally expensive. Thus these empirical methods allow thousands of quasi-random samples of the near-Sun solar wind to be propagated to Earth. The variation in the resulting ensemble provides an accurate measure of uncertainty in the deterministic MHD forecast (Owens & Riley, 2017).

4.3. *"Downscaling" of sub-grid and stochastic processes*

Finally, empirical models are also used to compensate for limited spatial and temporal resolution of numerical models. The Earth's magnetosphere responds to solar wind variations at time scales far below those currently reproduced by numerical solar wind models (Merkin *et al.* 2007). Future improvements to spatial and temporal resolution of models will not resolve this issue, as much of the high frequency variations in the solar wind are the result of stochastic processes such as turbulence. Thus a stochastic solution is required. Using empirical relationships between solar wind high-frequency variations and bulk properties, solar wind model results can be "downscaled" to include noise in a statistical sense prior to their use in driving magnetospheric simulations. This provides a more realistic representation of the magnetosphere (Owens *et al.* 2014).

5. Summary

Physics-based numerical models and simple, empirical models are not an either/or propositiom; they both have qualities which contribute to space-weather forecasting. As independent forecasts, empirical models add robustness to state-or-the-art methods, and can enable forecast lead times not achievable by numerical techniques. By relying on different observational inputs, empirical models produce resilience to interruptions in the observational data pipeline. Model validation/verification is aided by empirical models, as they provide a useful baseline against which numerical forecasts can be compared.

Perhaps most importantly, empirical models can add value to physics-based, numerical models in areas that are difficult to address from a first-principles approach. Empirical models can be used to parameterise physical processes that are not well represented in numerical models, to incorporate sub-grid and statistical processes, and to provide an accurate assessment of model uncertainty in a computationally inexpensive way.

6. Acknowledgements

M.O. is funded by Science and Technology Facilities Council (STFC) grant ST/M000885/1. We have benefited from discussions as part of the NASA LWS funded ProjectZed (NNX15AF39G).

References

Barnard, L., *et al.* 2011, *Geophys. Res. Lett.*, 381, doi:10.1029/2011GL048489
Bartels, J. 1934, *Terr. Magn. Atmos. Electr*, 39, 201, doi:10.1029/TE039i003p00201
Cranmer, S., Gibson, S. & Riley, P. 2017, *Space Sci. Rev.*, in press
Lionello, R., *et al.* 2013, *Astrophys. J.*, 777, 1, 76
Merkin, V., *et al.* 2007, *Space Weather*, 5, S12001, doi:10.1029/2007SW000335
Mikic, Z., Linker, J. A., Schnack, D. D., Lionello, R. & Tarditi, A. 1999, *Phys. Plasma*, 6, 2217
Odstrcil, D. 2003, *Adv. Space Res.*, 32, 497-506
Odstrcil, D., Riley, P. & Zhao, X. P. 2004, *J. Geophys. Res.*, 109, doi:10.1029/2003JA010135
Owens, M., *et al.* 2005, *J. Geophys. Res.*, 110, doi:10.1029/2005JA011343
Owens, M., *et al.* 2008, *Space Weather*, 6, doi:10.1029/2007SW000380
Owens, M., Challen, R., Methven, J., Henley, E. & Jackson, D. 2013, *Space Weather*, 11, 225-236, doi:10.1002/swe.20040
Owens, M., *et al.* 2014, *Space Weather*, 12, 395-405, doi:10.1002/2014SW001064
Owens, M., Riley, P. & Horbury, T. 2017, *Sol. Phys.*, 292, 5, 69, doi: 10.1007/s11207-017-1090-7
Owens, M. & Riley 2017, *Space Weather*, submitted
Riley, P., Linker, J. & Mikic, Z. 2001 *J. Geophys. Res.*, 106, A8, 15889-15901
Riley, P. & Lionello, R. 2011, *Sol. Phys.*, 270, 575-592, doi:10.1007/s11207-011-9766-x
Riley, P., Linker, J. A. & Arge, C. N. 2015, *Space Weather*, 13, 154-169, doi:10.1002/2014SW001144
Riley, P., *et al.* 2017, *Space Weather*, 15, 3, 526-540, doi:10.1002/2016SW001589
Siscoe, G., *et al.* 2004, *J. Atmos. Sol. Terr.*, 66, 1469-1480
Toth, G., *et al.* 2005, *J. Geophys. Res.*, 110, A12226, doi:10.1029/2005JA011126
Wang, Y. M. & Sheeley, N. R. Jr. 1990, *Astrophys. J.*, 335, 726

Space Weather of the Heliosphere: Processes and Forecasts
Proceedings IAU Symposium No. 335, 2017
C. Foullon & O.E. Malandraki, eds.

doi:10.1017/S1743921317011048

A New Technique to Provide Realistic Input to CME Forecasting Models

Nat Gopalswamy[1], **Sachiko Akiyama**[2], **Seiji Yashiro**[2] **and Hong Xie**[2]

[1] Solar Physics Laboratory, NASA Goddard Space Flight Center, Greenbelt, MD 20771, USA
email: nat.gopalswamy@nasa.gov

[2] Dept. of Physics, The Catholic University of America, Washington DC 20064, USA

Abstract. We report on a technique to construct a flux rope (FR) from eruption data at the Sun. The technique involves line-of-sight magnetic fields, post-eruption arcades in the corona, and white-light coronal mass ejections (CMEs) so that the FR geometric and magnetic properties can be fully defined in addition to the kinematic properties. We refer to this FR as FRED (Flux Rope from Eruption Data). We illustrate the FRED construction using the 2012 July 12 eruption and compare the coronal and interplanetary properties of the FR. The results indicate that the FRED input should help make realistic predictions of the components of the FR magnetic field in the heliosphere.

Keywords. coronal mass ejection, flux rope, reconnection.

---

1. Introduction

Coronal mass ejections (CMEs) are the most important players in space weather because they cause the severest of geomagnetic storms and accelerate energetic particles to GeV energies. The intensity of geomagnetic storms as measured by the *Dst* index primarily depends on the magnitude of the south-pointing out-of-the-ecliptic (B_z) component of the interplanetary magnetic field and the speed (V) with which the structure impacts Earth's magnetosphere: $Dst = -0.01V|B_z|-32$ nT (Wu & Lepping 2002; Gopalswamy *et al.* 2008). One of the vexing problems in space weather has been the prediction of southward (negative) B_z of the interplanetary field that reconnects with Earth's magnetic field to produce geomagnetic storms. While there has been reasonable progress in the prediction of the arrival time of CMEs, B_z prediction has been very limited due to the lack of realistic input to global MHD models that track CMEs into the heliosphere and provide asymptotic values of the CME parameters including the magnetic field. Since most CMEs arriving at Earth have a flux rope (FR) structure, it is inevitable that the global MHD models should use a FR input at the near-Sun boundary. Current models use a pressure pulse (which has no magnetic content) or ad hoc magnetic structures such as a spheromak (Odstrcil & Pizzo 1999; Jin *et al.* 2017; Shen *et al.* 2014; Shiota *et al.* 2014).

We construct a "flux rope from eruption data" (FRED) by combining two key results: (i) the reconnected (RC) flux during an eruption approximately equals the poloidal flux of the ejected flux rope (Longcope *et al.* 2007; Qiu *et al.* 2007; Hu *et al.* 2014; Gopalswamy *et al.* 2017a), and (ii) white-light or EUV coronal mass ejections (CMEs) can be fit to a FR to get its geometrical properties (see e.g., Temmer *et al.* 2011). The RC flux is computed from the area under post-eruption arcades (PEAs, McAllister and Martin 2000) and the underlying unsigned photospheric magnetic field strength. The poloidal flux of the FR is known from the RC flux; assuming that the FR is force free (Lundquist 1951) we can get the axial and azimuthal field components and the toroidal flux of

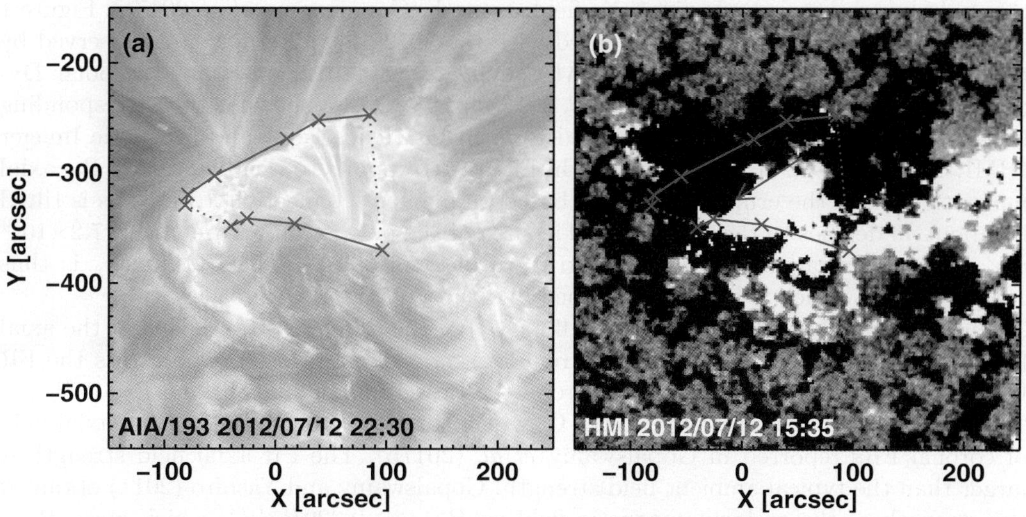

Figure 1. (a) PEA area (polygon) marked on an SDO/AIA 193 Å image. (b) SDO/HMI line-of-sight magnetogram with PEA and the filament magnetic field direction (arrow) overlaid.

the flux rope. Thus we have a fully-defined FR, complete with geometric and magnetic parameters that can be used as input to global MHD models. The axial and poloidal fields are essentially responsible for geoeffectiveness in high and low-inclination magnetic clouds at Earth, respectively and hence the FR input should lead to definite prediction schemes. The FRED technique is complementary to another one that uses the source-region relative magnetic helicity along with FR geometrical properties to estimate the FR magnetic properties (Patsourakos *et al.* 2016).

2. The Coronal Flux Rope

We illustrate the construction of FRED using the 2012 July 12 eruption that resulted in a large SEP event and a major geomagnetic storm (Dst = -139 nT). The event has been studied by many authors (see e.g., Gopalswamy *et al.* 2013; Gopalswamy *et al.* 2014; Hess & Zhang 2014; Möstl *et al.* 2014; Hu *et al.* 2016). Most of these papers were concerned with the kinematics and Sun-to-Earth propagation of the CME that ended up as a shock-driving magnetic cloud at Earth. Many of the authors also fit a flux rope using the graduated cylindrical shell (GCS) model (Thernisien 2011). Here we use the GCS model to get the geometric parameters of the FR.

Geometrical Properties. Fitting the CME observed by the three views provided by the SOHO and STEREO missions to the GCS model, we get the coordinates of the flux rope as S12W06, which is slightly different from the flare location in AR 11520 (S15W01, Gopalswamy *et al.* 2014). The ratio of the FR radius (R_0) to the leading edge distance from the Sun center (R_{tip}) is 0.26. At R_{tip} = 10 Rs, the R_0 = 2.6 Rs in the coronagraph field of view. The face-on and edge-on angular widths of the FR are 94° and 40°, respectively. The tilt angle of the FR axis at its apex with respect to the horizontal is about 53°, indicating a northwest-southeast orientation of the flux rope axis, consistent with the neutral line as inferred from filament location.

Magnetic Properties. We combine these geometrical information with the RC flux to get the magnetic properties of the FR. It was recently shown that the total RC flux during an eruption can be obtained from a snapshot of the post eruption arcade (PEA) and the underlying photospheric magnetic field strength (Gopalswamy *et al.* 2017a). Figure 1 shows a snapshot of the PEA associated with the 2012 July 12 eruption as observed by the Atmospheric Imaging Assembly (AIA, Lemen *et al.* 2012) on board the Solar Dynamics Observatory (SDO, Pesnell *et al.* 2012) at 193 Å. Also shown is the corresponding line-of-sight photospheric magnetic field from SDOs Helioseismic and Magnetic Imager (HMI, Scherrer *et al.* 2012). The arrow in Figure 1(b) points to the direction of the axial magnetic field of the erupted flux rope based on filament connectivity. The axis is tilted by about 46 degrees to the horizontal. The area of the polygon in Figure 1(a) is 7.2×10^{19} cm^2, and an average magnetic field strength of B of 392 G. The RC flux ϕ_r is thus 1.42×10^{22} Mx, which is half the unsigned flux through the PEA area.

Since ϕ_r roughly equals the poloidal flux (ϕ_p) of the erupted FR, we can get the axial field (B_0) of the coronal FR from the relation, $\phi_p = (L/x_{01})B_0R_0$, where L is the FR length and x_{01} (=2.4048) is the first zero of the Bessel function J_0. Taking $L = 2\ R_{tip} =$ 20 Rs, we get $B_0 = \phi_r x_{01}/LR_0 = 0.13$ G, which is near the higher end of the axial field of coronal FRs reported in Gopalswamy *et al.* (2017b). The FR axial field strength is larger than the typical ambient field strength. Gopalswamy and Yashiro (2011) obtained the strength of the ambient magnetic field as $B_{amb} = 0.329R^{-1.23}$, which gives B_{amb} $= 0.019$ G at a heliocentric distance R of 10 Rs. Clearly, the FR axial field strength is about 7 times greater than the ambient field, consistent with the fact that CME FRs are low-beta plasmas. The toroidal flux of the FR is $\phi_t = \phi_p(2\pi R_0/L)J_1(x_{01})$, where J_1 is the first order Bessel function. With the above numbers from the GCS fit, we get $\phi_t = 0.42\phi_p$ or 5.96×10^{21} Mx. The poloidal (B_p) and toroidal (B_t) field strengths at any distance r from the FR axis are given by $B_p = HB_0J_1(\alpha r)$ with $H = \pm1$ (helicity sign) and $B_t = B_0J_0(\alpha r)$. Here, $\alpha = x_{01}/R_0$ is the force-free parameter. In the present case, $H = +1$ because the axial field points mostly to the south and the azimuthal field goes from the positive to the negative side (see Figure 1b). The direction is consistent with a southern hemispheric eruption having right-handed helicity sign (e.g., Bothmer & Schwenn 1998). Finally, the relative helicity per unit length (H_r/L) is also well-defined for a Lundquist FR: $H_r/L = 0.7{B_0}^2{R_0}^3$ (e.g., Dasso *et al.* 2003). For the 2012 July 12 coronal FR, $H_r = 9.98\ \times10^{43}$ Mx^2, if we take the FR length to be $2R_{tip}$. Thus, the flux rope is fully defined both geometrically and magnetically. The three-dimensional CME speed was measured using STEREO-B as $V = 1548$ km s^{-1} (Gopalswamy *et al.* 2013). According to the empirical relation $V = 298{\phi_r}^{0.75}$ (Gopalswamy *et al.* 2017b), this speed indicates a ϕ_r of 9.0×10^{21} Mx, which is only about 36% lower than the observed value (1.42×10^{22} Mx).

3. Comparison with 1-AU Observations

A recent study showed that the axial field of the coronal FR derived from ϕ_r correlates with the axial field of the 1-AU flux rope (fitted) and also with the strength of the observed total magnetic field (B_{tot})(Gopalswamy *et al.* 2017b). In that study the white-light CME data were fit with the elliptical flux rope model of Krall and St. Cyr (2006). Figure 2 shows a scatter plot between ϕ_r and B_{tot} for a set of interplanetary CME (ICME) events from cycle 23. From the regression line $B_{tot} = 21.9\log\phi_r - 453.2$ nT, we get B_{tot} =31.9 nT with $\phi_r = 1.42\times10^{22}$ Mx for the 2012 July 12 eruption. The inferred B_{tot} is very close to the observed value at 1 AU: ~30 nT (see e.g., Hess and Zhang 2014). If the FR expands self-similarly, the 1-AU FR radius is expected to be 0.26 AU (since $R_0/R_{tip} = 0.26$ with $R_{tip} = 1$ AU). Therefore, one can get the 1-AU axial field strength from the coronal field strength ($B_0 = 0.13$ G) as: $B_{01AU} = B_0(10/214)^2 = 28.3$

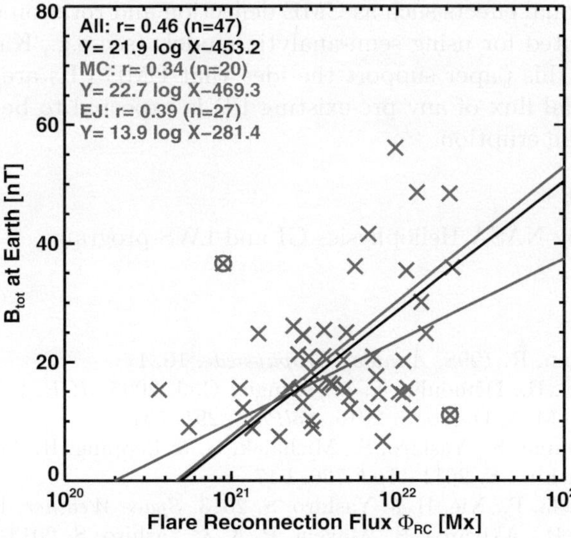

Figure 2. Scatter plot between observed B_{tot} and ϕ_r for a set of events from solar cycle 23 reported in Gopalswamy *et al.* (2017b). Two outliers are excluded from the correlation. Blue and red symbols correspond to magnetic clouds (MC) and non-cloud ejecta (EJ), respectively.

nT, again very similar to the observed value. The assumption of self-similar expansion seems to be valid because the FR radius is consistent with that obtained from Grad-Shafranov reconstruction using *in-situ* data (Hu *et al.* 2016). For the same reasons, H_r in the 1 AU FR is similar to that in the coronal FR. The 1-AU data also shows that the FR Y-component (azimuthal) of the magnetic field rotates from west to east, while the Z-component (axial) points southward throughout the cloud duration (Hess & Zhang 2014; Hu *et al.* 2016), consistent with the orientation of the coronal FR. In this event, the B_z is simply the axial field component.

4. Discussion and Summary

We have shown that the FRED technique is a viable starting point to infer the expected asymptotic magnetic structure in the heliosphere. It combines the line-of-sight photospheric magnetograms and PEA observations (X-ray, EUV, H-alpha, or microwave) with CME observations in the corona to obtain complete properties of coronal FRs. We have shown these properties at a distance of 10 Rs for the purpose of illustration, but FRED properties can be obtained at any initial height where the CME is observed. For example, the inner boundary of many MHD simulation models is located at $R_{tip} \sim 21$ Rs, where the axial field $B_0 = \phi_r x_{01}/LR_0$ will be weaker than the 10-Rs case because both L and R_0 will be larger. Expressing L and R_0 in terms of R_{tip} and noting that R_0/R_{tip} is invariant for each FR from the GCS model we get $B_0 = \phi_r x_{01}/[pq(R_{tip})^2]$, where p and q are constants: $L = pR_{tip}$ (assumed FR length) and $R_0 = qR_{tip}$ (from the GCS model). This expression shows that the axial field falls off as the square of the heliocentric distance. Note that we have taken $p = 2$ following Nindos *et al.* (2003). The length can be higher if the flux rope assumes the shape of a half torus, so $L = \pi R_{tip}$. Démoulin *et al.* (2016) recently reported a statistical value of $p = 2.6 \pm 0.3$, which is in between the two values noted above. The invariant toroidal flux can also be written in a general form as $\phi_t = \phi_p(2\pi q/p)J_1(x_{01})$. Global MHD models that use a realistic input such as FRED should be able to provide a realistic forecast of what to expect in the heliosphere. It must be noted that additional effects such as CME deflection and rotation can modify the FR, which can be accounted for using semi-analytic models (see e.g., Kay *et al.* 2015). The results presented in this paper support the idea that CME FRs are formed during the eruption. The poloidal flux of any pre-existing FR is expected to be a small fraction of that added during the eruption.

Acknowledgement

Work supported by NASA Heliophysics GI and LWS programs.

References

Bothmer, V. & Schwenn, R. 1998, *Annales Geophysicae*, 16, 1
Dasso, S., Mandrini, C. H., Démoulin, P. & Farrugia, C. J. 2003, *JGR*, 108, 1362
Démoulin, P., Janvier, M. & Dasso, S. 2016, *SolPhys*, 291, 531
Gopalswamy, N., Akiyama, S., Yashiro, S., Michalek, G. & Lepping, R. 2008, *JASTP*, 70, 245
Gopalswamy, N. & Yashiro, S. 2011, *ApJ*, 736, L17
Gopalswamy, N., Mäkelä, P., Xie, H. & Yashiro, S. 2013, *Space Weather*, 11, 661
Gopalswamy, N., Xie, H., Akiyama, S., Mäkelä, P. A. & Yashiro, S. 2014, *Earth, Planets, and Space*, 66, 104
Gopalswamy, N., Yashiro, S., Akiyama, S. & Xie, H. 2017a, *SolPhys*, 292, 65
Gopalswamy, N., Akiyama, S., Yashiro, S. & Xie, H. 2017b, *JASTP*, arXiv:1705.08912
Hess, P. & Zhang, J. 2014, *ApJ*, 792, 49
Hu, Q., Qiu, J., Dasgupta, B., Khare, A. & Webb, G. M. 2014, *ApJ*, 793, 53
Hu, H., Liu, Y. D., Wang, R., Möstl, C. & Yang, Z. 2016, *ApJ*, 829, 97
Jin, M., Manchester, W. B., van der Holst, B., *et al.* 2017, *ApJ*, 834, 173
Kay, C., Opher, M. & Evans, R. M. 2015, *ApJ*, 805, 168
Krall, J., St. Cyr, O. C. 2006, *ApJ*, 652, 1740
Lemen, J. R., Title, A. M., Akin, D. J., *et al.* 2012, *SolPhys*, 275, 17
Longcope, D., Beveridge, C., Qiu, J., *et al.* 2007, *SolPhys*, 244, 45
Lundquist, S. 1951, *Phys. Rev.*, 83, 307
McAllister, H. & Martin, S. F. 2000, *AdSpR*, 26, 469
Möstl, C., Amla, K., Hall, J. R., *et al.* 2014, *ApJ*, 787, 119
Nindos, A., Zhang, J. & Zhang, H. 2003, *ApJ*, 594, 1033
Odstrčil, D. & Pizzo, V. J. 1999, *JGR*, 104, 483
Patsourakos, S., Georgoulis, M. K., Vourlidas, A., *et al.* 2016, *ApJ*, 817, 14
Pesnell, W. D., Thompson, B. J. & Chamberlin, P. C. 2012, *SolPhys*, 275, 3
Qiu, J., Hu, Q., Howard, T. A. & Yurchyshyn, V. B. 2007,*ApJ*, 659, 758
Scherrer, P. H., Schou, J., Bush, R. I., *et al.* 2012, *SolPhys*, 275, 207
Shen, F., Shen, C., Zhang, J., *et al.* 2014, *JGR*, 119, 7128
Shiota, D., Kataoka, R., Miyoshi, Y., *et al.* 2014, *Space Weather*, 12, 187
Temmer, M., Veronig, A. M., Gopalswamy, N. & Yashiro, S. 2011, *SolPhys*, 273, 421
Thernisien, A. 2011, *ApJS*, 194, 33
Wu, C.-C. & Lepping, R. P. 2002, *JGR*, 107, 1346

Space Weather of the Heliosphere: Processes and Forecasts
Proceedings IAU Symposium No. 335, 2017
C. Foullon & O.E. Malandraki, eds.

doi:10.1017/S1743921317007396

Prospects for Modeling and Forecasting SEP Events with ENLIL and SEPMOD

J. G. Luhmann[1] M. L. Mays[2], D. Odstrcil[3], Yan Li[1], H. Bain[4], C. O. Lee[1], C. M. S. Cohen[5], R. A. Mewaldt[5], R. A. Leske[5] and Y. Futaana[6]

[1]Space Sciences Laboratory, University of California, Berkeley, CA, USA
email: jgluhman@ssl.berkeley.edu

[2]CCMC, NASA Goddard Space Flight Center, Greenbelt, MD, USA

[3]George Mason University, Fairfax, VA, USA

[4]NOAA Space Weather Prediction Center, Boulder, CO, USA

[5]California Institute of Technology, Pasadena, CA, USA

[6]IRF, Kiruna, Sweden

Abstract. One view of major Solar Energetic Particle (SEP) events is that these (proton-dominated) fluxes are accelerated in heliospheric shock sources created by Interplanetary Coronal Mass Ejections (ICMEs), and then travel mainly along interplanetary magnetic field lines connecting the shock(s) to the observer(s). This places a particular emphasis on the role of the heliospheric conditions during the event, requiring a realistic description of the latter to interpret and/or model SEP events. The well-known ENLIL heliospheric simulation with cone model generated ICME shocks is used together with the SEPMOD particle event modeling scheme to demonstrate the value of applying these concepts at multiple inner heliosphere sites.

Keywords. Space weather, solar energetic particles, interplanetary shocks

1. Introduction

The desire to interpret and forecast Solar Energetic Particle (SEP) events as part of space weather research has spawned a number of schemes for both source and particle transport descriptions (e.g. Marsh *et al.* 2015). One school of thought favors approaches with near-Sun sources and diffusive propagation, while another emphasizes the evolving geometry and properties of the coronal mass ejection driven shock sources and regards shock-connected interplanetary field lines as highways for their transport into the heliosphere (Heras *et al.* 1992; Kallenrode *et al.* 1997; Lario *et al.* 1997). Each of these have advantages and weaknesses, but the latter has the best prospects for describing the global nature of these events including those involving multiple shock sources. One well-exercised way of describing the necessary heliospheric conditions is the WSA-ENLIL-cone model (e.g. Odstrcil *et al.* 2005), which we refer to as ENLIL here. This MHD simulation bases its solar wind and interplanetary field description on inner boundary conditions set by global magnetic maps obtained with solar magnetographs (e.g. see Arge *et al.* 2004), while its coronal mass ejection (CME)-driven shocks are driven by localized injections of high pressure outflows into the solar wind with directions, stream widths, and initial speeds dictated by coronagraph images of CMEs (e.g. see Mays *et al.* 2015). Together, this combination can sometimes produce good matches to the plasma parameters observed at 1 AU, including high speed solar wind streams and interplanetary CME (ICME) shock arrivals (e.g. Taktakishvili *et al.* 2009).

SEPMOD takes the next step by using the results from ENLIL to time-integrate assumed SEP injections of energetic protons at the shocks onto magnetic field lines connecting to an observer located anywhere in the ENLIL domain (see Luhmann *et al.* 2017 for a more detailed description of SEPMOD). The result for any observer is a companion time series of SEP proton fluxes for the period of the ENLIL run, including all of the ICME shock source contributions present in ENLIL during that run. Sample results from this scheme are shown here, together with the corresponding available data, for SEP observer outposts at Earth, STEREO, Venus and Mars. The results demonstrate the extent to which a heliosphere-wide perspective of SEP events associated with sometimes highly complex heliospheric conditions can be obtained with the application of the combined ENLIL/SEPMOD approach.

2. Examples of ENLIL/SEPMOD Applications

Here we use a straw-man set of 5 inner heliosphere observers located at Earth, STEREO, Venus and Mars to illustrate the application of ENLIL with SEPMOD to the diagnosis of two cases of multipoint SEP observations. For 1 AU observations we use a combination of EPAM on ACE and EPS on GOES for Earths location (e.g. Gold *et al.* 1998), and LET and HET (Mewaldt *et al.* 2008, Von Rosenvinge *et al.* 2004) on STEREO-A and B at their changing heliocentric positions with respect to Earth. We also use an ASPERA IMA detector background from Venus Express (VEX) for Venus at 0.7 AU, and a similar detector background from Mars Express (MEX) at Mars at 1.5 AU (see Futaana *et al.* 2008).

Although the assumptions in SEPMOD of parallel (to the field), non-diffusive transport are best suited to 10-100 MeV energy protons, we apply the ENLIL/SEPMOD scheme to the energy range from a few MeV to 100 MeV to assess its usefulness at the lower energies. It should be mentioned here that the injected SEP protons are each initially isotropic bursts that have energy spectra determined by the ENLIL shock compression jump at the point of the observers magnetic connection, which change with the 5 minute cadence of SEPMODs adoption of ENLIL time-dependent field and shock geometries (see Luhmann *et al.* 2017). The injected SEP fluxes are determined using a semi-empirical formula derived by Lario *et al.* (1998) based on those authors observer field line backmapping to a modeled shock in a large number of SEP events. SEPMOD also includes an ad-hoc Energetic Storm Particle (ESP) component that is added to the injected flux when the observer is located within a few tenths of an AU of a strong shock. This component has a flux level that is a fixed fraction of the baseline flux and falls off with distance from the shock location. It also has a softened spectrum relative to the baseline flux. The construction of this component was based on a survey of ESP event phenomenology in the literature (e.g., Cohen 2006 and references therein).

The results shown illustrate a variety of issues and features that contribute to the complexity of SEP event analysis and modeling. For each case we first show the geometry of the various observers during the period under analysis. We then show the observed time series of SEP fluxes at each of the 5 observer sites, for the duration of the ENLIL run, and then the model counterparts to these time series calculated with the ENLIL results and SEPMOD. ENLIL runs of several weeks duration are typically used because in many cases multiple ICME shocks are present in the ENLIL domain at one time, and must all be considered in interpreting what is observed. In concert with this, it is found that ENLIL simulations out to 5 AU are desirable because of both contributions of shock

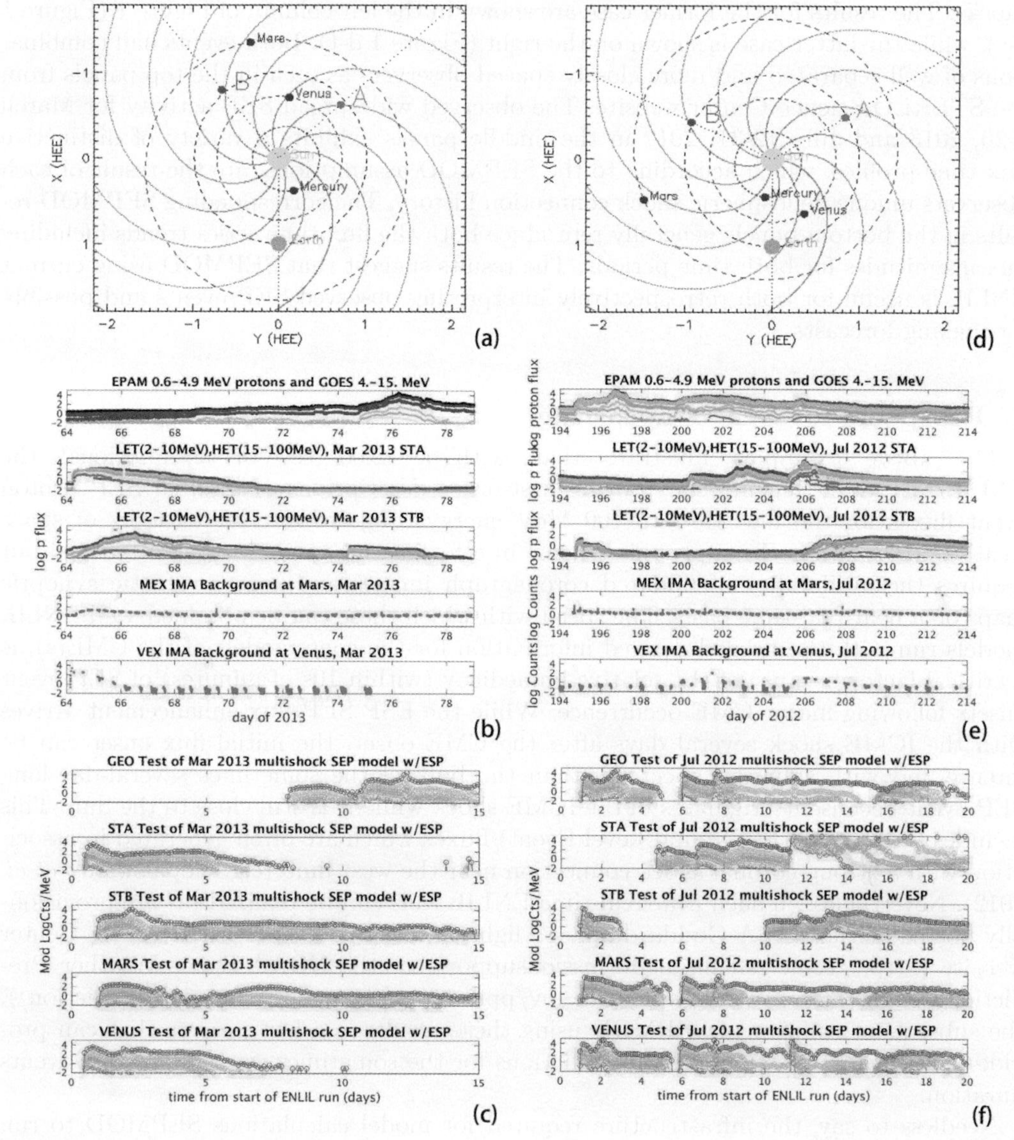

Figure 1. (a) Ecliptic plane locations of the observers at the Earth, STEREO-A and B sites, Venus and Mars from which SEP and supporting observations are available in March 2013. (b) Stacked SEP flux time profiles for days 64-79 (March 5-20) of 2013 from ACE/GOES at Earth, STEREO-A and B, and background in the Mars Express and Venus Express ASPERA-IMA detectors. (c) SEP flux time profiles calculated with ENLIL and SEPMOD for this period. (d-f) Same as (a-c) but for a period of high solar activity in July 2012 (July 12-31).

sources that have passed the observer, and SEP mirroring in the structured interplanetary magnetic field. Although many details of the observer shock connections could be shown, the modeled time series represent the ultimate test of the SEPMOD/ENLIL scheme, which is why we focus on them in this brief report.

We chose a period of relatively simple solar activity in March 2013, when only two major ICME shocks were present, well-separated in time and space, and another in July 2012 when there was exceptionally strong, complex activity with multiple overlapping shocks. The results for the former case are shown in the left column of Figure 1 (Figure 1 a-c), while the latter case is shown on the right (Figure 1 d-f). Both events had combinations of well-separated and more closely spaced observers as seen in the top panels from the STEREO Science Center website. The observed widespread SEP activity for March 5-20, 2013 and July 12-31, 2012 in the middle panels exhibits a variety of distinctive flux time profiles, which according to the SEPMOD assumptions, are the result of each observers unique heliospheric shock connection history. The corresponding SEPMOD results in the bottom panels generally reproduce both the flux time series trends including flux magnitudes for both time periods. The results suggest that SEPMOD using current ENLIL is useful for both retrospectively interpreting observed SEP events and possibly for making forecasts.

3. Prospects for SEP Forecasting

The above examples illustrate that with a fixed set of assumptions, the ENLIL/SEPMOD scheme can provide first-order descriptions of gradual SEP proton event fluxes for the few MeV to 100 MeV energy range of interest to users of space weather information. To apply this scheme in a forecasting mode is straightforward but requires the availability of updated coronagraph images and solar magnetic synoptic maps on a near-real-time basis. The speed with which these can be obtained, and ENLIL models run with the needed derived information for the cone models of the CME(s), is a critical factor because of the relative immediacy (within 10s of minutes) of SEP event onsets following major CME occurrence. While the ESP SEP flux enhancement arrives with the ICME shock several days after the CME onset, the initial flux onset can be intense and with a harder spectrum than the bulk of the sometimes several-day-long SEP event because it originates in the ICME shock while it is still close to the Sun. This includes the hard GLE (Ground Level Event) fluxes which are often generated in association with a strong coronal shock connection near the west limb (e.g. Gopalswamy *et al.* 2012). Nevertheless, if such near-real-time ENLIL runs can be realized (as are provisionally posted at the NASA Goddard Space Flight Center Space Weather Research Center website (http://ccmc.gsfc.nasa.gov/mssionsupport) or the NOAA Space Weather Prediction Center (http://www.swpc.noaa.gov/pproducts/wsa-enlil-solar-wind-prediction)), the subsequent running of SEPMOD using these results is a fast process that can provide related SEP flux time profile predictions for the sometimes several-day-long events duration.

Needless to say, the infrastructure required for model calculations SEPMOD to run must be a first priority of those invested in the goal of interplanetary radiation environment forecasting, including both the coronal and solar magnetic field observations and the continual improvement of heliospheric models like ENLIL. The availability of multiperspective imaging such as that possible with SOHO and ACE makes determination of CME properties much more realizable than images from only Earths perspective. In addition, while current ENLIL appears to provide a usable first order description of heliospheric shocks and fields, it does not include the region of strongest shocks inside 21 Rs (inside ENLILs inner boundary) -where GLE SEPs are generated. Nor does current ENLIL include the magnetic fields of the coronal ejecta in its cone model CMEs. The magnetic fields of the latter both influence the radial evolution and interactions of the ICME shocks, as well as alter the magnetic field line and hence connectivity descriptions in the post shock regions of the ICMEs. SEPMOD upgrades, in terms of physics included, can also be made, but these can only improve the outcomes if the underlying heliospheric simulations are also improved. Thus one has to regard any implementation or validation of the scheme described here in its entirety. These issues notwithstanding, modeled SEP flux time profiles like those above can be obtained for retrospective studies on a regular basis today. The CCMC at Goddard Space Flight Center is currently taking steps toward that goal.

Acknowledgement

The authors are grateful to the CCMC at NASA Goddard Space Flight Center for its provision of ENLIL model results for use in SEPMOD, and to all of the providers of the used solar coronagraph and magnetograph data, and multipoint SEP observations and their sponsors. This research is supported by the Heliophysics Division of NASA and the NSF Solar Terrestrial Research Program.

References

C. N., Luhmann, J. G., Odstrcil, D., Schrijver, C. J. & Li, Y. 2004, *J. Atmosph. Solar Terr. Phys.*, 66, 1295-1309

Cohen, C. M. S. 2006, *Geophys. Monograph Ser., 165, eds. N. Gopalswamy, R. Mewaldt and J. Torsti*, 275-282, AGU, Washington DC

Futaana, Y., *et al.* 2008, *Planet. Space Sci.*, 56, 873-880

Gold, R. E., Krimigis, S. M., Hawkins, S. E., Haggerty, D. K., Lohr, D. A., Fiore, E., Armstrong, T. P., Holland, G. & Lanzerotti, L. J. 1998, *Space Sci. Rev.*, 86, 541-562

Gopalswamy, N., Xie, H., Yashiro, S., Akiyama, S., Makela, P. & Usoskin, I. 2012, *Space Sci. Rev.*, 171, 23-60

Heras, A. M., Sanahuja, B., Smith, Z. K., Detman, T. & Dryer, M. 1992, *Astrophys. J.*, 391, 359-369

Kallenrode, M.-B. & Wibberenz, G. 1997, *J. Geophys. Res.*, 102, 22,311-11,334

Lario, D., Sanahua, B. & Heras, A. M. 1997, *Adv. Space Res.*, 20, 121-126

Lario, D., Sanahuja, B. & Heras, A. M. 1998, *ApJ* 509, 415

Luhmann, J. G., *et al.* 2017, *Space Weather*, 15, 1-21, doi:10.1002/2017SW001617

Mays, M. L., *et al.* 2015, *Sol. Phys.*, 290, 1775-1814

Marsh, M. S., Dalla, S., Dierckxsens, M., Laitinen, T. & Crosby N. B. 2015, *Space Weather*, 13, 386-394

Mewaldt, R. A., *et al.* 2008, *Space Sci. Rev.*, 136, 285-362

Odstrcil, D., Pizzo, V. J. & Arge, C. N. 2005, *J. Geophys. Res.*, 110, 2004JA010745

Taktakishvili, A., Kuznetsova, M., MacNeice, P., Hesse, M., Rastaetter, L., Pulkkinenen, A., Chulaki, A. & Odstrcil, D. 2009, *Space Weather*, 7, doi:10.1029/2008SW000448

Von Rosenvinge, T. T., *et al.* 2008, *Space Sci. Rev.*, 136, 391-436

Space Weather of the Heliosphere: Processes and Forecasts
Proceedings IAU Symposium No. 335, 2017
C. Foullon & O.E. Malandraki, eds.

doi:10.1017/S1743921317011012

Application of Test Particle Simulations to Solar Energetic Particle Forecasting

S. Dalla[1], B. Swalwell[1], M. Battarbee[1], M. S. Marsh[2], T. Laitinen[1] and S. J. Proctor[1]
[1]University of Central Lancashire, Preston, PR1 2HE, UK
[2]Met Office, Exeter, UK

Abstract. Modelling of Solar Energetic Particles (SEPs) is usually carried out by means of the 1D focused transport equation and the same approach is adopted within several SEP Space Weather forecasting frameworks. We present an alternative approach, based on test particle simulations, which naturally describes 3D particle propagation. The SPARX forecasting system is an example of how test particle simulations can be used in real time in a Space Weather context. SPARX is currently operational within the COMESEP Alert System. The performance of the system, which is triggered by detection of a solar flare of class >M1.0 is evaluated by comparing forecasts for flare events between 1997 and 2017 with actual SEP data from the GOES spacecraft.

Keywords. Sun: particle emission

---

1. The SPARX forecasting system

A number of physics-based forecasting models are currently being developed to predict the particle intensities and radiation risk associated with SEPs. The traditional approach is spatially 1D, meaning that particles are assumed to remain tied to the magnetic field line on which they were originally injected, with no propagation across the field (e.g. Aran *et al.* 2006; Luhmann *et al.* 2010). However, high energy SEPs have a strong influence on the radiation dose and these particles are affected by drifts due to the gradient and curvature of the Parker spiral, requiring a 3D description (Marsh *et al.* 2013; Dalla *et al.* 2013).

The SPARX forecasting system is based upon a 3D test particle model that accounts for drift and deceleration effects (Marsh *et al.* 2015). It is currently operational within the COMESEP alert system (www.comesep.eu, Crosby *et al.* 2012). In real time operation, SPARX is triggered by the detection of a solar flare of magnitude >M1.0. The system makes use of a database of runs of the test particle model to combine the contribution of many injection tiles near the Sun, describing an extended CME-driven shock structure. A profile of SEP intensities versus time at Earth for protons >10 MeV and >60 MeV is produced. Peak intensities are normalised by means of an empirical relationship between flare peak flux and particle peak intensities (Dierckxsens *et al.* 2015). Further details about SPARX may be found in (Marsh *et al.* 2015).

2. Evaluating model performance

To assess the performance of the SPARX model in forecasting SEP events and their parameters, we proceeded as follows. We started from a list of X-class solar flares that took place in the time range between 1 September 1997 and 30 April 2017, covering solar cycles 23 and 24 up to the present time. The initial list consisted of 169 flares.

hits = 20	false alarms = 27	47
misses = 20	correct negatives = 58	78
40	85	125

Table 1. Contingency table for SPARX forecasts for F_N threshold.

Bias	1.18
POD	0.5
FAR	0.57
POFD	0.32
CSI	0.30

Table 2. SPARX scores for F_N threshold.

For each flare, SPARX was run as it would have been in a forecast mode, i.e. with a fixed set of model input parameters which remained the same for all events. The output was then examined and a forecast constructed as follows: if the peak flux for protons >10 MeV over the entire duration of the event exceeded a specified threshold F, then a positive forecast of SEP event was made, otherwise a no event situation was predicted.

In the analysis below, we considered two thresholds: the first one is the standard NOAA threshold, F_N=10 pfu (where pfu is particles cm^{-2} s^{-1} sr^{-1}) and the second is a lower threshold F_1=1 pfu.

We then examined data from the >10 MeV proton channel of the GOES Energetic Particle Sensor (EPS) (Onsager *et al.* 1996) to verify whether an actual event took place following each of the flares in our list. To define the occurrence of an SEP event, we used in the first instance the threshold F_N and subsequently F_1. The methodology for associating an SEP event with a given flare is described in (Swalwell *et al.* 2017). At the start time of 36 flares in our list, the EPS detector already had an enhanced flux from a previous event, making it impossible to verify whether or not an SEP event occured: these flares were removed from the list. In addition, for 8 flares no EPS data was available. Therefore the total number of flares used in the analysis below is 125.

2.1. *NOAA threshold*

The contingency table for the NOAA threshold, F_N, is shown in Table 1. A correct negative is any X class flare which was not forecast to produce SEPs and which did not do so. From a contingency table, it is possible to derive a number of indicators that assess the performance of the model. They are defined as follows.

The bias score gives an indication of whether there is a tendency to under-forecast (bias < 1) or over-forecast (bias > 1) events. The perfect score is 1. The bias is given by:

$$\text{BIAS} = \frac{\text{hits} + \text{false alarms}}{\text{hits} + \text{misses}} \tag{2.1}$$

The probability of detection (POD) score gives the fraction of observed "yes" events which were correctly forecast. The perfect score is 1. The POD is given by:

$$\text{POD} = \frac{\text{hits}}{\text{hits} + \text{misses}} \tag{2.2}$$

The false alarm ratio (FAR) gives the fraction of predicted "yes" events which did not occur. The perfect score is 0. The FAR is given by:

hits = 40	false alarms = 31	71
misses = 12	correct negatives = 42	54
52	73	125

Table 3. Contingency table for SPARX forecasts for F_1 threshold.

Bias	1.37
POD	0.77
FAR	0.44
POFD	0.43
CSI	0.48

Table 4. SPARX scores for F_1 threshold.

$$\mathrm{FAR} = \frac{\text{false alarms}}{\text{hits} + \text{false alarms}} \tag{2.3}$$

The probability of false detection (POFD), also called false alarm rate, is the fraction of "no" events incorrectly forecast as "yes". The perfect score is 0. It is given by:

$$\mathrm{POFD} = \frac{\text{false alarms}}{\text{correct negatives} + \text{false alarms}} \tag{2.4}$$

Finally the critical success index (CSI) shows how well the forecast "yes" events correspond to the observed "yes" events. The perfect score is 1. The CSI is given by:

$$\mathrm{CSI} = \frac{\text{hits}}{\text{hits} + \text{misses} + \text{false alarms}} \tag{2.5}$$

For the SPARX model with F_N threshold, the values of the scores defined in eqs.(2.1)–(2.5) are given in Table 2.

2.2. *1 pfu threshold*

The contingency table is shown in Table 3 for the SPARX forecasts for the F_1 threshold. The various scores as defined in eqs.(2.1)–(2.5) are shown in Table 4 for the F_1 threshold.

3. Comparison between forecast and actual SEP profile characteristics

For those events which were correctly forecast by SPARX to occur, we can compare the properties of the forecast flux profiles with those of the observed GOES profiles. Figure 1 shows the peak fluxes forecast by the SPARX model versus actual peak fluxes from GOES EPS data. The colour scale of the symbols reflects the longitude of the flare. Figure 2 presents a comparison of times to maximum intensity, defined as the time between the start of the flare and the time of peak flux. SPARX times to maximum versus actual GOES EPS ones are plotted. As in Figure 1 the colour scale gives the flare longitude.

4. Discussion

SPARX performs much better if a lower threshold than that used by NOAA is used to define an SEP event. For both thresholds, the BIAS score is above 1, meaning that

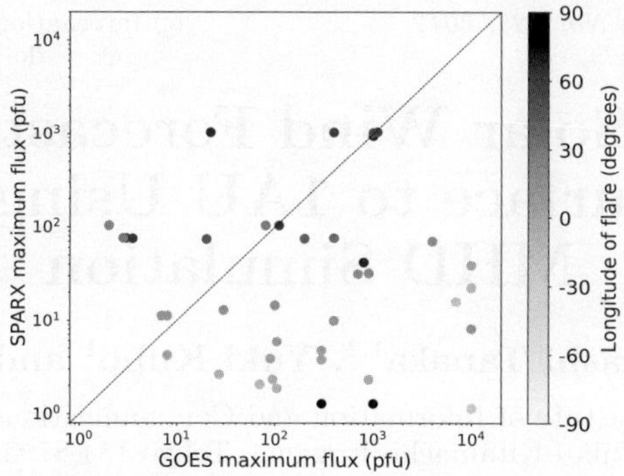

Figure 1. Peak fluxes forecast by the SPARX model versus actual GOES peak fluxes, for >10 MeV protons. The gray line is the 1:1 line.

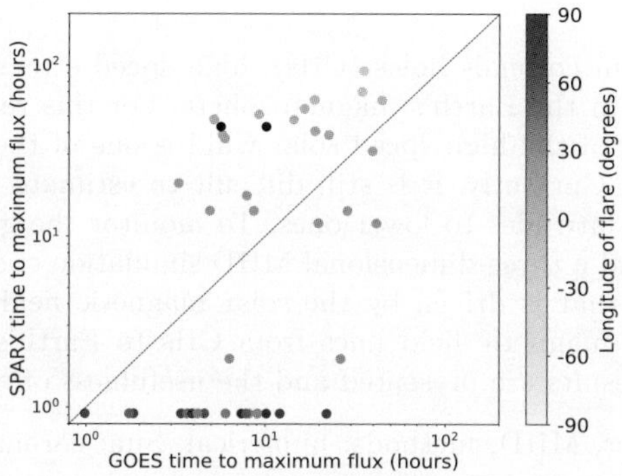

Figure 2. Time to peak intensity forecast by the SPARX model versus actual GOES time to peak intensity, for >10 MeV protons.

SPARX over-forecasts SEP events, more so for the lower threshold. All of the remaining scores are better for the lower threshold: the POD score for the lower threshold is much higher than for the lower (0.77 versus 0.5); the FAR is lower for the lower threshold (0.44 versus 0.57); and the CSI is higher for the lower threshold (0.48 versus 0.30).

Regarding the characteristics of the profiles of SEP fluxes, SPARX appears to significantly underestimate the time to peak intensity for well connected Western longitude flares, while for Eastern longitude flares reasonable agreement between forecast and actual times to peak are seen. Peak fluxes for Eastern longitude flares tend to be underestimated.

References

Aran, A., Sanahuja, B. & Lario, D. 2006, *Adv. Space Res.*, 37, 1240

Crosby, N. B., *et al.* 2012, in *AIP Conference Series*, vol. 1500, ed. Q. Hu, *et al.* p. 159

Dalla, S., Marsh, M. S., Kelly, J. & Laitinen, T. 2013, *J. Geophys. Res.*, 118, 5979

Dierckxsens, M., *et al.* 2015, *Solar Phys.*, 290, 841

Luhmann, J. G., *et al.* 2010, *Adv. Space Res.*, 46, 1

Marsh, M. S., Dalla, S., Kelly, J. & Laitinen, T. 2013, *Ap. J.*, 774, 4

Marsh, M. S., Dalla, S., Dierckxsens, M., Laitinen, T. & Crosby, N. B. 2015, *Space Wea.*, 13, 386

Onsager, T. G., *et al.* 1996, in *GOES-8 and Beyond, Proceedings of SPIE*, ed. E. R. Washwell, vol. 2812, p. 281

Swalwell, B., Dalla, S. & Walsh, R. W. 2017, *Solar Phys.*, 292, 173

Space Weather of the Heliosphere: Processes and Forecasts
Proceedings IAU Symposium No. 335, 2017
C. Foullon & O.E. Malandraki, eds.

doi:10.1017/S1743921317011607

High Speed Solar Wind Forecast Model from the Solar Surface to 1AU Using Global 3D MHD Simulation

Mitsue Den[1], Takashi Tanaka[1,2], Yuki Kubo[1] and Shinichi Watari[1]

[1]National Institute of Information and Communications Technology,
4-2-1 Nukui-Kitamachi, Koganei, Tokyo 184-8795, Japan
email: den@nict.go.jp
[2]Kyushu University,
744 Motooka, Nishi-ku, Fukuoka 819-0395, Japan

Abstract. Emanating from coronals holes (CHs), high speed streams (HSSs) cause recurrent geomagnetic disturbances in the Earth's magnetosphere. For this reason being able to predict the occurrence and timing of the high speed solar wind is one of the more important issues in space weather forecasting. Currently, it is still difficult to estimate the effect of a CH in case that it extends from high latitudes to lower ones. To monitor the global solar wind condition we have therefore developed a three-dimensional MHD simulation code, the REProduce Plasma Universe (REPPU) code, that is driven by the solar magnetic field from the solar surface to 1AU. The connectivity of magnetic field lines from CHs to Earth's orbit via HSSs has been investigated. Simulation results are presented and the usefulness of our model is discussed.

Keywords. Space weather, MHD, methods: numerical, Sun: corona (Sun:) solar wind, solar-terrestrial relations

1. Introduction

Coronal holes (CHs) are the origin of high speed streams (HSSs) and generally occur near the minimum of the 11-year solar activity cycle. When they interact with slow speed streams, it can result in co-rotating interaction regions that can cause temporary disturbances in Earth's magnetosphere. Thus, being able to reproduce the global solar wind structure is one of the important issues for space weather to consider when forecasting the space weather.

Several MHD models have been developed to simulate the solar wind structure ranging from the outer region of the sun (about 20 - 50 solar radii) to a distance of several AU (e.g., Shiota and Kataoka 2016, Odstrcil *et al.* 2005) or from the photosphere to interplanetary space (e.g., Nakamizo & Tanaka *et al.* 2009, Toth *et al.* 2005, Riley *et al.* 2001).

The REProduce Plasma Universe (REPPU code), developed by Tanaka (Tanaka 1994, Nakamizo & Tanaka *et al.* 2009) extends from the solar surface to about 200 solar radii. One of the distinguishing features of this simulation model is the grid structure used for the spherical surface, which has no polar singularity or no seam. This can avoid congested fine grids around the poles. In this way, we are able to reproduce both CHs and the global solar wind structure, thus allowing us to identify the connectivity of a CH and the associated HSS. It is expected that our results will be useful for assessing the potential severity of a CH. Our model is now being used for solar wind forecasting at our institute and provides a solar wind forecast every day. In this paper, we briefly describe the REPPU code and present several simulation results.

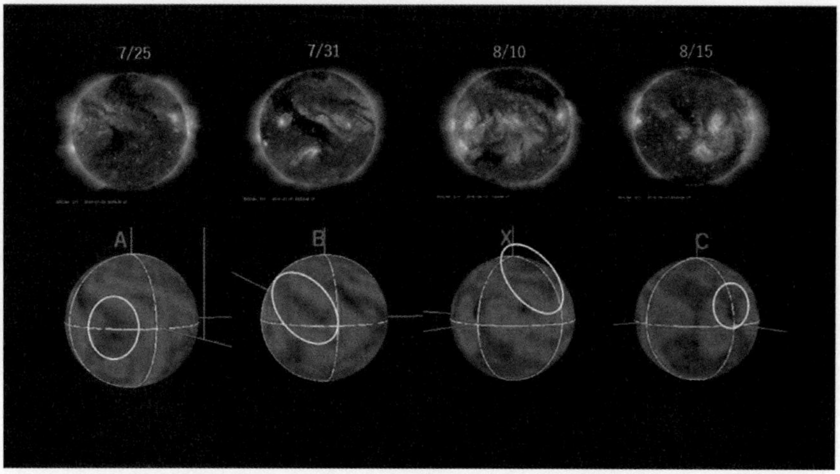

Figure 1. CH observations using SDO AIA211 data (upper panels) and corresponding color contour maps of CH plasma temperature obtained by simulation (lower panels).

2. Method

The simulation method and the model used in this work are described in Den *et al.* (2015) and Nakamizo & Tanaka *et al.* (2009). Here we depict different points from our previous papers. The simulation region extends to 217 solar radii. The number of grid points is 217 in the radial direction with a minimum resolution of 0.02 solar radii and a maximum resolution of 2.7 solar radii. We start by calculating the potential magnetic field of the solar corona using GONG observational data of the photospheric magnetic field and obtain a steady state representation of the solar wind.

We solve three dimensional MHD equations with heating terms including a volumetric heating function and Spitzer-type thermal conduction. Our heating function consists of the expansion factor of the solar magnetic field, the e-folding decay, and the intensity of the heating function. The decaying scale is here tuned to 0.7 solar radii.

3. Simulation Results

The period of GONG data used here is 2017/7/14-8/19. Figure 1 presents the CHs that occurred in this period on four different days. In Figure 1 (In the upper panels of Figure 1), CHs are observed as dark images in the SDO AIA211 data, and the lower panels present the corresponding simulation results as color contour maps of the plasma temperature. We label each simulated CH as A, B, X, and C to clarify the source of the solar wind observed at the L1 point. The CH in the third column (X) crossed the central meridian on Aug. 7: however, there are not observational data for the period Aug 7 - 9, so we used the data obtained on Aug. 10 instead. In each of the four cases, the simulation results reproduce almost all the features of the CHs.

In the upper panel figures of Figure 2 magnetic field lines are drawn starting from the identified CHs, and the same lines are mapped on the color contour plots of the solar wind speed on Earth's orbital plane (lower panels). It can be seen that the extended field lines correspond to the HSS region. These results present the magnetic connectivity between the CHs and the high speed solar wind region.

Figure 3 presents a plot of the simulated solar wind speed at the L1 point (top panel), the real-time solar wind velocity data observed by DSCOVR (middle panel) and the local geomagnetic K index, which quantifies the impact of space weather phenomena

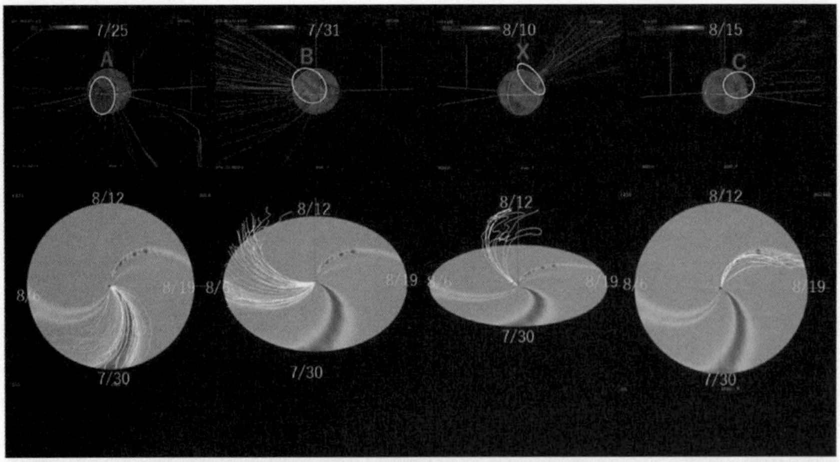

Figure 2. Magnetic field lines are drawn starting from the CHs (upper panel figures), where the red and blue lines represent an away and a toward direction respectively. The same field lines (in white) are also shown with the color contour of the solar wind speed on Earth's orbital plane (lower panel figures).

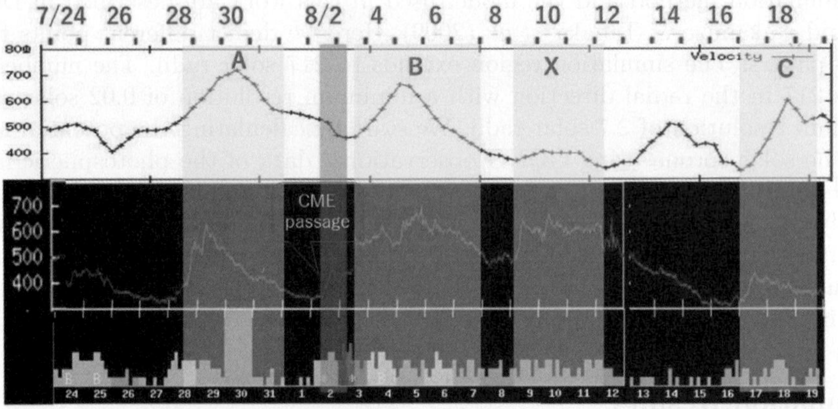

Figure 3. Simulated and measured solar wind velocity at the L1 point (upper and middle panels) and corresponding geomagnetic K-index (bottom panel) for the four cases.

on the Earth's magnetic field (bottom panel). The yellow and red bars of the K index imply significant disturbances in the horizontal component of Earth's magnetic field. The three yellow transparent vertical bars present high stream regions having corresponding simulated CHs (labeled A, B, C in Figures 1 and 2). The blue transparent bar represents the time interval when a CME passed, which is not an object of our study in this paper. The grey transparent bar also shows a high stream region: however, the simulated solar wind was not found to be high. This observed stream is thought to originate from the CH labeled X in Figures 1 and 2. The simulated magnetic field moved to a high latitude region and did not reach Earth's orbital plane (Figure 2). This discrepancy may be due to the evaluation of the expansion rate of the magnetic field, which is one of the problems of our model. It is interesting that the CH labeled C cannot be seen in the SDO data (Figure 1), but the simulation result and data observed by DSCOVR show that a medium speed solar wind originated from it (Figure 3).

4. Concluding Remarks

We developed a 3D MHD simulation model, the REPPU code, for simulating CHs and the global solar wind structure. We showed that our simulation can reproduce both CHs and associated HSSs in the solar wind and additionally identify the connectivity of magnetic field lines in HSSs to CHs. It is still difficult to assess the effect of a mid-latitude CH since this needs more precise modeling of the CHs, which is one of our future works.

Acknowledgements

We acknowledge the Global Oscillation Network Group (GONG) program, Solar Dynamics Observatory, NOAA DSCOVR satellite team, and Kakioka Magnetic Observatory for utilizing observational data obtained by the program or projects.

References

Den, M., Tanaka, T., Kubo, Y. & Watari, S. 2015, *Proceedings of Science, ICRC2015*, 184
Nakamizo, A., Tanaka, T., Kubo, Y., Kamei, S., Shimazu, H. & Shinagawa, H. 2009, *J. Geophys. Res.*, 114, A07109
Shiota, D. & Kataoka, R. 2016, *Space Weather*, 14, 56
Odstrcil, D., Pizzo, V. J. & Arge, C. N. 2005, *J. Geophys. Res.*, 110, A02106
Toth, G., Sokolov, I. V., Gombosi, T. I., *et al.* 2005, *J. Geophys. Res.*, 110, A12226
Riley, P., Linker, J. A. & Mikic, Z. 2001, *J. Geophys. Res.*, 106, 15889
Tanaka, T. 1994, *J. Comp. Phys.*, 111, 381

Space Weather of the Heliosphere: Processes and Forecasts
Proceedings IAU Symposium No. 335, 2017
C. Foullon & O.E. Malandraki, eds.

doi:10.1017/S1743921317007359

The HelMod Monte Carlo Model for the Propagation of Cosmic Rays in Heliosphere

Matteo J. Boschini[1,3], Stefano Della Torre[1], Massimo Gervasi[1,2], Davide Grandi[1,2], Giuseppe La Vacca[1,2], Simonetta Pensotti[1,2], Pier Giorgio Rancoita[1], Davide Rozza[1,2] and Mauro Tacconi[1,2]

[1]INFN Sezione di Milano Bicocca, I-20126 Milano, Italy
email: stefano.dellatorre@mib.infn.it

[2]Università degli Studi di Milano Bicocca, I-20126 Milano, Italy
[3]Cineca, Segrate, Italy

Abstract. The heliospheric modulation model HelMod solves the transport-equation for Galactic Cosmic Ray propagation through the heliosphere down to Earth. It is based on a 2-D Monte Carlo approach that includes a general description of the symmetric and antisymmetric parts of the diffusion tensor, thus properly treating the particle drift effects as well as convection within the solar wind and adiabatic energy loss. The model was tuned in order to fit 1) the data observed outside the ecliptic plane at several distances from the Earth and 2) the spectra observed near the Earth for both, high and low solar activity periods. Great importance was given to description of polar regions of the heliosphere. We present the flux for protons, antiprotons and helium nuclei computed for solar cycle 23-24 in comparison with experimental observations and prediction for the full solar cycle 24.

Keywords. Sun: activity, cosmic rays, diffusion, elementary particles, interplanetary medium, ISM: general, Sun: heliosphere, methods: numerical

---

1. Introduction

The intensity of Galactic Cosmic Rays (GCRs) observed at Earth changes with time as a result of solar activity. The overall effect of heliospheric propagation on the spectral flux of GCRs is called *solar modulation*. In this paper we present the HelMod Model for the description of the solar modulation, as reported by Boschini *et al.* (2017a), and its web interface. A stand-alone python module, fully compatible with GALPROP Porter *et al.* (2017), was developed for the calculation of solar modulation effects, resulting in a newly suggested set of local interstellar spectra as reported by Boschini *et al.* (2017b). Finally we present forecasts for GCRs intensities of Protons and Helium nuclei up to the next solar minimum that we estimate to occour in 2020.

2. HelMod Model for GCR Propagation in the Heliosphere

GCR propagation in the heliosphere was first studied by Parker (1965), who formalized the transport equation, also referred to as Parker equation (see, e.g., discussion in Bobik *et al.* 2012). The Parker equation can be solved by means of numerical techniques. The HelMod code computes the CR propagation from the Termination Shock (TS) down to the Earth orbit using a Monte Carlo approach (for details, see description in Bobik *et al.* 2016). The present form of diffusion parameter (as defined in Boschini at al. 2017a) includes a scale correction factor related to drift contribution. As discussed by Boschini *et al.* (2017b) this correction is evaluated for protons during positive HMF polarity period

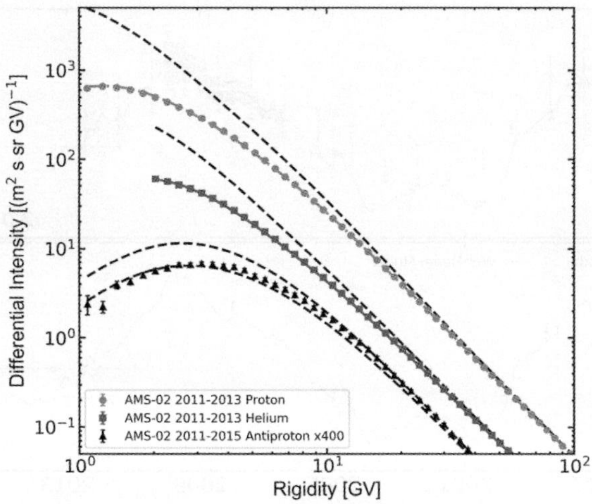

Figure 1. Differential intensity of galactic protons (red points), helium nuclei (blue squares) and Antiprotons (black triangle) measured by AMS-02 compared with modulated spectra from HELMOD. Antiproton spectrum is scaled multiplying by a factor 100. Black dashed lines are the GALPROP LISs. All presented simulated spectra are in close agreement with observations (see text).

to account the presence of latitudinal structure in the Galactic CRs spatial distribution. More details on the HELMOD model and its implementation can be found in Bobik *et al.* (2012), Bobik *et al.* (2013) and Boschini *et al.* (2017a).

As discussed by Boschini *et al.* (2017a), the validity of the HELMOD Model is verified down to 1 GV in Rigidity (1 GeV in equivalent Kinetic energy for electrons). Lower energies are not considered in the present work since they need further improvement in the description of solar modulation in the outer heliosphere – from TS up to interstellar space – as well as the inclusion of turbulence in drift coefficient evaluation.

2.1. *Local Interstellar Spectrum*

The local interstellar spectrum (LIS) is assumed to be isotropic along the heliosphere boundary. LIS parametrization is constrained by measurements mainly from Voyager probes, at low energy, and AMS-02, at high energy. As described by Boschini *et al.* (2017b), in order to derive the physically motivated LIS of CR species, an iterative procedure was developed to feed the GALPROP output into HELMOD to compare with AMS-02 data as observational constraints. A Markov Chain Monte Carlo (MCMC) interface to GALPROP was developed and described in Boschini *et al.* (2017b). Main propagation parameters were treated as free parameters in the scan using the 2D GALPROP model (Masi 2016). Parameters of the injection spectra, such as spectral indices and the break rigidities, were also free parameters, but their exact values depend on the solar modulation. As a matter of fact, the low energy part of the spectra are tuned together with the solar modulation parameters. In the scan we used all published AMS-02 data on protons (Aguilar *et al.* 2015a), helium (Aguilar *et al.* 2015b), B/C ratio (Aguilar *et al.* 2016a) and electrons (Aguilar *et al.* 2014), while antiproton data (Aguilar *et al.* 2016b) were explicitly excluded in order to make a prediction of the antiproton spectrum based on GCR primaries. The so obtained set of LISs was presented in Boschini *et al.* (2017b) and reported in Figure 1. In Figure 1 the differential intensity of galactic protons, helium nuclei and antiprotons, observed by AMS-02, are compared with spectra modulated by HELMOD.

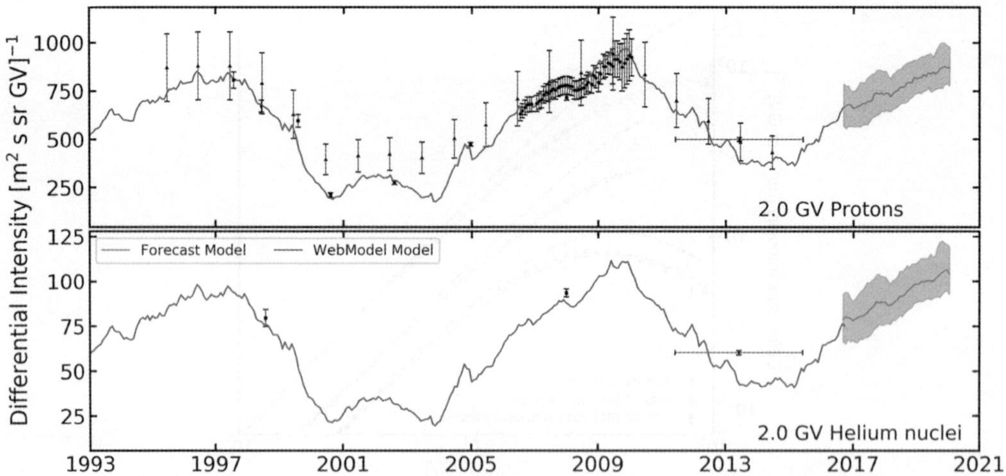

Figure 2. Differential intensity of galactic proton and Helium nuclei as a function of time. Experimental points are described in the text. Blue solid lines are obtained by HELMOD on-line solar modulator. Red solid lines are the forecast of protons and helium nuclei up to the end of 2020; shaded bands represents 20% uncertainty.

3. Long term solar modulation and forecasts

To simplify the calculations, we developed a python script reading the GALPROP output and providing the modulated spectrum for periods of selected experiments. The module, as described in Section 3.1 of Boschini *et al.* (2017b), uses pre-evaluated normalized probability functions from HELMOD thus this method dramatically accelerates the modulation calculations while maintaining the accuracy of the full-scale simulation.

The HELMOD python module can be downloaded from a dedicated website or used on-line†. The LISs showed in Figure 1 are provided as default choice, however the user can also specify their own LIS in GALPROP output or plain-text file formats. The on-line interface allows a direct comparison with selected experimental data sets or provide modulated spectra for each Carrington rotation between the years 1996 and 2016. In Figure 2 we show the differential intensity of protons and helium nuclei GCRs at 2 GV as a function of time (blue solid line) computed by means of on-line solar modulator. Simulations are compared with experimental observations at same rigidity provided by SOHO/EPHIN (black triangle, Kühl *et al.* 2016), BESS (black squares, Shikaze *et al.* 2007, Abe *et al.* 2016), PAMELA (black points, Adriani *et al.* 2013) and AMS (black stars, Aguilar *et al.* 2002, 2015a, 2015b). Simulations agree well with experimental data within one standard deviation (two standard deviations at solar maximum for SOHO/EPHIN observations only).

3.1. *Solar Modulation forecast*

HELMOD is also able to provide forecast of GCRs intensity. Using a procedure derived from Owens *et al.* (2011) and Owens *et al.* (2012), we are able to forecast the number of smoothed sunspot number (SSN), tilt angle of neutral sheet, magnetic field at Earth and Solar Wind velocity in forthcoming years. From the analysis of previous 10 solar cycles we derived the average variation of parameters along an ideal solar cycle of fixed average duration. The length and amplitude of the variation are then rescaled in order to fit measured values of SSN during Solar Cycle 24. The analysis found a cycle duration

† HELMOD (version 3.4), 2017, http://www.helmod.org.

that is consistent with 11 years, thus, since the beginning of Solar Cycle 24 is set to be in 2009, the next solar minimum is expected to occur at the end of 2020. In Figure 2 we report in red the forecast of protons and helium nuclei intensity from end 2016 to 2020. The shaded bands represent the uncertainties, i.e. $\sim 20\%$, that is dominated by uncertainties on sunspot number forecast.

4. Conclusion

We presented the 2-D Monte Carlo Heliospheric Modulation Model, i.e. HELMOD, to evaluate GCR modulated spectra at Earth. The model includes details on the individual processes occurring in the heliosphere particle propagation, thus providing a comprehensive description of the solar modulation. The model is tuned by using several proton observations at Earth in different solar activity periods, treating in a unified description the high and low solar activity periods. Moreover HELMOD, jointly with GALPROP and thanks to the precise observations of AMS-02, provided unprecedentedly tight constraints on proton and Helium LISs. To make the calculation available to the community, a python module and a dedicated website are developed to provide modulated spectra for selected GCR experiments in different solar activity periods. The model now includes also the forecast of future GCRs intensity until the end of cycle 24. The model estimate the next solar minimum will occur in 2020.

Acknowledgements

We wish to specially thank Pavol Bobik, Giuliano Boella, Karel Kudela, Marian Putis and Mario Zannoni for their support to the HELMOD code and many useful suggestions. This work is supported by ASI (Agenzia Spaziale Italiana) under contract ASI-INFN I/002/13/0 and ESA (European Space Agency) contract 4000116146/16/NL/HK.

References

Abe, K. & BESS Collaboration 2016, *ApJ*, 822, 65
Adriani, O. & PAMELA Collaboration 2013, *ApJ* 765, 91
Aguilar, M. & AMS Collaboration 2002, *Phys. Rep.*, 366, 331
Aguilar, M., *et al.* 2014, *Phys. Rev. Lett.*, 113, 121102
Aguilar, M., *et al.* 2015a, *Phys. Rev. Lett.*, 114, 171103
Aguilar, M., *et al.* 2015b, *Phys. Rev. Lett.*, 115, 211101
Aguilar, M., *et al.* 2016a, *Phys. Rev. Lett.*, 117, 231102
Aguilar, M., *et al.* 2016b, *Phys. Rev. Lett.*, 117, 091103
Bobik, P., Boella, G., Boschini, M. J., *et al.* 2012, *ApJ*, 745, 132
Bobik, P., Boella, G., Boschini, M. J., *et al.* 2013, *Advances in Astronomy*, 2013, 793072
Bobik, P., Boschini, M. J., Della Torre, S., *et al.* 2016, *J. Geophys. Res.*, 121, 3920
Boschini, M. J., Della Torre, S., Gervasi, M., *et al.* 2017a, *Adv. Space Res*, in press
Boschini, M. J., Della Torre, S., Gervasi, M., *et al.* 2017b, *ApJ*, 840, 115
Jokipii, J. R., Levy, E. H. & Hubbard, W. B. 1977, *ApJ*, 213, 861
Kühl, P., Gómez-Herrero, R. & Heber, B. 2016, *Solar Phys.*, 291, 965
Masi, N. 2016, Il Nuovo Cimento 39 C, 298
Owens, M. J., Lockwood, M., Barnard, L. & Davis, C. J. 2011, *Geophys. Res. Lett.*, 38, L19106
Owens, M. J., Usoskin, I. & Lockwood, M. 2012, *Geophys. Res. Lett.*, 39, L19102
Parker, E. N. 1965, *Plan. Space Sci.*, 13, 9
Porter, T. A., Jóhannesson, G. & Moskalenko, I. V. 2017, *ApJ*, 846,67
Shikaze, Y. & BESS Collaboration 2007, *Astrop. Phys.*, 28, 154

Space Weather of the Heliosphere: Processes and Forecasts
Proceedings IAU Symposium No. 335, 2017
C. Foullon & O.E. Malandraki, eds.

doi:10.1017/S1743921318000741

The Space Weather through a Multidisciplinary Scientific Approach

Zouhair Benkhaldoun[1], Malki Khalifa[1], Aziza Bounhir[1], Nicole Vilmer[2], Khaoula El bouyahyaoui[1], Mohamed Kaab[1], Amine Lagheryeb[1] and Ahmed Daassou[1]

[1]Oukaimeden Observatory, LPHEA, FSSM, Cadi Ayyad University, Marrakech, Morocco
[2]LESIA,Observatoire de Paris, 5 Place Janssen, 92195-Meudon Cedex, France

Abstract. Space weather encompasses understanding how the near-space environment responds to forces from lower-atmosphere weather systems as well as conditions on the Sun. Although the specific effects of space weather (including power grid failures, communication outages, and navigation errors when using space-based navigation systems such as GPS) are local in nature, understanding and predicting their occurrence requires a global view of the environment. Here we initiated a first attempt to link one solar event which occurred on 2014 February 25, and affected the Earth's upper atmosphere.

Keywords. Ionosphere, Airglow, Solar Eruption

---

1. Introduction

We adopted a multidisciplinary approach at the Cadi Ayyad University in Morocco to better understand the Sun-Earth interaction. Our objective is to identify the various correlations, which may exist between the phenomena observed by various instruments. For this we have deployed an experiment dedicated to the study of the Ionosphere - RENOIR - with the cooperation of the University of Illinois, USA. We also have installed a GPS station in cooperation with MIT and have started cooperation with the LESIA of Paris Observatory for the study of solar eruptions through the RHESSI satellite in particular. We present here a first attempt to combine the observations from these various facilities in the context of a solar event which occurred on February 25, 2014 in order to find a causal relationship between the solar activity and the disturbances of the upper layer of the terrestrial atmosphere. We explore here the ionospheric and thermospheric responses to the February 27-28 geomagnetic storm. We assume that this storm was caused by a flare-associated Coronal Mass Ejection (CME) that occurred on February 25 2014.

2. The Solar Event

The X4.9 flare of February 25, 2014 took place in a newly numbered sunspot region NOAA 11990 (S12^o E82^o) which was the sunspot region NOAA 11967 during the previous rotation studied in detail by Chen *et al.* (2014). Figure 1 shows shows the continuum image map of NOAA 11990 observed on February 25, 2014 with the Helioseismic and Magnetic Imager (HMI) instrument onboard Solar Dynamic Observatory (SDO; Schou *et al.* 2012). The Helioseismic and Magnetic Imager is a filtergraph with the full−disk coverage by two CCD cameras of 4096×4096 pixels. NOAA 11990 produced only one X−class flare during that day. The Geostationary Operational Environmental Satellite (GOES) Data indicates that the X-ray flare started at 00:39 UT, peaked at 00:49 UT, and ended at 01:03 UT.

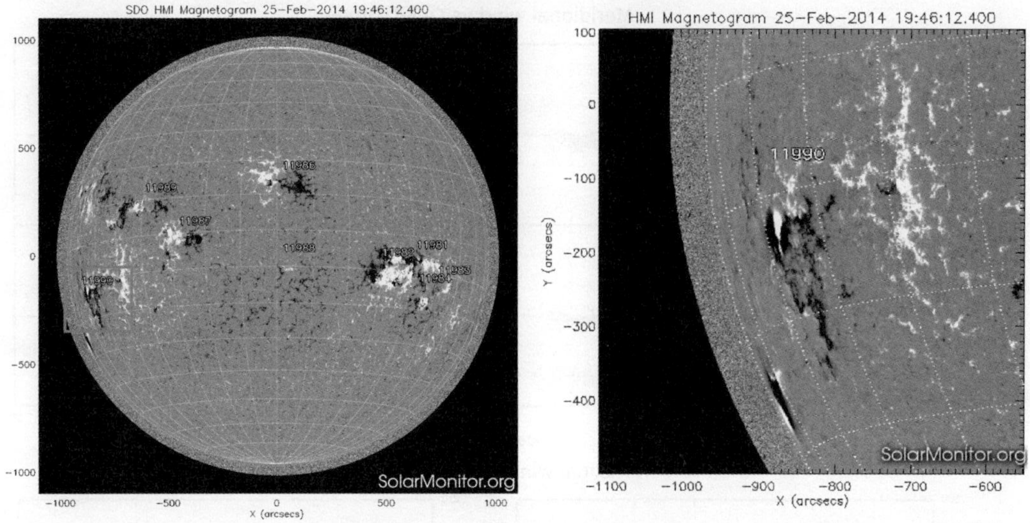

Figure 1. Left: SDO HMI Magnetogram of the solar disk obtained with HMI/SDO on February 25, 2014 at 19:46 UT. Active Region AR 11990 is indicated by the orange square. Right: Image centred on AR 11990 observed with HMI/SDO.

The X4.9 flare, corresponding to a 1−8 SXR produced an energy flux of 4.9×10^{-4} W/m^2 measured at Earth, assuming isotropic emission from the flare source. This represents an instantaneous luminosity of $\sim 1.1\times10^{26}$ ergs/s, or $\sim 3\times10^{-8}$ of the total solar luminosity from the SXRs alone.

3. Thermospheric and Ionospheric Response

Using GNSS measurements and prediction models Bergeot *et al.* (2014) classified the February 27, 2014 ionospheric storm as moderate based on the value of Dst index.

3.1. *RENOIR @ Oukaimeden*

The RENOIR (Remote Equatorial Nighttime Observatory of Ionospheric Region) experiment consists of a Fabry−Perot (FPI) interferometer and wide angle viewing camera. The FPI makes measurements of the thermospheric neutral wind velocities and neutral temperatures using observations of the 630.0 nm emission caused by the dissociative recombination of O_2^+. The wide−angle imaging system uses the same airglow emissions to provide measurements of ionospheric structures and irregularities. The main goal of this experiment is to characterize the mid-latitude ionosphere and thermosphere by establishing the climatology of both the neutral winds and the instabilities taking place in the ionosphere as well as the coupling between the ionosphere and thermosphere during quiet time conditions and during geomagnetic storms. The thermospheric storm, induced by high latitude heat input drives a global circulation of winds flowing from high to low latitudes creating large scale of atmospheric waves that propagates in a global scale. The local impression of this global picture of the storm was captured during the February 27th event. Figure 2 shows the meridional and zonal winds measured with the Fabry−Perot interferometer (FPI) during the 24, 25, 27 and the February 28th, 2014. The February 26 measurements are missing. The FPI winds are estimated from the 630 nm airglow emission caused by dissociative recombination of O_2^+. The red line refers to the monthly average of the nights with Kp strictly less than four. The blue line shows the meridional

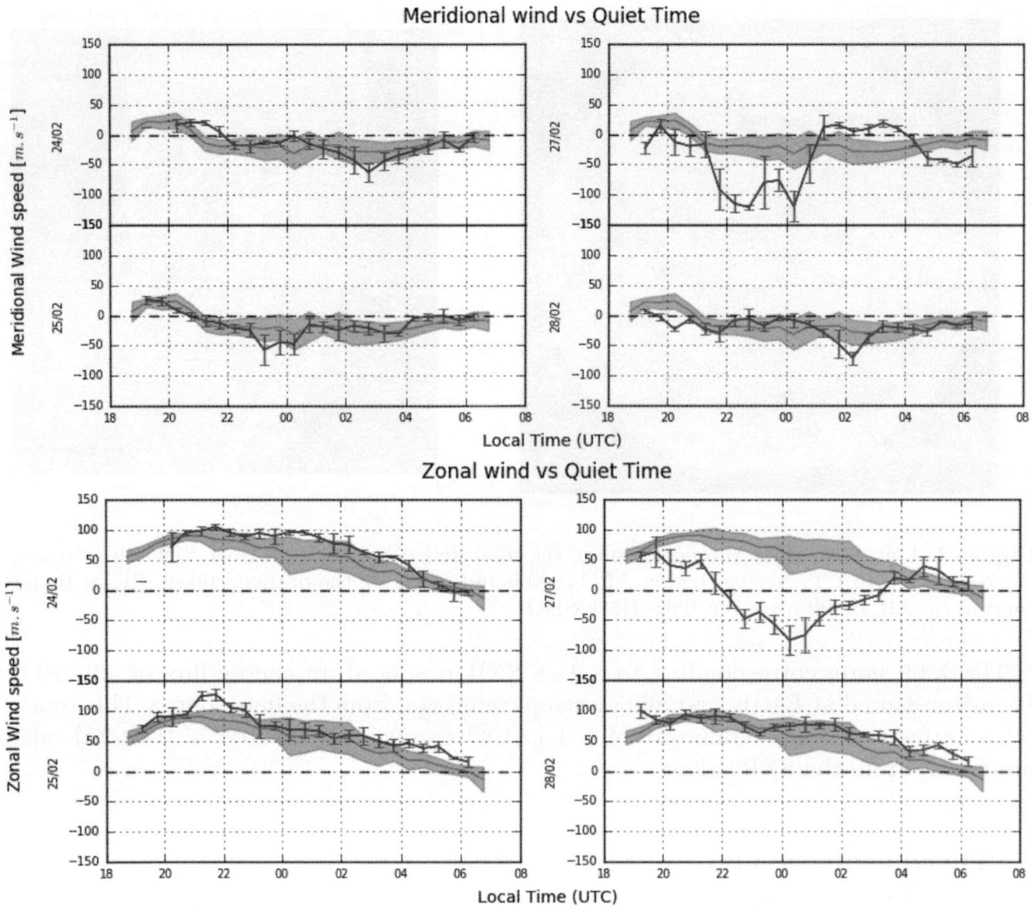

Figure 2. Thermospheric meridional and zonal winds measured by the the Fabry-Perot interferometer (FPI) over Oukaimeden Observatory (31°12′ 23.3″N, 7°51′ 58.8″ W).

(zonal) measurements during the night and the magenta area denotes the daily variability during February 2014. The average quiet time is in red color and the quiet time daily variability is illustrated as the shaded purple area.

3.2. *GPS Observations*

The estimated total electron content (TEC) measured from a GPS station installed in Rabat (33.998°N; 353.1457°E,geographic) is illustrated in Figure 3. This figure contains the TEC for the February 27-28, 2014 and the average of the quiet days ($Kp<3$). We can notice some oscillation of the 27th TEC between 14 LT and 00 LT and in the early hours of the 28th which correlate positively with NmF2 peak (the F2-layer peak electron density which is derived from the F2-layer critical frequency). As the meridional wind plays an important role in the electrodynamics of the ionosphere, we present in Figure 3 the evolution of the TEC along with the north and south component of the winds. We can notice as well that the positive storm on February 27 with a maximum TEC of 65 TECU (around 13 UT), remaining high for some hours and achieving 50 TECU at 00 UT. On the early hours of 28th, from 01 to 04 UT, the decrease is steep with an approximate rate of 8,3 TECU/hour. During the 28th, the TEC was still higher than

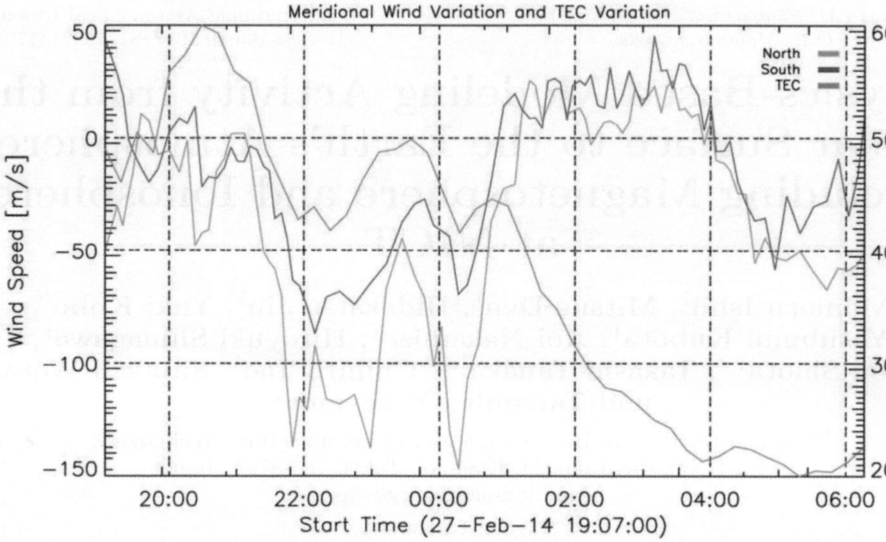

Figure 3. Temporal variation of the Total Electron Content measured over Rabat (33.998°N; 353.1457°E, geographic) overlay thermospheric meridional and of the thermospheric meridional and zonal winds measured by the the Fabry-Perot interferometer (FPI) over Oukaimeden Observatory (31°12′ 23.3″ N, 7°51′ 58.8″ W).

quiet time behavior. The observed gap during the early hours was about 40 TECU, and about 20 TECU during the day and diminishes to 10 TECU during the night.

4. Conclusion

The effect of the X4.9 flare of the February 25, 2014 solar event on the thermosphere and the ionosphere is investigated. Thermospheric winds and temperature were inferred from Fabry-Perot interferometer measurements deployed at Oukaimeden observatory. The ionospheric response of the storm was also analyzed with the VTEC (Vertical Total Electron Content) data and a wide-angle imaging camera. Attention was paid to the conditions on the Sun, interplanetary medium and geomagnetic indices that impacted the thermosphere-ionosphere coupling during that storm. This is the first time that a case study of a geomagnetic storm has been achieved in north Africa by using Fabry-Perot interferometer data of the thermosphere. Additionally, this is the first coincident study of a storm using a ground-based FPI and an all-sky imager in this sector. The African sector is lacking data and it is important to see how the thermosphere/ionosphere system reacts to geomagnetic storms in this area. Furthermore, the comparison with models is useful for validation or improvement. Our perspective is to address the variable ways in which the thermosphere-ionosphere system reacts to geomagnetic storms in this region. We are about to publish more detailed results concerning this event in a forthcoming paper.

References

Chen H., *et al.* 2014, *The Astrophysical Journal Letters*, 797, 2, L15
Bergeot C., *et al.* 2014, *Journal of Space Weather and Space Climate*, 4, A31, 2014
Schou J., *et al.* 2012, *Solar Phys.* 275, 229

Space Weather of the Heliosphere: Processes and Forecasts
Proceedings IAU Symposium No. 335, 2017
C. Foullon & O.E. Malandraki, eds.

doi:10.1017/S1743921317011668

Physics-Based Modeling Activity from the Solar Surface to the Earth's Atmosphere Including Magnetosphere and Ionosphere at NICT

Mamoru Ishii[1], **Mitsue Den**[1], **Hidekatsu Jin**[1], **Yuki Kubo**[1], **Yasubumi Kubota**[1], **Aoi Nakamizo**[1], **Hiroyuki Shinagawa**[1], **Daikou Shiota**[1,2], **Takashi Tanaka**[1,3], **Chihiro Tao**[1], **Shinichi Watari**[1] **and Tatsuhiro Yokoyama**[1]

[1]National Institute of Information and Communications Technology,
4-2-1 Nukui-Kitamachi, Koganei, Tokyo 184-8795, Japan
email: den@nict.go.jp

[2]Nagoya University,
Furo-cho, Chikusa-ku, Nagoya, Aichi 464-8601, Japan

[3]Kyushu University,
744 Motooka, Nishi-ku, Fukuoka 819-0395, Japan

Abstract. We present the development of physics-based models of solar-terrestrial regions from the solar surface to the Earth's atmosphere at NICT. Our models consist of three regions: (1) the solar surface and solar wind, (2) the Earth's magnetosphere-ionosphere, and (3) a model of the whole atmosphere from the troposphere to the ionosphere, called the Ground to Topside Model of Atmosphere and Ionosphere for Aeronomy (GAIA). We also have a solar wind and CME model, Space-weather-forecast-Usable System Anchored by Numerical Operations and Observations (SUSANOO). Furthermore, we have developed a high-resolution plasma bubble model. The coupling of these models is a future work.

Keywords. Space weather, MHD, methods: numerical, (Sun:) solar wind, solar-terrestrial relations

---

1. Introduction

Space weather is the electromagnetic condition in the near-Earth space which affects human social activities, such as radio telecommunications, broadcast, satellite positioning, power grid and aviation. The importance of space weather information is increasing in our society with a high ICT use. It is necessary to observe space weather with space- and ground-based instruments to monitor the present conditions. In addition, it is also important to develop and use models and simulation codes for space weather forecasting. The ultimate goal should be to establish a space weather forecast system with theoretical models using data assimilation; however, there are some difficulties in achieving this at present. The area we need to observe is very large and the observation points are few. There are still some important but unknown processes in space weather phenomena. Given this situation, it is meaningful to develop empirical models. For a long time, NICT has been developing both empirical and theoretical models of the sun, solar wind, magnetosphere and ionosphere. They have been used for research on the electromagnetic processes in space weather. Our physics-based models cover the solar surface to the Earth's atmosphere. The models for the ambient solar wind and the magnetosphere were developed using the same MHD simulation code called the REProduce Plasma Universe

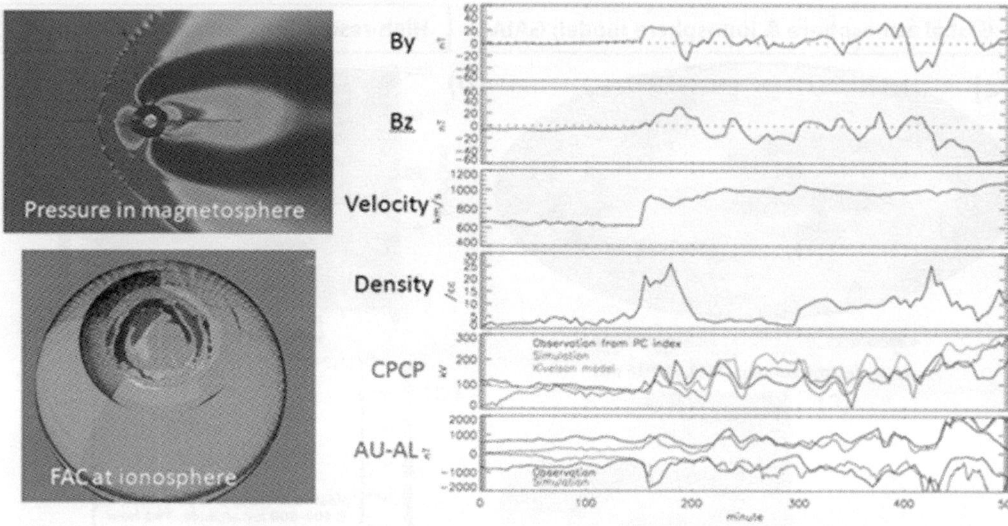

Figure 1. Simulation results of the Bastille Day event

(REPPU) code. (Den *et al.* 2015; Nakamizo *et al.* 2009; Moriguchi *et al.* 2008) One of the features of the REPPU code is that there is no singular point in the unstructured grid system, which makes it possible to simulate the global solar wind structure from about 1 solar radius to 1AU and to perform calculation with the uniform accuracy over not only the magnetosphere but also the ionosphere for the case of a strong solar wind such as Bastille event (Kubota *et al.* 2017). GAIA consists of atmosphere, ionosphere, and electrodynamics models, which are coupled with each other (Jin *et al.* 2011). The atmosphere model covers the whole region from the ground to the exobase and treats a full set of physical processes. The ionosphere model solves equations of mass, momentum, and energy for major ion species and electrons. The electrodynamics model treats the closure of global ionospheric currents induced by the neutral wind dynamo.

As REPPU for the solar version (Den *et al.* 2015) and SUSANOO (Shiota & Kataoka, 2016) are described in other papers (see Den *et al.*'s manuscript and Shiota *et al.*'s manuscript), we present information of the REPPU code for the magnetosphere and GAIA with a high-resolution plasma bubble model.

2. Magnetosphere

The numerical global MHD model developed by Tanaka (1994) self-consistently solves the SW-M-I coupling process. To achieve high resolution and accurately capture discontinuities, the MHD calculation employs a finite volume total variation diminishing (TVD) scheme with an unstructured triangular grid system, which was described by Moriguchi *et al.* (2008) and Nakamizo *et al.* (2009). The numbers of triangular grid points are 7682 in the horizontal direction and 240 in the radial direction. The outer and inner boundaries of the simulation are set at 200 and 2.6 Re, respectively.

Figure 1 shows simulation results of Bastille Day event (Kubota *et al.* 2017). The left panels are snapshots of the pressure distribution in the magnetosphere (upper) and the field-aligned current (FAC) in the ionosphere (lower panel). The right panels show the solar wind parameters (magnetic fields of By and Bz components, velocity, and density), the CPCP and the AE index. The horizontal axis (minute) is defined as the number of minutes starting from 12:00 UT on 15 July 2000. The red line indicates simulation

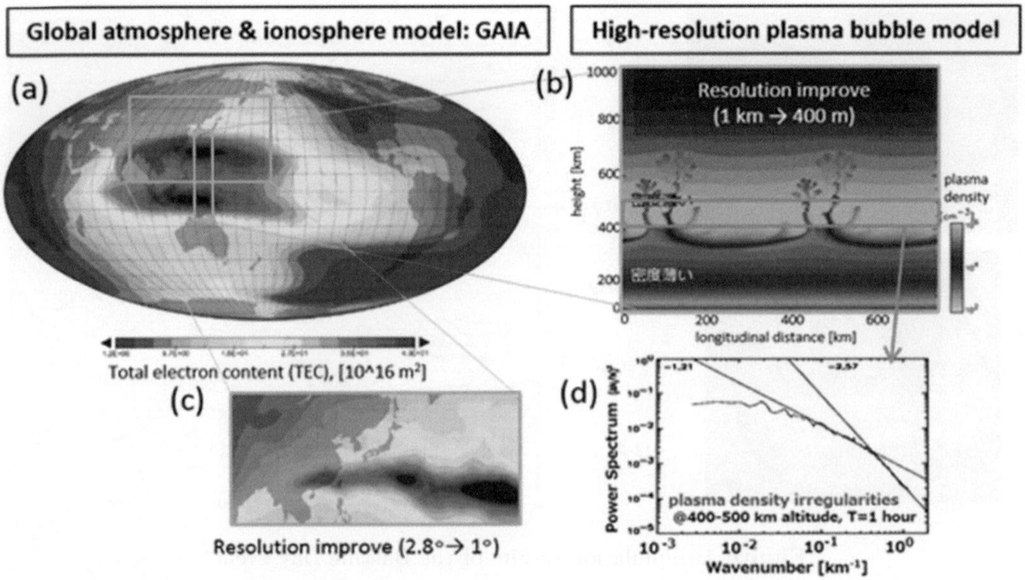

Figure 2. Simulation results using GAIA and high-resolution plasma bubble model

Table 1. Simulation methods of GAIA and Plasma bubble model.

	GAIA (ionosphere)	Plasma bubble model
Region	global	north-south 20 deg, east-west 3-5 deg
Resolution	2.5(lon.) × 1(lat.) deg, 1-2.8 deg (GCM)	0.4 - 1 km
Method	ion chemistry implicity & momentum balance combined with GCM	CIP (constrained interpolation profile) scheme
Referenes	e.g., Jin *et al.* 2011;Shinagawa 2009	e.g., Yokoyama *et al.* 2014

results, and the black line indicates observation values. Our simulation results of the CPCP and the AE index are consistent with the observation.

3. GAIA and plasma bubble model

Ionospheric disturbances affect radio propagation and communications, and cause significant errors in satellite positioning. Ionosphere modeling is one of the approaches to investigate and predict ionospheric disturbances. We have been developing two types of ionospheric model: global and local. The global model is the Ground-to-Topside Model of Atmosphere and Ionosphere for Aeronomy (GAIA), which solves the physical and chemical processes of the whole atmosphere from the troposphere to the exosphere under the interaction with the ionosphere (Jin *et al.* 2011) (Figure 2(a)). The local model is a high-resolution local model used to study the behavior and characteristics of the equatorial plasma bubble (e.g., Yokoyama *et al.* 2014) (Figure 2(b)). We have improved the spatial resolution of both the GAIA and bubble models. The high-resolution version of GAIA captures the ionosphere distribution with a horizontal scale of a few hundred kilometers (Figure 2(c)). The high-resolution plasma bubble model reproduces the ionospheric plasma structure with a scale of less than 1 km, showing the spectral breaking feature seen in previous observations (Figure 2(d)).

References

Den, M., Tanaka, T., Kubo, Y. & Watari, S. 2015, *Proceedings of Science, ICRC2015*, 184

Shiota, D. & Kataoka, R. 2016, *Space Weather*, 14, 56

Tanaka, T. 1994, *J. Comp. Phys.*, 111, 381

Moriguchi, T., Nakamizo, A., Tanaka, T., Obara, T. & Shimazu, H. 2008, *J. Geophys. Res.*, 113, A05204

Nakamizo, A., Tanaka, T., Kubo, Y., Kamei, S., Shimazu, H. & Shinagawa, H. 2009, *J. Geophys. Res.*, 114, A07109

Kubota, Y., Nagatsuma, T., Den, M., Tanaka, T. & Fujita, S. 2017, *J. Geophys. Res.*, 122, doi:10,1002/2016JA023851

Jin, H., Miyoshi, Y., Fujiwara, H., Shinagawa, H., Terada, K., Terada, M., Ishii, M., Otsuka, Y. & Saito, A. 2011, *J. Geophys. Res.*, 116, A01316

Shinagawa, H. 2009, *J. Natl. Inst. Inf. Comm. Tech.*, 56, 1–4, 199

Yokoyama, T., Shinagawa, H. & Jin, H. 2014, *J. Geophys. Res.*, 119, 10474

Space Weather of the Heliosphere: Processes and Forecasts
Proceedings IAU Symposium No. 335, 2017
C. Foullon & O.E. Malandraki, eds.

doi:10.1017/S1743921317007979

Numerical Effects on Wave Propagation in Atmospheric Models

Daniel J. Griffin[1] **and John Thuburn**[2]

[1,2]Department of Mathematics, University of Exeter,
North Park Road, Exeter, EX4 4QF, UK
emails: [1]djg211@exeter.ac.uk, [2]j.thuburn@exeter.ac.uk

Abstract. Ray tracing techniques have been used to investigate numerical effects on the propagation of acoustic waves in a non-hydrostatic dynamical core discretised using an Arakawa C-grid horizontal staggering of variables (Arakawa & Lamb 1977) and a Charney-Phillips vertical staggering of variables (Charney & Phillips 1953) with a semi-implicit timestepping scheme. It is found that the space discretisation places limits on resolvable wavenumbers and redirects the group velocity of waves towards the vertical. Wave amplitudes grow exponentially with height due to the decrease in the background density, which can cause instabilities in whole-atmosphere models. However, the inclusion of molecular viscosity and diffusion acts to damp the exponential growth of waves above about 150 km. This study aims to demonstrate the extent to which numerical wave propagation causes instabilities at high altitudes in atmosphere models, and how processes that damp the waves can improve these model's stability.

Keywords. atmospheric effects, diffusion, methods: numerical, waves.

1. Introduction

There is a current need for whole-atmosphere models to be developed as tools for space weather forecasting. The Met Office Unified Model dynamical core (Wood *et al.* 2014) is to be extended into a whole-atmosphere model, but initial attempts to raise the top boundary of this model give rise to instabilities, due to a combination of numerical effects and missing physical processes (Griffin & Thuburn 2018). The intended outcome of this study is to determine the effect of numerical wave propagation on the dynamical core, and to consider how the model's stability might be improved.

For propagation in a two-dimensional vertical cross section of the atmosphere, a wave packet with wavevector $\boldsymbol{k} = (k, m)$ and frequency ω approximately satisfies the local dispersion relation $\omega = \Omega(k, m; x, z, t)$, where the dependence on position $\boldsymbol{x} = (x, z)$ and time t can arise through variations in the background wind, temperature or stratification. In this work, any dependence on x or t is neglected. The wave packet then propagates along a space-time trajectory (a ray) at the group velocity $\boldsymbol{c}_g$ that depends on $\boldsymbol{k}$ and temperature T and the wave vector also evolves according to the ray tracing equations:

$$\frac{D_{\boldsymbol{c}_g}}{Dt}(\boldsymbol{x}) = \boldsymbol{c}_g(\boldsymbol{k}, T) = \boldsymbol{\nabla}_{\boldsymbol{k}}\Omega, \qquad \frac{D_{\boldsymbol{c}_g}}{Dt}(\boldsymbol{k}) = -\boldsymbol{\nabla}_{\boldsymbol{x}}\Omega, \tag{1.1}$$

(Lighthill 1978). The derivative operator $D_{\boldsymbol{c}_g}/Dt = \partial/\partial t + \boldsymbol{c}_g \cdot \nabla_{\boldsymbol{x}}$ represents the rate of change of a variable with time at a position moving with the group velocity $\boldsymbol{c}_g$. These equations are solved using a forward Euler timestep to describe the evolution of the position and wavevector of a wave packet. The wave energy conservation law:

$$\frac{D_{\boldsymbol{c}_g}}{Dt}(\log W) + \boldsymbol{\nabla}_{\boldsymbol{x}} \cdot \boldsymbol{c}_g + \frac{2}{\tau} = 0, \tag{1.2}$$

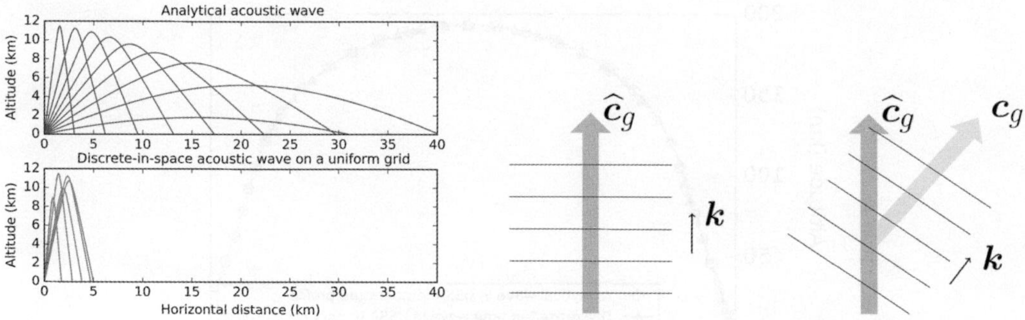

Figure 1. left: A comparison of ray plots of acoustic waves with a total wavenumber of 4×10^{-5} rad m^{-1} initiated with a range of wavevector angles measured from the horizontal, from 0 to $\pi/2$ (above: the analytical wave equation, and below: the spatially-discrete wave equation with $\Delta x = 100$ km, $\Delta z = 1$ km).Right: Illustrations of the directions of analytical group velocities $\boldsymbol{c}_g$ and spatially-discrete group velocities $\widehat{\boldsymbol{c}}_g$ of acoustic waves for varying wavevectors $\boldsymbol{k}$ (left: $k = 0$: the wavevector, analytical and numerical group velocities go in the same direction, and right: $k = \pi/\Delta x$: the largest k that can be considered for the wave to be resolved: $\boldsymbol{c}_g$ goes in the same direction as $\boldsymbol{k}$, but $\widehat{\boldsymbol{c}}_g$ is still vertical).

(Lighthill 1978) describes the evolution of the rate of growth of wave amplitudes, where W represents the wave energy per unit volume, and τ represents the timescale of any processes that act to damp the waves.

From the Navier-Stokes governing equations, a wave equation, and a dispersion relation $\widehat{\omega} = \widehat{\Omega}(\widehat{k}, \widehat{m}; x, z)$ (where the 'hats' denote the effective wavenumbers and frequency as seen by the continuous dispersion relation) and group velocity $\boldsymbol{c_g}$ can be found. The 'hat' variables and their derivatives change depending on whether the continuous equations or the spatially- and/ or temporally-discrete equations are being considered. However, the same ray tracing equations are used in all cases, so the analytical and numerical wave propagation can be directly compared.

2. Overview

The main effects of the space discretisation on wave propagation are twofold. Firstly, the coarse horizontal resolution ($\Delta x \sim 100$ km) compared with the finer vertical grid spacing ($\Delta z \sim 10$ km) means that waves are less well resolved in the horizontal direction than in the vertical. This prevents waves from propagating far horizontally, as demonstrated by Figure 1.

Secondly: there is a limit on horizontal wavenumbers k that may be used for waves to be resolved: $k < \pi/\Delta x$. For $k \sim \pi/\Delta x$, where the wavevector angle is low, the analytical group velocity points in the same direction, but the spatially-discrete group velocity points straight up, as seen in Figure 2. This causes excessive amounts of wave energy to be channelled upwards into the thermosphere in atmospheric models.

The time discretisation has the effect of slowing wave propagation, but maintaining the trajectory of the wave packet, as shown by Figure 3. This effect is much greater on high frequency acoustic waves than on gravity waves.

The wave amplitude grows exponentially with height due to the decrease in the background density, and varies with temperature. Molecular viscosity is a real physical process that reduces the growth of wave amplitudes in the thermosphere above ~ 130 km, as shown by Figure 4. Off-centering the semi-implicit timesteps can also artificially damp wave amplitudes throughout the whole atmosphere.

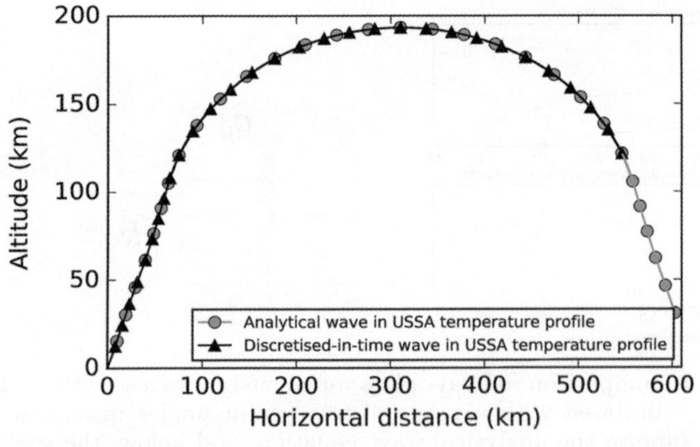

Figure 2. Ray plot of acoustic waves with a total wavenumber of 1×10^{-3} rad m^{-1}, initiated at an angle 0.3π using either the analytical wave equation or the temporally-discrete wave equation with a centred-in-time timestep of 3 seconds. The dots represent one-minute intervals and the simulation was run for 30 minutes.

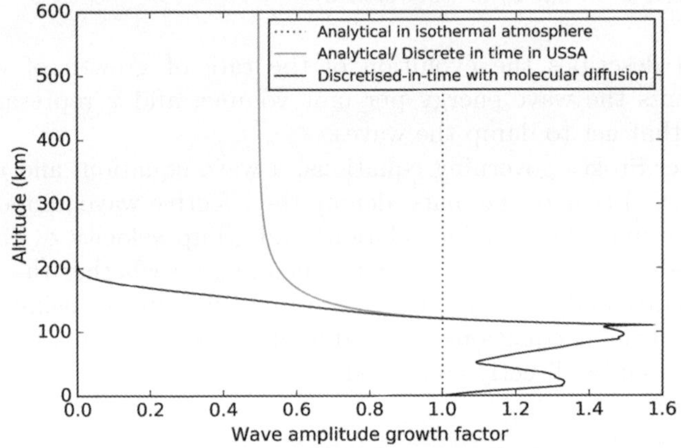

Figure 3. A comparison of the wave energy at each height for a vertically propagating acoustic wave with a total wavenumber of 1×10^{-4} rad m^{-1} for the cases with and without molecular viscosity and diffusion.

The improvement of the Met Office Unified Model dynamical core's stability with the inclusion of molecular viscosity and diffusion has been confirmed by experiments with a one-dimensional column version of the dynamical core. A small amount of off-centering is required in order to extend the top boundary up from 100km to the molecularly diffused region above 150km, whereupon molecular viscosity acts to regulate wave growth, keeping the model stable up to altitudes of 600km at the top of the thermosphere.

References

Arakawa, A. & Lamb, V. R. 1977, *Am. Met. Soc. J.*, 109, 1

Charney, J. G. & Phillips N. A. 1953, *J. Met.* 10, 2

Griffin, D. J. & Thuburn, J. 2018, *Q. J. R. Met. Soc.*, 144(711)

Lighthill, J. 1978, Waves in Fluids. *Cambridge University Press*

Wood, N., Staniforth, A., White, A., Allen, T., Diamantakis, M., Gross, M., Melvin, T., Smith, C., Vosper, S., Zerroukat, M. & Thuburn, J. 2014, *Q. J. R. Met. Soc.*, 140, 682

Space Weather of the Heliosphere: Processes and Forecasts
Proceedings IAU Symposium No. 335, 2017
C. Foullon & O.E. Malandraki, eds.

doi:10.1017/S1743921317008134

Comparison of Predictive Efficiency of Topological Descriptors and SHARP in Solar Flares Forecasting

Irina Knyazeva[1,2], Fedor Urtiev[1] and Nikolay Makarenko[1]

[1]Pulkovo Observatory, Saint-Petersburg
email: iknyazeva@gmail.com
[2]Saint-Petersburg State University
Saint-Petersburg, Russia
email: iknyazeva@gmail.com

Abstract. In the current paper, we investigate topological invariants, calculated by HMI LOS magnetograms as complexity descriptors of solar magnetic fields. We compared them with the physical parameters provided by the Space-weather HMI Active Region Patches (SHARP). We have repeated forecasting experiment of Stanford Solar Observatories Group with the same positive and negative active region patches database, but replace SHARP parameters with topological invariants of corresponding LOS magnetograms. The classification results turned out practically identical to those obtained by the Stanford Solar Observatory group. This means that using LOS magnetograms retains enough complexity for magnetic field description.

Keywords. methods: numerical, methods: statistical,Sun: magnetic fields, Sun: flares

---

1. Introduction

The growth of the quantity and quality of the solar magnetic field observational data and development of machine learning methods stimulate improved forecasting techniques for the solar flare forecasting. Most of techniques are based on the complexity of the photospheric magnetic field of the Sun's active regions. There are a large number of different characteristics that can be used for magnetic field complexity description. Due to many empirical assumptions during their calculation, they are hardly reproducible. For HMI/SDO vector magnetograms an automated active region tracking system exist called Spaceweather HMI Active Region Patch (SHARP). For each active region, key features called SHARP parameters were calculated and are available online. Computation of these features is based on SDO vector magnetograms. The line-of-sight component of the magnetic field (HMI/SDO LOS) includes only one projection of three-dimensional vector but is measured with higher quality. It is impossible to compute physically based parameters with only one component of a magnetic field but a large number of different proxy are known. We propose to use topological invariants based on a topological data analysis approach. They could be computed directly from the data without any assumptions and parameters. Using topological invariants as complexity description (in terms of critical points) is used by researchers in random field analysis Adler (2010). In this paper, we compared the efficiency of these features computed only by the LOS magnetograms with the SHARP parameters as solar flare forecasting model.

2. Methods

We perform a classification of positive and negative events that correspond with flaring and non-flaring active regions on the sun with two different sets of features. The first set of parameters are the physical parameters provided by the Space-weather HMI Active Region Patches (SHARP) calculated by vector magnetic data (Bobra *et al.* 2014). The second set of parameters is topological features calculated by corresponding LOS data. The number of possible combinations of flaring and especially non-flaring regions are very large. We reproduce exactly the same experiment made by Stanford Solar Observatories Group published openly with the data and active regions timing. With these two sets of features, we trained and tested a support vector machine classifier and evaluated its performance.

Topological features were calculated with the topological data analysis approach, see Edelsbrunner & Harer (2010). The main idea of TDA is to describe the shape of the data in term of some topological invariant and after that use these invariants as features. For random field, the basic research topic is understanding the distribution of geometric and topological features generated by a stochastic process. We could consider each magnetogram as a realization of some stochastic process drawn from some distribution, see Adler (2010). The structure of the random field is determined by the content of the local neighborhoods for the maxima and minima points, namely how many and at which level peaks appear that are close to the given maxima or minima. Also, we would like to know up to which level field maxima and minima are isolated in some local neighborhood. We can measure the lifetime of each isolated peak as the length of the interval during which it is separated from others. It is useful to draw it on the plane using the beginning and the end of the barcode as point coordinates. As the result, we obtain a set of points that lie above the diagonal that corresponds to barcodes of zero length. This graph is called a persistence diagram (PD). For each magnetogram we received two persistence diagram, roughly speaking one of them corresponds to minima neighborhood, the other to maxima neighborhood. In this work, we have used our own program with algorithm fully described in the paper Makarenko *et al.* (2014) and CAPD RedHom (Juda & Mrozek 2014) for PD estimation.

3. Results

The database was formed as follows. A positive event is an active region that flares with a peak magnitude above the M1.0 level, as defined by the GOES database. A negative event would be an active region that does not have such an event within a 24-hour time span. Two sets of features were formed. The first set was based on SHARP parameters that were taken from the page of the Stanford Solar Observatories Group. For the second feature set, PD for each LOS magnetogram were computed (see examples of AR and corresponding PD in Figures 1-4. For feature vector construction descriptive statistics of these PD were used. After that, these two data sets were run through a SVM. The performance of the classifier was esimated by using 6-fold cross validation and represented as ROC curve for each fold. The ROC curve is created by plotting the true positive rate or truly predicted flares events (TPR) against the false positive rate or predict flare when it was not (FPR) at various threshold settings. In the case of perfect classification ROC values equal to 1 in all points except 0. The closer the curve is to the diagonal the worse the forecasting ability. The cumulative area under the curve (AUC) measure summarize the quality of the ROC.

The result of classification for first (see Figure 5) and second set (see Figure 6) turned out to be practically the same.

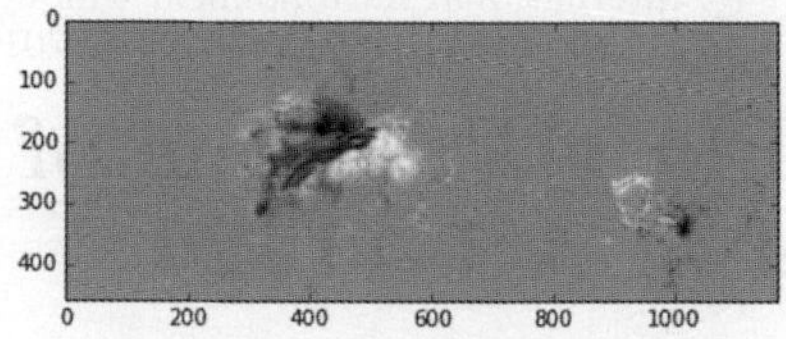

Figure 1. Positive AR

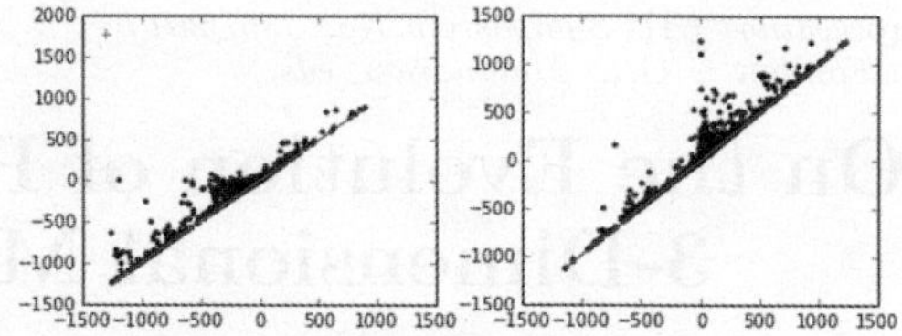

Figure 2. PD for AR of Figure 1

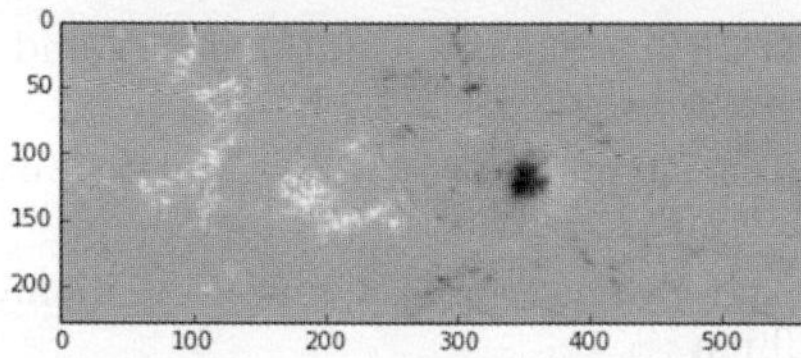

Figure 3. Negative AR

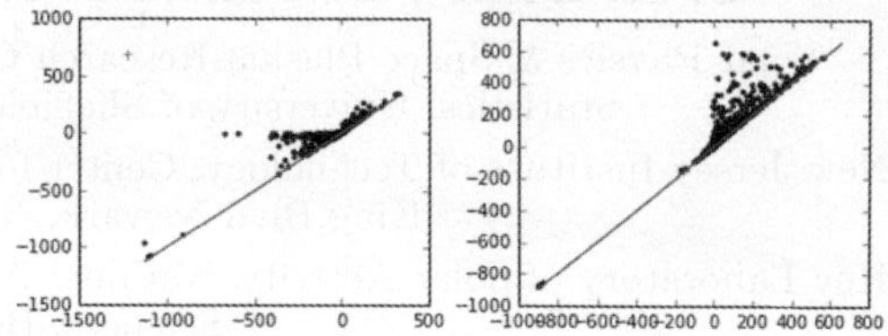

Figure 4. PD for AR of Figure 3

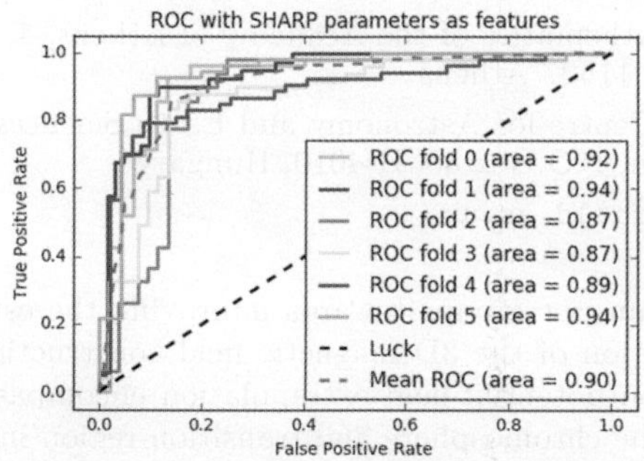

Figure 5. ROC curve based on SHARP parameters

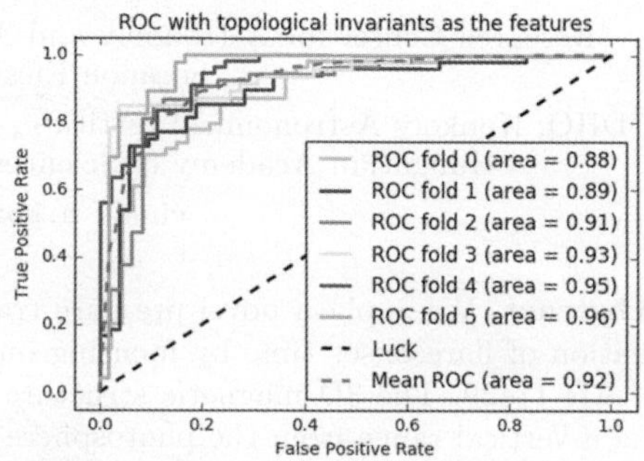

Figure 6. ROC curve based on topology

4. Conclusions and Forthcoming Research

We find that the forecasting efficiency of topological features computed only by the LOS magnetograms is practically the same as for vector magnetic data parameters, therefore LOS data could be useful in forecasting systems. Using distances between PD evolution of active region could be included. Also, PD depends only on the inner structure of the magnetogram and does not depend on absolute values, so potentially SOHO/MDI or other magnetic field data could be combined in one system.

Acknowledgments

This work was supported by RFBR grant number 16-32-00251

References

Bobra, M. G., Sun, X., Hoeksema, J. T., Turmon, M., Liu, Y., Hayashi, K., Barnes, G. & Leka K. D. 2014, *SoPh*, 289, 3549

Makarenko N. G., Malkova D. B., Machin M. L., Knyazeva I. S. & Makarenko I. N., 2014, *Fundam. Prikl. Mat. (Russian), translation in J. Math. Sci.* 203, 806

Edelsbrunner, H. & Harer, J. 2010, Computational topology: an introduction. *AMS*

Adler, Robert J. 2010, The geometry of random fields, volume 62 of Classics in Applied Mathematics. *SIAM*, Philadelphia, PA

Juda M. & Mrozek M. 2014, *ICMS* 160–166

Space Weather of the Heliosphere: Processes and Forecasts
Proceedings IAU Symposium No. 335, 2017
C. Foullon & O.E. Malandraki, eds.

doi:10.1017/S1743921317010432

On the Evolution of Pre-Flare Patterns of a 3-Dimensional Model of AR 11429

M. B. Korsós[1,6], S. Poedts[4], N. Gyenge[1,6], M. K. Georgoulis[5], S. Yu[2], S. K. Bisoi[3], Y. Yan[3], M. S. Ruderman[1] and R. Erdélyi[1]

[1]Solar Physics & Space Plasma Research Center (SP2RC), School of Mathematics and Statistics, University of Sheffield, Hounsfield Road, S3 7RH, UK

[2]New Jersey Institute of Technology, Center for Solar-Terrestrial Research, 323 Martin Luther King Blvd Newark, NJ 07102, United States

[3]Key Laboratory of Solar Activity, National Astronomical Observatories, Chinese Academy of Sciences, Beijing, China

[4]Center for Mathematical Plasma Astrophysics, KU Leuven, Celestijnenlaan 200B, 3001 Leuven, Belgium

[5]Research Center for Astronomy and Applied Mathematics of the Academy of Athens, 4 Soranou Efesiou Street, 11527 Athens

[6]DHO, Konkoly Astronomical Institute, Research Centre for Astronomy and Earth Sciences, Hungarian Academy of Sciences, Debrecen, P.O.Box 30, H-4010, Hungary

email: `m.korsos@sheffield.ac.uk`

Abstract. We apply a novel pre-flare tracking of sunspot groups towards improving the estimation of flare onset time by focusing on the evolution of the 3D magnetic field construction of AR 11429. The 3D magnetic structure is based on potential field extrapolation encompassing a vertical range from the photosphere through the chromosphere and transition region into the low corona. The basis of our proxy measure of activity prediction is the so-called weighted horizontal gradient of magnetic field (WG_M) defined between spots of opposite polarities close to the polarity inversion line of an active region. The temporal variation of the distance of the barycenter of the opposite polarities is also found to possess potentially important diagnostic information about the flare onset time estimation as function of height similar to its counterpart introduced initially in an application at the photosphere only in Korsós *et al.* (2015). We apply the photospheric pre-flare behavioural patterns of sunspot groups to the evolution of their associated 3D-constructed AR 11429 as function of height. We found that at a certain height in the lower solar atmosphere the onset time may be estimated much earlier than at the photosphere or at any other heights. Therefore, we present a tool and recipe that may potentially identify the optimum height for flare prognostic in the solar atmosphere allowing to improve our flare prediction capability and capacity.

Keywords. Solar flare - Magnetic field extrapolation

1. Introduction

The 3D magnetic field structure of an AR is a key to obtain a more accurate information about the pre-flare evolution of a flaring AR. However, to construct an accurate 3D magnetic skeleton of an AR is a challenging task, see e.g. Wiegelmann & Sakurai (2012). To accomplish this, one may need to measure the magnetic field at a number of heights. However, even nowadays, there are no routinely carried out direct measurements of the solar magnetic field in the solar atmosphere except only at the photosphere, see e.g. the magnetic field observations by SDO/HMI. Measuring routinely the magnetic field higher up in the solar atmosphere proved to be far too challenging because of a number of technical reasons, see e.g. Solanki *et al.* (2003) and the follow-up debate by Judge (2009). Therefore, we may need to develop other complementary theoretical tools, that capture the key features of the 3D magnetic structure of an AR in the (lower) solar atmosphere. A frequently used such tool is magnetic field extrapolation based on photospheric measurements, see e.g. Wiegelmann & Sakurai (2012). The method employs the approximation that the solar corona is considered to be in low-β plasma state, or in other words, the magnetic pressure dominates over the pressure in the plasma. From this, we may approximate the plasma being in force-free (FF) state satisfying

$$\mathbf{j} \times \mathbf{B} = 0. \tag{1.1}$$

Here, $\mathbf{j} = \frac{1}{\mu_0}\nabla \times \mathbf{B}$ is the electric current density, $\mathbf{B}$ the magnetic field and μ_0 is the magnetic permeability of free space. This approximation may be a good one for describing the state of the coronal magnetic field. However, many studies claim (for good reasons) that the FF approximation is not reliable below the transition region, especially in the photosphere. Equation (1.1) can be satisfied by $\mathbf{j} = \nabla \times \mathbf{B} = 0$, where the current density vanishes everywhere. This is called a potential field (PF). Another option is when $\mathbf{j} \parallel \mathbf{B}$, that can be rewritten as $\nabla \times \mathbf{B} = \alpha\vec{B}$ and $\mathbf{B} \cdot \nabla\alpha = 0$, where α is the FF parameter. When α is constant along $\mathbf{B}$ everywhere in a given volume, it is the linear FF field (LFFF) approximation, otherwise α may be a spatially dependent scalar function, in which case we have the nonlinear FF field (NLFFF) approximation.

It is elementary to understand the flare occurrence as the removal of free (i.e. non-potential) energy and helicity emerging through the Sun's surface, that is why the LFFF and NLFFF are more realistic (though not necessarily more pragmatic!) approximations than the PF extrapolation. Nevertheless, as a first step of our study, we use the PF extrapolation not only because of its mathematical simplicity but also to provide a first coarser view on how flare forecasting may be affected by tracking magnetic field structures in such 3D models of an AR in the lower solar atmosphere. Now, we only will investigate the typical pre-flare behaviour patterns as a function of height in the lower solar atmosphere.

2. Analysis

Korsós *et al.* (2015), hereafter K15, introduced a new parameter that may more accurately characterise the pre-flare evolution of the photospheric magnetic field of ARs. This is the weighted horizontal gradient of the magnetic field (WG_M) between nearby groups of spots of opposite magnetic polarities closest to the polarity inversion line. With an empirical analyses, for all the observed flare cases between 1996-2010 that satisfy certain selection criteria (for that see K15), they discovered two typical pre-flare patterns for higher than M5 GOES flare-classes: (i) The pre-flare behavior of the WG_M itself exhibits characteristic and unique patterns: steep rise, high maximum and a gradual decrease prior to flaring. (ii) There is a U-shaped typical pre-flare pattern during the compressing and diverging motion in the distance value of the area-weighted barycenters of opposite polarities preceding the flare onset. Also, K15 found, surprisingly, that if any one of these two typical pre-flare behaviours is missing then no flare occurs.

Based on the empirical study of K15, a further tool was developed for estimating the flare onset time: if one can identify the time when the distance between the two barycentres of the area-weighted opposite polarities begins to grow again after the compression phase of the evolution, then one may be able to estimate the onset time of the flare by Eq. (3) of K15. With the aim to improve the efficiency of the onset time estimation capability of the WG_M method that is based on the observed *photospheric* dynamics of pre-flare, we now propose to investigate the two pre-flare signatures not only at the

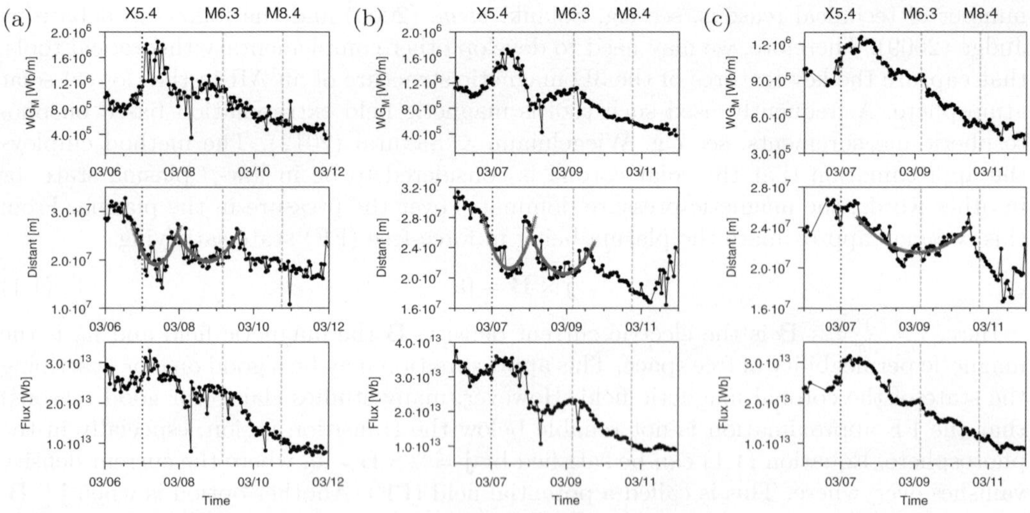

Figure 1. : Top panels are the variation of WG_M , middle panels depict the evolution of distance between the area-weighted barycenters of the spots of opposite polarities and the bottom panels show the unsigned flux of all spots at various heights as function of time. Column (a) is at photospheric level; (b) at 0.5 Mm and (c) at 1Mm levels, respectively, in the solar atmosphere. The red parabolae highlight the typical converging-diverging motion (U-shape) between the two area-weighted centres of the opposite polarities prior to flare(s). The blue/green vertical lines represent the X/two M-class flares onset times.

photosphere but higher up in the lower solar atmosphere, i.e. in the Interface Region and low corona in a *3D-constructed magnetic structure* for an AR. Implementing SDO/HMI line of sight magnetogram (and imaging) observations in the construction of the magnetic mapping of ARs may be a fruitful task. This procedure may also be cumbersome and computationally very expensive. Based on the WG_M method, we test now the feasibility of estimating the flare onset time on the magnetic mapping of AR 11429 using one of the simplest 3D magnetic constructions, namelly the PF exploration (see Figure 2a). This way, if it works and applied routinely, a new data catalogue may be established, for further analysis. This catalogue which is based on the 3D magnetic data would contain valuable information on position, area and magnetic field/flux for each sunspot across the lower solar atmosphere.

In order to test the concept, here, let us now apply the WG_M method and track the temporal variation of WG_M, distance of the area-weighted barycentre of the opposite polarities and the net flux at various heights for pre-flare dynamic studies for AR 11429 which produced one X5.4 (on 2012-03-07 00:24) and two M-class flares (M6.3 and M8.4 flares on 2012-03-09 03:53 and 2012-03-10 17:44). The investigation of the pre-flare dynamics is applied to the 3D construction of the AR between 6-12 March. Note, we cannot investigate the X5.4 flare because the limits of the WG_M method (e.g. the X5.4 flare occurred close to the East limb).

In the first step, as one ascends with a step-size of 100 km higher in the solar atmosphere and reaches the level of 2 Mm above the photosphere, following the application of the WG_M method detailed in K15, we now recognise at most of the heights the prominent typical pre-flare behavior of the WG_M (the increasing, maximum and after the decreasing phase of WG_M), and, the distance (evolving during the full converging-diverging motion of area-weighted barycenters) prior to flare(s). Furthermore, we also

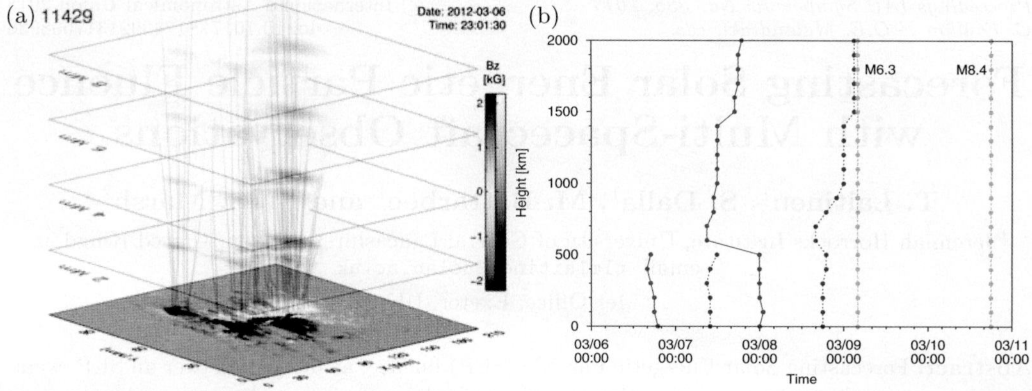

Figure 2. (a): The 3D visualisation of AR 11429 is based on PF extrapolation from the photosphere to 8 Mm in the solar atmosphere; colour bars represent magnetic polarities (b) The green lines are associated with the onset time of M8.4 and M6.3-class flares, respectively. The red lines are the actual moment of start times of approaching (beginning of the red parabola in Figure 1), times of moments of the closest approaching point between two barycenters are the blue lines.

find that the evolution of WG_M and distance changes remarkably as function of height (see Figure 1a-c).

In short, as we reach ∼0.4-0.6 Mm heights in the solar atmosphere, we see the beginning of the converging-diverging motion of the distance starts *earlier* than at the photospheric level.

3. Result

While applying the WG_M method to the PF extrapolated structure of AR 11429 we notice that the starting time of the approaching phase *changes as function of height*, as visualized in Figure 2. Here, we plot the start of the converging phase (red lines) and the moment of time of the minimum distance of barycentres (blue lines) as function of height. We notice that in the case of the M6.3 flare at about 0.4 Mm level, and, in the case of M8.4 flare at about 0.6 Mm, the converging phase starts much earlier and reaches the minimum point also earlier than it does at the photosphere or at other heights. From this behaviour we may conclude that the flare onset times (by using the empirical Eq. 3 from K15 or otherwise) could be estimated sooner at certain heights than at photosphere or in the corona. In summary, identifying this optimum height in the solar atmosphere for a given AR may increase the time window elapsing prior to a flare. If this conclusion is confirmed by using a larger, statistically significant sample of ARs containing flare eruptions this would be a significant progress.

Acknowledgements

MBK, NGY, MSR and RE are grateful to the U. of Sheffield, STFC and the Royal Society (UK) for the support. YY is supported by NSFC grant.

References

Judge, P. G. 2009, *A&A*, 493, 1121J
Korsós, M. B., Ludmány, A., Erdélyi, R. & Baranyi, T. 2015, *ApJ*, 802, L21
Solanki, S. K., Lagg, A., Woch, J., Krupp, N. & Collados, M. 2003, *Nature*, 425, 692-695
Wiegelmann, T. & Sakurai, T. 2012, *Living Reviews in Solar Physics*, 9 (5), 49

Space Weather of the Heliosphere: Processes and Forecasts
Proceedings IAU Symposium No. 335, 2017
C. Foullon & O.E. Malandraki, eds.

doi:10.1017/S1743921317008936

Forecasting Solar Energetic Particle Fluence with Multi-Spacecraft Observations

T. Laitinen[1], S. Dalla[1], M. Battarbee[1] and M. S. Marsh[2]

[1]Jeremiah Horrocks Institute, University of Central Lancashire, Preston, United Kingdom
email: tlmlaitinen@uclan.ac.uk

[2]Met Office, Exeter, UK

Abstract. Forecasting Solar Energetic Particle (SEP) fluence, as integrated over an SEP event, is an important element when estimating the effect of solar eruptions on humans and technology in space. Current real-time estimates are based on SEP measurements at a single location in space. However, the interplanetary magnetic field corotates with the Sun approximately 13° each day with respect to Earth, thus in 4 days a near-Earth spacecraft will have changed their connection about 60° from the original SEP source. We estimate the effect of the corotation on particle fluence using a simple particle transport model, and show that ignoring corotation can cause up to an order of magnitude error in fluence estimations, depending on the interplanetary particle transport conditions. We compare the model predictions with STEREO observations of SEP events.

Keywords. Sun: Particle Emission, interplanetary medium, diffusion

1. Introduction

Total dose due to Solar Energetic Particles (SEPs) is an important factor in evaluation of radiation risk for astronauts and spacecraft. SEP fluence is affected by acceleration at the Sun and in interplanetary space (e.g. Reames 1999), scattering of particles along and across the field lines (e.g. Jokipii 1966), cross-field propagation due to field-line meandering (Laitinen *et al.* 2016) and drifts (Dalla *et al.* 2013), and motion of the observer relative to the SEP source region, i.e. corotation (Giacalone & Jokipii 2012; Marsh *et al.* 2015). We study the effects of corotation on SEPs using a simple model and STEREO/LET (Mewaldt *et al.* 2008) SEP observations in 2013, when the spacecraft were at $\sim 60°$ heliolongitudinal distance from each other. Such separation mimics the recently suggested L1/L5 mission concepts (e.g. Trichas *et al.* 2015).

2. Effect of corotation of the SEP source: simple model

We estimate the effect of the corotation of the particle source on the SEP intensities with a simple 1D diffusion model. We use impulsive injection with Gaussian longitudinal injection distribution, which gives the particle intensity n at the observing spacecraft as

$$n(t,r,\phi) = n_0(\phi)\frac{1}{2\pi\left(\lambda v t/3\right)^{3/2}}\mathrm{e}^{-\frac{3r^2}{4\lambda v t}}, \qquad n_0(\phi) = n_{00}\,\mathrm{e}^{-(\Delta\phi)^2/(2\sigma_\phi^2)} \tag{2.1}$$

where ϕ and r are the observer's heliographic longitude and radial distance from the Sun, t is time from injection, λ the scattering mean free path, v the SEP speed, $\Delta\phi = \phi_F - \phi_{S/C,F}$ with ϕ_F and $\phi_{S/C,F}$ the SEP source and spacecraft footpoint longitudes, $\sigma_\phi = 40°$ (e.g. Richardson *et al.* 2014), and n_{00} a constant. We show in Figure 1 (a) the SEP intensities at 1 AU for different constant $\Delta\phi$ values.

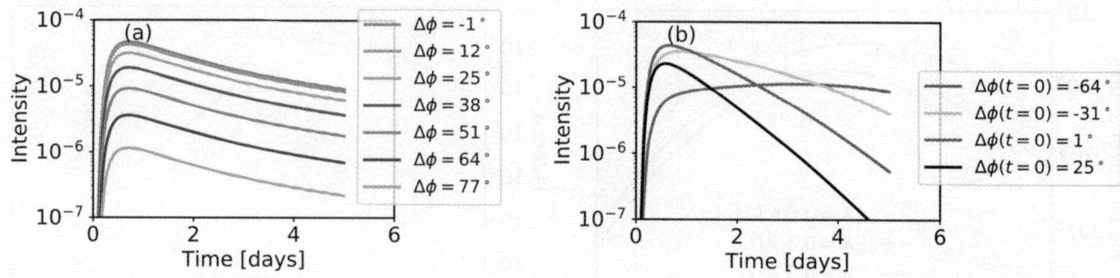

Figure 1. Time-intensity curves for 10 MeV protons given by Eq. (2.1) with $\lambda = 0.03$ AU at r =1 AU. (a): SEP intensities on field lines that corotate with the Sun. (b): SEP intensities observed with spacecraft at 1 AU moving relative to the corotating field lines.

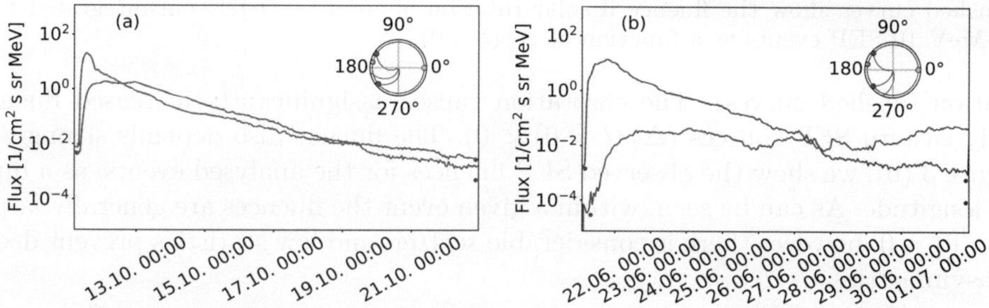

Figure 2. 10 MeV proton intensities on (a) 11 October and (b) 21 June 2013. In the inset, the red and blue circles show the STA and STB locations at the start of the event, respectively, and the red, blue and green spirals show the Parker spirals connected to the STA, STB and the SEP source location.

A spacecraft observing the SEPs at 1 AU rotates at rate $\Omega = -13.2°$/day with respect to the SEP surface, giving $\Delta\phi = \phi_F - (\phi_{S/C,F}(t = 0) + \Omega t)$. The spacecraft thus moves with respect to field lines that are connected to different-strength SEP injections at the Sun. The effect of this motion on SEP fluxes is demonstrated in Figure 1 (b), where the initially well-connected spacecraft (magenta curve) sees rapid SEP intensity onset followed by fast decay, whereas a spacecraft seeing the SEP event as central (green curve) sees the SEP intensities remaining high for several days. Similar features can be seen in SEP event profiles observed with different initial longitudinal separations (Cane *et al.* 1988). As seen in our simple model, corotation can produce these features without requiring a continuous, moving SEP source, as shown also by Marsh *et al.* (2015).

3. Multi-spacecraft observations

We analysed 10 MeV proton intensities during 10 events where SEPs were observed by both STEREO/A (STA) and STEREO/B (STB) LET instruments (Mewaldt *et al.* 2008) in 2013. The source longitude ϕ_F was analysed from STEREO and SOHO observations, and compared with previous studies (e.g. Richardson *et al.* 2014), while $\phi_{S/C,F}$ was obtained using the solar wind velocity at STA and STB.

Figure 2 (a) shows an event with the SEP source between the two spacecraft. As with all (4/10) events with this connectivity, the slower increase at STB is qualitatively consistent with STB propagating towards higher-intensity field lines. Figure 2 (b) shows an event where both spacecraft are advancing away from the source region. STA reaches $\Delta\phi = 180°$ on 27 June, which is seen as an increase in intensities. Overall, the SEP event profiles showed signatures of SEP source corotation.

In Figure 3 (a), we show 5-day fluences of 10 MeV protons, modelled with Eq. (2.1), as a function of the initial $\Delta\phi$, both for cases including corotation (solid curves) and ignoring

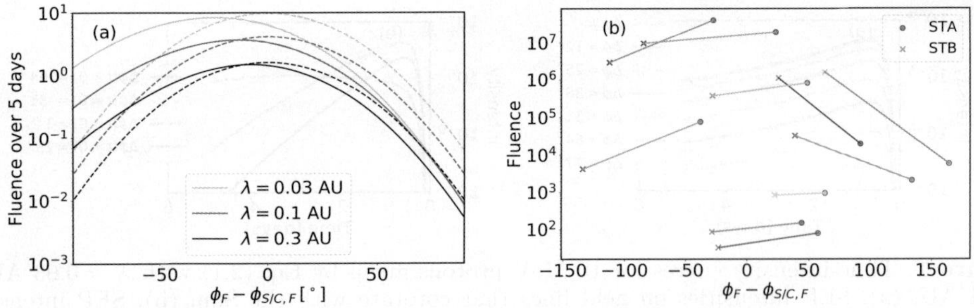

Figure 3. **(a)** Modelled 10 MeV proton fluence at 1 AU from the Sun as a function of $\Delta\phi(t = 0)$. The dashed curves show the fluence if solar rotation is ignored. **(b)** Event-integrated fluences for 10 MeV 10 SEP events as a function of $\Delta\phi(t = 0)$.

corotation (dashed curves). The corotation causes a significantly increased fluence for initially eastern SEP sources ($\Delta\phi(t = 0) < 0$). The fluence also depends strongly on λ. In Figure 3 (b), we show the observed SEP fluences for the analysed events, as a function of the longitude. As can be seen, within a given event the fluences are generally larger for $\Delta\phi(t = 0) \sim 0$, however there is considerable scatter, and low statistics prevent deducing clear asymmetries.

4. Conclusions

Propagation models show that the corotation of the SEP-filled field lines with the Sun affects the SEP intensities observed by interplanetary spacecraft. Our SEP event analysis shows that the time-intensity morphology is consistent with corotation affecting the event profiles. Low statistics prevent finding signatures of corotation in fluences without further, more involved modelling. Two-spacecraft observations bring more information, but require further study before their value can be estimated fully, including taking into account the varying interplanetary turbulence (e.g. Burlaga & Turner 1976), which has significant influence on SEP transport along and across the mean field, and SEP fluences.

Acknowledgements

We acknowledge the STEREO/IMPACT, STEREO/SECCHI and SOHO/LASCO teams for providing the data used in this study. TL and SD acknowledge support from the UK Science and Technology Facilities Council (STFC) (grants ST/J001341/1 and ST/M00760X/1) and MB from the Leverhulme Trust Grant RPG-2015-094.

References

Burlaga, L. F. & Turner, J. M. 1976, *JGR*, 81, 73
Cane, H. V., Reames, D. V. & von Rosenvinge, T. T. 1988, *JGR*, 93, 9555
Dalla, S., Marsh, M. S., Kelly, J. & Laitinen, T. 2013, J. Geophys. Res. (Space Physics), 118, 5979
Giacalone, J. & Jokipii, J. R. 2012, *ApJL*, 751, L33
Jokipii, J. R. 1966, *ApJ*, 146, 480
Laitinen, T., Kopp, A., Effenberger, F., Dalla, S. & Marsh, M. S. 2016, *A&A*, 591
Marsh, M. S., Dalla, S., Dierckxsens, M., Laitinen, T. & Crosby, N. B. 2015, Space Weather, 13, 386
Mewaldt, R. A., Cohen, C. M. S., Cook, W. R., *et al.* 2008, *Space Sci. Rev.*, 136, 285
Reames, D. V. 1999, *Space Sci. Rev.*, 90, 413
Richardson, I. G., von Rosenvinge, T. T., Cane, H. V., *et al.* 2014, *Sol. Phys.*, 289, 3059
Trichas, M., Gibbs, M., Harrison, R., *et al.* 2015, Hipparchos, vol. 2, Issue 12, pp. 25 - 31, 2, 25

Space Weather of the Heliosphere: Processes and Forecasts
Proceedings IAU Symposium No. 335, 2017
C. Foullon & O.E. Malandraki, eds.

doi:10.1017/S174392131700713X

Prediction of Ground Level Enhancements

Marlon Núñez[1], Pedro J. Reyes-Santiago[1] and Olga E. Malandraki[2]

[1]Department of Languages and Computer Sciences, University of Málaga,
Campus de Teatinos, 29071 Málaga, Spain,
email: mnunez@lcc.uma.es

[2]IAASARS, National Observatory of Athens, GR-15236 Penteli, Greece,
email: omaland@astro.noa.gr

Abstract. This paper summarizes the first tool that is able to predict Ground Level Enhancements (GLE). It makes real-time predictions of the occurrence of GLE events from the analysis of soft X-ray and differential proton flux measured by the GOES satellite network. Before the development of this tool, space weather systems have been warning users about evolving GLE events by processing neutron measurements recorded on ground level. This tool, called HESPERIA UMASEP-500, can predict GLE events before the detection by any neutron monitor (NM) station. The prediction performance measured for the period from 1986 to 2016 is presented for two consecutive periods, because of their notable difference in performance. For the 2000-2016 period, this prediction tool obtained a probability of detection (POD) of 53.8% (7 of 13 GLE events), a false alarm ratio (FAR) of 30.0%, and average warning times (AWT) of 8 min and 15 min with respect to the first NM station's alert and the GLE Alert Plus warning, respectively. This project has received funding from the European Union's Horizon 2020 research and innovation programme under agreement No 637324.

Keywords. Ground Level Enhancements, prediction

---

1. Introduction

Strong solar events may accelerate solar particles to near the speed of light reaching the Earth in a few minutes (Reames 2004). These particles may interact with the Earth's atmosphere to produce penetrating neutrons known as Ground Level Enhancements (GLEs) which are detected by neutron monitors (NMs) worldwide (Aschwanden 2012). These particles may irradiate astronauts in space and passengers and flight crews in commercial aircraft flying at polar latitudes (Shea & Smart 2012).

Currently, GLE warning systems can only provide a forewarning for an event that has already started (Souvatzoglou *et al.* 2014). In this work, we present the HESPERIA UMASEP-500 real-time predictor of the occurrence of GLE events constructed under the HESPERIA Horizon 2020 project (Malandraki *et al.* 2015), using the UMASEP scheme (Núñez 2011; Núñez 2015). Solar energetic proton (SEP) event and GLE predictors are of great significance because they offer the unique advantage of valuable added minutes to space weather users. Given a prompt and accurate warning message, pilots can re-route their planes at lower latitudes to be better protected by the Earth's shielding (Beck *et al.* 2005), and launch operators may postpone or re-schedule a launch (Shea & Smart 2012).

The UMASEP scheme makes predictions from the analysis of soft X-ray (SXR) and proton data. This approach has been used to develop other tools. The original UMASEP model was used to develop UMASEP-10, a tool for predicting > 10 MeV SEP events. Since 2010, NASA's integrated Space Weather Analysis system (iSWA) and the European Space Weather Portal redistributes UMASEP-10's forecasts, which have obtained successful results on an operational level (Tsagouri *et al.* 2013). The UMASEP-10 tool was also included as a real-time forecasting module in the European Space Agency's SEPsFLAREs system (García-Rigo *et al.* 2016). The HESPERIA project presented a tool, UMASEP-10mw (Zucca *et al.* 2017), that uses microwave instead of SXR data. In order to predict > 100 MeV SEP event, a High-Energy UMASEP model was proposed (Núñez 2015) to construct the UMASEP-100 tool from 5-min SXR and proton data.

This paper summarizes the HESPERIA UMASEP-500 GLE forecasting model and its results in Sections 2 and 3, respectively. Concluding remarks are presented in Section 4.

2. Model

The UMASEP scheme infers a magnetic connection, along which energetic protons are arriving in the near-Earth environment, by estimating a lag-correlation between SXR with differential proton fluxes at near-Earth. If the correlation is high and the associated solar flare is also strong, then the UMASEP scheme issues a SEP event prediction. To predict GLE events, we use the High-Energy UMASEP model, explained in Núñez (2015) with 1-min input data and a different set of thresholds.

The High-Energy UMASEP model generates a bit-based time series from the SXR time-derivatives and three bit-based time series from the time-derivatives of each of the P9-P11 channels of the GOES6-GOES15 satellites. The "1s" in each bit-based time series are set when an extreme rise is detected; more specifically, when a positive time derivative surpasses a percentage p of the maximum value of the time derivative in the present sequence of size L (beyond which no influence is assumed in the SEP event to be predicted); otherwise, the flux level is transformed into a "0". This forecasting approach creates a list of cause-consequence pairs as follows: it takes the first "1" of the SXR-based time series, and the first "1" of the proton-based time series, to create a pair; it then takes the second pair of "1s" in each time series, and thus successively, until all the "1s" of the SXR-based time series are inspected. For more information about this model consult Núñez *et al.* 2017).

The original purpose of the HESPERIA UMASEP-500 tool was to correlate SXR with neutron and proton data; however, we found that the use of neutron data provoked the generation of many false alarms due to some quality data problems caused by technical issues in the NMs (Souvatzoglou *et al.* 2014), such as problems in the sensor tubes and power supplies, temperature conditioning of the sensors, software disruptions, and bad communication links with the Neutron Monitor Data Base (NMDB). For this reason we decided not to use neutron data for making GLE predictions.

3. Results

The overall prediction performance of event occurrences for the analyzed period (1986-2016) was calculated in terms of probability of detection (POD), false alarm ratio (FAR) and average warning time (AWT). The forecasting results for the most recent half of the evaluation period (i.e. 2000-2016) may be summarized as follows: the POD was 53.8% (7 of 13 GLE events); the FAR was 30.0% (3/10); the AWT to the first NM's alert was 8 min; and, the AWT to the GLE Alert Plus's warning was 15 min. The GLE Alert Plus system (Souvatzoglou *et al.* 2014) is probably the cutting edge in operational GLE warning systems.

For the first half of the evaluation period (i.e. 1986-1999) the POD was 31.6% (6 of 19 GLE events), the FAR was 33.3% (3/9); and, the AWT to the first NM's alert was 13.3 min. For the whole evaluation period, the POD was 40.6% (13 of 32); the FAR was 31.6% (6 of 19); and, the AWT to the first NM's alert was 10.5 min. We do not know the reason for the better POD performance in the most recent period; nevertheless, we think that the use of a more recent and refined instrument technology and/or more experienced calibration procedures yields better forecasting performance.

We validated the HESPERIA UMASEP-500 GLE forecasting model by replacing the GOES/HEPAD proton data with SOHO/EPHIN proton data, and the results were similar in terms of POD, FAR and AWT (Núñez *et al.* 2017); this study shows that the predictions were issued at similar times using both EPHIN and HEPAD data. Previously, Kühl et at. (2016) had found that there is a possible > 10 MeV electron contamination of the EPHIN high energy proton measurements. Given the coincidence in prediction times using EPHIN and HEPAD data, we plausibly speculate that the HESPERIA UMASEP-500-based GLE predictions have possibly benefited from the early electron contamination in high-energy proton detectors.

4. Concluding remarks

The HESPERIA UMASEP-500 SEP forecasting tool makes real-time predictions of the occurrence of GLE events from the analysis of SXR and differential proton flux measured by the GOES satellite network. We assume that a prediction is successful when it is reported before the first GLE alert is issued by any NM station. Regarding the prediction of GLE events for the period 2000-2016, this tool had a POD of 53.8%, and a FAR of 30.0%. For this period, the tool obtained an AWT of 8 min and 15 min with respect to the first NM station's alert and the GLE Alert Plus warning, respectively.

The current GLE warning systems provide only a forewarning for an evolving GLE event. The main innovation is that the HESPERIA UMASEP-500 tool utilizes in-situ proton data measured in space to perform GLE predictions compared to existing tools based on ground-based neutron monitor measurements. Núñez *et al.* (2017) speculate that GLE predictions can possibly benefit from the early electron contamination in high-energy proton detectors.

This tool has been developed to improve the mitigation of adverse effects both in space and in the air from a significant solar radiation storm, providing valuable added minutes of forewarning to space weather users.

Real-time predictions are available via the HESPERIA web site (http://www.hesperia.astro.noa.gr/index.php/results/real-time-prediction-tools/umasep).

References

Aschwanden, M. 2012, *Space Sci. Rev.*, 171(1-4), p. 3-21

Beck P., Latocha M., Rollet S. & Stehno G. 2005, *Adv. Space Res.*, 16(9), p. 1627-1633

García-Rigo, A., Núñez M., Qahwaji R., Ashamari O., Jiggens P., Pérez G., Hernández-Pajares M. & Hilgers A. 2016, *J. Space Weather Space Clim.*, 6, A28

Kühl P., Dresing N., Heber B. & Klassen A. 2016, *Solar Physics*, 292(1)

Malandraki O., Klein K.-L., Vainio R., Agueda N., Núñez M., *et al.* 2015, Proceddings of the 34th International Cosmic Ray Conference, PoS (ICRC2015) 215

Núñez, M. 2011, *Space Weather*, 9, S07003

Núñez, M. 2015, *Space Weather*, 13, 11, p. 807-819

Núñez, M., Reyes-Santiago P. J. & Malandraki O. 2017, *Space Weather*, 15

Reames D. V. 2004, *Adv. Space Res.*, 34(2), p. 381-390

Shea M. A. & Smart D. F. 2012, *Space Science Review*, 171, p. 161-188

Souvatzoglou G., Papaioannou A., Mavromichalaki H., Dimitroulakos J. & Sarlanis C. 2014, *Space Weather*, 12, 11, p. 633-649

Tsagouri I., Belehaki A., Bergeot N., Cid C., Delouille V., Egorova T., Jakowski N., Kutiev I., Mikhailov A., Núñez M., *et al.* 2013, *J. Space Weather Space Clim.*, 3, 26

Zucca P., Núñez M. & Klein K.-L. 2017, *J. Space Weather Space Clim.*, 7, A13

Space Weather of the Heliosphere: Processes and Forecasts
Proceedings IAU Symposium No. 335, 2017
C. Foullon & O.E. Malandraki, eds.

doi:10.1017/S1743921317008122

A Probabilistic Approach to the Drag-Based Model

Gianluca Napoletano[1], Roberta Forte[2], Dario Del Moro[2], Ermanno Pietropaolo[1], Luca Giovannelli[2] and Francesco Berrilli[2]

[1]Dipartimento di Scienze Fisiche e Chimiche, Università degli studi dell'Aquila, Via Vetoio snc 67100, L'Aquila (AQ), Italy
email: gianluca.napoletano@graduate.univaq.it

[2]Department of Physics, University of Rome Tor Vergata, IT-00133, Rome, Italy

Abstract. Forecasting the time of arrival of a Coronal Mass Ejection at Earth is of critical importance for our high-technology society and for any future manned exploration of the Solar System. As critical as the forecast accuracy is the knowledge of its precision, i.e. the error associated to the estimate. Here a statistical approach to the computation of the time of arrival using the Drag-Based Model is proposed through the introduction of probability distributions, rather than exact values, as input parameters, thus allowing the evaluation of the uncertainty on the forecast.

Keywords. Sun: coronal mass ejections (CMEs).

1. Introduction

Coronal Mass Ejections (CMEs) are violent phenomena of solar activity with repercussions throughout the entire heliosphere and responsible for major geomagnetic storms; therefore predicting the arrival of Interplanetary Coronal Mass Ejections (ICMEs) at 1AU starting from remote observations is one of the primary subjects of the science of space-weather forecasting. Several forecasting methods have been proposed over the last two decades and are still under development. The Drag-Based Model of ICME propagation, a kinematical model, is rapidly revisited in Section 2, pointing out the observational quantities needed to make use of it for the purpose of forecasting. Section 3 describes the procedure to extend this model to compute the most probable ICME travel times by introducing the Probability Distribution Functions that are assumed for the input parameters. In closing, some comments about this approach to the forecasting are made in Section 4.

2. The Drag-Based Model

The Drag-Based Model (DBM) of heliospheric propagation of ICMEs is based on the hypothesis that beyond a certain distance from the Sun the ICME dynamics becomes governed solely by its interaction with the ambient solar wind, and that a hydrodynamical description may be employed. The Drag-Based Model with constant solar wind speed w and drag parameter γ is considered here, referring the reader to the papers from Cargill (2004) and Vršnak *et al.* (2012) for an extensive discussion about model assumptions and details. This model provides the following expression for the evolution of the ICME heliospheric radial distance with time:

$$r(t) = \pm\frac{1}{\gamma}\ln\left[1 \pm \gamma(v_0 - w)\,t\right] + w\,t + r_0 \tag{2.1}$$

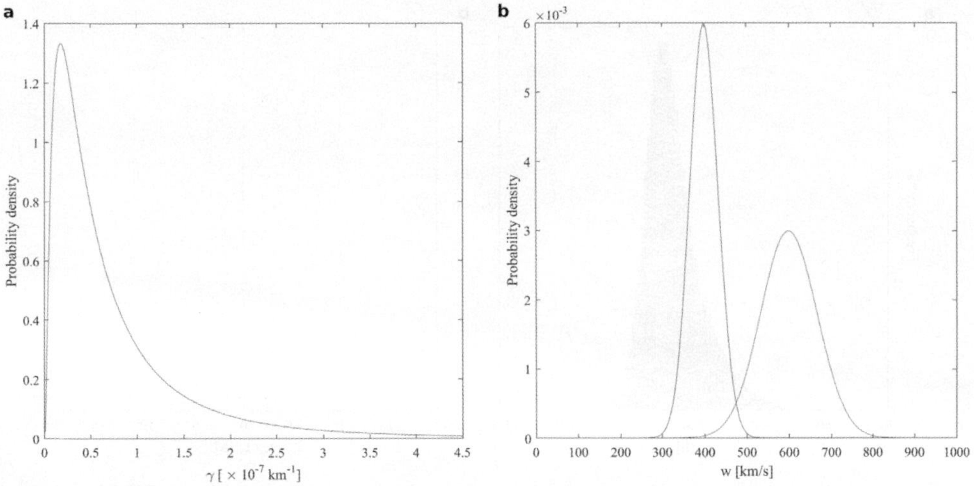

Figure 1. (a) PDF adopted for the random generation of γ in the P-DBM, modeled by a Log-Normal function with $\mu = -0.70$ and $\sigma = 1.01$. (b) PDF adopted for the random generation of w in the P-DBM, with the Slow w represented by a Gaussian PDF centered at 400km/s with σ = 33 km/s, and the Fast w represented by a Gaussian PDF centered at 600km/s with σ = 66 km/s.

Where r_0 is the CME initial radial distance from the Sun, and v_0 the initial radial speed. Such equation can be used to perform a forecasting of the ICME travel time to 1AU, provided that initial conditions r_0 and v_0, and the values for the solar wind speed w and the drag parameter γ are given from observations. While initial position and speed can be measured from coronagraphic images in time sequence (see Shi *et al.* 2012), the evaluation of the drag parameter relies on the knowledge of quantities far from being measurable with sufficient accuracy through coronagraphic images. Also, it is not immediate how to realistically make a choice of a unique, precise value for the solar wind speed accompanying the ICME from Sun to Earth.

3. The Probabilistic-Drag-Based Model

As discussed, the DBM needs four quantities for a travel time computation, namely (r_0, v_0, w, γ), and the last two are, in general, unknown. Following the method proposed by Vršnak *et al.* (2012), an evaluation of the drag parameter, through numerical inversion of the DBM equations for each event of the CME lists compiled by Schwenn *et al.* (2005), has been performed. A Log-Normal distribution, shown in Figure 1a, provides a good fit to the resulting distribution, capturing its properties in just two parameters. With regard to the choice of the solar wind speed, the common understanding is that there exist two different regimes: the so-called Slow solar wind (speed below 500 km/s), and the Fast solar wind, originated from the coronal holes. Therefore we adopt the two different Gaussian PDFs shown in Figure 1b, and choose the Fast w PDF in those cases where there is a prominent coronal hole in the center on the disk at the CME onset time, and the Slow w PDF otherwise. To complete, measurement errors on initial position and speed are generally described by Gaussian PDFs derived from the tracking algorithm of the CME leading edge in coronagraphic images (Shi *et al.* 2012).

The Probabilistic Drag-Based Model (P-DBM), performs an evaluation of the time of arrival of an ICME at a chosen distance from the Sun by generating N initial conditions sets (r_0, v_0, w, γ), each entry of which is randomly chosen from the relative PDF, and then

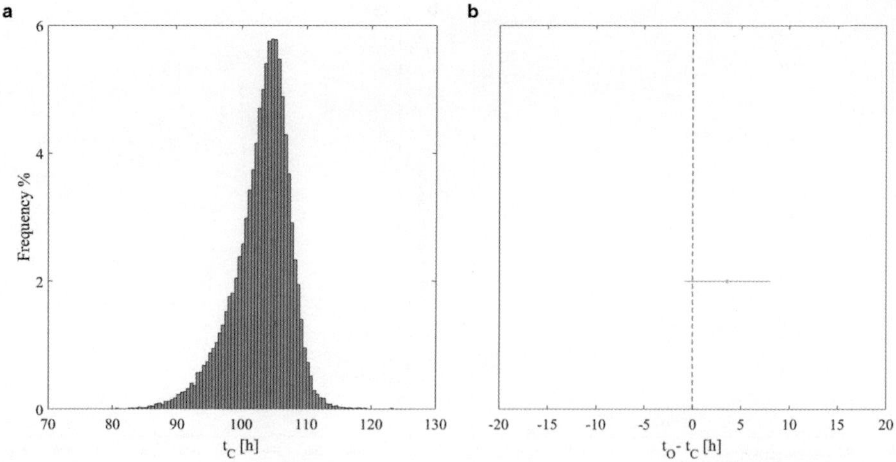

Figure 2. (a) Distribution of the transit times t_C calculated through the P-DBM generating $N = 50.000$ initial conditions. Mean value $< t_C >= 103.1$h and root mean square $\sigma = 4.4$h of this distribution are compared with the actual arrival time $t_O = 106.7$h in panel (b).

numerically computing via Eqn. (2.1) the transit time at 1 AU. This process produces in short computational times a distribution of values associated to the computed travel time t_C, which can be used to estimate relevant quantities statistically, in particular an error on the estimate of the travel time. Figure 2 shows the result from the application of the P-DBM to the CME occurred on 2008-12-12 08:37 UT and reaching the Earth about 113 hours later (see Shi *et al.* 2012). A graphical comparison of the P-DBM predicted travel time with the observed arrival time t_O is also shown.

4. Conclusions

This work proposed an approach to the evaluation of the forecast accuracy by evolving a previously well-established forecasting method based on a kinematical model. The error on the estimate evaluated in this way relies on general features obtained from a statistical analysis of a sample of CMEs, and takes into account our missing knowledge about ICME morphology, details of its dynamics and the precise values of the solar wind speed. Given that the lack of information about the onset parameters and characteristics of the ICME dynamics is a general issue for each forecasting model, the choice of putting together probabilistic and kinematical aspects certainly deserves further investigations, since it requires low computational facilities and certainly contributes to the important topic of the development of CME arrival prediction tools.

References

Cargill, P. J. 2004, *Solar Phys.*, 221, 135

Schwenn, R., Dal Lago, A., Huttunen, E. & Gonzalez, W. D. 2015, *Annales Geophysicae*, 23, 1033

Vršnak, B., Žic, T., Vrbanec, D., Temmer, M., Rollett, T., Möstl, C., Veronig, A., Čalogović, J., Dumbović, M., Lulić, S., Moon, Y.-J. & Shanmugaraju, A. 2013, *Solar Phys.*, 285, 295

Shi, T., Wang, Y., Wan, L., Cheng, X., M. Ding, M. & Zhang, J. 2015, *ApJ*, 806, 271

Space Weather of the Heliosphere: Processes and Forecasts
Proceedings IAU Symposium No. 335, 2017
C. Foullon & O.E. Malandraki, eds.

doi:10.1017/S1743921317009875

PFSS-Based Solar Wind Forecast and the Radius of the Source-Surface

Ljubomir Nikolić

Canadian Hazards Information Service, Natural Resources Canada,
2617 Anderson Road, Ottawa, ON, K1A 0E7, Canada
email: ljubomir.nikolic@canada.ca

Abstract. The potential-field source-surface (PFSS) model of the solar corona is a widely used tool in the space weather research and operations. In particular, the PFSS model is used in solar wind forecast models which empirically associate solar wind properties with the numerically derived coronal magnetic field. In the PFSS model, the spherical surface where magnetic field lines are forced to open is typically placed at 2.5 solar radii. However, the results presented here suggest that setting this surface (the source-surface) to lower heights can provide a better agreement between observed and modelled coronal holes during the current solar cycle. Furthermore, the lower heights of the source-surface provide a better match between observed and forecasted solar wind speed.

Keywords. Sun: corona, Sun: magnetic fields, solar wind

---

1. Introduction

The potential-field source-surface (PFSS) model is an important model of the solar coronal magnetic field. Due to its numerical simplicity, the model is widely used in the space weather research and operations. The PFSS model is based on a current-free approximation and uses synoptic solar magnetograms to derive the magnetic field between the surface of the Sun and the so-called source-surface (Altschuler & Newkirk 1969, Schatten 1969). At the source-surface, which is typically placed at 2.5 solar radii (R_0) in PFSS applications, the coronal magnetic field is purely radial, i.e. "open". An important application of the PFSS model is for solar wind forecasting (Arge *et al.* 2003, Hakamada *et al.* 2005). In the PFSS-based solar wind models, solar wind properties are empirically associated with the coronal magnetic field. These models can be used as stand-alone solar wind forecast models, or to provide boundary conditions to heliospheric MHD codes (e.g., Shiota & Kataoka 2016, McGregor *et al.* 2011). In these implementations of the PFSS model, the source-surface is typically fixed at 2.5 R_0 as well. For example, this is also the case in the Wang-Sheeley-Arge (WSA) solar wind and ENLIL MHD model configuration, which is used for operational forecasting of coronal mass ejection arrivals (Steenburgh *et al.* 2014). However, there is a question on how well the PFSS-based solar wind models with the fixed source-surface perform as the solar activity changes (Reiss *et al.* 2016). This question is particularly important since the current solar cycle (cycle 24) is relatively weak in comparison with previous solar cycles. It has been suggested by Lee *et al.* (2011) and Arden *et al.* (2014), that changing the height of the source-surface should be considered in order to satisfy agreement between modeled and observed open magnetic flux. However, different conclusions about the height have been reported. Here, agreement between modeled and observed coronal holes (CH) and forecasted and observed solar wind speed is investigated for different heights of the source-surface in the PFSS model.

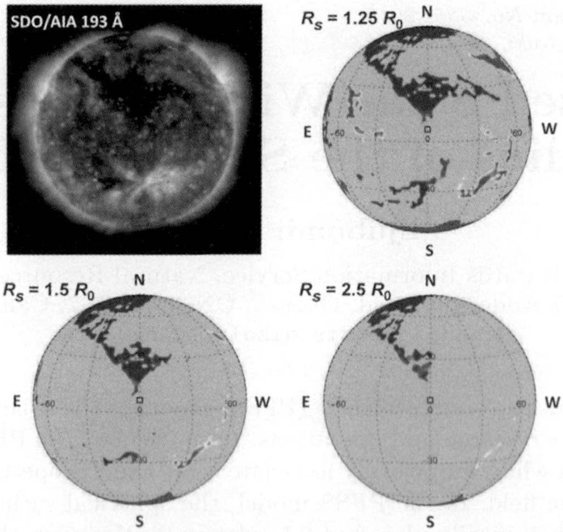

Figure 1. Observed (SDO/AIA 193 Å) and PFSS derived CH with R_s = 1.25, 1.5 and 2.5 R_0 for 22 August 2010. The red and blue color of modelled CH denote magnetic field lines which are pointed away from and toward the Sun, respectively.

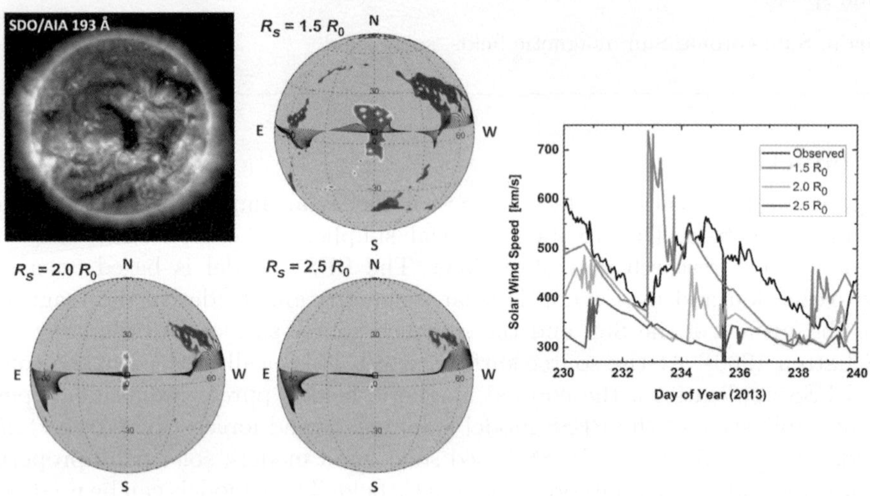

Figure 2. Observed (SDO/AIA 193 Å) and PFSS derived CH with R_s = 1.5, 2.0 and 2.5 R_0 for 18 August 2013 (230 day of year), and the observed and forecasted solar wind speed at Earth with different heights of the source-surface. The red and blue color of modelled CH denote magnetic field lines which are pointed away from and toward the Sun, respectively, while the black lines represent open magnetic field lines which connect to projected positions of Earth on the source-surface.

2. Results and Discussion

The PFSS and solar wind forecast code are developed at the Canadian Space Weather Forecast Center. Here, Global Oscillation Network Group (GONG) synoptic maps are used in the PFSS model to derive the coronal magnetic field for various heights of the source-surface.

To illustrate the dependence of the results on the height of the source-surface, in Figure 1 observed and PFSS modelled CH are shown for 22 August 2010. The observations are from the Solar Dynamics Observatory satellite (AIA 193 Å) and modelled CH are obtained with the source-surface radius (R_s) of 1.25, 1.5 and 2.5 R_0. Although identification of CH from observations is challenging (e.g., de Toma 2011), from Figure 1 we can see that the PFSS model with customary used $R_s = 2.5\ R_0$ does not provide a good resemblance to the large northern CH. The lower heights of the source-surface provide a better match. It is found that, in particular around the solar cycle 24 maximum, instead of 2.5 R_0, the source-surface with a radius of 1.5 - 1.8 R_0 provides better agreement between the model and observations. This can be seen also in Figure 2 where observed and modelled CH with $R_s = 1.5$, 2.0 and 2.5 R_0 are shown for 18 August 2013.

Since CH are the source of high-speed solar wind streams, the CH properties can be used to forecast the solar wind. Therefore, the fact that numerical modelling of CH can be improved by changing the source-surface height has implications on the PFSS-based solar wind models. In the WSA model, for example, the solar wind speed is empirically associated with open magnetic field lines, which form CH. In particular, in the WSA model, the solar wind speed depends on the flux tube expansion factor f_s, and the distance of the open magnetic field line foot-point from the CH boundary θ_b (Arge *et al.* 2003). Since the centrally located CH in Figure 2 is not represented well with $R_s = 2.5\ R_0$, the standard WSA solar wind relation (see Reiss *et al.* 2016) does not provide a good agreement between observed and derived solar wind speed. As we can see from Figure 2, using the same solar wind relation, but with lower heights of the source-surface, one could obtain an increased value of the solar wind speed. However, to obtain a better quantitative agreement, empirical coefficients used to associate the solar wind speed with f_s and θ_b should be tuned for lower source-surface heights. Since the tuning of the coefficients in the standard WSA relation for different heights of the source-surface poses a significant challenge, further research is needed to investigate and simplify the empirical relation between the coronal magnetic field and solar wind properties.

3. Summary

The obtained results suggest that changing the source-surface height in the PFSS model could improve agreement between observed and modelled CH. In general, during the current solar cycle, the source-surface location lies well below the customary used heights of 2.5 R_0. These results also suggest that the solar wind models with the solar activity dependent source-surface could provide better forecast of solar wind conditions.

References

Altschuler, M. D. & Newkirk, G. 1969, *Solar Phys.*, 9, 131

Arden, W. M., Norton, A. A. & Sun, X. 2014, *J. Geophys. Res. Space Physics*, 119, 1476

Arge, C. N., Odstrcil, D. & Pizzo, V. J. 2003, in: M. Velli, R. Bruno & F. Malara (eds.), *Solar Wind Ten* (IP Conf. Proc. 679), p. 190

de Toma, G. 2011, *Solar Phys.*, 274, 195

Hakamada, K., Kojima, M., Ohmi, T., Tokumaru, M. & Fujiki, K. 2005, *Solar Phys.*, 227, 387

Lee, C. O., Luhmann, J. G., Hoeksema, J. T., Sun, X., Arge, C. N. & de Pater, I. 2011, *Solar Phys.*, 269, 367

McGregor, S. L., Hughes, W. J., Arge, C. N., Owens, M. J. & Odstrcil, D. 2011, *J. Geophys. Res.*, 116, A03101

Reiss, M. A., Temmer, M., Veronig, A. M., Nikolic, L., Vennerstrom, S., Schngassner, F. & Hofmeister, S. J. 2016, *Space Weather*, 14, 495

Schatten, K. H., Wilcox, J. M. & Ness, N. F. 1969, *Solar Phys.*, 6, 442

Shiota, D. & Kataoka, R. 2016, *Space Weather*, 14, 56

Steenburgh, R. A., Biesecker, D. A. & Millward, G. H. 2014, *Solar Phys.*, 289, 675

Space Weather of the Heliosphere: Processes and Forecasts
Proceedings IAU Symposium No. 335, 2017
C. Foullon & O.E. Malandraki, eds.

doi:10.1017/S1743921317007293

Solar Flare Prediction Using Machine Learning with Multiwavelength Observations

Naoto Nishizuka, Komei Sugiura, Yuki Kubo, Mitsue Den, Shin-ichi Watari and Mamoru Ishii

National Institute of Information and Communications Technology,
4-2-1 Nukui-Kitamachi, Koganei, Tokyo, Japan 184-8795
email: nishizuka.naoto@nict.go.jp

Abstract. We developed a flare prediction model based on the supervised machine learning of solar observation data for 2010-2015. We used vector magnetograms, lower chromospheric brightening, and soft-X-ray data taken by Solar Dynamics Observatory and Geostationary Operational Environmental Satellite. We detected active regions and extracted 60 solar features such as magnetic neutral lines, current helicity, chromospheric brightening, and flare history. We fully shuffled the database and randomly divided it into two for training and testing. To predict the maximum size of flares occurring in the following 24 hours, we used three machine-learning algorithms independently: the support vector machine, the k nearest neighbors (kNN), and the extremely randomized trees. We achieved a skill score (TSS) of greater than 0.9 for kNN. Furthermore, we compared the prediction results in a more operational setting by shuffling and dividing the database with a week unit. It was found that the prediction score depends on the way the database is prepared.

Keywords. Sun: flares, magnetic fields; sunspots, methods: statistical, techniques; image processing

1. Introduction

Solar flares are an origin of space weather phenomena. They originate from the stored magnetic energy around sunspots, and the impulsive release of energy produces flares. Since the magnetic flux emerges from the bottom of the photosphere, solar flares can be predicted by monitoring the photospheric magnetic field in principle. Larger flares tend to occur when a sunspot is larger and more complex (e.g., Leka & Barnes 2003, Bloomfield *et al.* 2012, McCloskey et al. 2016). The repetition of flares is frequently observed. It has been shown by theory and observations that an emerging flux around magnetic neutral lines triggers a flare. Solar flares have been predicted by humans for a long time. However, the prediction accuracy has not significantly improved, despite the increase in the amount of high-resolution observation data available in near real-time.

On the other hand, supervised machine learning, which has recently been developed, is advantageous for dealing with a huge amount of data. New prediction models of solar flares using machine learning have been intensively studied (e.g., Ahmed 2013, Bobra & Couvidat 2015, Nishizuka *et al.* 2017, Raboonik *et al.* 2017, Liu *et al.* 2017, and references therein). Machine learning can tell us which feature is most effective for predicting flares in real-time operation. Several prediction models of flares have been developed (e.g., Barnes *et al.* 2017), but it is still difficult to compare the performance of each prediction model in the same way. There is no standard way of evaluating the prediction models in an operational setting. In this paper, we have attempted to solve these remaining problems.

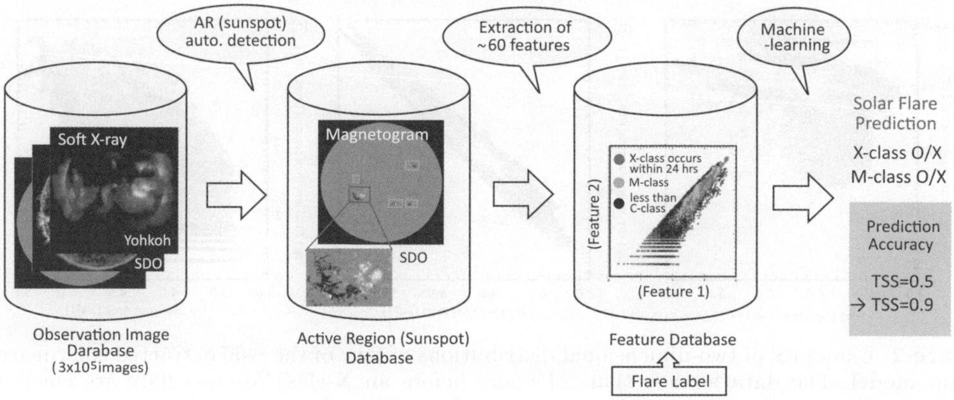

Figure 1. Flow chart of our model of flare prediction.

2. Flare Prediction Model and Observation Database

Figure 1 shows a flow chart of our flare prediction model. First, we compiled an original observation database for the period 2010-2015 by downloading the Solar Dynamics Observatory (SDO; Pesnell *et al.* 2012) data archives, i.e., the line-of-sight and vector magnetograms taken by the Helioseismic and Magnetic Imager (HMI; Scherrer *et al.* 2012) on board SDO, the lower chromospheric brightening taken by the 1600 Å filtergram of the Atmospheric Imaging Assembly (AIA; Lemen *et al.* 2012) on board SDO, and the soft-X-ray emission taken by the Geostationary Operational Environmental Satellite (GOES). Next, we automatically detected active regions from full-disk line-of-sight magnetograms using a threshold of 1.4×10^{-2} T (see Nishizuka *et al.* 2017 more in detail) and attached IDs to them. We identified a rotating active region in the time sequence with the same ID. The original data were reduced by employing one hour cadence. We attached flare labels to each active region in the time sequence, for which a larger flare than a GOES M-class flare occurred in the following 24 hours. Finally, we predicted the maximum size of flares occurring in the following 24 hours by using the following supervised machine-learning algorithms independently: the support vector machine (SVM; Cortes & Vapnik 1995), the k nearest neighbors (k-NN; Dasarathy 1991), and the extremely randomized trees (ERT; Geurts *et al.* 2006).

For each active region, we extracted ~60 features, including the sunspot area, the total magnetic flux, the length and number of magnetic neutral lines, the current helicity, the area of chromospheric brightening, the X-class and M-class flare histories, and their time evolutions (see Nishizuka *et al.* 2017 more in detail). We show some examples of two-dimensional distributions of four of the ~60 extracted solar features in Figure 2. The data for less than 24 hours before an X-class (M-class) flare are colored in red (green) and the non-flare data (including C-class flares) are colored in black. It was found that X-class flares tend to occur with a stronger magnetic field (>0.1 T), a larger sunspot area, a larger unsigned magnetic flux, and a larger number of magnetic neutral lines.

3. Evaluation Methods and Prediction Results

To evaluate the prediction results, we prepared the two datasets for training and testing in two ways (Figure 3). First, after standardizing the database of extracted features, we fully shuffled it and divided it into two for training and testing in the ratio of 7 to 3. Second, to investigate the performances of prediction models in a more operational setting, we shuffled the database with a one week unit. This is because while fully shuffling

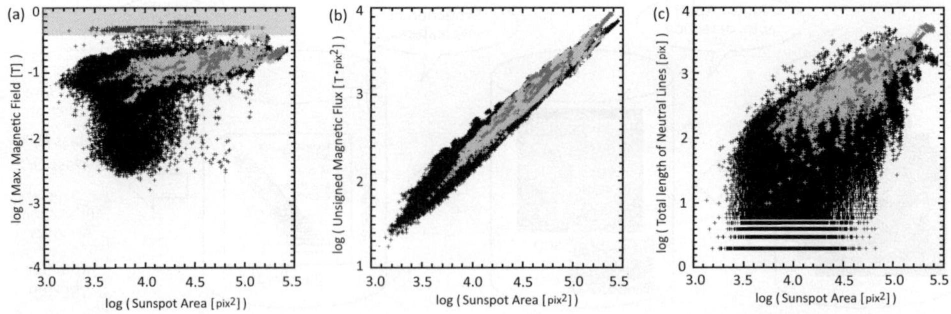

Figure 2. Examples of two-dimensional distributions of four of the ~60 extracted solar features in our model. The data for less than 24 hours before an X-class/M-class flare are colored in red/green. The non-flare data (including C-class flares) are colored in black. Data in the shaded area, which are affected by the observation fitting error, were removed from the database for machine learning. Here, the length unit of one pixel of SDO images is 0.5" or equal to about 362 km on the Sun.

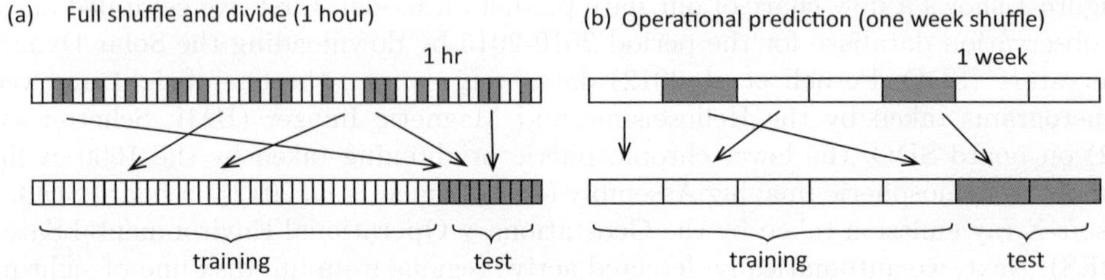

Figure 3. Two ways of producing the training and testing datasets: (a) full shuffle and divide and (b) one week shuffle and divide for investigation in a more operational setting.

and dividing the database, consecutive labeled data in a time sequence before a flare can be separated into both training and testing datasets, which results in an increase in the score.

The contingency tables of the prediction results for X-class flares are shown in Figure 4. We evaluated the prediction results using the skill score named the true skill statistic (TSS; Hanseen & Kuipers 1965). It is determined by $TSS = TP/(TP+FN) - FP/(FP+TN)$, where TP, FN, FP, and TN are the numbers of true positives, false negatives, false positives, and true negatives, respectively. Using the fully shuffled and divided database, all three methods achieved TSS greater than 0.8. It was found that, among the three algorithms, k-NN with k=1 exhibited the best performance with TSS greater than 0.9 (for reference, TSS=0.76 in Bobra & Couvidat 2015). In a more operational setting with a one week shuffle, k-NN still exhibited the highest score of greater than 0.9, but the scores of SVM and ERT decreased to 0.60 and 0.52, respectively.

4. Summary and Discussion

We developed a model of flare prediction employing machine learning using the multi-wavelength observation data of line-of-sight and vector magnetograms, the lower chromospheric brightening, and the soft-X-ray emission for the period 2010-2015. We applied three machine-learning algorithms for X-class flare prediction, i.e., SVM, k-NN, and ERT, and achieved TSS values of 0.93 for the fully shuffled datasets and 0.91 for the one-week-shuffled datasets in a more operational setting. k-NN exhibited the best score among the three methods, but in a more operational setting, the SVM and ERT scores decreased

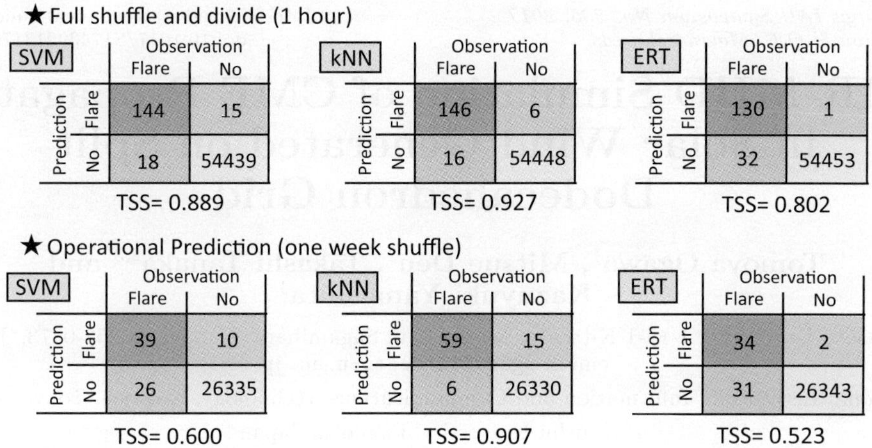

Figure 4. Prediction results of X-class flares occurring in the following 24 hours obtained by three machine-learning algorithms for the fully shuffled datasets and the one-week shuffled datasets in a more operational setting.

markedly. This indicates that the prediction scores strongly depend on the preparation method of the training and test datasets, as well as the evaluation method. Currently, there is no standard evaluation method for different prediction models and current operations (Crown 2012, Devos *et al.* 2014, Murray *et al.* 2017, Kubo *et al.* 2017). To ensure a fair comparison, we need to find a standard way of evaluating prediction models in a more operational setting.

References

Ahmed, O. W., Qahwaji, R., Colak, T., *et al.* 2013, *Sol. Phys.*, 283, 157
Barnes, G., Leka, K. D., Schrijver, C. J., *et al.* 2016, *Astrophys. J.*, 829, 89
Bloomfield, D. S., Higgins, P. A., McAteer, R. T. J. & Gallagher, P. 2012, *ApJ*(Letters), 747, L41
Bobra, M. G. & Couvidat, S. 2015, *Astrophys. J.*, 798, 135
Cortes, C. & Vapnik, V. 1995, *Mach. Learn.* 20, 273
Crown, M. D. 2012, *Space Weather*, 10, S06006
Dasarathy, B. V. 1991, *Nearest Neighbor (NN) Norms: NN Pattern Classification Techniques* (Los Alamitos, CA: IEEE Computer Society Press), 447
Devos, A., Verbeeck, C. & Robbrecht, E. 2014, *J. Space Weather Space Clim.*, 4, A29
Guerts, P., Ernst, D. & Wehenkel, L. 2006, *Mach. Learn.*, 63, 3
Hanssen, A. J. & Kuipers, W. J. 1965, *Meded Verhand*, 81, 2
Kubo, Y., Den, M. & Ishii, M. 2017, *J. Space Weather Space Clim.*, 7, A20
Leka, K. D. & Barnes, G. 2003, *Astrophys. J.*, 595, 1296
Lemen, J., Title, A. M., Akin, D. J., *et al.* 2012, *Sol. Phys.*, 275, 17
Liu, C., Deng, N., Wang, J. T. L. & Wang, H. 2017, *Astrophys. J.*, 843, 104
McCloskey, A. E., Gallagher, P. T. & Bloomfield, D. S. 2016, *Sol. Phys.*, 291, 1711
Murray, S. A., Bingham, S., Sharpe, M. & Jackson, D. R. 2017, *Space Weather*, 15, 577
Nishizuka, N., Sugiura, K., Kubo, Y., *et al.* 2017, *Astrophys. J.*, 835, 156
Pesnell, W., Thompson, B. J., Chamberlin, P. C., *et al.* 2012, *Sol. Phys.*, 275, 3
Raboonik, A., Safari, H., Alipour, N. & Wheatland, M. S. 2017, *Astrophys. J.*, 834, 11
Scherrer, P. H., Schou, J., Bush, R. I., *et al.* 2012, *Sol. Phys.*, 275, 207

Space Weather of the Heliosphere: Processes and Forecasts
Proceedings IAU Symposium No. 335, 2017
C. Foullon & O.E. Malandraki, eds.

doi:10.1017/S174392131700833X

AMR-MHD Simulation of CME Propagation in Solar Wind Generated on Split Dodecahedron Grid

Tomoya Ogawa[1], **Mitsue Den**[2], **Takashi Tanaka**[2,3] and **Kazuyuki Yamashita**[4]

[1]Kitasato University, 1-15-1 Kitasato, Minami-ku, Sagamihara, Kanagawa 252-0373, Japan
email: ogawa@kitasato-u.ac.jp

[2]National Institute of Information and Communications Technology, Koganei, Tokyo, Japan

[3]Kyushu University, Fukuoka, Japan

[4]University of Yamanashi, Kofu, Yamanashi, Japan

Abstract. We perform simulations of the interplanetary coronal mass ejections relating to the magnetic storm on 17 March 2015. A hierarchical mesh structure is used, which is controlled by an adaptive mesh refinement technique, with fine-scale cells where it matters, the structure of the running shock waves of the coronal mass ejections and co-rotating interactive regions. The initial and the inner-boundary conditions are derived from another simulation, which uses a split dodecahedron grid. The resulting shock-wave with the models adjusted to the observed ejection speed on the sky plane show delays by 20% in the arrival time at the Earth from the observed data. By contrast, the model adjusted to the observed arrival time at the Earth needs the ejection speed 30% higher than that in the above models.

Keywords. MHD, methods: numerical, Sun: coronal mass ejections (CMEs), solar wind

1. Introduction

A coronal mass ejection (CME) occurred on 15 Mar 2015, accompanied by an X-ray flare at S22W25. A shock wave generated by the CME, which subsequently reached the Earth, was unexpectedly strong and triggered a geomagnetic storm on 17 Mar 2015. In such a case, the interaction with the background solar-wind structure is considered to play an important role in the propagation of CME-caused shock waves. We have developed a magnetohydrodynamics (MHD) code, in which a hierarchical Cartesian mesh was used and an adaptive mesh refinement (AMR) technique was employed. The AMR technique enables us to trace running structures, e.g. CME-caused shock waves and co-rotating interaction regions (CIRs), with fine-scale cells. Simulations of propagation of interplanetary shock waves are performed and their results are compared with the observed data.

2. Method

First, the background solar wind structure should be determined. For the purpose, we employ an MHD code that uses a split dodecahedron grid, which is spherical and has no polar singularity (Nakamizo *et al.* 2009 and Den *et al.* 2015). We use the photospheric magnetic field data obtained by the Global Oscillation Network Group (GONG) and calculate the steady-state solar wind with the method by Den *et al.* (2015). The derived parameters of the solar wind are input to the AMR-MHD simulations as the initial and the inner-boundary conditions.

Second, we introduce the CME models. We focus on two CMEs: the halo CME on 15 March 2015, which triggered the geomagnetic storm, and the preceding partial halo CME on 14 March. The latter was very weak, but it may have affected the shock wave propagation, given that the subsequent CME seems to have caught up with it.

We model a CME as the cone velocity, $V_r(t, \xi) = T(t) \cdot V_0 cos(\xi/2)$, where ξ is an angle from the cone axis, $T(t)$ is a trapezoidal pulse characterized by the three parameters of duration times of rising d_1, maintaining maximum d_2, and decline d_3, and where V_0 is the maximum velocity on the cone axis (Odstrcil & Pizzo 1999). The values of d_1, d_2, d_3, and V_0 are determined from the observed values, i.e. the durations of relating X-ray flares and the CME speeds on the sky plane in the SOHO/LASCO CME Catalog; they are $d_1 = 85$min, $d_2 = 85$min, $d_3 = 240$min, and $V_0 = 1068$km/s for the halo CME, and $d_1 = 20$min, $d_2 = 10$min, $d_3 = 210$min, and $V_0 = 564$km/s for the partial-halo CME. We introduce three models: background solar-wind only (model 1), the halo CME input (model 2), and the two CMEs input (model 3). In addition, we use an extra model (model 4), in which the two CMEs with the same parameter values as in model 3 but $V_0 = 1388$km/s for the halo CME to adjust to the arrival time of the shock wave observed by the ACE space craft.

3. Results

Figure 1 shows our simulation results of the velocity distributions and mesh structures with model 3. The cells with the finest size of $1R_s$ successfully catch the running shock wave and CIRs. Note that we have made mesh refinement for the region only within the longitude of ±60deg and latitude of ±60deg and set the cell size finest near the sun.

Next, we compare the time profiles of velocity and density of 4 models with one another and the observed data by the ACE space craft (Figure 2). Several small gaps are identified in the profiles, which correspond to the points where the cell size changes. The results with models 2 and 3 show marginal differences only. It implies that the effect of the preceding partial halo CME is small. The shock waves with models 2 and 3 are found to be weaker and arrive later than the observed. By contrast, the peaks of velocity and density with model 4 show good agreement with the observation.

4. Discussion and conclusions

The steady-state solar wind derived on a split dodecahedron grid is input to the AMR-MHD simulations and it then generates the background structure. The dynamic mesh captures the CIRs with a high resolution. A running shock wave is also successfully captured with a high resolution and is found to maintain its sharp shape even at the arrival at the Earth. The results with models 2 and 3 show only a marginal difference from each other. The arrival times of the shock waves on the ACE's position with models 2 and 3, in which the CME velocities are determined by the observed value in the solar corona, are about 20% later than the observation. By contrast, that and the peaks of the velocity and density with model 4, in which the CME speed is 30% higher than the coronal observation, show good agreement with the ACE's observation. The after-shock velocities with models 2, 3 and 4 are underestimated. These results suggest that there is a margin for improvement in our CME model.

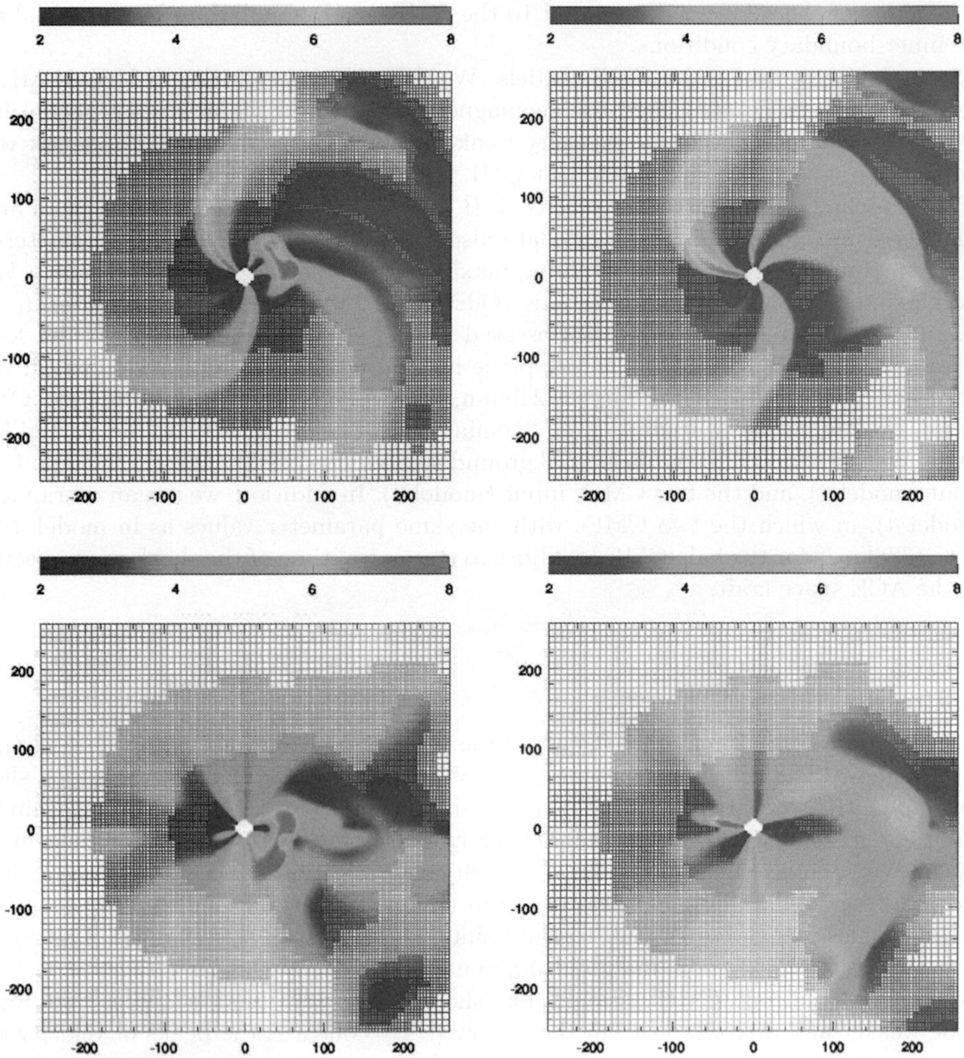

Figure 1. Two-dimensional color map of the simulated results of radial component of the velocity (color-bar scale in units of $81.52 km/s$) with model 3 on the (upper panels) equatorial and (lower panels) meridional planes at (left panels) 12 and (right panels) 60 hours after the halo CME occurs. The mesh structures in the simulations are indicated by grids.

5. Acknowledgement

This work utilizes the data obtained by the Global Oscillation Network Group (GONG) program, managed by the National Solar Observatory, which is operated by AURA, Inc. under a cooperative agreement with the National Science Foundation. The data were acquired by instruments operated by the Big Bear Solar Observatory, High Altitude Observatory, Learmonth Solar Observatory, Udaipur Solar Observatory, Instituto de Astrofísica de Canarias, and Cerro Tololo Interamerican Observatory.

We use the data from the SOHO/LASCO CME Catalog, which was generated and is maintained at the CDAW Data Center by NASA and The Catholic University of America in cooperation with the Naval Research Laboratory. SOHO is a project of international cooperation between ESA and NASA.

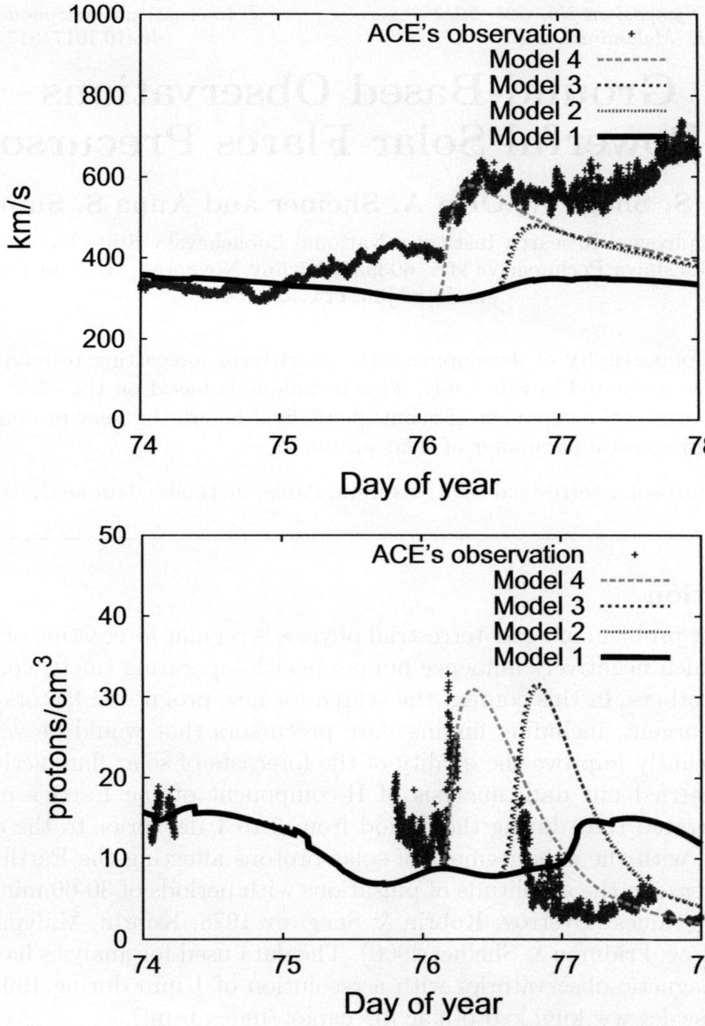

Figure 2. (Upper panel) Time profile of velocity in km/s for the 4 models at the ACE's position, compared with the observation by ACE. The horizontal axis is time in units of day of a year. (Lower panel) Proton number density in cm^{-3} for the same models.

We thank the ACE SWEPAM instrument team and the ACE Science Center for providing the ACE data.

References

Den, M., Tanaka, T., Kubo, Y. & Watari, S. 2015, *34th Int. Cosmic Ray Conf.*, 184

Khokhlov, A. M. 1998, *J.Comput.Phys.*, 143, 519

Nakamizo, A., Tanaka, T., Kubo, Y., Kamei, S. & Shimazu, H. 2009, *J. Geophys. Res.*, 114, A07109

Odstrcil, D. & Pizzo, V. J. 1999, *J. Geophys. Res.*, 104, 483

Space Weather of the Heliosphere: Processes and Forecasts
Proceedings IAU Symposium No. 335, 2017
C. Foullon & O.E. Malandraki, eds.

doi:10.1017/S1743921317008043

Ground-Based Observations of Powerful Solar Flares Precursors

Sergei S. Snegirev, Olga A. Sheiner and Anna S. Smirnova

Radiophysical Research Institute, National Lobachevsky State University,
25/12A Bol'shaya Pecherskaya str., 603950, Nizhny Novgorod, Russian Federation
email: rfj@nirfi.unn.ru

Abstract. The opportunity of development the short-term forecasting technique of geoeffective solar flares is presented in this study. This technique is based on the effect of growth the fluctuations of horizontal component of geomagnetic field before the solar proton flares, that is considered as a prognostic parameter of solar proton flares.

Keywords. (Sun:) solar-terrestrial relations, Sun: flares, methods: data analysis

1. Introduction

An important problem of Solar-terrestrial physics is regular forecasting of solar activity phenomena, which negatively influence human health, operating safety, communication, radar sets and others. In this context the search for new prognostic factors and patterns becomes more urgent, including finding flare precursors that would be very important and can significantly improve the quality of the forecasts of solar flare activity.

Previously carried out data analysis of H-component of the Earth's magnetic field values demonstrated that during the period from 3 to 1 day prior to the onset of solar flare associated with the enhancement of solar protons affecting the Earth environment there is an increase in the amplitude of pulsations with periods of 30-60 min compared to "magnetocalm" values (Bystrov, Kobrin & Snegirev 1978, Kobrin, Malygin & Snegirev 1985 and Snegirev, Fridman & Sheiner 2000). The data used for analysis have been taken from 100 geomagnetic observatories with a resolution of 1 min during 1991, 2001, 2005 –2007 (http://swdcwww.kugi.kyoto-u.ac.jp/ caplot/index.html).

2. Overview

The phenomenon of growth of the long-period fluctuations of the horizontal component of geomagnetic field before the solar proton flares is global in nature (Figure 1). See, for example, data for 19-23.03.91, obtained from 52 stations (solar flare 22.03.91, X9, starts at 22:43E UT; it is marked with a red triangle). As one can see from the left part of picture, at stations located on close geomagnetic latitudes, pre-flare pulsations are identical and occur simultaneously: their power differs insignificantly. The correlation coefficient of the spectral densities (averaged over the periods of 30-60 min) is very high, which indicates the reliability of the observed effect. For geomagnetic stations located on close geomagnetic meridian, the power of long-period oscillations decreases with decreasing latitude. At medium-latitude stations the oscillations appear several hours ahead of the same at higher latitudes' stations.

A possible reason for the enhancement of long-period pulsations-precursors is the relationship between oscillatory processes in the solar atmosphere before the flare events and in the Earth's atmosphere. Fluctuations in the physical parameters of the solar atmosphere must modulate the emission in a wide range of electromagnetic waves.

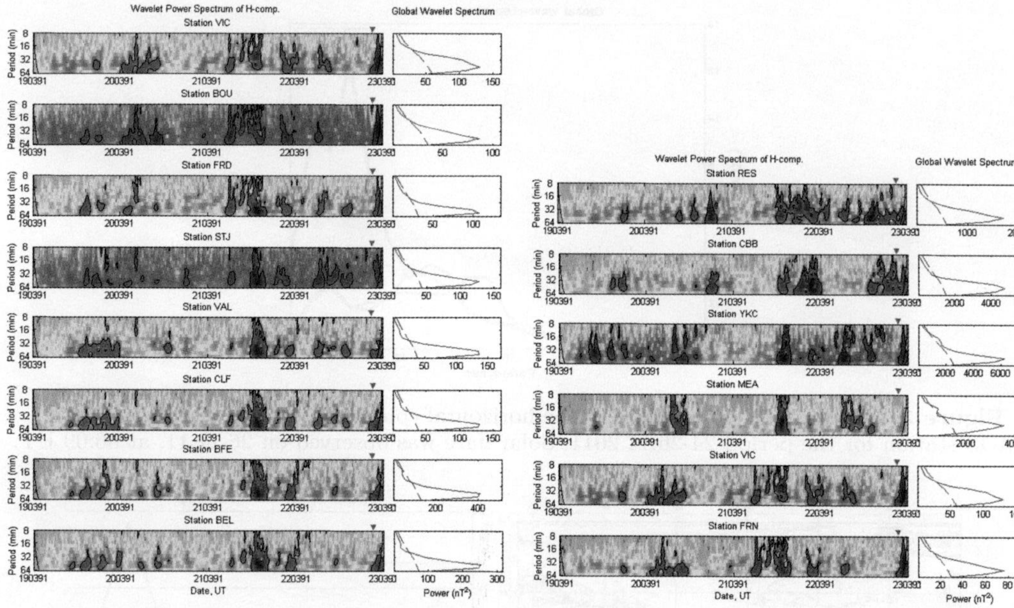

Figure 1. Spatial distribution of pulsations in latitude (left) and in longitude (right).

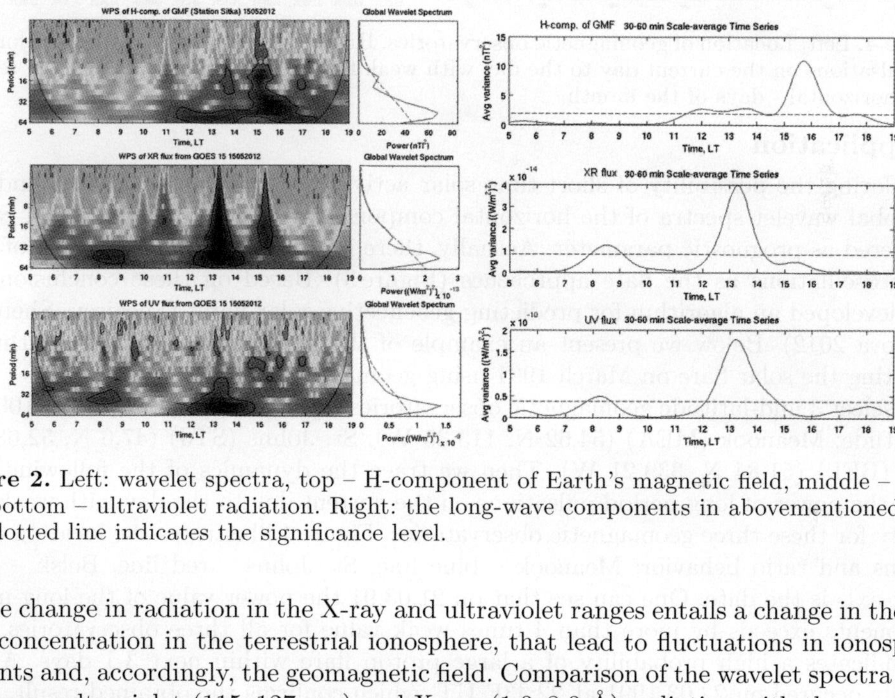

Figure 2. Left: wavelet spectra, top – H-component of Earth's magnetic field, middle – X-ray, and bottom – ultraviolet radiation. Right: the long-wave components in abovementioned data. The dotted line indicates the significance level.

The change in radiation in the X-ray and ultraviolet ranges entails a change in the electron concentration in the terrestrial ionosphere, that lead to fluctuations in ionospheric currents and, accordingly, the geomagnetic field. Comparison of the wavelet spectra of the long-wave ultraviolet (115–127 nm) and X-ray (1.0–8.0 Å) radiation with the spectra of the horizontal component of the Earth's magnetic field illustrates the similar behavior of long-period spectral components of X-ray and ultraviolet emission prior to proton flares (Figure 2). For example, 2 days prior to proton flare on 17.05.2012 (starts at 01:25 UT, M5) there are the power increase in the spectrum of ultraviolet, X-ray radiation and H-component of geomagnetic field. Increasing of preflare pulsations in long-wave ultraviolet and X-ray radiation is observed a little ahead than in the H-component.

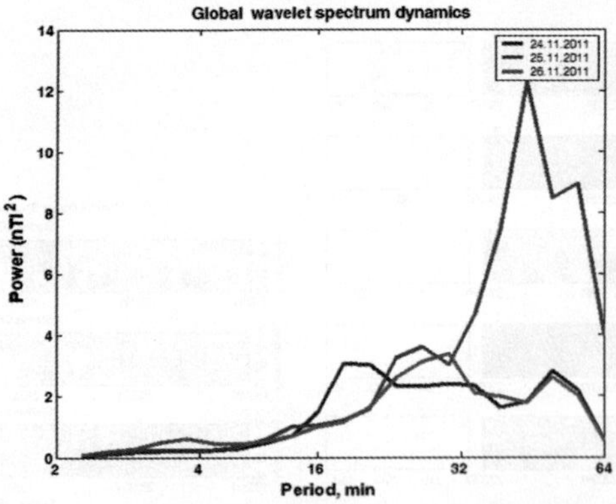

Figure 3. Global wavelet spectrum of the horizontal component of the geomagnetic field, MOS station for the period 24-26.11.2011. Solar flare was observed on 26.11.11, at 06:09 UT.

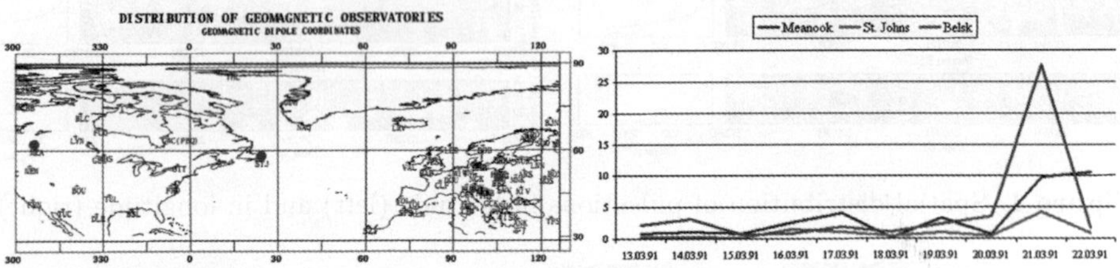

Figure 4. Left: Location of geomagnetic observatories. Right: The ratio of the powers of long-period pulsations on the current day to the day with weak flare activity; vertical axis is in relative units, horizontal - days of the month.

3. Application

Exploring the possibility of short-time solar activity forecasting, we have found that the global wavelet spectra of the horizontal component of the geomagnetic field can be considered as prognostic parameter. Actually, there is the increase in the power of long-period oscillations as the flare approaches (Figure 3). Based on these conclusions, we have developed an algorithm for predicting geoeffective solar flares (Snegirev, Sheiner & Smirnova 2012). Below we present an example of the application of this algorithm for predicting the solar flare on March 1991 using geomagnetic data.

We select 3 mid-latitude geomagnetic observatories separated by not less than 1000 km in latitude: Meanook (MEA) (54,62 N, 113,35 W), St. Johns (STJ) (47,6 N, 52,68 W), Belsk (BEL) (51,84 N, 339,21 W). Then we trace the dynamics of the following ratio value: the power of long-period pulsations on the current day to the day with weak flare activity for these three geomagnetic observatories. Figure 4 illustrates the location of the stations and ratio behavior: Meanook – blue line, St. Johns – red line, Belsk – green line; x-axis is the date. One can see that on 21.03.91 the power value of the long-period components exceeds by more than 4 times weak value for all three observatories. This fact indicates a high probability of a large proton flare within next 3-1 days. A flare actually occured on 22.03.1991 at 22:43E UT, which confirms the obtained result.

References

Bystrov, M. V., Kobrin, M. M. & Snegirev, S. D. 1978, *Soviet Astronomy Letters* 4, 76

Kobrin, M. M., Malygin, V. I. & Snegirev, S. D. 1985, *Planetary Space Science* 33, 1251

Snegirev, S. D., Fridman, V. M. & Sheiner, O. A. 2000, *The Sun and Space Weather, 24th IAU Meeting, JD 7, Manchester*

Snegirev, S. D., Sheiner, O. A. & Smirnova, A. S. 2012, *Invention RU 2491583 C1*, 1

Space Weather of the Heliosphere: Processes and Forecasts
Proceedings IAU Symposium No. 335, 2017
C. Foullon & O.E. Malandraki, eds.

doi:10.1017/S1743921317008055

Solar Radio Emission as a Prediction Technique for Coronal Mass Ejections' Detection

Vladimir M. Fridman and Olga A. Sheiner
Radiophysical Research Institute, National Lobachevsky State University,
25/12A Bol'shaya Pecherskaya str., 603950, Nizhny Novgorod, Russian Federation
email: rfj@nirfi.unn.ru

Abstract. In this report we present a possible scheme of short-term CME detection forecasting developed on the basis of statistical analysis of solar radio emission regularities prior to "isolated" solar Coronal Mass Ejections registered in 1998, 2003, 2009-2013.

Keywords. Sun: radio radiation, Sun: coronal mass ejections (CMEs), methods: data analysis

1. Introduction

It is generally accepted that solar activity is the driver of space weather. Space weather forecasting is an emerging science that is facing many tasks. Some of them are the definition, classification and representation of solar features, and the establishing an accurate correlation between the occurrence of solar activities (e.g., solar flares and coronal mass ejections (CMEs)) and solar features observed in various wavelengths. The concept of CME as a global phenomenon of solar activity caused by magnetohydrodynamic processes is generally accepted. These processes occur in different wavelength ranges of emission and manifest in the microwave emission. The usage of radio-astronomical data for CME prediction is convenient and promising. It is so because the majority of the proceeding processes, as a rule, is reflected in the radio emission; spectral measurements cover all heights of solar atmosphere, and sensitivity and accuracy of measurements make it possible to record even small energy changes.

It should be noted that for the development of a prediction technique 2 ways of using the experimental data are possible: a) to be based on the totality of all available data, obtained by different methods, different tools, including of those not intended for the solution of the problems of forecast; b) the use of data obtained by one apparatus, and with uniform procedure. We use the first way and present several facts to be the basis of CME prediction during different time intervals.

2. Overview

We have used the data of solar microwave emission obtained by ground based round-the-clock monitoring of solar radio emission (data of the worldwide network of solar observatories in the radio-frequency band). Our examined set of CMEs consists of "isolated" CMEs: the time-nearest Coronal Mass Ejections are not recorded for least 8 h before and 6 h after the considered event. Selection of "isolated" CMEs is related with the ability to identify uniquely the registered radio emission with the sources of formed CMEs. We also use the data of X-ray and Hα flares location on the solar disk to determine possible sources of radio precursors of Coronal Mass Ejections.

Obviously, phenomena in the radio range that precede powerful energy releases, primarily, Coronal Mass Ejections, should be observed during different time intervals: from

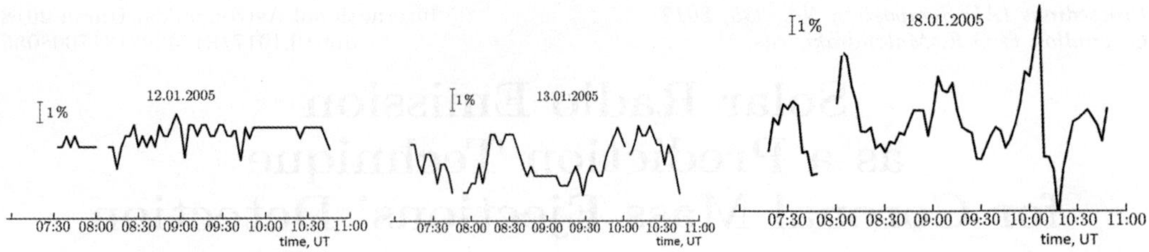

Figure 1. Dynamics of the radio emission flux during the January 2005 extreme events at the frequency $f = 9\,114$ MHz according to the RAS NIRFI "Zimenki" data.

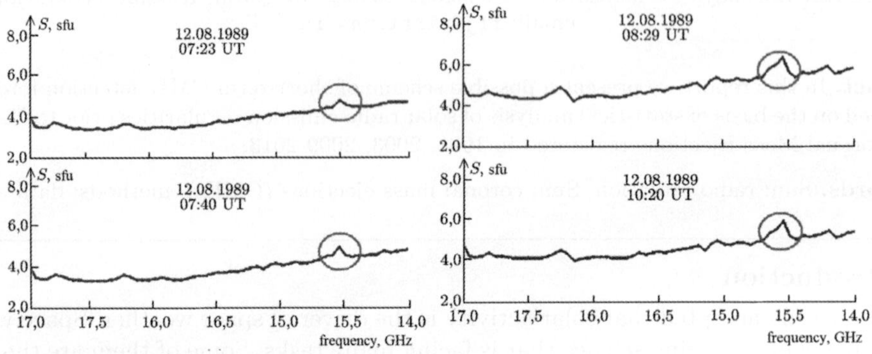

Figure 2. Radio emission spectra of active region on August 12, 1989 (NOAA/USAF 5638) at fixed times before CME's detection on coronagraph (12:59 UT) according to the sweeping spectrograph data (1 spectrum / sec, 100 MHz spectral resolution).

several days (which is typical for the evolution of the active region as a whole, and development of solar activity centers and complexes) to several hours and tens of minutes.

The study of the evolution of radio emission flux in the centimeter range showed that a few days before CME's detection, an increase in the amplitude of long-period ($T \geqslant 20$ min) pulsations of radio flux was observed (Fridman & Sheiner 2009).

This effect being observed before solar extreme events on January, 2005 (15/01/2005, 06:30:05 UT; 19/01/2005, 08:29:39 UT) is illustrated below (Figure 1). As can be seen from the figure, before the Coronal Mass Ejection's detection on January 15, 2005, the amplitude of long-period pulsations increased, on January 13, 2005, compared with January, 12. On January 18, 2005, the pulsation amplitude also increased in the day before the CME's detection on January, 19. After the events, the observed oscillations are at the level of the noise characteristics of the equipment used.

At closer times, before the detection of Coronal Mass Ejections at the coronagraph, a steady narrow-band spectral structure is detected in the intensity of microwave radio emission, 5-8 hours prior to the event. For example, approximately 5 hours before the CME's detection on August 12, 1989, one can see very small spectral features on the right side of spectra (Figure 2). The closer the time of CME detection the larger the amplitude of the feature is. The value of its flux is about 1 sfu and the spectral interval varies in 200-400 MHz just 1.5 hours before CME's detection. According to our ideas, the formation of this special feature is the reflection of reconstruction of the structure of the active region. Its stability is explained by the large three-dimensional scale of the phenomenon.

The third observed phenomena, preceding the appearance of a CME, are the presence of microwave precursors in the two-hour interval prior to CME's detection (Sheiner & Durasova 1994, Sheiner & Fridman 2010). This two-hour interval is especially important

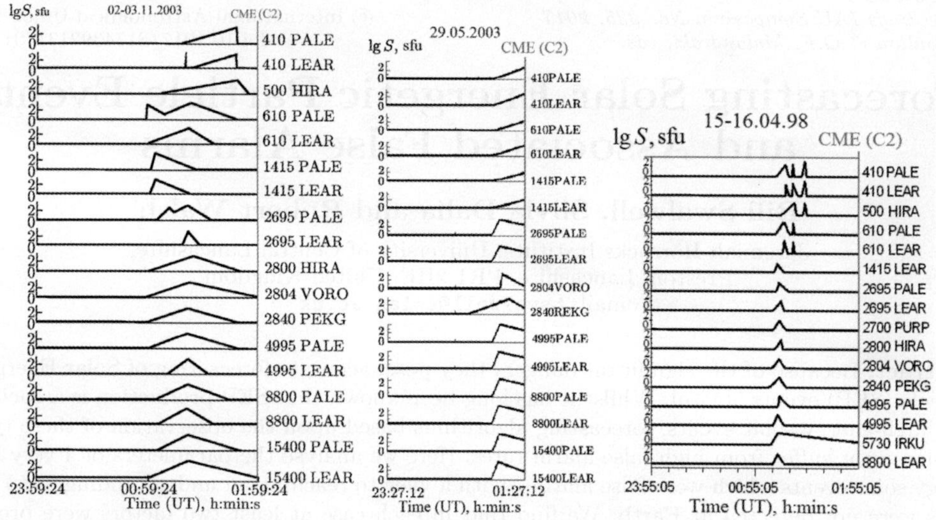

Figure 3. Schematic time profiles of microwave precursors. Left Y-axis is the observed flux on a logarithmic scale (in sfu). Right Y-axis (in red) is the moment of CME's detection on coronagraph LASCO, indication of the frequencies of the observations and abbreviations of observatories itself.

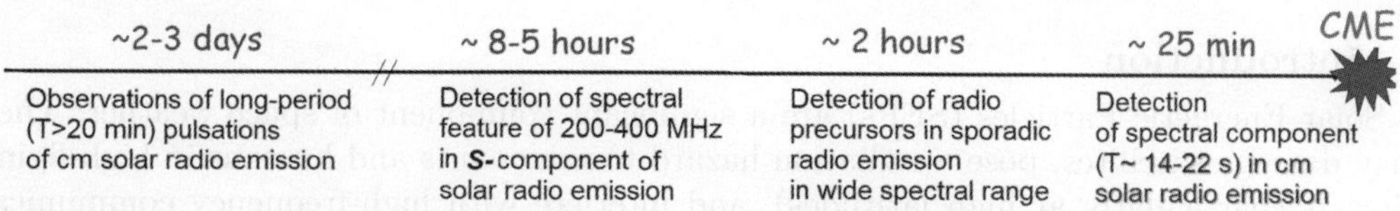

Figure 4. Preliminary qualitative forecasting scheme.

in studies, because it clearly exceeds the estimated propagation time of CMEs from the surface to the coronagraph FOV. It covers the period of the direct formation of CMEs, their separation and initial propagation in the lower corona of the Sun. These phenomena are illustrated in Figure 3. In the time interval of about 25 minutes before the Coronal Mass Ejections' detection periodic oscillations of the intensity of solar microwave emission are observed, with the periods of 14-22 seconds at wavelengths 3 and 10 cm. Moreover after the passage of the CME at the heights of the solar atmosphere where radio emission at these wavelengths is formed, such fluctuations on $\lambda = 3$ cm disappeared, and they were weakened considerably on $\lambda = 10$ cm. It is worthwhile to note that in several cases, when observed radiobursts were not connected with CMEs, periodic oscillations of intensity with these periods were not observed in the time intervals indicated.

The proposed way of short-term forecast of CMEs is based on the analysis of time and spectral data of solar microwave emission in the period preceding the phenomenon of CME (Figure 4).

Hereby it is possible to use the data of solar radio emission monitoring belonging to the Solar Data Service, for analysis and developing a procedure for short-term and very short-term CME forecasting.

References

Fridman, V. M. & Sheiner, O. A. 2009, *Geomagnetism & Aeronomy* 49, 1133
Sheiner, O. A. & Durasova, M. S. 1994, *Radiophysics & Quantum Electronics* 37, 575
Sheiner, O. A. & Fridman, V. M. 2010, *Radiophysics & Quantum Electronics* 53, 281

Space Weather of the Heliosphere: Processes and Forecasts
Proceedings IAU Symposium No. 335, 2017
C. Foullon & O.E. Malandraki, eds.

doi:10.1017/S1743921317011036

Forecasting Solar Energetic Particle Events and Associated False Alarms

Bill Swalwell, Silvia Dalla and Robert Walsh

Jeremiah Horrocks Institute, University of Central Lancashire,
Preston, Lancashire, PR1 2HE, United Kingdom
email: bswalwell@uclan.ac.uk

Abstract. Because of the significant dangers they pose, accurate forecasting of Solar Energetic Particle (SEP) events is vital. Whilst it has long been known that SEP-production is associated with high-energy solar events, forecasting algorithms based upon the observation of these types of solar event suffer from high false alarm rates. Here we analyse the parameters of 4 very high energy solar events which were false alarms, with a view to reaching an understanding as to why SEPs were not detected at Earth. We find that in each case at least two factors were present which have been shown to be detrimental to SEP production.

Keywords. Sun: flares, coronal mass ejections, particle emission

1. Introduction

Solar Energetic Particles (SEPs) are a significant component of space weather. They may damage satellites, pose a radiation hazard to astronauts and humans in high-flying aircraft (particularly at high latitudes), and interfere with high-frequency communications' systems (Desai & Giacalone 2016). Therefore, accurately forecasting their arrival at Earth has become vital.

It has long been known that the detection of SEPs at Earth is associated with solar flares which exhibit high emission in soft X-rays (SXR) and fast Coronal Mass Ejections (CMEs) (*e.g.* Dierckxsens *et al.* 2015). The difficulty for SEP forecasting algorithms is, however, that SEPs are not detected at Earth following all such large solar events. For example, Klein *et al.* 2011 investigated all X class flares between longitudes W0° and W90° in the period 1996 to 2006, and found that 30% did not produce an enhancement of >10 MeV protons above the background level.

Figure 1 shows two plots of proton intensity as measured by the ~40-80 MeV energy channel of the Geostationary Orbital Environmental Satellites' (GOES) Energetic Particle Sensor (EPS) instruments (Onsager *et al.* 1996). In Figure 1a, a steep rise is seen following a magnetically well-connected large solar event which occurred on 17 May 2012 as may be expected; by contrast Figure 1b shows that a similar large event which occurred on 18 March 2003 produced no rise at all. Such an event, which might reasonably have been expected to produce SEPs at Earth, may be termed a "false alarm". Here we examine 4 sample false alarm events with a view to determining why they failed to produce SEPs at Earth.

2. False alarms for simple forecasting algorithms

In Swalwell *et al.* 2017 we defined two simple SEP forecasting algorithms: the first, A.1, is based upon the observation of magnetically well-connected CMEs with a speed reported by the CDAW catalogue (Gopalswamy *et al.* 2009) to be greater than 1500 km/s

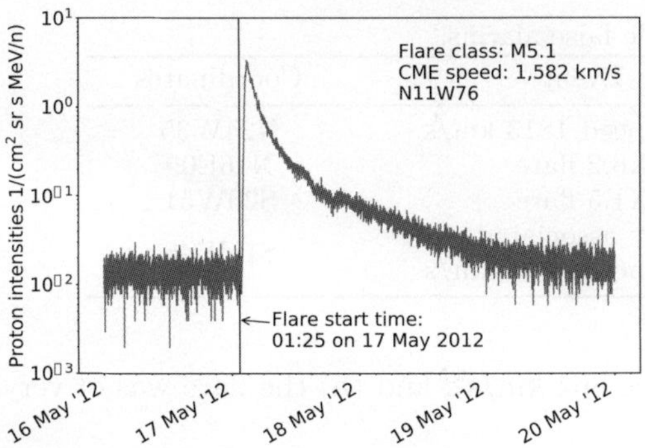

(a) A steep rise in energetic proton intensity is seen following a magnetically well-connected M5.1 large solar event.

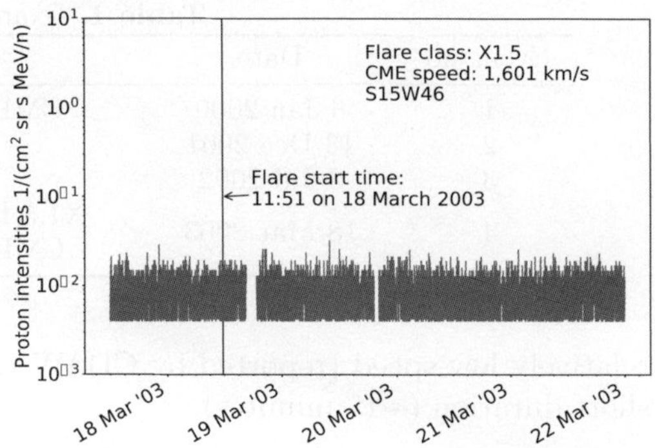

(b) A similarly large solar event produced no rise in energetic proton intensity. This event may be termed a "false alarm".

Figure 1. A comparison of 40-80 MeV proton intensity following two different magnetically-well-connected large solar events.

("fast CMEs"); the second, A.2, is based upon the observation of well-connected flares of class X. We compared the forecasts of each with historical data sets between January 1996 and March 2013.

Algorithm A.1 had a lower false alarm rate (28.8%) than A.2 (50.6%) but both missed a significant number of SEP events (53.2% and 50.6% respectively). We determined that an algorithm which was based upon the parameters of both CMEs and flares produced better results than one based upon the observation of only one type of solar event.

We identified a number of factors which are important to SEP production. Fast CMEs were less likely to produce SEPs if they were associated with a flare of class <M3, if their associated flare was of relatively short duration (< ~30 minutes), and if they were not reported to be a halo. X class flares were less likely to produce SEPs if either they were not associated with a CME or were associated with a CME slower than 500 km/s, and if they were of relatively short duration (< ~30 minutes) (Swalwell *et al.* 2017).

3. Examples of false alarm events

Table 1 gives four examples of high-energy solar events which were expected to produce SEPs at Earth, but which failed to do so. Examination of some of their parameters sheds some light on why they resulted in false alarms.

3.1. *Event 1: 1,813 km/s CME from N24W35 on 6 Jan 2000*

This was a very well magnetically-connected, very fast, CME. However, it was associated with a flare which was both short (~21 minutes) and of relatively low class (C5.8). Furthermore, the angular width of the CME was also reported by CDAW to be small (67°).

3.2. *Event 2: X6.2 flare at N16E09 on 13 Dec 2001*

The event originated from the edge of the best magnetically-connected region, and included a large flare which was expected to produce at least some enhancement in energetic protons. The fact that it did not may be due to two parameters: (a) the CME was of

Table 1. Example false alarms.

Event no	Date	Event	Coordinates
1	6 Jan 2000	CME speed 1813 km/s	N24W35
2	13 Dec 2001	X6.2 flare	N16E09
3	3 Jul 2002	X1.5 flare	S20W51
4	18 Mar 2003	X1.5 flare associated with CME speed 1601 km/s	S15W46

relatively low speed (reported by CDAW to be 864 km/s), and (b) the flare was of very short duration (~15 minutes).

3.3. *Event 3: X1.5 flare at S20W51 on 3 Jul 2002*

This very well magnetically-connected flare was associated with a very slow (265 km/s) non-halo (width 261°) CME, and was of very short duration (~8 minutes).

3.4. *Event 4: X1.5 flare with 1601 km/s CME from S15W46 on 18 Mar 2003*

This is the event for which the energetic proton intensity is shown in Figure 1b. It was extremely well magnetically-connected to Earth, and in this instance both the flare class and CME speed were high. Flare duration, however, was relatively short (at ~ 29 minutes) and the CME was reported by CDAW to have an angular width of 263°.

4. Conclusions

Some very high-energy solar events may fail to produce SEPs at Earth even if they are very well magnetically-connected. Understanding why such events are false alarms may provide an insight as to which of their associated parameters are important for SEP production.

Swalwell *et al.* 2017 reported that fast CMEs associated with flares of class <M3 or of relatively short duration (<~ 30 minutes), and fast CMEs which were not reported to be a halo were more likely to be false alarms. X class flares not associated with a CME, or associated with a CME slower than 500 km/s were more likely to be false alarms, as were those of relatively short duration.

Here we considered 4 sample false alarms. In each case 2 or more of the factors which were found to be detrimental to SEP production by Swalwell *et al.* (2017) were found to exist. In event number 1 there were 3: the associated flare was relatively short (~21 minutes); it was of class <M3 (C5.8); and the CME was not a halo (the width was 67°).

Event number 2 was a poorly magnetically-connected flare (from E09) but at X6.2 it was very intense; Event 3 was a lower class flare (albeit still large at X1.5) but very well connected. Neither produced SEPs but both had factors likely to result in a false alarm: each flare was short (~15 minutes and ~8 minutes respectively), and each was associated with a CME of relatively low speed (864 km/s and 265 km/s respectively); and in the case of event 3, the associated CME was not a halo.

Event number 4 illustrates that even a fast CME associated with a high intensity flare may sometimes be a false alarm. In this case, too, however the fact that the CME was not a halo and the short duration of the flare may explain why SEPs were not detected.

The full results and a more detailed analysis are presented in Swalwell *et al.* 2017.

References

Bentley, R. D., Csillaghy, A., Aboudarham, J., Jacquey, C., Hapgood, M. A., Bocchialini, K., Messerotti, M., Brooke, J., Gallagher, P., Fox, P., Hurlburt, N., Roberts, D. A. & Duarte, L. S. 2011, *Advances in Space Research*, 47, 2235-2239

Desai, M. & Giacalone, J. 2016, *Living Reviews in Solar Physics*, 13, 3

Dierckxsens, M., Tziotziou, K., and Dalla, S., Patsou, I., Marsh, M., Crosby, N., Malandraki, O. & Tsiropoula, G. 2015, *Solar Physics*, 290, 841-874

Gopalswamy, N., Yashiro, S., Michalek, G., Stenborg, G., Vourlidas, A., Freeland, S. & Howard, R. 2009, *Earth Moon and Planets*, 104, 295-313

Klein, K.-L., Trottet, G., Samwel, S. & Malandraki, O. 2011, *Solar Physics*, 269, 309-333

Onsager, T., Grubb, R., Kunches, J., Matheson, L., Speich, D., Zwickl, R. W. & Sauer, H. 1996, *Proceedings of SPIE - The International Society for Optical Engineering*, 2812

Swalwell, B., Dalla, S. & Walsh, R. 2017, *Solar Physics*, 292, 173

Space Weather of the Heliosphere: Processes and Forecasts
Proceedings IAU Symposium No. 335, 2017
C. Foullon & O.E. Malandraki, eds.

doi:10.1017/S1743921318000613

METU Data Driven Forecast Models: From the Window of Space Weather IAU Symposium 335

Yurdanur Tulunay[1] **and Ersin Tulunay**[2]

[1]ODTÜ-METU Dept. of Aerospace Eng., 06800 Ankara, Turkey
email: ytulunay@metu.edu.tr
[2]ODTÜ-METU Dept. of Electrical and Electronic Eng., 06800 Ankara, Turkey
email: ersintul@metu.edu.tr

Abstract. Space weather processes, in general, are non-linear and time-varying. In such cases 'data driven models' such as Neural Network, Fuzzy Logic and Genetic Algorithm based models were proved promising to be used in parallel with the mathematical models based on first physical principles. In particular, with the recent developments in 'big data' systems, one of the urgent issues is the development of new signal processing techniques to extract manageable, representative data out of the 'relevant big data' to be employed in 'training', 'testing' and validation phases of model construction. Since 1990, under the EU Frame Work Program Actions, we have developed such models for nowcasting, forecasting, warning and also for filling the data gaps on space weather cases including prediction of orbital spacecraft parameters. In particular, some typical, illustrative examples include the forecasting of the ionospheric critical frequencies foF2, during disturbed conditions, such as solar storms and extreme events; GPS total electon content(TEC); solar flare index during solar maximum and the construction of solar EUV flux variations. The associated input data organisation and the typical errors which have been within the acceptable operational expectations are summarised in terms of absolute values, percent and RMS. The aim of the paper is to show that the data driven approaches are promising for the forecasting of space weather.

Keywords. Near Earth Space, Space Weather, Data Driven Models, Forecast

---

1. Introduction

Near Earth Space (NES) Processes are highly complicated in that they are non-linear and time varying. Therefore, mathematical modelling, based on the 1st Physical Principles is usually very difficult or impossible. For such cases, we recommend the data driven modelling methods to be employed in parallel with the mathematical approach. Because non-linear systems are usually too complicated where exact analytical relationships based on first physical principles are prohibitively difficult to construct. We, a small group at the Middle East Technical University (METU), had worked on the data driven generic models of the Near Earth Space (NES) during the EU Actions of COST 238; 251; 271; 296, IHY and, ES 0803 between 1990 and 2015. The data driven models have been designed and developed. Briefly, we have demonstrated that the models based on Neural Network (NN), Fuzzy Logic (FL) and Genetic Programming (GP) have been promising both in modelling the Space Weather (SpW) effects on the ionosphere parameters and in filling the data gaps of the missing data and mitigation processes. The basic requirement for this is the availability of the representative data for the phenomena under the investigation. NN models have the capability of "learning", in other words, the capability of 'generalization'; FL models have the capability of inbedding the 'expert knowledge' into

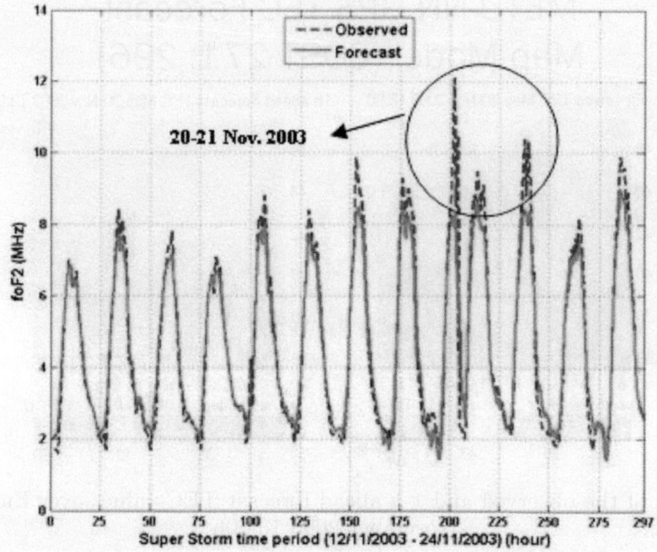

Figure 1. Superimposed are the observed and 1-h ahead forecast values of the Chilton f_oF2 versus the number of hours for the period of 12–24 Nov. 2013.

the model (see Altınay *et al.* 1997, Tulunay *et al.* 2004a, 2004b and 2005). Generalization capability of NNs are better than the generalization capability of numerical methods. At the present time, however, with the recent developments in the 'big data' systems, one of the urgent issues have been the development of new signal processing techniques to extract manageable representative data to be employed in 'training', 'testing', 'validation' phases during the process of model construction (Tulunay & Tulunay 2016). Some typical examples and results are presented in Section 2.

2. Some Typical Examples and Results

2.1. *AE NFN F_oF2 Forecast Model (COST 296)*

Figure 1 shows the forecast of the f_oF2 during a super storm of 2003 for the Chilton ionosonde critical frequencies. The sharp variations in these frequencies due to the super storm of 2003 are followed closely by the 1-h ahead forecast model.

2.2. *METU-NN GPS Total Electron Content (TEC) Forecast Model*

The 10-minute vertical TEC data evaluated from GPS data are employed in the model to forecast vertical TEC 1 hour ahead for 16–29 November 2003 which includes the space weather events. Figure 2 shows the observed and 1h ahead forecast TEC map for 20 November 2003 13:00h. Figure 2 (left) presents the observed map of TEC values over Europe and Figure 2 (right) presents the 1-h ahead forecast map of TEC values for the same area. It shows that the value distributions are considerably similar. This similarity demonstrates the satisfactory performance of METU-NN GPS TEC Forecast Map Model. Table 1 shows the performance parameters for the GPS forecast TEC values.

3. Conclusions

Data driven models have been developed and constructed to forecast the space weather events and to mitigate their associated probable effects (e.g. Earth System Science and

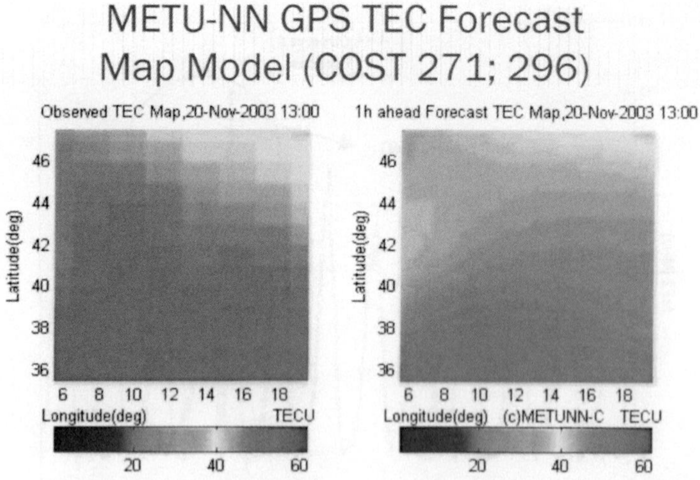

Figure 2. Map of the observed and 1-h ahead forecast TEC values over Europe for the 20 November 2003 13:00h.

Table 1. GPS TEC Forecast Performance Parameters for 16–29 November 2003

(COST 296 8-th MCM 13–15 March 2008, WG 1 Report)

LOCATION	Absolute error (TEC units) ($\times 10^{16}$ electron/m^2)	Cross correlation coeff. (%)
11.5°E; 38.5°N	1.58	98
13.5°E; 41.5°N	1.49	97
15.5°E; 44.5°N	1.52	97
Overall TEC Map	1.65	97

biological systems). Some typical examples are presented to give some idea concerning the performance of data driven models when used in space weather applications. In short, the data driven models are efficient not only in problem solving, but also in understanding the natures of the associated processes.

4. Acknowledgment

We acknowledge and thank all the co-authors of the topics mentioned herewith. Last but not the least we acknowledge the devoted care of Oktay Sipahigil, MS EE in the formatting of the proceeding.

References

Altınay O., Tulunay E. & Tulunay Y. 1997, *Geophysical Research Letter* 24 (12), 1467-1470

Tulunay Y., Tulunay E. & Senalp E.T. 2004, *Advances in Space research* 33 (6), 983-987

Tulunay Y., Tulunay E. & Senalp E.T. 2004, *Advances in Space research* 33 (6), 988-982

Tulunay E., Senalp E.T., Radicella S.M. & Tulunay Y. 2005, *Radio Science* 41 (4), RS4016, http://dx.doi.org/10.1029/2005 RS 003285

Tulunay Y. & Tulunay E. 2016 From the Near Earth Space (NES)/Space Weather Window META/BIG DATA: Some Turkish Initiatives *(2016 COSPAR Istanbul – cancelled) and European Space Weather Week 13 (ESWW 13), Oestende, Belgium, 14-18 November 2016*

The Editors Claire Foullon and Olga Malandraki.

Part 8. Space Weather Monitoring, Instrumentation, Data and Services

Space Weather of the Heliosphere: Processes and Forecasts
Proceedings IAU Symposium No. 335, 2017
C. Foullon & O.E. Malandraki, eds.

doi:10.1017/S1743921318000029

The MOTH II Doppler-Magnetographs and Data Calibration Pipeline

Roberta Forte[1], Stuart M. Jefferies[2,3], Francesco Berrilli[1], Dario Del Moro[1], Bernhard Fleck[4], Luca Giovannelli[1], Neil Murphy[5], Ermanno Pietropaolo[6] and Wayne Rodgers[7]

[1]Universitá di Roma Tor Vergata, Roma, Italy
via della Ricerca Scientifica 1, 00133 Roma
email: roberta.forte@roma2.infn.it

[2]2Georgia State University, Atlanta, USA

[3]University of Hawaii, Maui, USA

[4]European Space Agency

[5]Jet Propulsion Laboratory, USA

[6]Universitá degli Studi dell'Aquila, L'Aquila, Italy

[7]Eddy Company, Apple Valley, USA

Abstract. The calibration pipeline of the level zero images obtained from the Magneto-Optical filters at Two Heights (MOTH II) instrument is presented. MOTH II consists of two 20 cm aperture instruments, each using a Magneto-Optical Filter (MOF): one at 5896 Å (Na D2-line), the other one at 7700 Å (K I-line). MOTH II instruments thus provide full disk line-of-sight Doppler velocity and magnetic field measurements at two heights in the solar atmosphere. The developed MOTH II pipeline employs a set of standard calibration corrections, a correction for signal leakage, due to the non-ideal behavior of the polarizers, and the geometrical registration between the eight images acquired by four CMOS cameras, relative to two components of the signal in two circular polarization states, in each of the two channels. MOTH II data are used to investigate atmospheric dynamics (e.g., internal gravity waves and magneto-acoustic portals) and Space Weather phenomena. Particularly, flare forecasting algorithms, based on the detection of magnetic active regions (ARs) and associated flare probability estimation, are currently under development. The possible matching of MOTH II data with SDO/HMI and SDO/AIA images into a flux rope model, developed in collaboration between Harvard-Smithsonian CfA and MIT Laboratory for Nuclear Science, is being tested.

Keywords. polarimetric, spectrographs, solar atmosphere, helioseismology, magnetic fields

1. Introduction

The MOTH II (Magneto Optical filters at Two Heights) telescope consists of a dual channel equipment, mounting MOFs (magneto-optical filters) at either 5896 Å (Na D2-line) or 7700 Å (K I-line), respectively. It thus provides full disk line-of-sight Doppler velocity and magnetic field measurements at two heights in the solar atmosphere, 300-400 km and 600-700 km respectively. This allows to characterize the solar atmospheric dynamics, as the dynamics and topology of coronal magnetic loops is controlled by photospheric emergent magnetic field and plasma flow. It also allows to characterize the evolution of solar magnetic activity, in order to identify signature parameters of Space Weather events (e.g. flares).

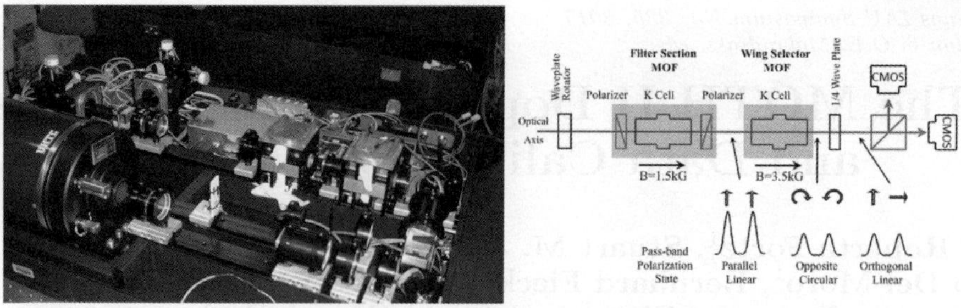

Figure 1. Picture of one channel of the MOTH II instrument (left) and optical scheme of the main section of the instrument (right).

2. The Instrument

The current version of the MOTH II instrument (Figure 1, left) mounts filters in K and Na lines, but two more channels in Ca and He (Murphy *et al.* 2005) are under testing, in order to probe the solar atmosphere at 4 altitudes in total, and to combine with HMI data, which would allow to obtain a 3D reconstruction of the magnetic and velocity fields.

The scheme in the right panel of Figure 1 shows the main elements of the current instrumental setup in one single channel (K or Na), based on Finsterle *et al.* (2004). The scheme shows the two MOF cells, both containing vaporized gas (of K or Na, according to the channel) in a uniform longitudinal magnetic field. The first one is the pre-filter, within two crossed linear polarizers, while the second one is the wing selector. This setup filters the incoming light at the typical wavelength of absorption of the vaporized element contained in the cell; the use of linear polarizers and of half wave retarder along the optical path allow to encode wavelength information into two circular polarization states ($\sigma+$ and $\sigma-$). The light beam is then split into a blue and red component, in order to get the blue and the red wing of the line, acquired by two different CMOS cameras. So in the end a total of 4 frames (R$\sigma+$, R$\sigma-$, B$\sigma+$, B$\sigma-$) are obtained in each channel (Na and K), which means 8 frames during each acquisition.

For each channel we can derive the magnetic field and velocity field maps along the line of sight by inserting the values of the 4 components into equations 2.1 and 2.2, according to Cacciani *et al.* (1990).

$$B_{LOS} = \frac{R^+ - B^+}{R^+ + B^+} - \frac{R^- - B^-}{R^- + B^-} \tag{2.1}$$

$$V_{LOS} = \frac{R^+ - B^+}{R^+ + B^+} + \frac{R^- - B^-}{R^- + B^-} \tag{2.2}$$

The MOTH II data have a pixel scale of 1.46 arcsec/pixel and a cadence of 5 seconds.

The MOTH II instrument is normally located at the Mees Observatory in Hawaii, where in particular the test phase was carried out and the pipeline developed. In November 2016 it was deployed to the South Pole for long observation campaigns, the first campaign was conducted during the Antarctic summer 2016-2017 (November 2016-January 2017), and the second one has started in November 2017.

3. Calibration Pipeline

The MOTH II instrument requires a solid pipeline for the automatic calibration and data reduction. A summary of the pipeline is shown in the flowchart in Figure 2.

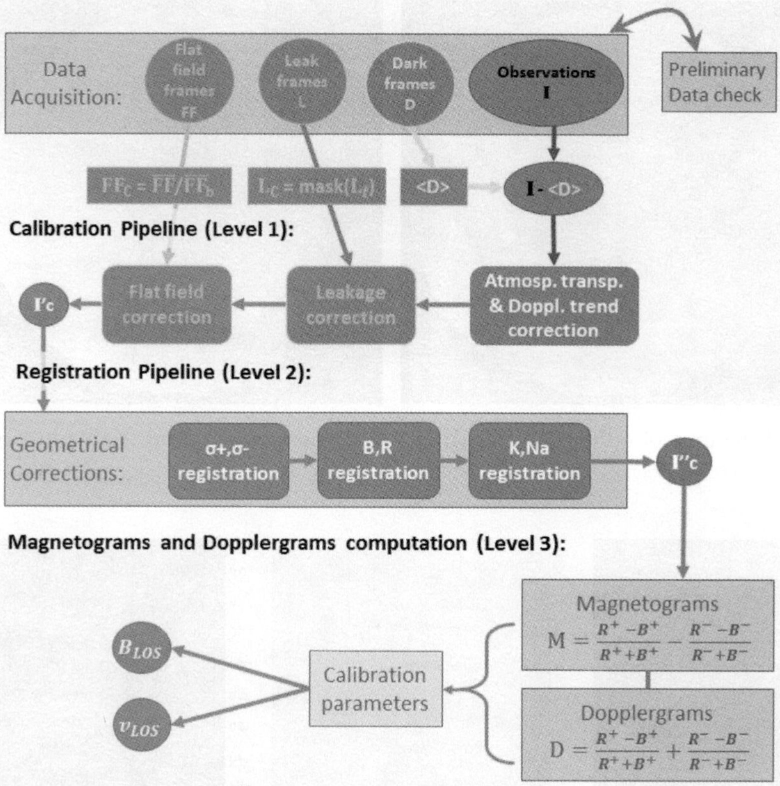

Figure 2. Scheme of the MOTH II calibration pipeline.

<u>Zero Level:</u> acquisition of scientific data and calibration images (flat field, dark and leak), and preliminary data check, performed through a Graphic User Interface to control temperatures, image tracking, dark, flat field, image statistics, etc.

<u>First Level:</u> standard photometric corrections, such as subtraction of the dark current, correction for the flat field and for the variation due to the atmospheric transparency and to the Doppler trend. The main issue of this step of the pipeline is the reconstruction and correction for the unpolarized signal leakage, due to the non-ideal polarizers which constitute MOFs.

<u>Second Level:</u> as the instrument provides 4 images per channel, it is extremely important to geometrically register them pixel-by-pixel in order to compute precise magnetograms and dopplergrams. In particular: there is a translation that has to be removed between images acquired by the same camera ($\sigma+$ and $\sigma-$ components); a translation and a rotation between images acquired by the 2 cameras (blue and red); translation, rotation and scaling between images from the 2 channels (K and Na).

<u>Third Level:</u> computation of the full-disk magnetograms and dopplergrams relative to the two channels of the instrument (K and Na), applying the formulas in equations 2.1 and 2.2 to the calibrated images. A sample of a resulting magnetogram and dopplergram is shown in Figure 3.

MOTH II data, together with data from other instruments (i.e. IBIS, ALTEA, PAMELA) will be made available in the near future in the Space Weather at TOr Vergata (SWERTO) database located on the SWERTO service website (www.spaceweather.roma2.infn.it).

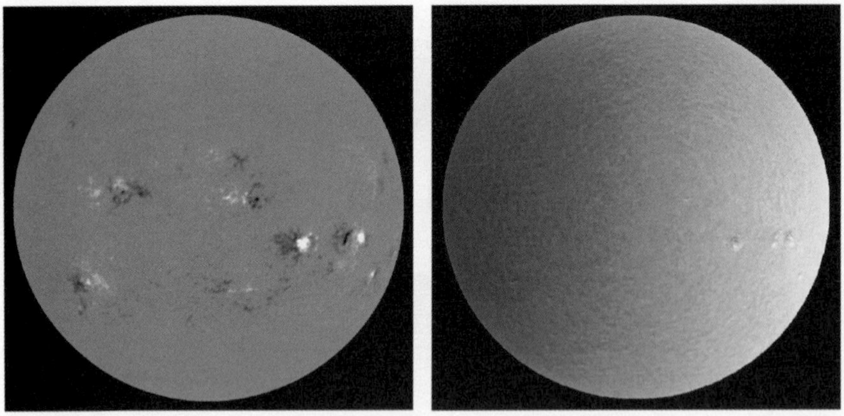

Figure 3. Sample of a final Magnetogram (left) and Dopplergram (right) in K channel.

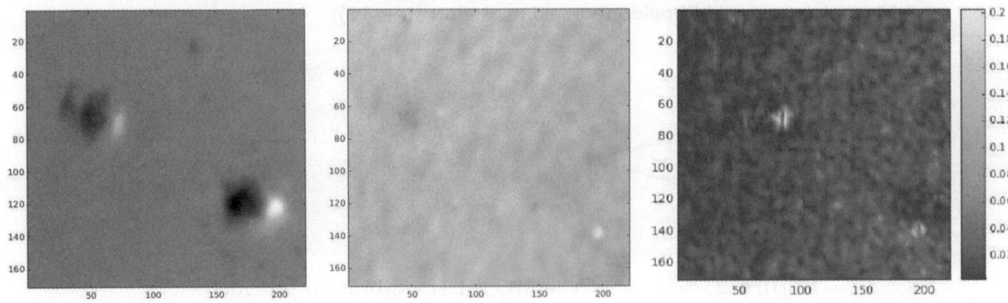

Figure 4. Magnetogram (left), Dopplergram (center) and power spectra displayed in color scale at 7 mHz (right) of the 175x225 pixels area, containing both the flaring AR 12628 (top-left) and the flare-quiet AR 12627 (bottom-right), on 21st January 2017.

4. Preliminary Data Analysis

A preliminary data analysis has been carried out on the recent dataset of magnetograms and dopplergrams acquired during the South Pole campaign. In particular several C-class flares were registered on the 21st January 2017, mostly from AR 12628. It is interesting to analyze the magnetic and doppler signal measured in the area containing both the flaring AR 12628 and the flare-quiet AR 12627. A power spectra analysis of the doppler signal in that area, in every pixel, has been carried out in a 6 hours interval, computing the power spectra of the signal with a moving window over intervals of 42 minutes with an overlap of 10 minutes. Flaring AR 12628 shows high power signal compared to the flare-quiet AR 12627. The max power of the spectra occurs at a frequency of 7 mHz (Figure 4). This enhancement may be attributed to power in an acoustic halo observed in the penumbral regions. Further studies will be carried out in the near future to investigate this result.

Other applications of MOTH II data are currently under development, such as flare forecasting algorithms, based on the detection of magnetic ARs and associated flare probability estimation, and possible matching of MOTH II data with SDO/HMI and SDO/AIA images into models for magnetic loops in ARs.

References

Cacciani, A., Ricci, D., Rosati, P., Rhodes, E. J., Smith, E., Tomczyk, S. & Ulrich, R. K. 1990, *Nuovo Cimento C*, 13:125–130

Murphy, N., Smith, E., Rodgers, W. & Jefferies, S. 2005, *ESA Special Publication*, vol 592, p. 511

Finsterle, W., Jefferies, S. M., Cacciani, A., Rapex, P., Giebink, C., Knox, A. & Dimartino, V. 2004, *Solar Physics*, 220:317–331

Space Weather of the Heliosphere: Processes and Forecasts
Proceedings IAU Symposium No. 335, 2017
C. Foullon & O.E. Malandraki, eds.

doi:10.1017/S1743921317008584

The Inner Coronagraph on Board ADITYA-L1 and Automatic Detection of CMEs

Banerjee, D.[1,2], Patel, R.[1], Pant, V.[1] and ADITYA team

[1]Indian Institute of Astrophysics, Bangalore-560034, India
[2]Center of Excellence in Space Sciences, IISER Kolkata, India

Abstract. Visible Emission Line Coronagraph (VELC) on board ADITYA-L1 is an internally occulted coronagraph with mirror as its primary objective element. It has a field of view (FOV) starting from 1.05 $R_\odot$ – 3 $R_\odot$. It will observe the corona in continuum centered at 5000 Å and will perform spectroscopic observations of inner corona in two visible (5303 Å and 7892 Å) and one infrared (10747 Å) wavelengths. VELC will be capable of observing **the** corona with high spatial and temporal resolutions. We present an overview of the inner coronagraph (VELC) design and introduce the concept of an on-board automated coronal mass ejections (CMEs) detection logic proposed for this payload.

Keywords. Sun, Coronal Mass Ejections, Coronagraph

1. Introduction

The outer atmosphere of the Sun, corona, has brightness million times fainter than that of the solar disk. Hence it is too faint to be seen unless solar disk is occulted by moon during a total solar eclipse. However, total solar eclipses are rare events and last only for few minutes giving observers/researchers very little time to perform experiments. This problem was resolved when Lyot invented coronagraph (Lyot 1930) to observe corona by creating an artificial solar eclipse. Since ground-based coronagraphs suffer from atmospheric scattered intensity, coronagraphs were sent to space to overcome atmospheric effects. A white light coronagraph having FOV from 3 $R_\odot$ – 10 $R_\odot$ on board OSO-7 (Koomen *et al.* 1975) observed large scale coronal transients for the first time from space in 1971 (Hansen *et al.* 1971). White light coronagraphs on board Skylab (MacQueen *et al.* 1974) and Solwind (Michels *et al.* 1980) having FOV from 1.5 $R_\odot$ – 6 $R_\odot$ and 2.6 $R_\odot$ – 10 $R_\odot$, respectively, have added to our understanding of CMEs. Solar Maximum Mission (SMM) (MacQueen *et al.* 1980) had a white light coronagraph with FOV from 1.5 $R_\odot$ – 6 $R_\odot$ and spatial resolution better than 10 arcsec. Coronagraph onboard SPARTAN 201 (Guhathakurta *et al.* 1993) with FOV of 1.25 $R_\odot$ – 6 $R_\odot$ flew five times between April 1993 and November 1998 and provided valuable data. Among the existing space based coronagraphs, Large Angle Spectroscopic COronagraph (LASCO) (Brueckner *et al.* 1995) on board Solar and Heliospheric Observatory (SOHO) has FOV starting from 2.5 $R_\odot$ whereas Sun Earth Connection Coronal and Heliospheric Investigation (SECCHI) (Howard *et al.* 2008) on board Solar Terrestrial Relations Observatory (STEREO) has FOV starting from 1.2 $R_\odot$. The only coronagraph which observed corona closest to the Sun is LASCO C1 with FOV starting from 1.1 $R_\odot$ up to 3 $R_\odot$. However, it stopped observations in 1998 two years after its launch. MkIII (Fisher *et al.* 1981) in Mauna Loa Solar Observatory (MLSO) observed the inner solar corona from 1.2 $R_\odot$ to 2.2 $R_\odot$. Later Mk4 (Elmore *et al.* 2003) recorded the polarization brightness (pB) images of corona from 1.12 $R_\odot$ to 2.8 $R_\odot$. It was succeeded by K-COR coronagraph (de Wijn *et al.* 2012) in 2013 having FOV same as that of Visible Emission Line Coronagraph (VELC). However, ground-based coronagraphs suffer from atmospheric conditions and limited observing time. Though our understanding and knowledge of solar outer corona has increased over the years, the observations of inner corona is not yet fully explored with high spatial and temporal resolutions. VELC ([Prasad *et al.* 2017, Singh *et al.* 2013, Singh *et al.* 2011) which is expected to launch in 2020, will provide an opportunity to study the inner corona and dynamics of CMEs in this region with unprecedented details having field of view starting from 1.05 $R_\odot$.

With the rise in the volume of data generated, automated CME detection techniques are also planned to be implemented on board so that the telemetry load is reduced and only images containing CMEs are downloaded. A number of on-ground automated CME detection techniques have been developed (Hess & Colaninno 2017). However, on ground automated CME detection algorithms are memory intensive and hard to be implemented on-board for CME detection. Therefore, a simple algorithm is designed to detect CMEs in the coronagraph images onboard VELC. Among the upcoming solar missions, Multi Element Telescope for Imaging and Spectroscopy (METIS) for Solar Orbiter mission scheduled to be launched in 2018 is also planning to have an on-board automated CME detection algorithm (Bemporad *et al.* 2014). This algorithm is primarily based on intensity threshold only.

In this proceedings we discuss the capabilities and uniqueness of VELC in section 2. The performance of the proposed algorithm on the synthetic images is briefly discussed in section 3.

2. Visible Emission Line Coronagraph

Visible Emission Line Coronagraph (VELC) is one of the seven payloads on-board ADITYA-L1 which will observe solar corona from Lagrangian 1 (L1) point. L1 point is located at 1.5×10^6 km from Earth at sun-earth line which provides continuous observations of Sun throughout the year. In addition to this, L1 point is devoid of Earth's magnetic field, thus accurate magnetic field measurements are possible at this location.

VELC has four channels which include one continuum channel that will have field of view from 1.05 - 3 $R_\odot$ and will take images of corona at 5000 Å with passband of 10 Å, while the other channels will perform spectroscopic observations from 1.05 $R_\odot$ to 1.5 $R_\odot$ using multi-slit spectrograph to study the solar corona at emission wavelengths of 5303 Å, 7892 Å, 10747 Å. While combining imaging and spectroscopic observations, we will be able to study the origin of coronal transients and their dynamics in the inner corona.

2.1. *Optical Design of VELC*

The optical layout of VELC can be referred to Figure 1 of Prasad *et al.* (2017). VELC has an entrance aperture of 147 mm. The primary mirror is an off-axis parabolic mirror located at a distance of 1570 mm from entrance aperture which reflects light to a spherical concave mirror having a hole at its center. The surface roughness of this mirror has to be in the range of 1.5-1.7 Å to keep the scattered light minimum and achieve the required signal to noise. It rejects the light up to 1.05 $R_\odot$ from center of solar disk through a hole perpendicular to entrance aperture. This serves as an internal occulter by rejecting the disk light and allowing only coronal light into the instrument. The reflected coronal light is collimated using a doublet lens and passed through a dichoric beam splitter which reflects the light with wavelength less than 5100 Å to the imaging channel and passes the rest of the light to the spectroscopic channels.

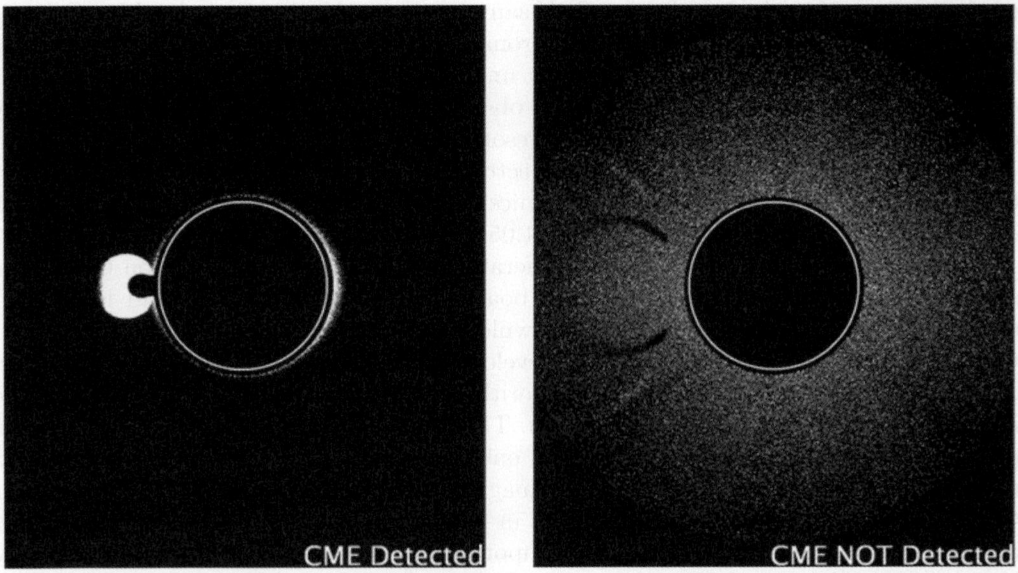

(a) (b)

Figure 1. (a) CME detected in synthetic coronal image when the intensity and size of the CME is more than chosen intensity and area threshold, (b) CME not detected in synthetic coronal image when CME merges with the background

In the imaging channel a narrow filter of pass band 10 Å centered at 5000 Å images the solar corona through a four element imaging lens on a 2560 × 2160 CMOS detector with pixel size of 6.5 μm.

VELC is unique amongst the existing space-based coronagraphs. It will take images of corona to height as low as 1.05 $R_\odot$ with cadence as high as 1 s. It should be noted that the existing space-based coronagraphs have cadences no faster than 5 min. VELC will perform simultaneous spectroscopic and imaging observations. The complete description of other channels of VELC can be found in Prasad *et al.* (2017).

3. On-board Detection of CMEs in VELC

The capability of VELC to image the corona at high cadence will generate more than 200 Gbits of data per day for VELC continuum channel alone. To identify and store the most useful data and download them using the available telemetry a simple algorithm has been designed which can be easily implemented in the on-board electronics. This algorithm will be the first of its kind.

The algorithm is implemented by first building the signal in coronagraph images. In order to increase the signal, 10 images of 100 ms exposure are added to make a 1 s equivalent exposure image. 10 images of 1 s exposure are further averaged to build the signal. These images are re-binned to lower resolution and difference is taken at an interval of 120 s. Intensity threshold is applied on the difference images with threshold value proportional to the mean of each image. It is accompanied by area threshold i.e. convolving the binary image with a kernel of appropriate size depending on the science requirements. The maximum value of convolved image is compared with an area threshold. The use of area threshold separates the CMEs structures from the noise which can be otherwise detected. The aim of this algorithm is to detect the CMEs in the images and not to track them. Once the CME is detected the observations will continue at a higher cadence. The images containing CMEs will then be downloaded and rest will be discarded.

In order to test the performance of this algorithm, synthetic coronal images with pixel resolution of 2.51 arcsec were prepared for VELC field of view, details of which will be presented in the forthcoming paper (Patel *et al.* 2018). The synthetic images contain contribution from scattered intensity (Venkata *et al.* 2017) and photon noise. CMEs of different sizes and intensities were launched on these images.

Figure 1a demonstrates that the algorithm has detected a CME in a simulated image while Figure 1b shows that the detection has failed when the intensity of the CME became faint and almost merged with the background. The value of convolution threshold for this case is chosen as 0.8 with kernel of size 10×10.

4. Conclusion

VELC will provide an unique opportunity to understand the dynamics of large scale coronal transients in inner corona by combining observations from imaging and spectroscopic channels. We have demonstrated that the detection algorithm based on intensity and area thresholding can detect CMEs in synthetic coronagraph images on-board VELC/ADITYA-L1. A thorough analysis has been planned for studying the detection efficiency of the CME detection algorithm for various cases of CMEs and modifying values of free parameters suitable for our requirements. The detailed analysis will be presented in Patel *et al.* (under preparation).

References

Bemporad, A., Andretta, V., *et al.* 2014, *Proceedings of the SPIE*, Vol. 9152

Brueckner, G., Howard, R., *et al.* 1995, *Solar Physics*, 162, 357

de Wijn, A. G., Burkepile, J. T., Tomczyk, S., Nelson, P. G., Huang, P. & Gallagher, D., 2012, *Proceedings of SPIE*, 8444, 84443N

Elmore, D. F., Burkepile, J. T., Darnell, J. A., Lecinski, A. R. & Stanger, A. L., 2003, *Proceedings of SPIE*, 4843, 66-75

Fisher, R. R., Lee, R. H., MacQueen, R. M. & Poland, A. I., 1981, *Applied Optics*, 20, 1094-1101

Guhathakurta, M., Fisher, R. R., Holzer, T. E. & Sime, D. G. 1993, *Amer. Astron. Soc*, 25, 1213

Hansen, R. T., Garcia, C. J., Grognard, R. J.-M. & Sheridan, K. V. 1971, *Proceedings of the Astronomical Society of Australia*, 2, 57

Hess, P. & Colaninno, R. C. 2017, *The Astrophysical Journal*, 836, 9

Howard, R. A., Moses, J. D., Vourlidas, A., *et al.* 2008, *Space Science Reviews*, 136, 67

Koomen, M. J., Detwiler, C. R., Brueckner, G. E., Cooper, H. W. & Tousey, R. 1975, *Applied Optics*, 14, 743

Lyot, B. 1930, *C. R. Acad. Sci. Paris* ,191, 834

MacQueen, R. M., Csoeke-Poeckh, A., Hildner, E., House, L., Reynolds, R., Stanger, A.,Tepoel, H. & Wagner, W 1980, *Solar Physics*, 65, 91

MacQueen, R. M., Eddy, J. A., Gosling, J. T., *et al.* 1974, *The Astrophysical Journal*, 87, L85

Michels, D. J., Howard, R. A., Koomen, M. J., Sheeley, N. R. J.r.: In Kundu, M. R. & Gergely, T. E. 1980, *Radio Physics of the Sun*, p.439, D. Reidel, Hingham MA

Prasad, B., Banerjee, D., Singh, J., *et al.* 2017, *Current Science*, 113(4), 613

Singh, J., Bayanna, R. & Sankarasubramanian, K. 2013, *Journal of Optics*, 42, 96

Singh, J., Prasad, B. Venkatakrishnan, P., *et al.* 2011, *Current Science*, 100, 167

Thompson, William *et al.* 2003, *SPIE Proceedings*, 4853

Venkata, S. N., Prasad, B. R., Nalla, R. K. & Singh, J. 2017, *Journal of Astronomical Telescopes, Instruments, Systems*, 3, 14002

Space Weather of the Heliosphere: Processes and Forecasts
Proceedings IAU Symposium No. 335, 2017
C. Foullon & O.E. Malandraki, eds.

doi:10.1017/S1743921318000510

Data Handling and Assimilation for Solar Event Prediction

Petrus C. Martens[1] **and Rafal A. Angryk**[2]

[1]Department of Physics & Astronomy, Georgia State University,
25 Park Place, 6th Floor, Atlanta, GA 30303, USA
email: `martens@astro.gsu.edu`

[2]Department of Computer Science, Georgia State University,
25 Park Place, 7th Floor, Atlanta, GA 30303, USA
email: `angryk@cs.gsu.edu`

Abstract. The prediction of solar flares, eruptions, and high energy particle storms is of great societal importance. The data mining approach to forecasting has been shown to be very promising. Benchmark datasets are a key element in the further development of data-driven forecasting. With one or more benchmark data sets established, judicious use of both the data themselves and the selection of prediction algorithms is key to developing a high quality and robust method for the prediction of geo-effective solar activity. We review here briefly the process of generating benchmark datasets and developing prediction algorithms.

Keywords. Astronomical data bases: miscellaneous, Sun: flares, catalogs

---

1. Introduction: Flare Prediction With Machine Learning

Forecasting the timing and magnitude of geo-effective flares, CMEs, and Solar Energetic Particle (SEP) events is of great practical significance for our increasingly technologically dependent society, and hence research in the field of space weather is strongly supported in many developed nations. For example the US has developed a National Space Weather Strategy and Space Weather Action Plan† that gives priority to research on the forecasting of solar flares, eruptions, and particle storms, while also emphasizing the need to develop benchmarks, the subject of this paper.

One approach to flare and SEP forecasting is using the methods of Big Data Mining, also called Machine Learning. The method involves gathering large amounts of relevant data, then cleaning up and balancing these data, as described in detail below, to produce a "benchmark data set" that algorithms of data analytics can be safely applied on without generating spurious results. That procedure, developed and widely used in Computer Science, is still only sparsely used in Solar Physics. The purpose of this paper is to specify and explain this procedure for the solar physics community.

The record for this method in the prediction of flares and SEP events gives us reason to be optimistic. First of all, experienced human flare forecasters at NOAA's Space Weather Prediction Center have used a similar method for decades with good outcomes. Secondly, recent research by Falconer *et al.* (2014), Bobra & Couvidat (2015), as well as our own work (Angryk *et al.* 2018, in preparation), using data from SDO's Helioseismic and Magnetic Imager (HMI; Scherrer *et al.* 2012), has yielded strong results for flare forecasting.

† https://goo.gl/r0Vcd5

Figure 1. Time difference in minutes between metadata entries produced by our trainable CBIR module for each AIA channel for three years of AIA data. Yellow shows a gap of over 12 h. Figure from Schuh, Angryk & Martens (2015).

2. Development of a Benchmark Data Set

The data repositories for most solar observatories are what we call "Operational Archives", meaning that all the data collected are archived, to be sorted out for usefulness by researchers as they analyze them. For most observatories the images undergo substantial processing – e.g. background subtraction, North up, etc. – prior to release for analysis. For the data from NASA's Solar Dynamics Observatory (SDO), also extensive metadata are derived, mostly from automated modules developed by the SDO Feature Finding Team (FFT; Martens *et al.* 2012) and stored in the Heliophysics Events Knowledge base (HEK; Hurlburt *et al.* 2012). These metadata too usually become part of a benchmark data set; in fact the analysis from most benchmarks is performed on metadata only.

Since the benchmark is being developed for *automated* analysis, it is important first to *clean* the data, i.e. remove all the images that are compromised in any way, for example by CCD bleeds, proton storm hits on the CCD, or from electronic garbling. Operational archives often have been labeled for such cases. The second step is to register all data gaps. Figure 1 (Schuh *et al.* 2015) shows the identification of data gaps in the metadata derived by our trainable Content Based Image Recognition (CBIR) module (Martens *et al.* 2012), using all channels of the Atmospheric Imaging Assembly (AIA; Lemen *et al.* 2012) on SDO. The data gaps are either in the AIA data or from the module itself not operating properly. Note the bi-annual gaps when SDO is eclipsed by Earth.

Why is it necessary to catalog these gaps for the benchmark data set? We'll give one example. Say, one does a study of the occurrence of sigmoids for the prediction of flares. For automated sigmoid detection we have the so-called "sigmoid-sniffer" (Martens *et al.* 2012), developed by Dr. Georgoulis, and for flares we use the X-ray data from the GOES satellites†. For the study results to be reliable one must exclude the periods when there are no AIA, or only low cadence 94 Å data that the "sigmoid sniffer" uses, as well as

† https://satdat.ngdc.noaa.gov/sem/goes/data/new_avg/

any intervals for which no GOES flare data are available in the pertinent X-ray channel, and also any timespans for which the "sigmoid sniffer" itself has no recorded metadata. Without that knowledge any classifier algorithm would be misled and produce incorrect results. Solar physicist conducting their own studies do this data cleaning by themselves and by hand, which is one reason why such studies often are limited to several dozen events – more would require a far too large effort.

The purpose then of our group in developing benchmark datasets is twofold. First by reviewing large amounts of data, as exemplified for AIA in Figure 1, we can develop clean, properly labeled datasets with thousands of data points. This entails an enormous amount of work – more than a year by a graduate student in the Schuh, Angryk & Martens (2015 and 2016) and Kucuk, Banda & Angryk (2017) papers. The latter released a dataset of automatically generated records of ARs, Coronal Holes, Sigmoids, and Flares observed by AIA, together with their image data, with 270,000 event records. It would be truly a wasteful multiplication of effort to repeat such a program over and over again for different research projects, hence our benchmark datasets.

Secondly, and at least as important, our benchmark can then be used to make fair comparisons between different methods to answer the same research question. Consider the prediction of solar flares: the "skill" scores that any prediction algorithm achieves will obviously depend on the dataset that one uses to train and test the prediction algorithm on. Hence comparisons between different algorithms honed on different data sets are rather meaningless; a fair comparison is achieved only by comparing the "skill scores" of different algorithms when applied to the same dataset. This is one of the reasons why the US Space Weather Action Plan, cited above, specifically asks for benchmarks to be developed. Obviously benchmarks must be "Static Archives", that is, once established they should not be changed to make the analyses carried out with them comparable.

Since the launch of SDO up to July 2017 we found 580 detected X- and M-class flares, leaving us with a still fairly small dataset to train and test predictions on, but still enough to obtain promising results as Bobra & Couvidat (2015) have shown, and as our own work indicates (Angryk *et al.*, in preparation). The number of geo-effective SEP events is less then one hundred, and hence we use a somewhat different approach here, that of *instance based learning*; see Filali *et al.* (2017).

3. The Need for Balanced Training Sets

When developing a flare prediction algorithm one needs to consider not only recorded flare events but also a similar amount of *non-events*, that is, instances of ARs that did *not* flare in the forecast window. Solar ARs do not have X- and M-class flares most of the time. There have been ≈ 600 in the six years of the SDO mission, that is roughly one every four days over the entire disk, and a considerably longer cadence per individual AR. Hence there are far more non-flaring daily intervals for each AR than there are flaring instances. Considering all these instances together will lead to a very uneven distribution. Any sophisticated flare prediction algorithm may then simply fall back to an "All Clear" prediction all of the time for each AR, to get the correct outcome in the great majority of cases. This is meaningless of course since the whole point of the exercise is to predict these rare flares.

One standard solution is to reduce the number of non-flaring intervals in the training data derived from the benchmark dataset by a large factor, for example by picking a random subsample. It is important to be very careful here to not introduce a systematic bias. A truly random sampling of non-flaring intervals may be the safest solution here, but it leaves us with a rather small dataset for training and testing, only about 1200 events, 600 flaring and 600 non-flaring intervals.

An alternative method to address imbalance is to identify forecasting evaluation measures that are resistant to imbalanced data. One can use the Heidke Skill Score (HSS) (e.g. Barnes & Leka, 2008) or the Gilbert Skill score (GS; e.g. Mason & Hoeksema, 2010).

4. Using Classifiers

The next step is applying machine learning methods to the benchmark dataset, in our case to find the best possible flare prediction. Almost all data mining papers on flare forecasting thus far fall into the category of parameter-based classification, which transforms spatio-temporal solar image data into a time series representation. Usually a group of expert-picked parameters is extracted from the images, that summarize the evolution of these images. For example, Bobra & Couvidat (2015) use 25 physical parameters derived from vector-magnetograms. Thus the project is reduced to the study of time series of metadata, sometimes already published online by the instrument teams.

One then applies conventional classification methods, such as regression models, decision trees, support vector machines, neural networks, k-nearest neighbor, etc. on the extracted parameters. The classifier looks back over a given time-interval to come up with a prediction for a second time interval, the prediction window. Once these classifiers are trained on one part of the benchmark dataset (usually of the order of 2/3 of the data), they are tested for their forecasting capability on the remaining part of the benchmark data. After that one picks the best, or a combination of the best algorithms, to employ them in an operational setting to predict real-time solar activity. That is the second and decisive test of the capabilities of the system.

A key aspect of this procedure is that one has to have a strong understanding of how the classifiers that one applies work, to figure out when true machine learning, i.e. data generalization, occurs. So some attempt needs to made to understand what the black box is actually doing - *look inside it!* Otherwise fundamental mistakes will be made, for instance using the same data for the training and predicting. Even a basic undergraduate textbook like *Principles of Data Mining* (Bramer 2013, second edition) can be very helpful. Another, even more effective approach is the close collaboration between solar physicists and computer scientists, as we have in the recently established interdisciplinary Solar-stellar Informatics Cluster at Georgia State University.

References

Barnes, G. & Leka, K. D. 2008, *ApJ Letters*, 688, L107

Bobra, M. G. & Couvidat, S. 2015, *ApJ*, 798, 135

Bramer, M. 2013, *Principles of Data Mining (Undergraduate Topics in Computer Science)*, Second Edition (Secaucus, NJ, USA: Springer-Verlag New York, Inc.)

Falconer, D. A., Moore, R. L., Barghouty, A. F. & Khazanov, I. 2014, *Space Weather*, 12, 306, 2013SW001024

Filali Boubrahimi, S., Aydin, B., Martens, P. & Angryk, R. 2017, *IEEE Proceedings*, in press

Hurlburt, N., Cheung, M., Schrijver, C., *et al.* 2012, *Sol. Phys.*, 275, 67

Kucuk, A., Banda, J. M. & Angryk, R. A. 2017, *Nature: Scientific Data*, 4, data Descriptor

Lemen, J. R., Title, A. M., Akin, D. J., *et al.* 2012, *Sol. Phys.*, 275, 17

Martens, P. C. H., Attrill, G. D. R., Davey, A. R., *et al.* 2012, *Sol. Phys.*, 275, 79

Mason, J. P. & Hoeksema, J. T. 2010, *ApJ*, 723, 634

Scherrer, P. H., Schou, J., Bush, R. I., *et al.* 2012, *Sol. Phys.*, 275, 207

Schuh, M. A., Angryk, R. A. & Martens, P. C. 2015, *Astronomy and Computing*, 13, 86

—. 2016, *Journal of Space Weather and Space Climate*, 6, A22

Space Weather of the Heliosphere: Processes and Forecasts
Proceedings IAU Symposium No. 335, 2017
C. Foullon & O.E. Malandraki, eds.

doi:10.1017/S1743921318000054

SWERTO: a Regional Space Weather Service

Francesco Berrilli[1,2,3], Marco Casolino[1,2], Dario Del Moro[1], Roberta Forte[1], Luca Giovannelli[1], Matteo Martucci[1,2], Matteo Mergé[1,2], Gianluca Napoletano[4], Livio Narici[1,2], Ermanno Pietropalo[4], Giuseppe Pucacco[1], Alessandro Rizzo[1,2], Stefano Scardigli[1] and Roberta Sparvoli[1,2]

[1]Department of Physics, University of Rome Tor Vergata, IT-00133, Rome, Italy
email: francesco.berrilli@roma2.infn.itl

[2]INFN - National Institute for Nuclear Physics, IT-00044 Frascati (Rome) Italy [3]INAF - National Institute for Astrophysics, IT-00136 Rome, Italy [4]Department of Physical and Chemical Science, University of L'Aquila, IT-67100, L'Aquila, Italy

Abstract. The Space WEeatherR TOr vergata university (SWERTO) service is an operational Space Weather service based on multi-instrument data from space-based (PAMELA, ALTEA) and ground-based (IBIS, MOTHII) instruments. The service (*spaceweather.roma2.infn.it*) is located at the Physics Department of the University of Rome Tor Vergata, Italy (UTOV) and will allow registered users to access scientific data from instrumentation available to UTOV researchers through national and international collaborations. It will provide intuitive software for the selection and visualization of such data and results from prototype forecasting codes for flare probability and Solar Energetic Particle (SEP) fluxes. The service is designed to promote access to technical and scientific information by the regional industries which employ technologies vulnerable to Space Weather effects. Basically, SWERTO aims to: *i)* design and construct a data-base with particle fluxes recorded by space missions and spectro-polarimetric measurements of the solar photosphere; *ii)* allow an *Open Access* to the data-base and to prototype forecasts to regional industries involved and exposed to Space Weather effects; *iii)* implement a tutorial and a FAQ section to help decision makers to became aware of and evaluate the risks from Space Weather events; *iv)* outreach and customer products. SWERTO has been financed by the Regione Lazio FILAS-RU-2014-1028 grant.

Keywords. Space weather instrumentation, space weather data, space weather services

1. Introduction

The effects of Space Weather (SW) on human high-tech activities, infrastructures and health on the ground, in the air, and in space is well documented in a large number of technical and scientific reports (e.g., Baker *et al.* 2004, Narici 2008). The *Regione Lazio*, situated in central Italy, hosts a large number of high-tech industries and national critical infrastructures whose operators and decision makers must have situational awareness of SW events in order to be able to take actions to mitigate possible SW impacts. The SWERTO service aims to provide to registered users a series of scientific data from instruments operating within national and international programs in which UTOV researchers are involved. Moreover, it offers a user web interface consisting of a dashboard designed to present SW information in a way that is easy to read for final operational users and decision makers. The SWERTO service is constructed from a database containing space-based and ground-based data which are relevant for characterizing the space environment and solar activity. The database contains mainly particle flux data from two satellite-borne experiments (Anomalous Long Term Effects on Astronauts (ALTEA) and Payload for Antimatter- Matter and Light-nuclei Astrophysics (PAMELA)) as well as

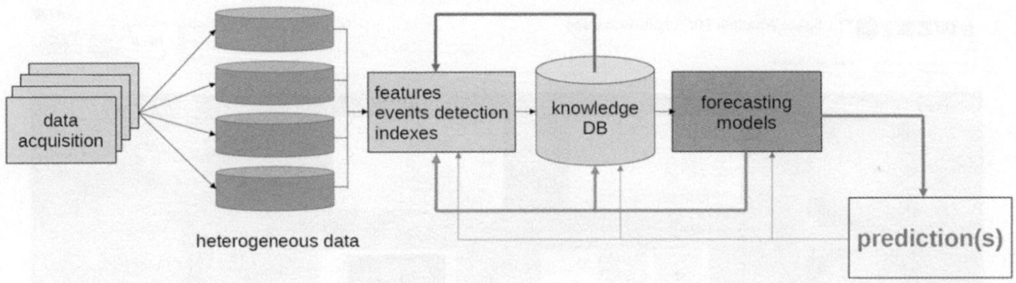

Figure 1. Scheme of the SWERTO database concept.

solar data from ground-based measurements (Interferometric BIdimensional Spectropolarimeter (IBIS) instrument, updated Magneto-Optical filters at Two Heights (MOTH II) suite of instruments). Data from these instruments feed into the second component of the SWERTO service which is the operational analysis and forecasting system. A Flowchart of the SWERTO basic concept is shown in Figure 1.

2. The database and the web interface

The SWERTO data component, the heterogeneous data layer in Figure 1, currently collects data from the four above-mentioned space-based (PAMELA and ALTEA) and ground-based (IBIS and MOTH II) observing systems. The analysis and forecasting models from these observing systems and additional instruments (e.g., SDO/HMI magnetograms or GOES X-ray fluxes) feed into the second SWERTO component which is the operational SW service. The flowchart in Figure 1 presents the different types of procedures performed in the second SWERTO component (features event detection and indexes, knowledge Data Base, forecasting models and predictions). The user web interface consists of a dashboard designed to organize and present information in a way that is easy to read for final operational users and decision makers. On the SWERTO service dashboard (Figure 2) the following information is shown:

(*a*) The latest H_a image acquired by the Global Oscillation Network Group (GONG).

(*b*) The latest magnetogram from Helioseismic and Magnetic Imager (HMI) instrument on board of Solar Dynamics Observatory (SDO).

(*c*) Proton and Helium counts from PAMELA. PAMELA, originally built to study antimatter in cosmic rays (Picozza *et al.* 2007), has over the years specialized in the study of the solar component of cosmic rays. The instrument measures populations of particles within the cosmic radiation, from energies of a few tens of MeV up to a few TeV.

(*d*) Averaged flux or averaged dose rate from ALTEA. ALTEA measure the radiation field (e.g., Di Fino *et al.* 2014) inside the International Space Station (ISS). It has been on-board the ISS since 2006, recording data till 2012 in different configurations. ALTEA returned to Earth in Feb. 2015 to be upgraded with a new detector entitled LIDAL (Light Ion Detector for ALTEA) and it will be back on-board the ISS in 2018.

(*e*) High resolution solar velocity or magnetic maps from the IBIS instrument, a spectro-polarimetric imager based on capacitance-stabilized piezoelectric scanned Fabry-Perot etalons. The instrument can rapidly and precisely scan solar photospheric and chromospheric spectral lines from 550 to 860 nm (e.g., Viticchié *et al.* 2009).

(*f*) Full disk line-of-sight (LOS) Doppler velocity and LOS magnetic field strength from MOTH II. MOTH II consists of two telescopes observing in the K I (769.9 nm) and Na I (589.0 nm) lines and provides observations of the velocity and magnetic filed

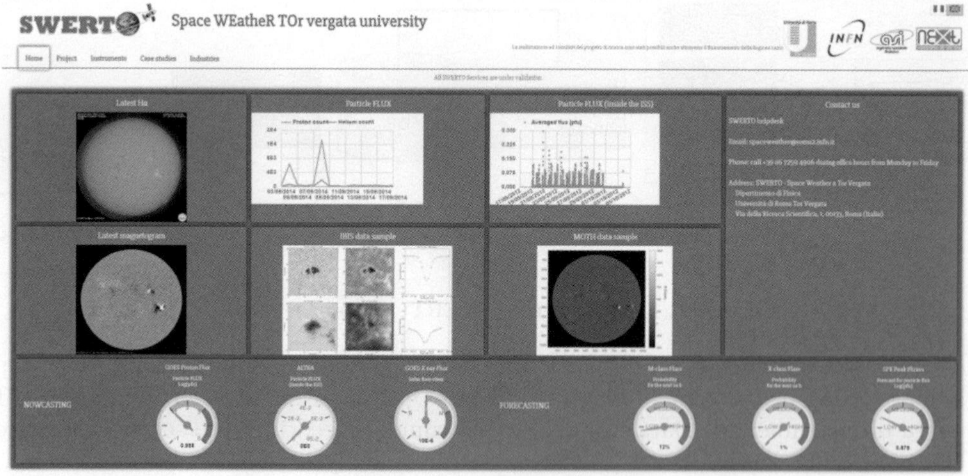

Figure 2. The SWERTO service dashboard.

at two heights of the solar atmosphere. The core of MOTH II is a tandem of magneto-optical filter (MOF) constituted by a cell containing vapours of K (or Na) immersed in a longitudinal magnetic field. The telescope is managed by the University of Hawaii's Institute for Astronomy (IfA), Georgia State University (GSU) and JPL.

Moreover, needle gauges (see lower panel on Figure 2) provide an easy graphical interface to easily obtain *nowcasting* and *forecasting* information. More in detail, they show: *i)* recent proton flux near Earth, measured for three different particle energy ranges-5-minute GOES-13 prepared by the U.S. Dept. of Commerce, NOAA, Space Weather; *ii)* particle flux inside the ISS. ALTEA will be back in orbit in 2018 (upgraded with LIDAL); *iii)* recent GOES X-ray flux near Earth; *iv)* M- and X-class flare probability for the next 24h, currently based on HMI magnetograms using the algorithm reported in Schrijver (2007), *v)* SEP flux created making use of detected CME parameters: some characteristics of the proton spectrum can be forecast in LEO orbit, using the correlation between the SEP characteristics and the CME velocity (Papaioannou *et al.* 2016, Napoletano *et al.* 2017).

3. Acknowledgments

We acknowledge the use of SDO/HMI full disk magnetograms and GONG H_a images. SDO data are courtesy of the NASA/SDO HMI science team. We acknowledge the support from Regione Lazio FILAS-RU-2014-1028 grant (November 1st, 2015 - October 31st, 2017).

References

Baker, D. N., Daly, E., Daglis, I., Kappenman, J. G. & Panasyuk, M. 2004, *Space Weather*, 2, S02004

Di Fino, L., Zaconte, V., Stangalini, M., Sparvoli, R., Picozza, P., Piazzesi, R., Narici, L., Larosa, M., Del Moro, D., Casolino, M., Berrilli, F. & Scardigli, S. 2014, *JSWSC*, 4, A19

Napoletano, G., Forte, R., Del Moro, D., Pietropaolo, E., Giovannelli, L. & Berrilli, F. 2018, *JSWSC*, 8, A11

Narici, L. 2008, *New J. of Phys.*, 10, 075010

Papaioannou, A., Sandberg, I., Anastasiadis, A., Kouloumvakos, A., Georgoulis, M., Tziotziou, K., Tsiropoula, G., Jiggens, P. & Hilgers, A. 2016, *JSWSC*, 6, A42
Picozza, P., Galper, A. M. & Castellini, G., *et al.* 2007, *Astropart. Phys.*, 27, 296
Schrijver, C. J. 2007, *ApJ*, 655, L117
Viticchié, B., Del Moro, D., Berrilli, F., Bellot Rubio, L. & Tritschler, A. 2009, *ApJ*, 700, L145

Space Weather of the Heliosphere: Processes and Forecasts
Proceedings IAU Symposium No. 335, 2017
C. Foullon & O.E. Malandraki, eds.

doi:10.1017/S174392131800025X

The Ionosphere Prediction Service

Carlo Albanese[1], Filippo Rodriguez[1], Roberto Ronchini[1], Stefano Di Rollo[1], Francesco Berrilli[2], Alice Cristaldi[2], Dario Del Moro[2], Roberta Forte[2], Giorgiana De Franceschi[3], Claudio Cesaroni[3], Luca Spogli[3], Vincenzo Romano[3], Marcio Aquino[4], Sreeja Vadakke Veettil[4], Osman Kalden[5], Michael Hutchinson[6] and Eric Guyader[7]

[1]Telespazio, Rome, Italy
[2]University of Rome "Tor Vergata", Rome, Italy
[3]Istituto Nazionale di Geofisica e Vulcanologia, Rome, Italy
[4]University of Nottingham, Nottingham, UK
[5]Telespazio Vega DE, Darmstadt, Germany
[6]Nottingham Scientific Ltd, Nottingham, UK
[7]European Commission, Bruxelles, Belgium

Abstract. The aim of the Ionosphere Prediction Service (IPS) project is to design and develop a prototype platform to translate the prediction and forecast of the ionosphere effects into a service customized for specific GNSS user communities. The project team is composed by Telespazio (coordinator), Nottingham Scientific Ltd, Telespazio Vega Deutschland, the University of Nottingham, the University of Rome "Tor Vergata" and the Italian Istituto Nazionale di Geofisica e Vulcanologia (INGV). The IPS development is conceived of two concurrent activities: prototype service design and development & research activity that will run along the whole project. Service design and development is conceived into four phases: user requirements collection, architecture specification, implementation and validation of the prototype. A sub-activity analyses also the integration feasibility in the Galileo Service center, located in Madrid. The research activity is the scientific backbone of IPS that will provide the models and algorithms for the forecasting products.

Keywords. Sun: CMEs, Sun: activity

1. Introduction

Changes of the environmental conditions in the near-Earth space (space weather) due to solar transient events, such as flares or Coronal Mass Ejections (CMEs), can affect the performance and longevity of space-borne and ground-based technological systems. Space weather monitoring and forecasting is thus of crucial importance to prevent harmful effects on the technological society. Within this context, the Ionosphere Prediction Service (IPS) aims to develop specific methodologies and techniques in order to monitor and predict changes in the physics of the ionosphere that can affect the Global Navigation Satellite System (GNSS) performance.

2. User Requirements

IPS is designed on the basis of GNSS users requirements. Indeed, representatives of the user communities (Aviation, High Accuracy, Mass Market and the Emerging Applications) have been contacted to fill a questionnaire about IPS objectives, and provided a collection of requirements used to define the service components, based on the evaluation and prediction of ionospheric effects on GNSS receivers. Specifically, the Aviation

community highlighted the importance of an ionospheric prediction service since GNSS is critical to performance-based navigation operations, including both en-route and runway approach operations. High Accuracy GNSS service providers expressed their need to account for ionospheric effects in order to achieve the levels of positioning accuracy required by their customers. Liability Critical Applications users stressed the consequences on navigation of undetected GNSS miss-performances, which can generate significant legal or economic consequences. Furthermore, Smart grids and Micro grids, where timing accuracy and continuity is more challenging, are particularly threatened by GNSS service outages. Finally, it is worth mentioning the impact of space weather on the financial services. Some operations in this sector depend on accurate timing, provided by GNSS and commonly used to generate time stamps for financial transactions.
All those infrastructures can be affected by ionospheric events.

3. Sensor Networks

High-quality data recorded from both ground-based and satellite-based instruments allow to monitor and forecast the ionospheric changes and solar eruptive events.
We provide here a list of the sensor networks employed to monitor the ionospheric and near-Earth environment:

• The Ionospheric Scintillation Monitor Receivers (ISMR) network, composed by 12 GNSS stations with standard dual frequency receivers (NovAtel GSV4004) or with special scintillation high rate receivers (50 Hz Septentrio PolaRxS).

• The RING geodetic network, composed by ≃180 GPS dual-frequency receivers distributed over Italy. L1 and L2 signals from GPS satellites are acquired by RING receivers at 30 s sampling period and sent to 2 main servers located at INGV premises in Rome and Grottaminarda (in Southern Italy).

• GONG (Global Oscillation Network Group), managed by the National Solar Observatory (NSO), H-Alpha worldwide Network composed by 6 sites equipped with a Fourier tachometer.

• MOTH (Magneto-Optical filters at Two Heights), a 20 cm telescope equipped with 2K x 2K CMOS cameras providing magnetogram, intensity and velocity maps at two different layers of the solar atmosphere.

• HMI (Helioseismic and Magnetic Imager, Schou *et al.* 2012), an instrument designed to study oscillations and the magnetic field at the solar surface, aboard the SDO (Solar Dynamic Observatory, Pesnell *et al.* 2012) satellite.

• AIA (Atmospheric Imaging Assembly, Lemen *et al.* 2012), aboard the SDO, which observes the solar corona by taking images that span at least 1.3 solar diameters in multiple wavelengths, at a resolution of about 1 arcsec.

• LASCO (Large Angle and Spectrometric Coronagraph, Brueckner *et al.* 1995), one of the instruments aboard the Solar and Heliospheric Observatory satellite, which consists of three solar coronagraphs with three nested fields of view.

• PAMELA (Payload for Antimatter Matter Exploration and Light-nuclei Astrophysics) satellite-borne experiment, which has been measuring particle (100 MeV - 100 GeV) fluxes in polar orbit. Due to the unavailability of PAMELA data since the beginning of 2017, the project will continue using the GOES-13 satellite proton fluxes.

4. Research Activities

Current studies related to flares and CMEs predictions, ionospheric changes and disturbances are carried out by the University of Rome "Tor Vergata" (flare, CME and energetic particles now- and fore-casting; see, e.g. Berrilli *et al.* 2014, Di Fino *et al.* 2014, Romano *et al.* 2015), the INGV (ionspheric plasma morphology and dynamics, scintillation mapping and modelling on global and/or regional scale; see, e.g., Spogli *et al.* 2013, Spogli *et al.* 2016, Cesaroni *et al.* 2017), and the University of Nottingham (modelling and prediction of GNSS errors, detection of travelling ionospheric disturbances in collaboration with INGV; see, e.g., Sreeja *et al.* 2011, Sreeja *et al.* 2012, Aquino *et al.* 2013).

5. IPS Architecture

The IPS architecture is based on the following elements:

• Sensors: this layer collects all the elements used to gather raw data for the space weather and ionospheric products generation.

• Remote Processing Facilities (RPFs): these processing elements generate space weather and ionospheric products starting by collected raw data or intermediate products generated by other RPFs. They interacts with both the remote sensors for the collection of the needed input data and the central storage to save the generated products, retrieve and process data from other RPFs or trigger one or more processes implemented in the central unit. In particular, each RPF deals with:

◦ RPF-1: Sun activity, such as sunspots, solar flares, CMEs, solar energetic particles.

◦ RPF-2: ionosphere activity, such as Total Electron Content variation, scintillation and travelling ionospheric disturbances (RPF-2 products are input to both RPF-3 and RPF-4).

◦ RPF-3: forecast of disturbance at receiver level: loss of lock, position error, tracking error.

◦ RPF-4: monitor and forecast of the ionospheric effects on the performance of GNSS systems at service level for several classes of users. It also provides ABAS (Aircraft-based Augmentation System) and SBAS (Satellite-based Augmentation System) forecast services to civil aviation users, including integrity, accuracy, availability and continuity.

• Central Storage and Processing Facility (CSPF): this is the IPS central facility that implements all the functionalities related to the collection and distribution of the products and the interactions with the GNSS user communities.

Aknowledgements

IPS has been financed by the EC through the Tender contract No: 434/PP/GRO/RCH/15/8381.

References

Aquino, M. & Sreeja, V. 2013, *Space Weather*, 11
Berrilli, F., Casolino, M., Del Moro, D., *et al.* 2014, *JSWSC*, 4, A16
Brueckner, G. E., Howard, R. A., Koomen, M. J., *et al.* 1995, *Solar Phys.*, 162, 357
Cesaroni C., Alfonsi L., Pezzopane M., *et al.* 2017, *JGR: Space Physics*, 122
Di Fino, L., Zaconte, V., Stangalini, M., *et al.* 2014, *JSWSC*, 4, A19
Lemen, J. R., Title, A. M., Akin, D. J., *et al.* 2012, *Solar Phys.*, 275, 17
Pesnell, W. D., Thompson, B. J. & Chamberlin, P. C. 2012, *Solar Phys.*, 275, 3
Romano, P., Zuccarello, F., Guglielmino, S. L., *et al.* 2015, *A&A*, 582, A55
Schou, J., Scherrer, P. H., Bush, R. I., *et al.* 2012, *Solar Phys.*, 275, 229
Spogli, L., Alfonsi, L., Romano, V., *et al.* 2013, *JASTP*, 105, 199-206
Spogli, L., Cesaroni, C., Di Mauro, D., *et al.* 2016, *JGR: Space Physics*, 121(12)
Sreeja, V., Aquino, M. & Elmas, Z. G. 2011, *Space Weather*, 9, S10002
Sreeja, V., Aquino, M., Elmas, Z. G. & Forte, B. 2012, *Space Weather*, 10, S06005

Space Weather of the Heliosphere: Processes and Forecasts
Proceedings IAU Symposium No. 335, 2017
C. Foullon & O.E. Malandraki, eds.

doi:10.1017/S1743921317008353

The Belgian Space Weather Observatory in Dourbes

D. Sapundjiev, T. G. W. Verhulst, S. M. Stankov and J. C. Jodogne
Royal Meteorological Institute (RMI), Ringlaan 3, B-1180 Brussels, Belgium
email: danislav.sapundjiev@meteo.be

Abstract. The paper outlines the space-weather-related observations, instrumentation, research and service developments carried out at the Geophysical Centre of the Royal Meteorological Institute (RMI) in Dourbes, Belgium.

Keywords. Space weather, solar-terrestrial relations, cosmic rays

1. Introduction

The Geophysical Centre (http://dourbes.meteo.be) of the Royal Meteorological Institute (RMI) was established in the 1940s in Dourbes (50.1°N, 4.6°E), a remote location in the south of Belgium, providing a suitable environment for accurate observations of the geomagnetic field, ionosphere, and cosmic ray activity. Regular observations started in the early 1950s, long before the term "space weather" was first coined, and continue to this day. Fundamental and applied research was carried out by several generations of scientists, who also develop services for users from the industry, academia, and public. This paper provides an overview of the observations, research and development activities.

2. Instrumentation and measurements

The geomagnetic observatory, in operation since 1953 and following on the previous century-old tradition of magnetic observations in the Brussels area, consists of variometers for measuring the variations of the field components about the baseline and magnetometers for the absolute measurements needed to establish the baseline values (Verhulst *et al.* 2014a). The current data acquisition cadence is 1 s for time, with a precision of 0.1 nT for the induction and 0.001° for both, the declination and the inclination.

The main instrument utilised for ionospheric observations is the digital ionospheric sounder (a Lowell Digisonde-4D), a state-of-the-art equipment using high-frequency radar principles of remote sensing to evaluate, with high accuracy and precision, the local ionospheric plasma characteristics (Stankov *et al.* 2012). Ionospheric observations are also helped by several GNSS signal receivers providing high-resolution measurements of the total electron content (TEC) and the small-scale variability of the ionosphere.

Cosmic rays are observed with a 9NM-64 neutron monitor providing measurements of the secondary neutron component of the cosmic rays on the ground, which (after pressure correction) mirrors the primary cosmic ray intensity (Sapundjiev *et al.* 2014). Thus, one can reliably detect proton events (due to energetic solar flares) and Forbush decreases (due to the occurrences of strong solar winds).

RMI supports the Belgian Open Data policy, readily shares data, and provides (in real time) all measurements to international databases, such as the International Magnetic Observatory (http://www.intermagnet.org), Global Ionosphere Radio Observatory (http://giro.uml.edu), International Neutron Monitor Data Base (http://www.nmdb.eu).

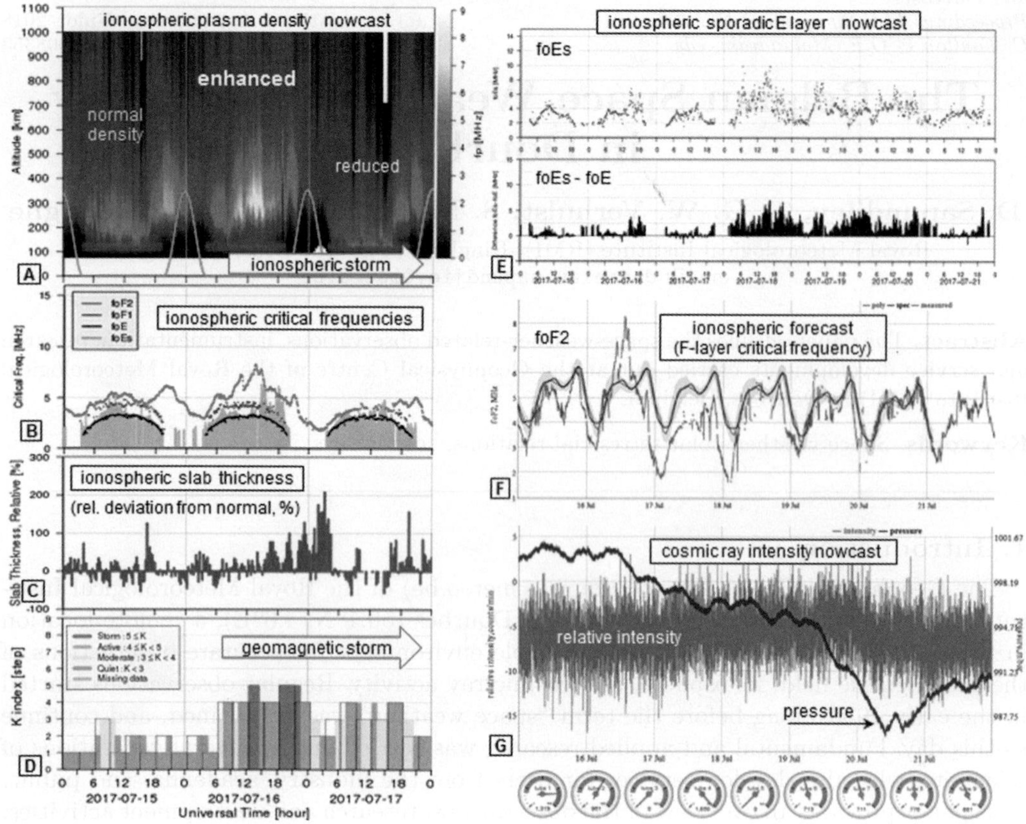

Figure 1. Space weather monitoring services (http://ionosphere.meteo.be): (A) ionospheric plasma density (plotted as plasma frequency) nowcast, (B) ionospheric critical frequencies, (C) ionospheric slab thickness, (D) geomagnetic activity nowcast, (E) ionospheric sporadic E-layer activity nowcast, (F) ionospheric forecast (climatological), and (G) cosmic ray intensity nowcast.

3. Research

Research is focused on local and regional space weather, including ionospheric plasma structures and dynamics, ionospheric irregularities, geomagnetic disturbances (Jodogne & Stankov 2002; Stankov *et al.* 2010; Verhulst *et al.* 2014a), and their effects on communications and navigation system performance (Warnant *et al.* 2007; Lejeune & Warnant 2008; Stankov *et al.* 2009; Warnant *et al.* 2009; Lejeune *et al.* 2012). Also designed, developed, and programmed are robust algorithms for nowcast and short-/long- term forecast of geospace plasma parameters by ground-based (ionospheric incidence sounding, geomagnetic, cosmic rays) and space-based (GNSS, LEO satellite) measurements (Spits & Warnant 2008; Stankov & Warnant 2009; Spits & Warnant 2011; Stankov *et al.* 2011a; Stankov *et al.* 2011b; Stankov *et al.* 2012; Verhulst & Stankov 2013; Verhulst & Stankov 2014b; Sapundjiev *et al.* 2014; Verhulst & Stankov 2015; Sapundjiev & Stankov 2016). RMI is also actively involved in the research activities of the Belgian Solar-Terrestrial Centre of Excellence (http://www.stce.be). For example, a joint recent multi-instrument observation campaign focused on the solar eclipse effects on the geospace environment (Verhulst *et al.* 2016; Stankov *et al.* 2017). Over the years, the team has participated in and managed several national and international research projects funded by the Belgian Government, European Commission, ESA, NATO, etc.

4. Services

All instruments, used as a finely-tuned observatory, yield high-resolution concurrent and collocated measurements of key characteristics of the space weather and its effects. Several services (http://ionosphere.meteo.be) were developed and are now available to the users (Figure 1). A nowcast system for operational (1-hour cadence) calculation of a proxy K-type geomagnetic index (K-LOGIC) was installed (Stankov *et al.* 2011a), estimating the local magnetic field activity and issuing an alert when storm levels are indicated. The local ionospheric specification and visualisation system LIEDR (Stankov *et al.* 2011b) acquires and processes GNSS TEC and ionosonde measurements, and deduces the vertical electron density distribution in the local ionosphere. It provides nowcast of the critical frequencies, peak densities and heights of the ionospheric layers, and the ionospheric slab thickness. The ionospheric sporadic layer Es is also monitored and alert is issued when high levels of the Es critical frequency, foEs, are reached. Ionospheric forecast (climatological) is offered, based on long-term statistics of the local ionospheric critical frequency foF2 (Sapundjiev & Stankov 2016). For cosmic-ray monitoring, an automatic data quality control system for real-time correction of data from the neutron monitor was developed and implemented (Sapundjiev *et al.* 2014).

5. Future developments

Space weather research demands high accuracy and precision of instruments, high resolution of measurements, prompt and convenient dissemination of data, and continuous availability of services. Such demands necessitate regular upgrades of the instruments and become even more challenging when a service utilises different types of instruments. The NM64 neutron monitor proved to be a reliable tool for cosmic-ray monitoring, so a second 9-tube monitor is under construction. Initially, a producer-free concept will be exploited for verification of several theoretical simulations and investigation of the monitor's yield functions. A multi-directional muon telescope is being developed, which will improve the characterisation of interplanetary space conditions. Also planned is the installation of a radio telescope for solar activity monitoring. Research and developments will be focused on high-resolution nowcast and more reliable forecast of space weather.

References

Jodogne, J. C. & Stankov, S. M. 2002, *Ann. Geophys.*, 45, 5, 629
Lejeune, S. & Warnant, R. 2008, *J. Atm. Sol.-Terr. Phys.*, 70, 6, 889
Lejeune, S., Wautelet, G. & Warnant, R. 2012, *GPS Solutions*, 16, 1, 105
Sapundjiev, D., Nemry, M., Stankov, S. M. & Jodogne, J. C. 2014, *Adv. Space Res.*, 53, 1, 71
Sapundjiev, D. & Stankov, S. M. 2016, *Acta Geophys.*, 64, 3, 810
Spits, J. & Warnant, R. 2008, *J. Atm. Sol.-Terr. Phys.*, 70, 15, 1885
Spits, J. & Warnant, R. 2011, *Adv. Space Res.*, 47, 2, 296
Stankov, S. M. & Warnant, R. 2009, *Adv. Space Res.*, 44, 11, 1295
Stankov, S. M., Warnant, R. & Stegen, K. 2009, *Adv. Space Res.*, 43, 9, 1314
Stankov, S. M., Stegen, K. & Warnant, R. 2010, *Adv. Space Res.*, 46, 10, 1318
Stankov, S. M., Stegen, K. & Warnant, R. 2011a, *Ann. Geophys.*, 54, 3, 285
Stankov, S. M., Stegen, K., Muhtarov, P. & Warnant, R. 2011b, *Adv. Space Res.*, 47, 7, 1172
Stankov, S. M., Jodogne, J. C., Kutiev, I., *et al.* 2012, *Ann. Geophys.*, 55, 2, 283
Stankov, S. M., Bergeot, N., Berghmans, D., *et al.* 2017, *J. Space Weather Space Clim.*, 7, A19
Verhulst, T. & Stankov, S. M. 2013, *Adv. Space Res.*, 51, 11
Verhulst, T., Stankov, S. M. & Rasson, J. 2014a, *KBVE Rev. E Tijdschrift*, 30, 2, 1

Verhulst, T. & Stankov, S. M. 2014b, *Radio Sci.*, 49, 3, 181
Verhulst, T. & Stankov, S. M. 2015, *Adv. Space Res.*, 55, 8, 2058
Verhulst, T. G. W., Sapundjiev, D. & Stankov, S. M. 2016, *Adv. Space Res.*, 57, 11, 2407
Warnant, R., Lejeune, S. & Bavier, M. 2007, *J. Lilensten (ed.), Space Weather (Springer)*, 129
Warnant, R., Foelsche, U., Aquino, M., *et al.* 2009, *Ann. Geophys.*, 52, 3-4, 373

Space Weather of the Heliosphere: Processes and Forecasts
Proceedings IAU Symposium No. 335, 2017
C. Foullon & O.E. Malandraki, eds.

doi:10.1017/S1743921317011334

The Coronal Solar Magnetism Observatory

Michael J. Thompson[1], Steven Tomczyk[1], Sarah E. Gibson[1], Scott W. McIntosh[1] and Enrico Landi[2]

[1]High Altitude Observatory, National Center for Atmospheric Research, P.O. Box 3000, Boulder, CO 80307, USA
email: mjt@ucar.edu, tomczyk@ucar.edu, sgibson@ucar.edu, mscott@ucar.edu

[2]Department of Climate and Space Sciences and Engineering, University of Michigan, Ann Arbor, MI 48109-2143, USA
email: elandi@umich.edu

Abstract. The Coronal Solar Magnetism Observatory (CoSMO) is a proposed new facility led by the High Altitude Observatory and a consortium of partners to measure magnetic field and plasma properties in a large (one degree) field of view extending down to the inner parts of the solar corona. CoSMO is intended as a research facility that will advance the understanding and prediction of space weather. The instrumentation elements of CoSMO are: a white-light coronagraph (KCor), already operational at the Mauna Loa Solar Observatory (MLSO); the Chromosphere and Prominence Magnetometer (ChroMag), due for deployment to MLSO next year; and the CoSMO Large Coronagraph (LC) which has completed Preliminary Design Review.

Keywords. Sun: corona, Sun: chromosphere, Sun: magnetic fields

---

1. Introduction

Measurements of coronal and chromospheric magnetic fields are arguably the most important observables required for advancing our understanding of the processes responsible for coronal heating, coronal dynamics, and the generation of "Space Weather" that affects communications, GPS systems, space flight, and power transmission. The Coronal Solar Magnetism Observatory (CoSMO) is a proposed ground-based facility that will be home to a suite of instruments designed for routine study of coronal and chromospheric magnetic fields and their plasma environment (Tomczyk *et al.* 2016).

CoSMO will provide measurements of the magnetic field and thermodynamic properties of coronal, chromospheric and prominence plasmas over a range of timescales from minutes to years. These measurements are required to understand the physics of coronal heating, solar wind acceleration, and the genesis and evolution of coronal mass ejections (CMEs), prominences and related activity. CoSMO measurements will lead to a dramatic improvement in our knowledge of the coupling of the Sun's magnetic field with the plasma physical properties from the photosphere through the heliosphere and how the field evolves over solar cycle timescales. As the facility matures, these measurements will be made available to the operational space weather community in a concerted effort to improve the predictability of ongoing space weather and also of extreme space weather events.

2. CoSMO Facility and Instrumentation

The **CoSMO facility** will be built and operated by a consortium consisting of the National Center for Atmospheric Research (NCAR), University of Michigan, University of Hawaii and George Mason University, on or adjacent to the site of the current NCAR

Mauna Loa Solar Observatory (MLSO), the best possible coronal observing site in Hawaii. The dome that will house the 1.5m apperture f/5 refractive Large Coronagraph (see below) will be 12.2 meters in diameter, and will be positively pressurized with HEPA filtered air to maintain cleanliness of the objective lens.

The white-light polarized-brightness (pB) **K-Coronagraph ("KCor")** (de Wijn *et al.* 2012) observes the column density of electrons in the corona. KCor is currently operational at MLSO. This instrument allows high temporal cadence observations of CMEs and shocks very close to the solar limb.

The **Chromospheric Magnetometer ("ChroMag")** (de Wijn *et al.* 2014) is devoted to polarimetric measurements on the disk and above the limb. Vector magnetic field observations will be made using the chromospheric lines of He 587nm, H 656nm, Ca 854nm and He 1083nm, as well as the photospheric line of Fe 617nm.

The **CoSMO Large Coronagraph ("LC")** (see Tomczyk *et al.* 2016) is devoted to obtaining the highest quality polarimetric data of coronal emission lines at visible and IR wavelengths. It will operate as a "light bucket" far from the diffraction limit. The LC can make detailed polarization measurements of the extended corona, in addition to making Doppler maps critical for event detection. Once operational, the LC will also be the only available instrument capable of providing near-simultaneous 2D measurements of the thermodynamic properties of coronal plasmas from near the photosphere to the extended corona (2 solar radii or more).

As a critical part of its make-up, CoSMO will also provide a unique community testbed for new instrumentation concepts to be built and deployed in the pristine observing environment on Mauna Loa.

For more details about the CoSMO instrument suite, see Tomczyk *et al.* (2016) and references therein.

3. Proof of Concept: CoMP

The concept for the observations to be performed by the LC has already been demonstrated by the Coronal Multi-Channel Polarimeter (CoMP) instrument, which is operational on a daily basis at MLSO, albeit with a very much smaller aperture (and therefore sensitivity) than will be the case for the LC. Figure 1 shows coronal properties measured by the CoMP instrument in October 2005 on the 20 cm aperture OneShot coronagraph at the National Solar Observatory's Sacramento Peak Observatory in New Mexico. The CoMP is a filter-based polarimeter observing the FeXIII emission line at 1074.7 nm. These observations have a spatial sampling of 4.5 arcseconds per pixel, and required 1.5 hours of integration time due to the small coronagraph. Nonetheless they do demonstrate that plasma velocities and magnetic field direction and strength can be measured in the corona.

The CoSMO/LC by contrast will allow magnetic fields of strength 1-2 Gauss to be measured at, e.g., 2 arcsecond resolution with a temporal resolution of 15 minutes, or 4 arcsecond resolution with a temporal resolution of 5 minutes.

4. Coming Developments

With the US's national strategy gearing up to improve space weather forecasting capability, a facility like CoSMO is a critical piece of the puzzle. We are in active development of a ChroMag prototype and building an upgraded CoMP ("UCoMP") instrument as scientific pathfinders towards CoSMO.

UCoMP will serve as a proof of concept for the plasma diagnostic capabilities of CoSMO/LC, by observing the solar corona in a number of coronal, transition region

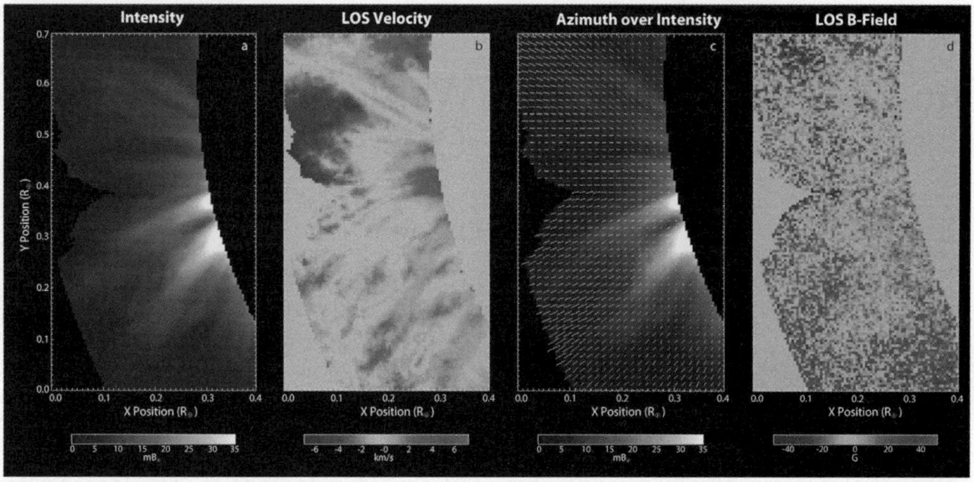

Figure 1. Inferred coronal plasma and magnetic field properties from CoMP observations. a) The intensity of FeXIII 1074.7 nm emission. b) The line-of-sight velocity of coronal plasma. c) Vectors showing the plane-of-sky direction of the magnetic field (the length of the vectors being proportional to the degree of linear polarization), superimposed on the intensity of FeXIII 1074.7 nm emission. d) The line-of-sight strength of the coronal magnetic field. (Adapted from Tomczyk *et al.* 2008.)

and chromospheric lines that will measure the plasma properties of both the quiescient and active corona, as well as CMEs. It will provide near-simultaneous 2D measurements of line intensities, Doppler velocities and widths, as well as plasma properties such as electron density, temperature distribution, and elemental composition everywhere from the photosphere to 2 solar radii (see Landi *et al.* 2016).

We hope that both UCoMP and the ChroMag prototype will be deployed to MLSO by summer 2018 and data will become available as soon as possible. Another interesting development is the "CoSMO Community Spar" – an eight-sided sun-pointing spar where we hope to work with sponsors to create a community instrument development program for space weather and synoptic observations that will be supported by the existing MLSO infrastructure on the mountain.

Acknowledgements

The National Center for Atmospheric Research is sponsored by the National Science Foundation (NSF). UCoMP development is funded by the NSF through grant AGS-1408789.

References

de Wijn, A. G., Burkepile, J. T., Tomczyk, S., Nelson, P., Huang, P. & Gallagher, D. 2012, *Proc. SPIE*, 8444, doi:10.1117/12.926511

de Wijn, A. G., Bethge, C., Tomczyk, S. & McIntosh, S. W. 2014, *Proc. SPIE*, 8446, doi:10.1117/12.926395

Landi, E., Habbal, S. R. & Tomczyk, S. 2016, *J. Geoph. Res. Space Phys.*, 121, 8237

Tomczyk, S., Card, G. L., Darnell, T., Elmore, D. F., Lull, R., Nelson, P. G., Streander, K. V., Burkepile, J., Casini, R. & Judge, P. G. 2008, *Solar Phys.*, 247, 411

Tomczyk, S., Landi, E., Burkepile, J. T., Casini, R., DeLuca, E. E., Fan, Y., Gibson, S. E., Lin, H., McIntosh, S. W., Solomon, S. C., de Toma, G., de Wijn, A. G. & Zhang, J. 2016, *J. Geoph. Res. Space Phys.*, 121, 7470

Space Weather of the Heliosphere: Processes and Forecasts
Proceedings IAU Symposium No. 335, 2017
C. Foullon & O.E. Malandraki, eds.

doi:10.1017/S174392131700881X

The Forecast of Space Weather According to Ground-Based Observatories

Andrey Tlatov[1,2]

[1]Kislovodsk mountain astronomical station of the Pulkovo observatory,
357700, Gagarina str. 100, Kislovodsk, Russia
email: tlatov@mail.ru

[2]Kalmyk State University, Elista, Russia

Abstract. The space weather forecast can be conditionally divided into three components: 1) The forecast of recurrent slowly varying events which are associated with the topology of a large-scale magnetic field; 2) Evaluation of fluxes of ultraviolet and extreme radiation; 3) Detecting and determining the parameters of eruptions and CME. All three components can be observed and evaluated by ground-based telescopes. In this article we describe the telescopes of the space weather service in Kislovodsk Mountain Astronomical Station. Telescope-magnetograph STOP for the forecast of recurrent events and patrol telescopes-spectrographs for detecting CME and estimating hard radiation fluxes.

Keywords. Sun, space weather, ground-based observation

1. Overview

At present one of the most urgent tasks in the study of solar-terrestrial relationships is the study of the nature of space weather variations caused by solar activity. Space weather effectively affects space vehicles, also on Earth, its strongest manifestations are localized in the polar regions. From the solar corona continuously expires corpuscular stream in the form of solar wind (SW). In the case of low activity when there are no flare disturbances on the Sun leading to sporadic coronal mass ejections (CMEs) into the interplanetary medium, the so-called quiet solar wind forms a quasi-stationary pattern in the interplanetary space. From the point of view of the prediction of magnetic disturbances on Earth as an important part of space weather, a stationary SW state is favorable, because it has a tendency toward 27-day event recurrence (recurrent forecast).

The Sun is also a source of radiation in the X-ray and ultraviolet ranges, which can have a negative impact on technological systems. The estimation of these flows, as well as the forecast of their intensity for the next few days, is the subject of space weather service (SWS). The SWS should be based on continuous observations of solar activity, and also include an operative process of primary data analysis and the formation of activity indices. There are two main components of the SWS forecast. Firstly, this definition of the parameters of SW, when there are no flares and CME on the Sun, the so-called quiet solar wind. The second component is the accounting of flares and CMEs. It should be noted that in the case of solar activity, the CME plasma interacts with the plasma of the quiet solar wind. The spread in this case depends on the background state interplanetary medium, which must be considered when modeling disturbance.

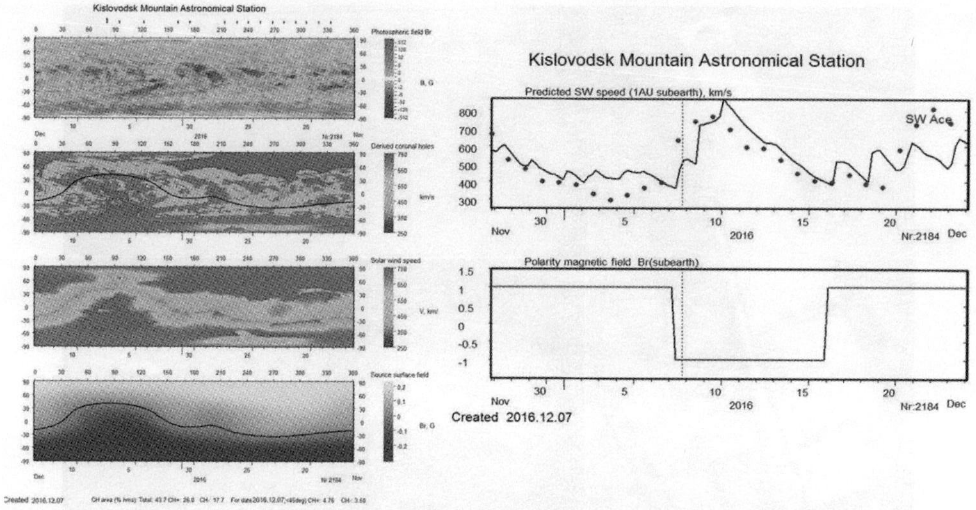

Figure 1. Example of operational calculations of the photospheric magnetic field, positions of open magnetic flux structures, solar wind speed at the source surface, and the polarity of magnetic field on the source surface according to Kislovodsk data.

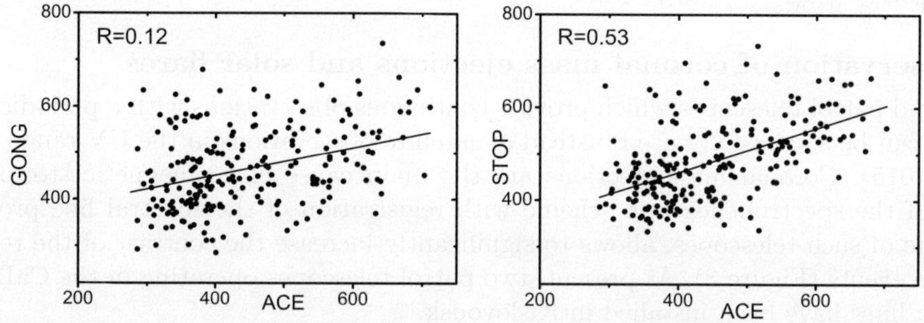

Figure 2. Comparison of the reliability of solar wind speed in km/sec forecasts according to Kislovodsk and GONG data for 2016 with ACE data. Kislovodsk data have significantly better correlation.

2. Forecast of Solar Wind Parameters According to STOP Magnetograph Observations

Large-scale solar magnetic fields have been regularly observed at Kislovodsk mountain astronomical station (KMAS) from July 1, 2014 (Tlatov *et al.* 2016). The telescope is mostly used to obtain full disk magnetograms of longitudinal largescale solar magnetic fields with the angular resolution of $\sim$ 30$\times$6 arc.sec. The coronal magnetic field structure can vary significantly within one synoptic rotation because of the emerging active regions. For this purpose, series of daily updated synoptic maps are constructed at the KMAS, i.e., every day newly received magnetograms are integrated into the existing weather map. These maps have a resolution $2.5^o \times 2.5^o$(degrees of latitude and longitude). These maps are used to compute the coronal magnetic field in the PFSS and WSA models (Figure 1).

Different input data can provide different extrapolations of coronal structures in the photosphere and source surface as well as influence the results near the Earth. It is necessary to note that the simulation results depend on the spatial resolution of the magnetograms, which are not the same for different observatories (Figure 2).

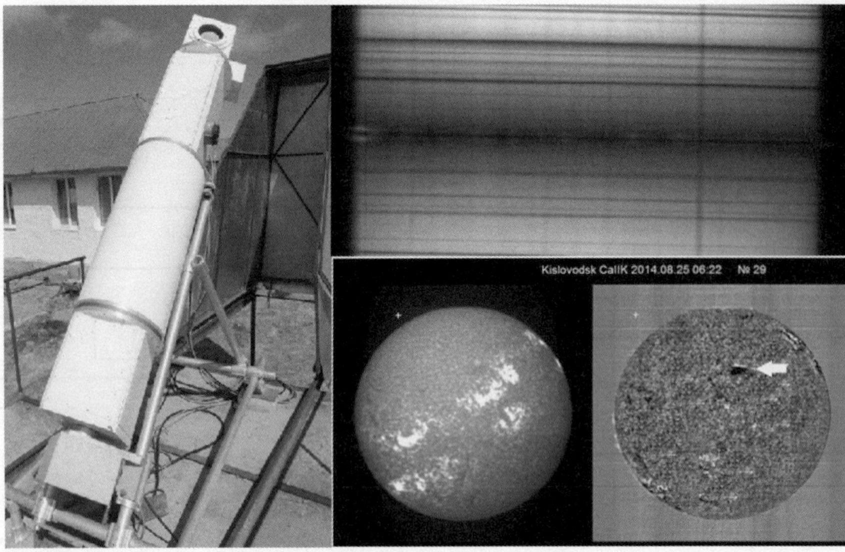

Figure 3. The Patrol telescopes observations on UV fluxes and coronal mass ejections. The use of an intensity difference in the wings of the spectral lines makes it possible to effectively detect the CME (see arrows).

3. Observation of coronal mass ejections and solar flares

Ground patrol telescopes which provide continuous observations with a periodicity per minute can be successfully used instead of satellite observations in the UV range Tlatov *et al.* (2015). Coronal mass ejections are the main cause of geomagnetic storms. The choice of the spectroheliograph scheme with registration of the spectral line profile, as a scheme of such telescopes, allows to significantly increase the contrast of the recorded eruptive events (Figure 3). At present, two patrol telescopes operating in the CaIIK and H-alpha lines have been installed in Kislovodsk.

4. Conclusions

This article discussed the main tasks of the space weather forecasting service, as well as the experience of using ground-based observations in Kislovodsk. Daily magnetographic observations are necessary to determine the parameters of quasi-stationary fluxes of solar wind forward by 3-7 days and the specification of physical conditions for models of CME propagation.

For predicting effects of solar flares and coronal mass ejections, a continuous 24-hour observation regime is ideal. For implementation, it is necessary to install patrol telescopes at observation posts, including the western hemisphere, with a total number of about 8. Patrol telescopes-spectrographs allow to determine the direction and speed of coronal mass ejections and to estimate the fluxes of cruel and UV radiation.

This work was supported by of the RSF N 15-12-20001.

References

Tlatov, A. G., Dormidontov, D. V., Kirpichev, R. V., Pashchenko, M. P. & Shramko, A. D. 2015, *Ge&Ae*, 55, 961

Tlatov, A. G., Pashchenko, M. P., Ponyavin, D. I., Svidskii, P. M., Peshcherov, V. S. & Demidov, M. L. 2016, *Ge&Ae*, 56, 969

Space Weather of the Heliosphere: Processes and Forecasts
Proceedings IAU Symposium No. 335, 2017
C. Foullon & O.E. Malandraki, eds.

doi:10.1017/S1743921317008328

The High-Energy Particle Detector (HEPD) on Board the CSES Mission

Vincenzo Vitale[1], for the CSES/HEPD collaboration

[1]Agenzia Spaziale Italiana (ASI) Science Data Center
and INFN, Sezione di Roma "Tor Vergata"
I-00133 Roma, Italy
email: vincenzo.vitale@roma2.infn.it

Abstract. The High-Energy Particle Detector (HEPD) will measure electrons, protons and light nuclei fluxes, in low Earth orbit. This detector consists of a high precision silicon tracker, a versatile trigger system, a range-calorimeter and an anti-coincidence system. It is one of the instruments on board the China Seismo-Electromagnetic Satellite (CSES). HEPD can detect multi-MeV particles trapped within the geomagnetic field. When operated at large latitudes HEPD can also detect un-trapped solar particles and low energy cosmic rays. A detailed description of the HEPD will be given.

1. Introduction

The High-Energy Particle Detector (HEPD) is a space experiment for the detection of electrons from 3 to 100 MeV, protons from 30 to 300 MeV and light nuclei up to few hundreds of MeV. HEPD will be flown on the CSES satellite (Shen *et al.* 2011), together with several other instrument, for the study of the low Earth orbit (LEO) electromagnetic environment (ionosphere, magnetosphere and Van Allen belts). The CSES satellite will have a 98 degrees inclination Sun-synchronous circular orbit, an altitude of 500 km and an expected lifetime is 5 of years. At low latitude HEPD will detect particles which are trapped within the geomagnetic field, while when operated at large latitudes, with a lower geomagnetic cutoff, HEPD will be able to access also un-trapped particles, such as solar energetic particles (SEPs).

SEPs consist of electrons, protons and nuclei with larger Z, with energies from tens of keV to a few GeV. They might be accelerated via magnetic reconnection during flaring events, or by shocks related to coronal mass ejections (CME). The following SEPs transport within the heliosphere washes out the characteristic imprints of the acceleration process, making more difficult to determine which one is at work. SEPs are a major factor in determining the radiation environment outside Earth. During intense events, solar particles can also penetrate the magnetosphere and reach the atmosphere, also due to sudden variations of the ionospheric currents. Then SEPs are an hazard factor for space (and high altitude) activities. Rarely, solar particle emission is intense enough to cause ground level enhancement (GLEs) of the radiation on Earth.

Solar activity effects on the Earth magnetosphere can be studied with the HEPD, which can also obtain SEPs fluxes and spectra. HEPD will work in a peculiar energy range, below the one covered by ground-based instruments and at the lower end of the energy range of the space borne cosmic ray detectors. Operating at similar energies (>80 MeV), the PAMELA experiment performed observations of the 2012 May 17 SEP (Adriani *et al.* 2015), which was associated to GLE. The PAMELA team reported two distinct SEP protons populations: - a first one below 1GeV, which undergoes important scattering effectes; - another one above 1GeV, which reaches the Earth without strong disturbances

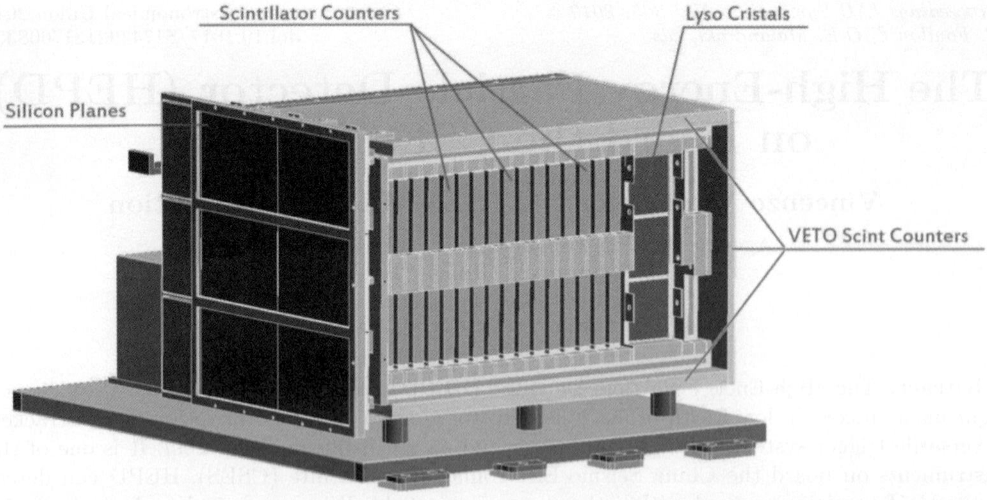

Figure 1. A detailed view of the High-Energy Particle Detector and its sub-systems. Starting from the left we see the the silicon tracker, in blue. Then it is located the trigger plane (not seen here because it is covered by the other elements). In light blue are drawn the 16 plastic scintillators of the upper calorimeter. Then going to the right, there are three of the LYSO crystals. Around the calorimeter three counters of the VETO can be seen. The power supply and electronics are drawn in green.

and is well consistent with neutron monitor measurements. Further PAMELA measurement of low energy protons, both trapped and solar particles, are reported in Bruno *et al.* (2017).

2. HEPD Components and Measurement Methods

The sub-systems composing the HEPD instrument are: (a) a **Tracker** (TRK) which is located on the top of the HEPD and consists of two silicon detector planes (213×213 mm^2). Si detectors have 192 micron read-out (p and n) for a resultant resolution of 40-50 micron; (b) a **Trigger System** (TS), which consists of one thin plane of 6 counters (dimension 200×30×5mm^3), dubbed bars, made of plastic scintillator EJ-200 and read out by PMTs Hamamatsu R9880U; (c) a **Range-Calorimeter** (CALO), which is divided in two sections. Its upper part is a tower made of 16 plastic scintillator EJ-200 planes while the lower part is a 3×3 matrix of an inorganic scintillator LYSO, located on the bottom of the tower. Each plane measures 150×150×10 mm^3 , and is read out by two PMTs at two opposite corners. The lower calorimeter consists of 9 crystals. Each crystal has dimensions 48×48×40 mm^3 and is read out by one PMT located on the bottom face; (d) a **Veto System** (Veto), which consists of 5 plastic scintillator counters (four lateral and one at the bottom of the instrument) similar to those of the CALO. (e) the **Electronics and Power Supply Sub-System** is composed by the Electronics Subsystem (ELS), made by all front-end electronics and four boards, which are the DAQ board, the PMT/Trigger board, the CPU board and the Power Control board and the Power Supply Subsystem (PWS), which is made with the Low Voltage Power Supply (LVPS) and High Voltage Power Supply (HVPS). An illustration of the HEPD sub-systems is given in Figure 1.

The instrument dimensions are 530 ×382×404 mm^3 with a mass budget of 45 kg and a power consumption limited to 27 W. HEPD reaches a geometrical acceptance of few hundreds cm^{2}sr, relatively large for this type of instruments. Preliminary indications of the effective acceptance for electrons and protons were given in Sparvoli *et al.* (2015),

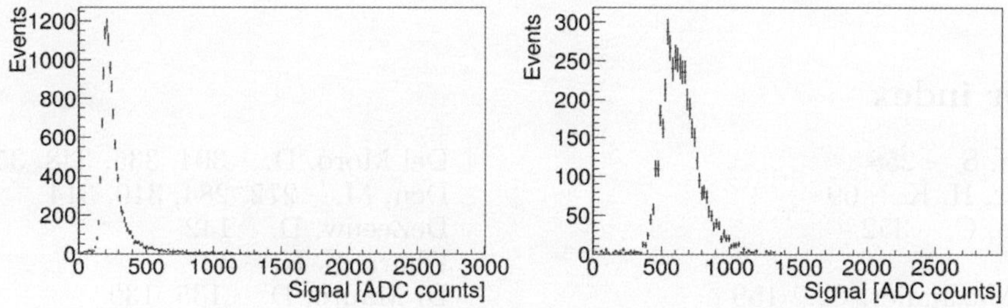

Figure 2. Energy loss within the Trigger System: on the left atmospheric muons; on the right 228 MeV protons acquired during a dedicated beam test.

with peak acceptance in the order of hundred of $cm^2 sr$, depending on the particle type and trigger conditions. In facts the trigger logic is a key factor in defining the instrument effective acceptance. The main trigger logic requires to have in coincidence: (a) a signal within one of the trigger bars; (b) a signal from one of the calorimeter planes, usually the first from above. The acceptance and the energy threshold can be modulated by changes in the trigger logic: - the acceptance can be reduced by allowing only one or few of the trigger bars to form the main trigger, or by requiring no signal in the lateral veto; - the energy threshold can be raised by associating to the trigger logic a calorimeter plane which is closer to the instrument bottom, so that a particle needs more energy to reach it.

This instrument can achieve a good particle identification. In the energy range of interest while electrons are always relativistic, protons are much slower. Then it is possible to used the ΔE vs E_{total} method for the electron vs. proton discrimination. The ΔE is the energy loss within a thin detector layer, and can be obtained with the silicon tracker measurement, or the with the trigger bars and the individual calorimeter planes. An example of energy loss within the trigger bars is in Figure 2. The E_{total} is the particle energy and can be measured by the calorimeter, for particles which are fully contained within the detector. Discrimination of higher Z particles can be achieved also with the energy release within the silicon tracker.

3. Conclusions

The HEPD, a space detector for particles up to few hundred of MeV, is scheduled to be launched during late 2017-early 2018. For large part of the 24th solar cycle, it will be able monitor SEPs and the effects of solar activity on the Earth magnetosphere.†

References

Adriani, O., *et al.* 2015, *ApJS*, Volume 801, Number 1

Bruno, A., *et al.* 2017, *Adv. Sp. Res.*, Volume 60, 788–795

Shen, X., *et al.* 2011, *Earthquake Science*, 24(6):639–650

Sparvoli, R., for the CSES/HEPD Collaboration, 2015, *Proceedings of the 34th International Cosmic Ray Conference*, 34, 567S

† V. Vitale is very grateful to Dr. M. Martucci for the interesting discussion on the SEPs and GLEs.

Author index

Some illustrations of the parallel education program with public lecturers L. Green and S. Limaye, the Norman Lockyer archive (Old Library), public and conference participants attending the Tuesday lectures, G. Provan's planeterrella experiment (Monday school workshop) and magnet demonstrations (Tuesday public fun fair).

PARALLEL EDUCATION PROGRAM

The recent IAU Symposium 335 on 'Space Weather of the Heliosphere: Processes and Forecasts' held at the University of Exeter, UK in July 17-21 2017, was the occasion to promote our field of astronomy and space weather science, by giving participants the opportunity to engage about space weather with local young people (students), teachers and the general public in an active 3-day parallel education/public outreach program.

The following events took place from Monday 17th to Wednesday 19th of July, reaching out to schools, teachers and the general public (~300 people).

- Monday of workshops and poster competition for schools (with Poster judging and Prize ceremony).
- Tuesday CPD session for teachers
- Space Weather: public lectures and fun fairs
  - Tuesday, Exeter: Two public lectures, followed by a 'fun fair' with some hands-on demos for children and adults to use.
  - Wednesday, Sidmouth: Demonstrations at the Norman Lockyer Observatory (NLO) followed by a public lecture, Kennaway House.

In addition were History of Science and Outreach Exhibitions that were specially arranged by the University Old Library & Bill Douglas Cinema Museum on the occasion of the IAU Symposium. These exhibitions were open to the general public for several weeks together (13-31 July), including the week of the conference.

I take this opportunity to thank all guest public lecturers, demonstrators, and judges from across the UK but also from abroad, who contributed to share their enthusiasm about space weather: C. Beggan, S. Bingham, L. Green, M. Kutzenova, D. Knipp, S. Limaye, H. Mason, K. Muglach, G. Provan, P. Reiff, S. Strawbridge, Y. Tulunay, E. Verwichte, M. Weber. This outreach program was facilitated by N.B. Crosby, A. Mills, M. Berger, D. Jackson and D. Strange from SOC/LOC, and an extensive number of Met Office and NLO staff associated to members of the LOC, as well as many Exeter staff across several departments and Colleges.

Financial support from the Royal Astronomical Society and the University of Exeter and donations from the DK publisher supported these activities and are gratefully acknowledged.

Post-conference materials including photo albums and education resources are posted on the website www.exeter.ac.uk/iaus335.

Dr Claire Foullon, Chair of IAUS335 LOC & SOC
Exeter, 20 November 2017